MANAGE WEEDS ON YOUR FARM

A GUIDE TO ECOLOGICAL STRATEGIES

Charles L. Mohler, John R. Teasdale and Antonio DiTommaso

SARE handbook series 16

SARE is supported by the National Institute of Food and Agriculture, U.S. Department of Agriculture.

www.sare.org

This book was published by Sustainable Agriculture Research and Education (SARE), supported by the National Institute of Food and Agriculture (NIFA), U.S. Department of Agriculture under award number 2019-38640-29881. USDA is an equal opportunity employer and service provider.

To order:
Visit www.sare.org/manage-weeds-on-your-farm or call (301) 374-9696. Discounts are available for orders in quantity.

Library of Congress Cataloging-in-Publication Data:
Names: Mohler, Charles L., 1947-2021, author. | Teasdale, John R., author. | DiTommaso, Antonio, author. | Sustainable Agriculture Research & Education (Program), issuing body.
Title: Manage weeds on your farm : a guide to ecological strategies / Charles L. Mohler, John R. Teasdale, Antonio DiTommaso.
Other titles: SARE handbook ; 16.
Description: College Park : Sustainable Agriculture Research & Education (SARE), [2021] | Series: SARE handbook series ; 16 | Summary: "Manage Weeds on Your Farm: A Guide to Ecological Strategies provides you with in-depth information about dozens of agricultural weeds and the best ways to manage them. The book begins with a general discussion of weeds and then describes the strengths and limitations of the most common cultural management practices, physical practices and cultivation tools. Part Two is a reference section that describes the identification, ecology and management of 63 of the most common and difficult-to-control weed species found in the U.S"-- Provided by publisher.
Identifiers: LCCN 2021042413 | ISBN 9781888626209 (paperback)
Subjects: LCSH: Weeds--Handbooks, manuals, etc. | Weed control.
Classification: LCC QH545.W43 M65 2021 | DDC 581.6/52--dc23
LC record available at https://lccn.loc.gov/2021042413

Production credits:
Authors: Charles L. Mohler, John R. Teasdale and Antonio DiTommaso
Production Manager: Andy Zieminski
Copy Editing: Lizi Barba
Graphic Design: Peggy Weickert, University of Maryland Design Services
Cover Photos: Main photo courtesy of Practical Farmers of Iowa. Front cover photo strip (from left): Oregon NRCS; University of Kentucky Extension; Laura McKenzie, Texas A&M AgriLife; Matt Ryan, Cornell University. Back cover photo strip (from left): Matt Ryan, Cornell University; Lynn Betts, NRCS; Lucila De Alejandro, Suzie's Farm; Practical Farmers of Iowa.
Indexing: Linda Hallinger
Printing: University of Maryland Printing Services

In Memoriam – Charles L. Mohler

We dedicate this book to our dear friend, colleague, and coauthor, Charles "Chuck" Mohler. Unfortunately, Chuck passed away in April 2021 and was not able to witness his 15-year-long book project to its deserved culmination. Chuck was a unique individual in that he was not only a brilliant scientist able to produce some of the most innovative weed science research, but he could translate this often highly technical research into practical and useful information and advice for growers. Seeing or hearing of growers using his advice was undoubtedly one of the most satisfying aspects of his position as a weed scientist. Chuck lived a simple life and genuinely cared for people. He offered his help to anyone who asked for it and this assistance did not come with preconditions or expectation of repayment.

Chuck was in the vanguard of scientists that brought ecology to the forefront of a sustainable-based approach to weed science. He valued rigorous science and the scientific approach to solving applied aspects of weed science. Chuck was a great role model to so many aspiring and well-established scientists, and was generous with his time and knowledge.

Despite his failing eyesight during the latter years of this book project, Chuck never once mentioned that he could no longer finish this project, which was so important to him. Instead, he would just mention that his parts of the book might take "just a little longer to complete, but they will get done!" We certainly wish that Chuck could be with us to witness firsthand the very book he spent so many years thinking about and working on. We know that he would be most proud of this book and the positive impact that it will have for growers not only in the United States, but across the world, who are interested in ecological strategies for managing weeds!

Chuck, this is YOUR book and we are most grateful to have had the opportunity to assist you in this journey!

Antonio DiTommaso, John Teasdale

Contents

About the Authors

Charles ("Chuck") Leon Mohler was a senior research associate in the Section of Soil and Crop Sciences, School of Integrative Plant Science, at Cornell University's College of Agriculture and Life Sciences. Dr. Mohler was born in Salem, Ore., graduated from the University of Oregon in 1971 with a bachelor's degree in biology, and obtained a doctorate in ecology and evolutionary biology from Cornell in 1979. Subsequently he became a close colleague and 20-year member of Professor Antonio DiTommaso's Weed Ecology and Management Laboratory at Cornell. His research there produced new and innovative approaches to integrated weed management. His studies of the ecology of agricultural weeds and ecological methods of weed management made him a key innovator in this area of weed science. He examined the effects of tillage, cultivation and crop residue on weed control, testing his methods to make them hypothesis driven yet still practical. He wanted growers to be able to use the methods, and they did. He used some of these same approaches in the management of weeds in his extensive vegetable garden at home.

Dr. Mohler was instrumental in moving the field in this new direction of integrated weed management. Over time, younger researchers followed and improved on what he had done and on the standards he had set. These young students are now some of the top weed ecologists, and they were all influenced by his work. Dr. Mohler co-authored four books, including the highly acclaimed Ecological Management of Agricultural Weeds (2001) with co-authors Matt Liebman and Charles Staver. He also produced dozens of refereed scientific articles as well as a number of media presentations. He mentored numerous graduate and undergraduate students and visiting scholars. His significant contributions to the field of weed science were recognized when he received the Outstanding Researcher Award from the Northeastern Weed Science Society in 2014. Dr. Mohler died in 2021 at the age of 73.

John Teasdale is retired from the USDA Agricultural Research Service (ARS) at Beltsville, Md. He was born and raised in St. Paul, Minn., and received a doctorate in agronomy from the University of Wisconsin-Madison. After joining the ARS Weed Science Laboratory at Beltsville, he established a research program developing integrated weed management systems with an emphasis on understanding the interactions between cover crop management and weed emergence and growth dynamics. He developed a novel cover-crop-based, minimum-tillage system for vegetable production that served as a model for defining environmental, microbial community, physiological and molecular responses to conservation soil management practices. He was instrumental in the design and analysis of long-term cropping system experiments at Beltsville that established the importance of phenologically diverse crop rotations for organic farming. He played a leadership role in establishing the sustainable agriculture research program at

Beltsville and served as the founding research leader for the ARS Sustainable Agricultural Systems Lab for 11 years until retirement. He currently lives in Bowie, Md., where he is the coordinator of a community garden that supports the Bowie Food Pantry, and he serves on city committees that advise municipal staff and elected officials on environmental and sustainability issues.

Antonio ("Toni") DiTommaso is a professor of weed science and chair in the section of Soil and Crop Sciences in the School of Integrative Plant Science at Cornell University, where he has research and teaching responsibilities. He was born in southwestern Italy, but his family immigrated to Montreal when he was 9 years old. He received a bachelor's degree in environmental sciences (1986) and a doctorate in weed ecology (1995) from McGill University in Montreal, as well as a master's degree in plant ecology (1989) from Queen's University in Kingston, Ontario. He joined Cornell as an assistant professor in 1999. His primary areas of scholarship focus on the effects of biotic and abiotic factors on the biology and ecology of important agricultural weeds (e.g., common ragweed, pigweeds) and introduced invasive plant species of natural and semi-natural areas (e.g., swallowworts) in the northeastern United States and southern Canada. His most recent work focuses on the impact of climate change, including drought and elevated temperatures on the ecology and evolution of weedy and invasive species and their potential geographic distributions. He has published extensively in scientific journals and has served as editor of the scientific journal Invasive Plant Science and Management since 2015. He served as the president and is a fellow of the Northeastern Weed Science Society. He is also co-author of the soon-to-be-published revised version of the popular weed identification guide, Weeds of the Northeast. DiTommaso, an avid gardener, currently lives with his family in Dryden, N.Y., where he enjoys growing fig plants (in pots) and multiple fruit trees.

Acknowledgments

We gratefully acknowledge the assistance of several people who helped assemble the information on which this book is based: S. Morris, N. Hubert, D. Glabau, S. Green, J. Ligai, P. Mukergee, B. Bundran, P.O. Pentrarlow, M. Williams, B. Chi, C. Platzer, G. Moniz, B. Morgan, K. Ellis, and V.J. Dagun. E. Gallandt, E.B. Rosen, R. Kersbergen, M. Schonbeck, A. Seaman, E. Stockman and L. Perkins provided helpful comments on early drafts of the manuscript. K. Ellis and L. Mohler drafted some of the figures. Y. Li matched references with citations and checked the validity of web addresses. R. Padilla helped convert online versions of the taxonomy chapters to the form that appears in this book. E. Buck, J. Kerr, S. Morris and V. Wikel were instrumental in developing the botanical descriptions. We are especially grateful to the many colleagues who provided technical reviews and photographs for the taxonomy chapters: P. Bhowmik, R. Boydston, D. Brainard, J. Byrd, J. Cardina, S. Clay, W. Curran, A. Dille, F. Forcella, E. Gallandt, R. Gulden, C. Johnson, M. Leblanc, J. Leeson, R. Leon, J.L. Lindquist, R.F. Norris, A. Price, E. Regnier, K.A. Renner, D. Robinson, M.J. Simard, L. Sosnoskie, D. Stoltenberg, K. Tørresen, M. VanGessel, G. Wehtje, C. Willenborg and M. Williams. Any errors are, however, the responsibility of the authors.

About SARE

Sustainable Agriculture Research and Education (SARE) is a grant-making and outreach program. Its mission is to advance—to the whole of American agriculture—innovations that improve profitability, stewardship and quality of life by investing in groundbreaking research and education. Since it began in 1988, SARE has funded more than 7,600 projects around the nation that explore innovations—from rotational grazing to direct marketing to ecological pest management—and many other best practices. Administering SARE grants are four regional councils composed of farmers, ranchers, researchers, educators and other local experts. SARE-funded Extension professionals in every state and island protectorate serve as sustainable agriculture coordinators who run education programs for agricultural professionals. SARE is funded by the National Institute of Food and Agriculture, U.S. Department of Agriculture.

SARE GRANTS
www.sare.org/grants
SARE offers several types of competitive grants to support the innovative applied research and outreach efforts of key stakeholders in U.S. agriculture. Grant opportunities are available to farmers and ranchers, scientists, Cooperative Extension staff and other educators, graduate students, and others. Grants are administered by SARE's four regional offices.

RESOURCES AND EDUCATION
www.sare.org/resources
SARE Outreach publishes practical books, bulletins, online resources and other information for farmers and ranchers. A broad range of sustainable practices are addressed, such as cover crops, crop rotation, diversification, grazing, ecological pest management, direct marketing and more.

SARE REGIONS

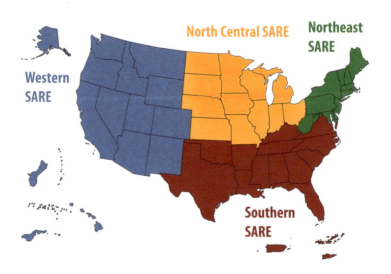

SARE's four regional offices and outreach office work to advance sustainable innovations to the whole of American agriculture

PART I

Concepts of Ecological Weed Management

CHAPTER 1

Introduction

PURPOSE AND PHILOSOPHY

The purpose of this book is to provide you with information about the biology of agricultural weeds, including identification, management strategies and ecological facts that will help you understand and manage them. The book is focused on the weeds of arable cropping systems. It does not discuss the management of weeds in forests, turf, permanent pastures or perennial bioenergy crops. Weed management issues in forage production are discussed to some extent since forages are often rotated with other crops.

The basic philosophy behind the book is that understanding the biology of weeds is critical to ecological weed management. Ecological management of weeds is information intensive rather than input intensive. This book is intended to provide the information you need to grow crops without synthetic herbicides, great expense or backbreaking work. By understanding how weeds work as organisms in the context of your farm ecosystem, the task of weed management becomes easier. By learning about the biology of weeds that cause you problems and then exploiting their weaknesses you can make weed management an integral part of your overall management effort. That does not mean that learning about particular weed control practices is useless. On the contrary, ecological weed management depends on a large bag of tricks. The key to success, however, lies in knowing when to apply a particular tactic, and that requires an understanding of how weeds operate, both in general and as particular species.

The premise of this book is that although weeds can be useful as food and as protection for the soil, most farmers will prefer to strictly limit their abundance. Weeds are neither always bad nor always good, but usually they tend to reduce yield, hinder harvest and cause a variety of other problems.

Some readers may wonder why we refer to the system of weed management described in this book as "ecological" rather than as "organic." There are two reasons. First, the principles described here apply equally well whether you are fully committed to organic management or whether you still use some chemical fertilizers and pesticides. They are adequate to provide successful weed management on an organic farm, but they can be useful on any farm. Second, we wish to emphasize that the principles and methods are based on careful scientific investigation. Ecology provides the theoretical basis for the applied science of weed management, much as physics provides the theoretical basis for the applied science of engineering. To build a sewer system one does not need to be a physicist, but some knowledge of loads, stresses and fluid mechanics will reduce the cost of materials and prevent the drains from backing up. Similarly, a minimal knowledge of some specific ecological ideas will help you minimize your weed management effort and improve your success.

Finally, thinking about your farm from an ecological perspective has an additional bonus. Ecology is the science of how organisms interact with their environment. Its subject encompasses the workings of the everyday biological world that you see and touch. We expect that understanding something about the small corner of this science that deals with agricultural weeds will enrich your life. The great ecologist Richard Root once commented that the best preparation one could have for the serious study of ecology is to grow a vegetable garden and become fascinated by a particular group of organisms, e.g., plants or insects, that are encountered and how their behavior can change from season to season. We hope that this book will help you better understand weeds and how to manage them.

HOW TO USE THIS BOOK

The book is divided into two main sections. The initial chapters contain general information on the ecology of weeds and methods of ecological management. Some of the information in Chapter 2 ("How to Think About Weeds") may not seem immediately relevant to weed management. As you learn more about the application of ecological management methods and about individual weed species, however, you will see that understanding how weeds work as organisms is the key to effective management. Chapter 3 ("Cultural Management of Weeds") and Chapter 4 ("Mechanical Weed Management") describe the many and varied methods of ecological weed management. Often a method will work well to help control one species but not another. Thus, your management will be most effective if you combine the general information in chapters 3 and 4 with specific information about the particular weed species present in your fields. Additional documentation of most statements made in chapters 2, 3 and 4 can be found in the book *Ecological Management of Agricultural Weeds* by Liebman, M., C.L. Mohler and C.P. Staver. 2001. Cambridge University Press: New York. Additional references have been added where needed. Chapter 5 gives case studies of how ecological management methods have been applied on farms around the United States.

Because we believe that effective ecological weed management requires a multi-year approach that involves a varied sequence of crop types, this book takes a weed-centered rather than a crop-centered approach. Thus, although the book contains much information you will find useful for the management of weeds in wheat, corn, carrots or other specific crops, the book does not contain sections addressing management of individual crops.

The second section of the book contains species-by-species information on most of the common agricultural weeds of the United States and Canada. Each species entry is divided into three sections. The first, **Identification**, gives a description of the weed, photographs of various stages in the weed's life cycle and tips on distinguishing the species from similar looking species. The second, **Management**, gives a deliberately brief summary of major control strategies. Use chapters 3 and 4 to get more information on particular procedures mentioned in the management section. The third, **Ecology**, provides a series of short but specific statements about various aspects of the species' ecology. This information provides much of the basis for the recommendations in the management section. Farms differ greatly in soils, climate, economic resources and the management style of the farmer. Understanding the ecological behavior of your weeds will help you manage them in cases where the general recommendations in the management section do not apply. The relevance of the information will only be apparent, however, through study of the management section and chapters 2, 3 and 4.

We have attempted to keep technical terminology to a minimum throughout the book. Descriptions of plants and their ecological behavior become longwinded or even inaccurate, however, without the introduction of a few technical terms. These are defined in the Glossary. Also, some specialized tools have been developed to facilitate weed management, and many readers will be unfamiliar with some of these. Rather than describe the tool each time it is mentioned, we refer to it by its usual name and describe it in the Glossary. Pictures of implements are provided in appropriate places in the text, and the page reference for these is given in the Glossary.

Few people ever read a book of this sort in its entirety. We believe that you will find study of chapters 2, 3 and 4 highly useful, and Chapter 5 may give you some ideas about how to combine various tactics into an overall control plan. Then read about the weeds that are actually giving you problems. Finally, combine the specific information on those weeds with the general principles and methods from chapters 2, 3 and 4 to derive a weed management strategy that works with your particular mix of soils, crops and resources. We, the authors, would greatly appreciate you contacting us with your stories of success and failure using this approach to weed management.

CHAPTER 2

How to Think About Weeds

Understanding the biology and behavior of weeds is key to managing them. As explained in this book, weeds are very good at colonizing and persisting in your fields. They are, however, short on brain power and, with a little effort, you can outsmart them. Outsmarting them requires understanding how they operate as organisms.

WHAT IS A WEED?

Weeds are commonly defined as plants growing in places where they are not desired. That definition, however, does little to further understanding of how weeds operate as species or how to use an understanding of weed biology to manage them.

From an ecological point of view, weeds are "plants that are especially successful at colonizing disturbed, but potentially productive sites, and at maintaining their abundance under conditions of repeated disturbance" (Liebman, Mohler and Staver 2001). Weeds have adapted to highly disturbed conditions, and these adaptations are revealed by their biological characteristics. Those characteristics reveal how to manage them.

For example, the seeds of annual weed species are generally small because these species must produce many seeds so that some seedlings survive repeated disturbance. Similarly, the seeds of most weed species survive many years in undisturbed soil. This, plus the ability of the seeds of many weed species to recognize cues associated with tillage, like light and large fluctuations in soil temperature, allows weeds to emerge after many years of absence when sod is turned to begin the annual cropping part of a crop rotation.

There are many ways to be a weed: Different combinations of characteristics allow various plant species to thrive in the same sort of disturbed habitat. Annuals, for example, complete their lifespan within less than a year and survive between growing seasons as seeds in the soil. Common lambsquarters and giant foxtail are good examples of these.

Stationary perennials, such as dandelion and curly dock, are fixed in place and unable to spread by vegetative reproduction, except by human intervention. As explained in the section "Vegetative Propagation of Perennial Weeds," few of the weeds found in recently tilled ground are stationary perennials. These species are common problems, however, in established pastures and hay fields.

Finally, some of the worst weeds are creeping perennials, which spread by horizontal roots or underground shoots (rhizomes) from which daughter plants emerge aboveground. Examples of these species include quackgrass and field bindweed. They are notoriously hard to kill, and because they renew themselves by vegetative reproduction, individual clones are essentially immortal if left alone. Table 2.1 summarizes some of the contrasting characteristics of these three types of weeds.

THE ORIGINS OF WEEDS

Most species of agricultural weeds that propagate by seed can be found in naturally disturbed or open environments, and these were probably their habitats prior to the development of agriculture. These habitats include stream flood plains, cliffs, beaches and locations disturbed by animals. Both the spectacular increase in growth rate many weed species show in response to the addition of nutrients and the frequent ability of weed seeds to pass through mammalian digestive tracts without harm make their association

Table 2.1. Characteristics of Three Types of Weeds

Characteristic	Annuals	Fixed Perennials	Creeping Perennials
Vegetative lifespan	< 1 year	Two to a few years	Long, indefinite
Vegetative reproduction	No	Accidental[1]	Yes
Seed longevity	Years to decades	Years to decades	A few years
Energy allocated to seed production	High	Medium high	Low
Establishment	Seeds	Seeds	Mainly vegetative
Usual means of dispersal	By wind, in soil, by livestock in manure, with crop seed	In soil, by livestock in manure, with crop seed, a few species by wind	In soil, a few by wind-blown seeds
Examples	Common lambsquarters, barnyardgrass	Dandelion, curly dock	Quackgrass, field bindweed

[1]Normally, fixed perennials do not propagate vegetatively, but they may do so if chopped up by a tillage implement or cultivator.

with animal disturbances seem particularly likely. The extensive modification of Eurasia by human settlement makes tracing the original habitats of most Eurasian weeds impossible, but the original habitats of some North American weeds are relatively clear. For example, giant ragweed is most frequently found in river bottom fields, stream banks and drainage ditches, and before agriculture, it probably inhabited soil deposited along rivers by floodwater. Waterhemp, as the name implies, can be found growing in the mud at the edges of lakes and ponds, and it seems to have moved into agricultural fields relatively recently.

Many of the creeping perennials were probably minor components of prairies and natural wet meadows prior to agriculture. Their ability to propagate by division of rhizomes or roots, or their ability to emerge from deep rhizome or root systems, allowed them to expand in agriculturally disturbed soils when perennial plants more sensitive to disturbance were killed by tillage.

A third group of weeds has evolved from our crops. Representatives of this class of weeds important in North America include wild-proso millet, shattercane and common sunflower. In most grain crops, mutant forms that shatter rather than retain the seeds until harvest arise spontaneously. These may subsequently undergo natural selection that adapts them to life as weeds. For example, the black seeded forms of wild-proso millet not only shatter but also have evolved increased dormancy and a more resistant seed coat. Consequently, these seeds persist in the soil much longer than seeds of domesticated proso millet. Weedy forms of crops also frequently cross with the wild progenitors when the latter are present. For example, domesticated sorghum, shattercane and their wild progenitor have

intercrossed repeatedly in the savannah zone of Africa (de Wet 1978). The resulting hybrids have been selected for desirable traits within the crop, while human management has selected weedy hybrids that resist that management. Some of the latter have subsequently been introduced into North America and constitute the weediest forms of shattercane.

WEED POPULATIONS ARE DYNAMIC

The population of each weed species in a field changes constantly. A weed population consists of not only the green, growing plants present in a field but also the seeds present in the soil. Seeds are essentially tiny dormant plants, and management of seeds in the soil seed bank is a critical component of ecological weed management. The number of green plants and seeds of a weed species changes with the season and from year to year as seeds germinate, seeds and green plants die, and more seeds are produced. Both management and natural processes like weather and consumption by insects affect the number of seeds and growing plants in a weed population.

New weeds are born into a population when plants produce seeds or daughter plants form on a perennial plant by vegetative reproduction. The total number of individuals of a species is the result of the balance between birth and death. If births exceed deaths, then the population increases. If deaths exceed births, then the population declines. This is a simple idea, but many complex factors can determine how birth and death rates actually balance out in your fields. Understanding some of the complexity in the balance between birth and death provides insight into why management succeeds or fails.

Plants die from a variety of causes, and the causes of death change as a weed moves from one stage in its life cycle to the next and grows from a little seed to a large mature plant (Figure 2.1). Fungi and a wide range of animals consume weed seeds. Other seeds die after aging or from problems during germination. Tiny seedlings die of desiccation, fungal disease and attack by invertebrates both before and after they emerge from the soil. In contrast, large, nearly mature weeds are rarely killed by any of these factors. Large weeds, however, are more likely to be noticed and killed when hand hoeing a vegetable crop than are tiny seedlings. Both the probability that the weed will flower and the number of seeds it will produce also change with the size of the plant. Figure 2.1 summarizes some of the life-stage and size-dependent processes acting on a weed population. Note that as weeds get larger, the strategies for control tend to shift from the management of biological processes to direct attack on the weed.

Ecological weed management attempts to increase the death rate of multiple life stages and sizes of weeds, both by direct attack and by modifying the physical and biological environment of weeds. It should also aim to reduce weed reproduction. As explained in succeeding chapters, however, the small, early stages in the weed life cycle are easier to manage than later stages.

- *The fundamental principles of weed management are thus: 1) to ensure that weeds die before moving into the next size class and 2) to prevent mature plants from producing seeds and vegetative daughter plants.*

WEED DENSITY AFFECTS WEED DEATH AND REPRODUCTION

The density of a weed population matters. A small increase in weed density has its greatest effect on crop yield when weeds are sparse (Cousens 1985). When weed density is low, however, the absolute effect on yield of those few weeds may not be obvious. On the other hand, when weed density is high, the absolute effect on yield will be considerable, but additional small increases in density will have little further effect.

The density of a weed species also affects what proportion of individuals will die or reproduce during the course of a week or of a growing season (Garcia de Leon 2014). Some factors kill a greater *proportion* of weeds when the weed species is dense. Other factors kill a smaller proportion of the species when it is dense. Still other factors kill about the same proportion of weeds regardless of the species' density. Factors that influence reproduction, like flower-eating insects and competition from crops, also vary in how dependent they are on weed density. Thus, as a weed population increases, the balance between birth and death rates changes, leading either to further increase or to decline.

Your management enters into this balance. As a weed problem becomes more acute, you are likely to take additional measures to bring the population down. Similarly, if you can drive a species to low density in a field, natural processes may be sufficient to keep it in check, and you can relax your control efforts. Ecological weed management manipulates natural

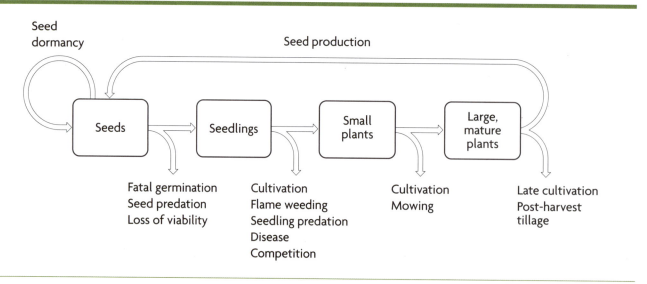

Figure 2.1. The life cycle of an annual weed species showing mortality factors that affect the population at various life stages.

processes like seed germination and beneficial organisms in concert with physical management methods like cultivation.

- *The goal of ecological weed management is to arrive at a balance between birth and death that keeps the density of weed populations low most of the time and reduces them quickly when density starts to increase.*

Some examples will help explain how density affects weed mortality and reproduction.

As the density of weed seeds in the soil increases, the number of seedlings that emerge increases, but the *proportion* of seeds that produce seedlings decreases (Grundy et al. 2003). This process tends to slow population growth as weeds proliferate. Similarly, hand weeding kills a larger proportion of weeds as density increases. In the first pass, a weeding crew may hoe out 95% of the weeds in half a day. A second half-day of hoeing immediately after the first will certainly remove far less than 95% of the remaining weeds because the ones that are left are smaller, or are hiding under crop leaves, or appear to have been killed by the previous hoeing. Conversely, the higher the initial density, the greater the *proportion* of weeds removed is likely to be, although the *number* of weeds remaining is also likely to be higher.

Many of the mortality factors that have an effect on weed populations are independent of weed density; an example is blind cultivation, which is disturbance of the entire surface soil, including the crop row. Organic corn and soybean growers commonly use a rotary hoe or tine weeder for blind cultivation before or shortly after the crop emerges. The objective of blind cultivation is to kill tiny, shallowly rooted weed seedlings, both those that are newly emerged and those that are still in the "white thread" stage of development in the soil. Blind cultivation will kill more seedlings if more are present, but it usually kills the same *proportion* of seedlings regardless of density.

In contrast, a row crop cultivator will kill a smaller proportion of weeds when the weeds become very dense. At low to moderate density, the cultivator will likely kill the same proportion of weeds regardless of how many are present. At very high densities, however, the weed roots tangle together and hold onto soil. The cultivator increasingly throws up slabs of weeds and soil as density increases. In these slabs, the roots are not exposed to drying, so the weeds live on to reroot

and continue growing. Thus, as weed density increases, a row crop cultivator kills the same proportion of weeds until they become very dense, and then an ever-smaller proportion is killed with further increases in density.

Factors that affect reproduction are also influenced by weed density. Competition from the crop normally decreases seed production to the same degree regardless of weed density. In contrast, fungal diseases often kill an increasing proportion of flowers and immature seeds as weed density increases because dense vegetation keeps humidity higher, thereby promoting infection, and the close proximity of plants increases movement of spores onto not-yet-infected individuals.

If you have a problem weed, any practice that kills that species or reduces its reproduction is desirable. Note, however, that when weeds are dense, many can be killed easily, but this will not necessarily result in control. You will not achieve long-term control of the weed unless your overall management kills a greater proportion of the weeds as their density increases. Ecological management uses many "little hammers," but just any collection of hammers will not do. Your overall management program needs to create a higher death rate and a lower birth rate as the population increases, or the population will continue to increase.

VEGETATIVE PROPAGATION OF PERENNIAL WEEDS

Perennial weeds potentially live for more than one growing season, and many are nearly immortal. All have buds on roots or underground stems that sprout when the plant is damaged. These dormant buds allow quick recovery following disturbance, and they may result in multiplication of the weed if it is fragmented. Understanding the behavior of buds and how they tap into the weed's stored energy reserves is critical for successful management of these weeds.

All temperate zone perennial weeds have someplace to store carbohydrates and other nutrients through the winter; these stored nutrients allow the weed to sprout vigorously in the spring. In creeping perennials, the storage organ is either a thickened, horizontal storage root (for example, field bindweed), a rhizome (for example, quackgrass) or a tuber (for example, purple nutsedge). A rhizome is simply an underground stem, though most rhizomes look more like roots than like stems. Roots and rhizomes form buds at short intervals as they grow, and these can turn into new shoots if conditions are right. Most buds remain dormant due to hormones that are released from the growing tip of the rhizome or storage root. This phenomenon is known as *apical dominance*. As the rhizome or storage root grows, the distance from the tip to a particular bud becomes longer, and the concentration of the hormone at a particular bud decreases. Eventually, hormone concentration drops so low that the previously suppressed bud can sprout and become a new shoot. The new shoot also produces hormones that suppress the sprouting of neighboring buds. The result is a semi-regular spacing of shoots along the length of the storage organ.

When the root or rhizome system of a creeping perennial is cut up into small pieces during tillage, hormonal suppression of the lateral buds stops and the last bud on each segment becomes capable of sprouting (Figure 2.2). Usually sprouting will occur immediately, but if the season is not suitable for growth, the buds may remain dormant until soil conditions become favorable again. In any case, tillage often increases the number of shoots of a creeping perennial. This is not necessarily bad, however, because each of the shoots has smaller storage reserves on which to draw. Hence, although the number of individual shoots increases, each one has more difficulty growing up out of the soil, growing through a crop canopy, and recovering from subsequent damage by cultivators.

The smaller the piece of storage organ, the weaker an individual will be, and the more sensitive it will be to other management practices. The perennial storage organs of a few species are tough enough that they can be worked to the soil surface and physically removed or subsequently killed by drying (see "Remove Perennial Storage Organs" and "Dry Out Perennial Storage Organs" in Chapter 4). Longer pieces make working the storage organs to the surface easier.

For most species, however, tillage aimed at controlling perennials should attempt to cut or break up the pieces as small as possible. A disk, rotary tiller or spading machine will be more effective than a chisel or moldboard plow at chopping up roots and rhizomes into small pieces. The former machines are also less likely to spread the weed around the field. Note, however, that if the pieces are buried deep with a moldboard plow after being cut up, then energy reserves will be heavily depleted as the shoot elongates out of

the ground. This will make the weed easier to manage with subsequent cultivation or with a competitive crop. Management of perennials with tillage is discussed more fully in Chapter 4.

When a piece of root or rhizome sprouts, energy drains out of the storage tissue and into the shoot. This process continues even after the shoot has emerged from the soil (Figure 2.3). Eventually, the shoot develops leaves and becomes self-sufficient, after which it begins to replenish the depleted underground reserves, and new rhizomes or storage roots begin to form. The ideal point at which to destroy the plant by removing the shoot occurs when the storage organs reach their minimum weight. This point varies among weed species. For quackgrass, the minimum is usually reached when the plant has formed three to four leaves, whereas for Canada thistle it occurs when the plant is about 12 inches tall. Probably, burial of the shoot is more harmful to the weed than simply removing it, since a buried shoot will continue to draw on root or rhizome reserves by respiration, whereas a severed shoot will not. This requires testing, however, and in any case, the shoot must be completely buried, or it will continue to grow.

Many stationary perennials like dandelion rely on a taproot for overwinter storage. Taproots of stationary perennial weeds have dormant shoot buds or tissues

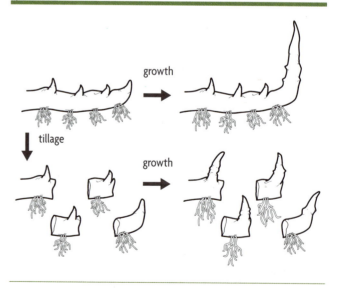

Figure 2.2. The growing tip of a storage root or rhizome produces hormones that prevent nearby buds on the root or rhizome from sprouting. Cutting up the root or rhizome during tillage releases the buds from hormonal suppression and allows them to sprout. Each of the resulting shoots has smaller reserves to draw on and is thus less vigorous than the shoot from an uncut root or rhizome.

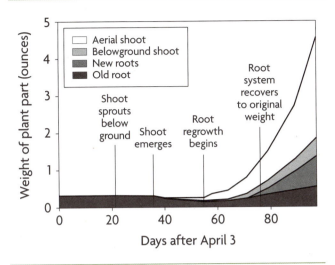

Figure 2.3. Recovery of fragments of storage roots of perennial sowthistle, a typical creeping perennial. When a root fragment sprouts, the new shoot begins drawing energy from the root fragment, thereby reducing its dry weight. Even after the shoot emerges above the soil surface, it continues to grow at the expense of the storage root. The plant is weakest and most susceptible to further disturbance when the root system reaches minimum weight, but several weeks may be required before the storage roots have fully recovered. Most herbaceous perennials go through these stages following fragmentation, but the time of course varies among species. Five-inch root segments were buried in November at a 2-inch depth, and the first observations were made on April 3. This figure was plotted from data in Håkansson (1969).

Figure 2.4. A dandelion that has recovered from burial of the taproot. Note the long underground shoots that have formed to act as bases for the production of new rosettes of leaves.

that can differentiate into buds. They thus behave similarly to creeping perennials in that breaking the taproot into pieces produces multiple new individuals. Because the taproots of stationary perennials are generally much larger in diameter than the storage organs of creeping perennials, breaking the root into sufficiently small pieces to create loss of vigor in the daughter plants is difficult. Moreover, the taproots of stationary perennials are often so tough that they are not broken up much by tillage. Partial damage to the root may stimulate sprouting of dormant buds. This results in a tight clump of shoots arising from the same piece of root. The same effect occurs when the top of a taproot is cut off by a cultivator.

Deep burial by tillage is often an effective first step for controlling taprooted perennials despite their substantial storage reserves. Unlike the storage roots and rhizomes of creeping perennials that recover well from a change in orientation, a taproot prefers to sit vertically in the soil. If the root is inverted or laid on its side deep in the soil, it must expend substantial energy developing a new vertical axis. Moreover, most taprooted perennial weeds form a whorl of leaves arising at ground level (a rosette) before the plant forms a flowering stalk. Thus, the new vertical root or underground shoot must be substantial enough to support development of the rosette (Figure 2.4). Development of a new, thick vertical underground root or shoot drains the original taproot and leaves the resulting plant susceptible to further management.

Some stationary perennial weeds (for example, the various species of plantain) overwinter as a rosette. In these species, the leaf bases, root crown or short underground stem act as the storage organs. Most such species do not propagate vegetatively: Fragmentation of the root crown usually kills the plant rather than multiplying it. Few stationary perennials without taproots survive well in tilled fields, because turning the soil buries the root crown and kills it. However, individuals that end up at the soil surface may occasionally survive and re-establish, even after losing most or all of its leaves. In general, stationary perennials, both with and without taproots, are primarily weeds of grasslands (e.g., lawns, pastures, hayfields and roadsides). Although poorly adapted for surviving tillage, they are well adapted to survive mowing and grazing because the shoot arises near the soil surface and thus below the height of cutting.

Characteristics of Weeds That Affect Their Management

The following sections are arranged approximately in order of plant development, from seed to mature plant. They correspond to the topics listed under "Ecology" in the sections on individual weed species that are in Part Two of this book.

SEED WEIGHT

Seed weight indicates much about the biology of a weed species and hence its response to a wide range of management practices. Seed weight is important because it indicates the resources available to the seedling during establishment. Consequently, seed weight correlates with the depth from which the seedling can emerge from the soil, how well it can grow around obstacles in the soil, its ability to grow up through organic mulch (Mohler and Teasdale 1993), and how quickly it will begin competing with young crop plants.

Since seed weight governs the species' ability to establish in adverse situations, it also indicates how likely the species is to respond to environmental cues associated with favorable conditions for establishment. Small seeded weeds commonly germinate in response to multiple environmental signals associated with near-surface conditions, recent soil disturbance and bare soil (see "Seed Germination: Why Tillage Prompts Germination"). The small seeded species balance these cues physiologically to determine whether to germinate. In contrast, large seeded weed species are usually insensitive or only weakly sensitive to many of these cues, probably because their greater seed reserves allow them to successfully establish from deeper in soil regardless of surface soil conditions. Unlike most small seeded species, germination of many large seeded species requires weathering of the seed coat. As with all plant species, germination of large seeded weeds requires appropriate soil temperature and moisture conditions.

The seed size of a weed species affects how you can manage the species. Weeds are always easier to control by cultivation when they are young, but the smaller the seed, the easier the seedlings are to kill. Thus, shallow cultivation with a tine weeder or rotary hoe often provides good control of small seeded weeds like common

lambsquarters and the pigweeds, since these can only emerge successfully from near the soil surface. In contrast, shallow-working machines are usually less effective for controlling large seeded weeds like velvetleaf or the morningglories because the seedlings have sufficient resources to emerge from below the implement's operating depth.

Seed weight indicates the species' ability to grow up into and through the canopy of an established crop. Species that grow from tiny seeds tend to stagnate in the shade under a crop. Those from larger seeds can rely on seed reserves, however, to get above the lower leaves of the crop into partial sunlight and to continue growth. The ultimate extension of this pattern is to perennial species that rely not on seed reserves but on substantial tubers, rhizomes or storage roots for their early growth. Enhancing crop competitiveness helps control all sorts of weeds. A highly competitive crop can devastate the seedlings of most small seeded species. It may also check the growth of a seedling from a large seed or slow new shoot growth of a perennial weed. If the crop does not establish dominance quickly enough, however, even a small seeded weed species can outgrow the crop and become highly competitive. In general, the smaller the seed size of a weed, the more rapidly its size multiplies each day (see "Growth and Competition for Light").

Finally, smaller seeded weeds are more easily controlled by organic mulch than large seeded species. First, fewer seeds of the small seeded species will receive the cues they need for germination if covered with organic mulch (see "Seed Germination: Why Tillage Prompts Germination"). Second, those that do emerge from the soil will be more likely to starve or become so weak that they succumb to disease before growing through the mulch.

This discussion has described general trends across many species. Exceptions to any of these generalities can arise from the unique growth form or physiology of individual weed species. Nevertheless, seed size tells much about the biology of a weed and how to manage it.

Most crops have seeds that are much larger than

the weeds with which they compete. Corn, soybeans, wheat, peas and pumpkins all have seeds that weigh 50–1,000 times more than many common agricultural weeds like the pigweeds, giant foxtail and common lambsquarters (Figure 2.5). Potato seed pieces and onion sets dwarf weed seeds even more. This difference in seed size gives large seeded crops a strong head start in competition with weeds, and it makes possible several organic weed management techniques. Transplanting small seeded crops like lettuce, tomatoes and cabbage gives these species a competitive head start similar to that of the large seeded crops. Conversely, the notorious difficulty of controlling weeds in small seeded crops that do not transplant well (for example, carrots and parsnips) is not a coincidence.

In this book, including in the accounts of individual species, we use the term "seed" in the loose sense used by most farmers and gardeners rather than the strict botanical sense. For example, the grain of a grass is technically a type of fruit called an "achene," in which the wall of the fruit adheres tightly to the seed proper. Moreover, in some species, including large crabgrass and common lambsquarters, the dispersal unit usually includes some flower parts that also cling to the seed. Although the most ecologically relevant part of the dispersing unit is the germ plus the endosperm, which is the seed's food store, data on the weight of the seed proper are available for few weed species. Consequently, the seed weight values reported in the accounts of individual species are the weights of the dispersing units. Usually, most of the weight is that of the seed proper, since thin fruit skins and tightly clinging chaff are usually light relative to the seed. Exceptions include cocklebur and the sandbur species, and for these we give the weights of the shelled-out seeds.

SEED GERMINATION: WHY TILLAGE PROMPTS GERMINATION

Most weed species have very small seeds. Hence the newly emerged seedlings are tiny and incapable of competing with established vegetation. Consequently, these species have been naturally selected to respond to environmental cues that indicate the nearness of the soil surface and the absence of competing plants. In natural situations, usually plants are only absent if

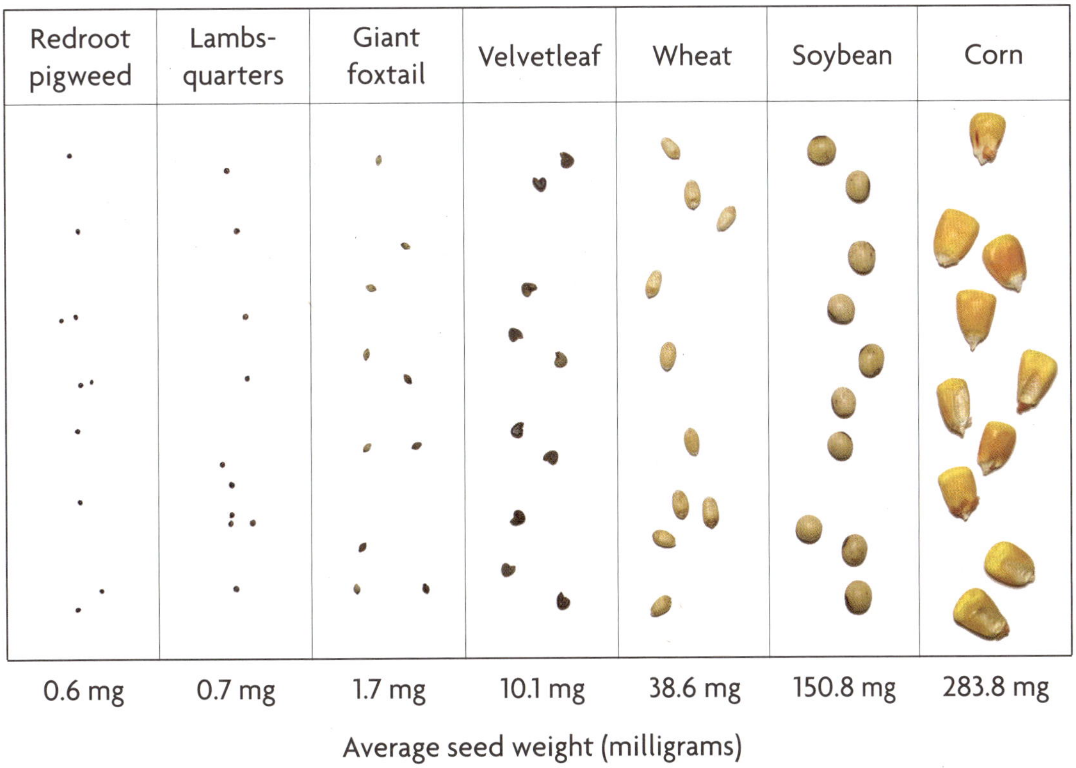

Redroot pigweed	Lambs-quarters	Giant foxtail	Velvetleaf	Wheat	Soybean	Corn
0.6 mg	0.7 mg	1.7 mg	10.1 mg	38.6 mg	150.8 mg	283.8 mg

Average seed weight (milligrams)

Figure 2.5. Comparison of the seed size of some common weeds and crops.

the soil has been recently disturbed, for example, by animals or flooding. Thus, weed seeds often respond to cues associated with soil disturbance, and this means that they germinate in response to tillage. Cues that prompt germination of many weed species include light, high soil temperatures, fluctuation in temperature between day and night, the presence of nitrate in the soil, and the absence of volatile substances that are released by soil organisms when oxygen is in short supply.

High light levels at the soil surface occur when competing plants and plant residues are removed by tillage. Also, tillage rolls the soil around, briefly exposing buried seeds to light, even if they are again covered. Exposure to white light increases germination of most weed species, particularly if the seeds have been previously buried (Wesson and Waring 1969). In contrast, light that has been depleted in red wavelengths by passage through a plant leaf canopy inhibits germination of many weed species (Taylorson and Borthwick 1969).

Bare soil exposed to direct sunlight becomes much warmer during a sunny day than soil that is covered by plants and plant residues. Germination of some species, like redroot pigweed and common purslane is prompted by high soil temperatures. In addition, bare soil radiates more heat to the night sky than covered soil, and therefore it cools more rapidly. The daily alternation between high daytime and low nighttime temperatures prompts germination of many weed species, including common lambsquarters and curly dock. Note that high soil temperatures and large daily variation in soil temperature only occur near the soil surface, and thus they indicate to the seed that it is in a safe location for germination.

Inside soil aggregates, oxygen may become depleted due to respiration by roots and microbes. Low oxygen concentrations inhibit germination of some species until air is stirred into the soil by tillage. Normally, however, enough oxygen is present within the plow layer for seed germination. Venting of volatile organic compounds during tillage may be more important than increased oxygen in prompting weed seed germination. Ethanol, acetone and aldehydes can accumulate in undisturbed soil particles due to insufficient oxygen for the complete breakdown of carbon compounds. Dormant seeds can release these substances if oxygen is in short supply. When soil is stirred by tillage, these volatile substances are vented into the atmosphere and some species, including velvetleaf and tall morningglory, germinate in response to their sudden absence (Holm 1972).

Finally, the increase in warmth and oxygen associated with tillage stimulates microbes to consume organic matter and thereby release nitrogen-containing compounds into the soil. Specialized bacteria convert these to nitrate. Normally, plant roots absorb nitrate so quickly that concentrations in the soil remain very low. After tillage, however, no living plants are present to take up the nitrate, and it accumulates in the soil. Even though the concentrations of nitrate in the soil usually remain low, seeds of some weed species, like common lambsquarters and common chickweed detect the slight increase in nitrate concentration following tillage and germinate in response to it.

Many species respond to several of the germination cues discussed above. Thus, for example, a few common lambsquarters seeds will germinate in the dark at constant temperature and no nitrate. More will germinate if any one of these three cues (light, alternating temperature or nitrate) is present, still more will germinate if two cues are present, and most will germinate if all three are present. Seeds of many weed species add up cues in this way to assess the suitability of the environment for establishment.

As a result of the changes in soil properties following tillage and the variation in soil properties with depth, the seeds of many weed species are able to detect 1) when they are near the soil surface and 2) when competing vegetation and dead organic materials have been removed. Consequently, a flush of weed emergence usually follows tillage, provided the soil is warm and moist enough for seed germination and the seeds are not in a seasonal dormancy state (see the next section).

The response of weeds to soil disturbance and cues associated with a bare soil surface provides a means for controlling them when they germinate. Tillage can induce a flush of new seedlings that can then be killed with cultivation. Conversely, if the weeds are killed without further soil disturbance, for example by flaming when establishing a stale seedbed, then relatively few additional seedlings will establish. Alternatively, keeping the soil cool and dark with a mulch or with a dense crop canopy will suppress most additional germination.

SEASON OF WEED EMERGENCE

Most weed species have a particular season of the year in which they emerge most abundantly. The seeds of many weed species are dormant when shed from the parent plant. These must go through some period of aging (after-ripening), a period of cold or some other process before they are capable of germination. Dormancy in weeds is generally either 1) a physiological process in which biochemical changes must occur within the seed to ready it for germination or 2) the result of a hard seed coat that prevents water from entering the seed. Often, weeds with relatively large seeds like hedge bindweed and velvetleaf rely on hard seed coats to maintain dormancy, whereas small seeded weeds like common lambsquarters and common ragweed rely on physiological mechanisms. However, many exceptions to this pattern also occur.

Physiological Dormancy

Many weed species experience an annual dormancy cycle in which dormancy is broken by some mechanism but is then re-established if the seed does not receive appropriate cues for germination or if soil conditions are unfavorable (Baskin and Baskin 1985). Spring germinating weeds usually either require a cold treatment to release their seeds from dormancy (for example, common ragweed) or seeds undergo a gradual loss of dormancy called after-ripening (for example, the foxtails). If they do not germinate, eventually the higher temperatures of summer may restore them to a dormant state. They then must pass through another winter before they can germinate. Other species may be held in a dormant state by high temperatures but become capable of germinating again in the fall when temperatures are lower (for example, shepherd's-purse). These species tend to establish in both the spring and fall. Species that germinate only in the fall typically require a period of high temperatures to release them from dormancy and then germinate when temperatures again lower (for example, purple deadnettle). However, few common agricultural weeds behave in this way. Summer germinating species, like common purslane, have high temperature thresholds for germination and may germinate directly after falling from the parent plant if the soil is warm enough. But except in regions with long, hot summers, the temperature will often have dropped by the time the plants shed seeds.

Hard Seed Coat Dormancy

In a few species with hard seed coat dormancy, water can enter the seed, but the hard seed physically restricts growth of the embryo (for example, mayweed chamomile). Most species with hard seed coats, however, maintain dormancy by preventing entry of water into the seed. They have a specific area, usually near the point where the seed attached to the parent plant, which changes to allow water entry (Baskin and Baskin 2000). Aging of the seed or repeated wetting and drying of the outer layers of the seed coat change the structure of this region so that a crack forms or the tissue softens to allow water to enter. Damage to the seed coat by insects, fungi or abrasion can also allow the seed to take up water and germinate. But under field conditions, these mechanisms often result in death of the seedling due to attack by pathogens or germination at an inappropriate time or depth in the soil (Baskin and Baskin 2000). Since winter provides a long period in which the processes that open the pore can act on the seed, often species with hard seed coat dormancy show a peak of germination in the spring. These species also usually continue to germinate sporadically throughout the growing season as the seed coats of additional individual seeds become permeable.

Consequences of Seasonal Germination for Crop Rotation

Crops that are planted in any given season compete primarily with weeds that characteristically germinate during that season. Earlier germinating weeds are wiped out during tillage and seedbed preparation. Species that germinate later, after the crop is well established, usually pose few problems since the crop is more competitive, and if it is a non-competitive crop, at least it will be sufficiently well rooted to stand aggressive cultivation.

Thus, different sets of weed species are typical of spring, summer and fall planted crops. For example, common lambsquarters, common ragweed and foxtail grasses are often abundant in spring planted crops; purslane and pigweeds plague summer planted vegetable crops; and downy brome, mayweed chamomile and shepherd's-purse are common in fall planted grains. An important consequence of this variation in the dominant weed species associated with the variation in season of crop planting is that *rotation between spring, summer and fall planted crops tends to interrupt most weed life cycles and prevent any one suite of species from becoming extremely abundant.*

SEED LONGEVITY

If conditions are not suitable for germination, a seed may remain alive in the soil for years. How long the seed survives in the soil depends both on the characteristics of the species and on soil conditions. Since soil disturbance tends to prompt germination of most weed species (see "Seed Germination: Why Tillage Prompts Germination"), the nature and frequency of soil disturbance is particularly critical in determining how long seeds survive in the soil.

Stating how long the seeds of a species survive in the soil is difficult. Seed longevity is usually reported in literature sources as the percentage of seeds alive after some number of years. However, such numbers do not reflect the nature of seed mortality in the soil. Unlike healthy humans that mostly die of old age, seeds in the soil tend to die at a constant rate (Roberts and Dawkins 1967). That is, the same percentage of the seeds that are still left die each year, regardless of how many years they have been in the soil (Box 2.1). This means that if the soil seed bank is not replenished by reproduction, the number of seeds in the soil declines relatively rapidly at first, but it also means that some seeds will persist for many years. This is the primary

reason complete eradication of most weeds from a field is so difficult.

Because seed mortality is best expressed as the percentage of seeds that die in a year, we report seed mortality in this way in the discussion of individual species whenever the data make such a computation reasonable. Seed survival depends on weather (particularly for recently shed seeds), the presence or absence of seed predators, soil conditions and management. Consequently, all types of seed survival data should be used only as a general indicator of a species' persistence in the soil.

Weed seeds usually die in one of three ways: 1) they begin to germinate in conditions that do not allow establishment, 2) they are eaten by seed predators or 3) they die from physiological breakdown. The relative importance of the three mechanisms may be in the order just listed. Most of the species that have been studied systematically, however, have been grasses with relatively large seeds; small seeded and broadleaf species may behave differently.

Although most weed species possess mechanisms for determining appropriate seasons and conditions for germination (see "Season of Weed Emergence" and "Seed Germination: Why Tillage Prompts Germination"), these mechanisms do not work perfectly. Consequently, many seeds germinate too deep in the soil

Box 2.1. Seeds usually die in the soil at a constant rate. The number of seeds left after one year can be determined by multiplying the initial number in the soil by the proportion that survive for a year. The proportion surviving is equal to (1 - (percentage mortality / 100)). Repeated multiplication by the proportion surviving gives the number left after any given number of years.

Year	20% Annual Mortality		50% Annual Mortality	
	Number/ft^2	Computation	Number/ft^2	Computation
Initial	1,000		1,000	
After 1 year	800	=1,000 x (1 - 0.2)	500	=1,000 x (1 - 0.5)
After 2 years	640	=800 x (1 - 0.2)	250	=500 x (1 - 0.5)
After 3 years	512	=640 x (1 - 0.2)	125	=250 x (1 - 0.5)
After 4 years	410	=512 x (1 - 0.2)	62	=125 x (1 - 0.5)
After 5 years	328	=410 x (1 - 0.2)	31	=62 x (1 - 0.5)
After 6 years	262	=328 x (1 - 0.2)	16	=31 x (1 - 0.5)
After 7 years	210	=262 x (1 - 0.2)	8	=16 x (1 - 0.5)
After 8 years	168	=210 x (1 - 0.2)	4	=8 x (1 - 0.5)
After 9 years	134	=168 x (1 - 0.2)	2	=4 x (1 - 0.5)
After 10 years	107	=134 x (1 - 0.2)	1	=2 x (1 - 0.5)

for successful emergence. Others begin to germinate and then dry out and die, or they begin to germinate and are then killed by soil organisms.

In general, seed predators are most effective against seeds that are on or near the soil surface. Consequently, if weeds have gone to seed in a field, fall tillage will bury and protect the weed seeds. Ground foraging birds and mice are major predators on seeds greater than 2–5 milligrams, whereas ground beetles of the carabid family are among the major predators of smaller seeds. Earthworms consume grass seeds and digest a substantial proportion of the ones they eat. The action of earthworms on broadleaf weeds has not been studied much, but large night crawlers drag seeds into burrows where many germinate, die, rot and are subsequently eaten by the worms.

Although seeds are inanimate creatures, they are metabolically active. In the soil, most species of weed seeds persist in a moist condition and are capable of metabolic repairs (Villiers and Edgecombe 1975). Eventually, however, damage to membranes, genetic mistakes and toxins accumulate and cause loss of vigor. This happens most quickly at high temperatures and when the seeds are damp but not fully moistened (Villiers and Edgecombe 1975). Since both of these conditions occur most often near the soil surface, seeds near the surface that remain dormant deteriorate quickly. In contrast, seeds buried deep in the soil remain cold and wet, and retain viability for longer periods.

Thus, seed survival tends to improve with the depth of burial in the soil (Figure 2.6). The soil surface is a deadly environment for weed seeds, and the top few inches are an unhealthy environment. In contrast, seeds tend to survive well deep in the soil where they stay cool and moist and are relatively safe from animal activity. Some species, like velvetleaf and common lambsquarters, survive well when buried. Others, like the galinsogas and downy brome, survive poorly when buried, but no species survives well on the soil surface.

Although weed seeds die most rapidly near the soil surface, tillage that incorporates seeds into the soil often reduces the number of weeds that establish in the next crop after the seeds are produced. Whether or not those seeds will return to the surface during subsequent tillage events depends on the death rate of the seeds deep in the soil and how easily seedlings can reach the surface when the seeds are more than slightly buried. The complexity of managing seed banks with tillage is discussed more fully in Chapter 4.

A few generalizations about the persistence of different taxa are possible. Few grass species persist well in the soil for more than a few years, whereas many broadleaf species form long-lasting seed banks. Wetland species often survive well in the soil and then establish profusely when a dry period exposes the mud to light and air. Finally, large seeds and elongated seeds tend to survive more poorly than small, round seeds (Thompson et al. 1993). This last generalization may stem from the difficulty large and elongated seeds have in entering untilled soil. Mechanisms allowing long-term survival would provide little advantage for seeds stuck on or very near the soil surface. Many exceptions to these generalizations exist. For example, the small seeds of broadleaf galinsoga species survive poorly in the soil, whereas the large seeds of velvetleaf persist well in the soil provided they are hard when they are shed. Nevertheless, these rules of thumb may be useful when managing a species that is not covered in the later chapters of this book.

DEPTH OF SEEDLING EMERGENCE FROM THE SOIL

In the discussion of individual species, we give the optimal and maximum depths of emergence for each species for which this has been measured. Emergence

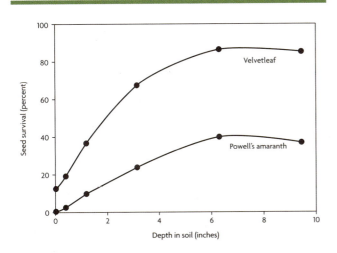

Figure 2.6. Survival of seeds of velvetleaf and Powell amaranth (a pigweed species) at different depths in the soil over the course of one year (spring to spring). Data are for the survival of seeds that did not produce emerged seedlings. The apparent slight decline in survival at the maximum depth tested is not statistically significant (Mohler, unpublished data).

Table 2.2. Growth Rates of Some Weeds and Crops in Relation to Seed Size (from Seibert and Pearce 1993)

Species	Seed Weight (mg)	Initial Growth Rate (mg/day)	Relative Growth Rate (mg/mg/day)
Common lambsquarters	0.41	0.14	0.35
Velvetleaf	7.8	1.9	0.24
Cocklebur	38	7.1	0.19
Sunflower	61	12	0.2
Soybean	158	24	0.16

from various depths is affected by soil properties (Gardarin et al. 2010), and thus studies sometimes vary in the proportion of seedlings emerging from a given depth. For most species, emergence is poor to moderate at the soil surface, increases rapidly to a peak with shallow soil covering and then decreases exponentially as burial depth increases. The increased emergence with slight burial is due to increased germination from improved soil-seed contact (Grundy et al. 2003).

The ability of seedlings to emerge from various depths is a major factor in how the species responds to tillage following seed production. If the weed seeds are incorporated to the optimal depth for emergence, then you will likely see more seedlings than if the seeds are left too shallow or too deep for good emergence. Subsequent tillage events redistribute seeds that were previously buried, however, and as explained in the "Seed Longevity" section, depth of burial also affects seed survival. The interaction of seed survival, depth of emergence and the redistribution of seeds by tillage implements is discussed in the section "Tillage Effects on Weed Seedling Density" in Chapter 4.

The ability of weed seedlings to reach the soil surface is roughly related to seed size, although other factors are also involved (Grundy et al. 2003, Gardarin 2010). The giant seeds of bur cucumber emerge well from anywhere in the plow layer, whereas the little seeds of the two galinsoga species can barely emerge from a depth of 0.25 inch. Most species with seeds less than 1 milligram emerge poorly from depths greater than 1 inch, and only a few weed species have seeds that allow a high rate of emergence from deeper than 2 inches.

The relationship between seed size and depth of emergence is far from exact, however. Grass seeds of a given size often emerge from deeper in the soil than broadleaf weeds with seeds of the same size, though

this has not been studied systematically. Also, long seeds and seeds with an uneven surface tend to germinate exceptionally poorly on the soil surface due to poor soil-seed contact.

GROWTH AND COMPETITION FOR LIGHT

The growth rate of a plant depends on its size. For annual weeds and annual crops (plants that complete their life cycles in less than one year), larger plants usually grow faster than small ones, at least until they begin to mature and redirect resources to reproduction. Since weeds tend to have small seeds (see "Seed Weight"), their seedlings are usually much smaller than crop seedlings and transplants. Consequently, the growth of small seeded weeds (weight gain per day) as seedlings is usually slow compared to crops, simply because the crops have more leaf area per plant (Table 2.2). This gives the crops an initial advantage. If managed carefully, the crops can grow up above the weeds and shade them, thereby slowing further growth.

The tiny plants that emerge from little seeds, however, tend to have a higher *relative* growth rate (weight gain/unit weight/day) (Table 2.2). This is largely because they have more photosynthetic area relative to the weight of non-photosynthetic tissues, like roots and the inside of stems. Due to their small seed weight and to other factors (see "Nutrient Use"), weed seedlings have some of the highest relative growth rates recorded for any plant species. Consequently, small seeded weeds start out life weak and non-competitive, but if given a chance, they rapidly grow large enough to compete with crops.

If a plant already occupies a piece of ground, other plants will have difficulty displacing it. Its roots already fill the most opportune crevices in the soil and have already multiplied to tap tiny pockets of high nutrient concentration. Furthermore, during dry spells, an established root system can tap moisture in deeper layers of soil that will·be unavailable to newly germinated seedlings. Even more importantly, an established plant casts shade. It intercepts light and uses it for further growth. In the process, it deprives newly established plants of the light they need to grow. Thus, plant competition is governed by the rules of "first come, first served" and "those who have will get more." Thus, *two critical principles of weed management are 1) always maintain a positive size difference between the crop and the weeds, and 2) ensure that crops occupy as much of the space available as possible.* If you can implement these two principles, the crop will do most of your weed management for you.

PHOTOSYNTHETIC PATHWAY

Various plant species perform photosynthesis by different metabolic pathways. The most common of these are referred to as the C_3 and C_4 pathways. The biochemical details of these photosynthetic pathways do not require discussion here but can be found in any botany text and in most general biology texts. The two pathways do, however, give plants contrasting ecological behaviors. Thus, knowing whether a species uses the C_3 or C_4 pathway can help you understand its growth requirements.

In general, C_3 plants thrive under cool, moist conditions, although some species tolerate warmer, drier conditions as well. Their rate of photosynthesis reaches a maximum at some fraction of full daylight. In contrast, C_4 plants reach peak performance at high temperatures and are often drought tolerant. Their photosynthetic rate does not reach a peak even at full sunlight intensity. Thus, photosynthetic pathway is one indicator of the season and habitat favored by a species. Most broadleaf weeds and cool season grass weeds like quackgrass use the C_3 pathway. Most warm season grasses like bermudagrass and the foxtails use the C_4 pathway. In addition, a few broadleaf weeds, including the pigweeds and common purslane, use the C_4 pathway.

C_3 plants substantially increase their photosynthetic rates and growth rates as the concentration of atmospheric carbon dioxide increases. Experiments show that C_3 weeds like common lambsquarters and Canada thistle will likely become worse pests as carbon dioxide concentrations continue to rise due to the burning of fossil fuels and clearing of tropical forests. In contrast with C_3 weeds, carbon dioxide concentration has little

Table 2.3. Effect of Increased Atmospheric Carbon Dioxide on the Relative Competitive Ability of Crops and Weeds as Affected by Photosynthetic Pathway (Ziska 2001)

	C_3 Crop	C_4 Crop
C_3 weed	Depends on the species	Favors the weed
C_4 weed	Favors the crop	Depends on the species

effect on C_4 species. Although C_4 weeds will probably not become less vigorous with increased carbon dioxide, competition between C_4 weeds and the many C_3 crops like wheat, soybeans, potatoes and pumpkins will likely tip in favor of the crops (Table 2.3).

Competition, however, of C_4 crops like corn and sorghum with C_3 weeds will likely favor the weeds. The future competitive balance between C_3 crops and C_3 weeds is difficult to predict in general, since it depends on the relative responsiveness of the particular crop and weed to increased carbon dioxide (Table 2.3).

The shifting competitive balance between C_3 and C_4 plants due to increased carbon dioxide will be affected by changes in climate. For example, the warmer, drier climate predicted for the Southeast may favor C_4 weeds and crops relative to C_3 crops and weeds despite the fertilization effect of carbon dioxide on the C_3 species. In contrast, the more modest changes predicted for the West Coast may allow the carbon dioxide fertilization effect to be expressed.

SENSITIVITY TO FROST

No species tolerates a rapid change from summer-like temperatures to sub-freezing conditions. Most temperate species, however, adapt to temperatures that are low but above-freezing by a variety of physiological changes that can include modification of the structure of membrane lipids, an increase in soluble carbohydrates that essentially act as antifreeze, and, in some species, an increase in the amino acid proline (Hughes and Dunn 1990). In some weed species and winter hardy crop cultivars, these changes allow the plant, after a period of acclimation to low but above-freezing temperatures, to tolerate freezing with little or no damage.

Weed species vary greatly in their tolerance to frost. Some, like hairy galinsoga, die with even a mild frost, whereas others, like field pennycress, tolerate freezing at 7°F. Such species can persist through even a harsh northern winter, particularly if protected from drying winds by snow cover, crop debris or a cover crop. Many other species, like wild mustard, can tolerate temperatures a few degrees below freezing, but not much lower. Knowing the frost sensitivity of your various weed species allows you to judge whether late emerging individuals will be controlled by frost before they set seed or whether some cleanup action is required.

Most perennials survive low temperatures by dying back to the ground and persisting in the soil as rhizomes or storage roots. These belowground structures may be sensitive to freezing but escape it below the freezing depth of the soil. If the storage organs are primarily in the plow layer, sometimes partial control can be achieved by working them to the surface where they will freeze, or by plowing frozen ground to expose them to sub-freezing soil temperatures.

DROUGHT TOLERANCE

In irrigated cropping systems, withholding water for a few days before and after cultivation will help prevent weeds from rerooting. Similarly, in rainfed systems, timing cultivation with weather and soil conditions in mind can improve cultivation effectiveness. With the exception of drought stress on uprooted weeds, drought is rarely a viable weed management strategy since crops also require adequate water for good growth. Nevertheless, understanding your weeds' response to drought can help inform your management decisions when drought does occur.

A few, mostly cool-season species like common chickweed and henbit, are quite sensitive to dry conditions and typically die back, often completely, after a few warm, dry periods of a week or more. Most weeds, however, like many crops, can survive a few weeks of dry weather and then recover when water is again available. A few weeds, like Palmer amaranth and common purslane, are physiologically adapted to withstand high temperatures and prolonged drought. Their anatomy, enzyme systems and gas exchange processes allow them to maintain growth when experiencing water stress. Since drought usually occurs during hot weather with intense sunlight, the characteristics that adapt these species to drought cause them to grow poorly during cool weather and make them susceptible to dense shade. Other species, like shattercane and field bindweed, are susceptible to drought as seedlings but develop exceptionally deep root systems that allow

them to access water that is unavailable to most crops. Both types of drought tolerant weeds are likely to be highly competitive during dry conditions. Although they always grow faster if given adequate water, the difference in growth rate between the crop and a highly drought tolerant weed is likely to be less under moist conditions. This understanding can help in allocating limited irrigation water among fields with different weed problems. Although deliberate drought is rarely useful on a field-wide basis, as explained in Chapter 3, benefits can be obtained from applying water in ways that maximize uptake by the crop rather than by the weeds.

MYCORRHIZAE

Mycorrhizal fungi are species that live in close and usually mutually beneficial association with plant roots. Several types of mycorrhizal associations exist, but most mycorrhizal crop and weed species form arbuscular mycorrhizal associations. In these, the fungi penetrate the cells of the plant roots as well as spread into the soil beyond the roots. Carbohydrates are passed to the fungus from the plant, and the fungus passes nutrients and water to the plant. This exchange is believed to occur in special bodies, arbuscules, that the fungus makes in cells of the plant root. Mycorrhizal fungi should not be confused with rhizobium bacteria that fix nitrogen in the roots of legumes.

Although the great majority of flowering plants form mycorrhizal associations, many weed species and some crops do not. In particular, plants in the mustard, goosefoot, pigweed, smartweed, pink and sedge families usually do not (Jordan et al. 2000). In contrast, both crops and weeds in the grass, aster, legume and nightshade families usually form extensive mycorrhizal associations.

The ability of mycorrhizae to assist host plants by obtaining nutrients, especially phosphorus, and by reducing stress from drought, non-optimal pH, salinity and soil toxicity is well established. Mycorrhizal fungi also help prevent infection by soil pathogens. Nevertheless, if the soil is in good condition and the plant can obtain all the nutrients and water it needs without the help of the fungus, then supporting the fungus may be detrimental. This may be why many weeds adapted to highly fertile conditions, like those frequently found in agriculture, have lost their fungal associates during evolution.

Some agricultural practices are detrimental to mycorrhizal fungi. These include tillage, bare fallow periods, chemical fertilizers, manure-based amendments that are high in P, and rotations with non-mycorrhizal crops (Jordan et al. 2000). Prolonged use of such practices can lead to low populations of mycorrhizal fungi. In contrast, reduced tillage systems and most cover crops promote dense populations of mycorrhizal fungi. When conditions are favorable to mycorrhizal fungi, they can make a mycorrhizal crop more competitive against non-mycorrhizal weeds and can greatly reduce weed growth (Jordan et al. 2000).

NUTRIENT USE

When you disturb the soil, say by spring tillage, you incorporate oxygen into the soil and release the carbon dioxide that has built up in the soil profile through the respiration of plant roots and soil organisms. Soil temperatures also increase because plants and dead plant residues are worked into the soil and no longer shade the soil surface. The increases in oxygen and soil temperature stimulate microorganisms that decompose soil organic matter. As they consume organic matter, they release the nutrients in the plant tissues. Weeds are well adapted to take advantage of the pulse of nutrients that accompanies the breakdown of organic matter following soil disturbances.

Small seeded species, including most agricultural weeds, have small diameter roots. In contrast, the roots of large seeded crops usually have a greater diameter. This difference in root diameter has important consequences. First, the small diameter of weed roots allows them to grow longer more quickly than large seeded crops. Essentially, the energy gathered by the leaves is packaged into great lengths of fine roots instead of into a shorter length of thick roots. As one example, redroot pigweed at emergence had a root length to weight ratio eight times higher than that of domestic sunflower, and after 28 days of growth, pigweed plants had a total root system twice as long as the sunflower plants (Seibert and Pearce 1993). The difference in root length means that weeds are better at thoroughly exploring the soil for nutrients that are released from decaying organic materials. Second, the greater length of smaller diameter roots provides a greater surface area for nutrient uptake.

Several studies have examined the concentration of mineral nutrients in weeds. Weeds commonly have one to three times higher concentrations of N, P, K, Ca and Mg in their tissues than the crops with which they compete (Vengris et al. 1955). This deprives the crop and gives the weed a competitive advantage. High nutrient concentrations are probably one factor contributing to the generally high relative growth rate of weeds, since those nutrients supply the metabolic machinery of the weed with the materials needed for rapid growth. High tissue concentrations of nutrients stored early in the life of a weed may also allow it to continue growth later in the season when soil nutrients may become scarcer.

Whereas small seeded weeds rely on rapid acquisition of soil nutrients to achieve their extraordinarily high relative growth rates, large seeded crops and transplants can achieve good growth rates at moderate soil fertility levels because they are partially supported by nutrients stored in the large seed or in the transplant plug. Consequently, *nutrient sources that release steadily as the crop grows tend to favor crops, whereas a large pulse of nutrients at planting tends to favor weeds.*

The high mineral content of agricultural weeds has two other interesting consequences. First, many weeds are highly nutritious food for humans and livestock. Second, however, the nitrate content of some weed species, for example, redroot pigweed and black nightshade, can reach levels that are toxic to livestock when growing on highly fertile ground.

Not only do weeds concentrate nutrients, but they also often continue to respond to additional nutrients with increased growth even after crops have reached their maximum yield. For example, following incorporation of clover and alfalfa, several important weed species responded to rates of composted chicken manure far beyond the point at which corn ceased to increase dry weight (Figure 2.7). This result highlights that when compost is applied in excess of crop needs, increased weed problems may result. Even though lettuce continued to respond to high application rates in this case, the increased yield might not be worth the additional cost of keeping the crop free of weeds (Figure 2.7).

In another experiment, poultry litter composted with wood fiber was applied at various rates in an organic corn-soybean-spelt rotation. The spelt was overseeded with red clover, and in the fourth year the clover was incorporated to supply nitrogen for the next corn crop. No compost was applied for the corn that year. Corn yield did not respond to the various levels of residual nutrients in the soil from applications the previous three years, but several annual weed species grew significantly larger at the higher nutrient rates (Figure 2.8). This experiment illustrates the danger of creating lasting weed problems through over fertilization with organic nutrient sources like compost and manure. Unlike mineral salts that are often quickly leached or converted to unavailable forms in the soil, organic materials can continue to supply nutrients to

weeds for several years after application. Phosphorus is particularly prone to buildup with repeated heavy applications of manure or compost, and many weed species respond strongly to high rates of P (Blackshaw et al 2004). We believe that many growers are creating severe weed problems through over-application of manure and compost.

ALLELOPATHY

Plants compete for light, mineral nutrients and water, but sometimes they also "cheat" and slip some poison

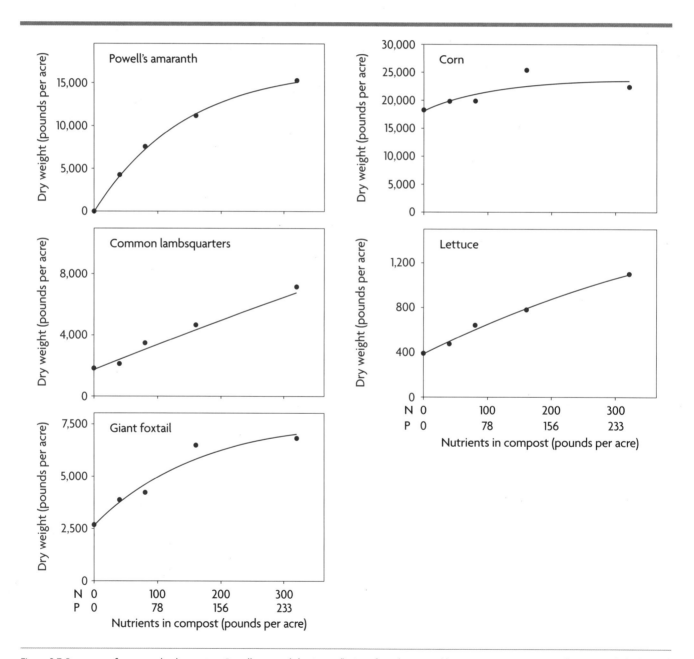

Figure 2.7. Response of common lambsquarters, Powell amaranth (a pigweed), giant foxtail, corn and lettuce to increasing rates of composted chicken manure. Rates of composted chicken manure are expressed as the amount of N and P_2O_5 contained in the compost (redrawn from Little et al. 2015).

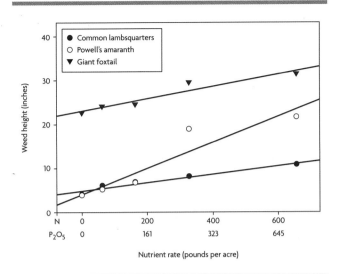

Figure 2.8. Response of weed size to residual nutrients (Mohler, Bjorkman, Martens and DiTommaso, unpublished data). Nutrients were applied as composted poultry litter over a three-year corn–soybean–spelt crop rotation. Spelt was overseeded with red clover, and the red clover was plowed under prior to planting corn in the fourth year. No compost was applied in the fourth year, but weed size responded strongly to increasing residual nutrient levels from the previous years. In contrast, corn yield did not differ across the nutrient treatments. Nutrient rates are the total applied over the first three years of the crop rotation.

to potential competitors. This is known as allelopathy.

Most plants are full of chemical compounds that are not involved in the processes of growth and reproduction. These compounds are referred to as "secondary" because they play no direct role in plant metabolism. Many of these substances are, however, toxic to a variety of other organisms. Sometimes they are toxic to the plant itself and must be stored in special organs or compartments within the plant's cells where they will not cause harm. The best explanation for the presence of these compounds is that they help defend the plant against insects and diseases. When tissues are broken by an insect's chewing or by the invasion of a fungus, the toxins poison the attacker. You are familiar with many secondary plant compounds because they provide the pungent odor of onions and the characteristic taste of cabbage and carrots. In most vegetable crops, however, the levels of secondary compounds have been reduced by centuries of plant breeding to make the produce more palatable.

Some species release such large amounts of secondary compounds into the environment that they poison adjacent plants. Weeds that do this include shattercane and common milkweed. Many crops and cover crops, including rye, barley, sorghum-sudangrass and buckwheat, have allelopathic properties, and that activity probably contributes to their effectiveness in suppressing weeds. Farmers who use straw mulch in vegetable production employ allelopathy in their weed management. Transplants and most large seeded crops are relatively immune to the allelopathic compounds from straw. The allelopathic substances move only a fraction of an inch in the soil. Since large seeded crops are planted below this toxic zone, their germination and root growth are not affected (Figure 2.9). The shoot quickly pushes up through the toxic layer. In contrast, most small seeded weeds arise from the top inch or less of soil, where the allelopathic compounds are concentrated. In addition, large seeded seedlings have a greater capacity to metabolize toxins than do seedlings of small seeded species. Consequently, most crops are relatively immune to allelopathic toxins from straw compared to the majority of weed species.

Incorporated cover crops can also release allelopathic compounds that kill weed seedlings. These compounds are released from tissues as they decompose and are generally most active during the first one to two weeks after incorporation, after which they are degraded. They can also inhibit crop growth, though some crops, like corn, seem to be more sensitive than others, like soybeans. Allelopathy is one of several reasons to consider delaying planting for one to two weeks after incorporating a heavy stand of a cover crop.

Allelopathy is notoriously unpredictable. The effect will depend on weather, soil type, soil organic matter content and the particular species and growth stage involved. Researchers often find no effect in field trials, even when the same material shows potent effects in the laboratory. Researchers and farmers are still working out the rules for using allelopathy for weed management.

RESPONSE TO SOIL PHYSICAL CONDITIONS

All plant species have habitats in which they grow best. For agricultural weeds, the principal habitat condition required is soil disturbance, and physical and chemical characteristics of the soil generally are only secondarily important. Thus, most agricultural weeds grow well on a wide range of soil textures from sand to clay, provided moisture and fertility (see "Nutrient Use") are available. Nevertheless, a few species do thrive better on particular soil textures. For example, the sandburs do best on sandy soils, and giant ragweed grows most

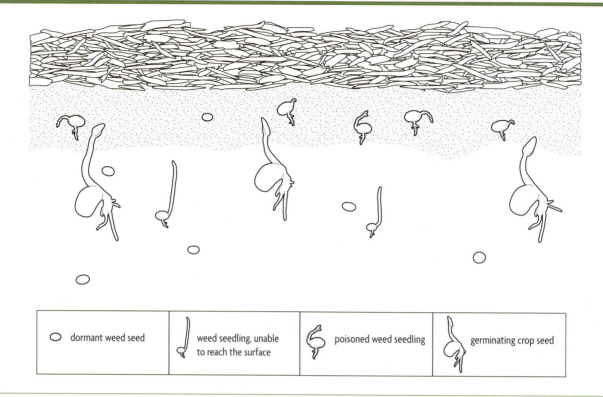

| dormant weed seed | weed seedling, unable to reach the surface | poisoned weed seedling | germinating crop seed |

Figure 2.9. Allelopathic effect of straw mulch on weeds. Allelopathic compounds leaching from the mulch are concentrated near the soil surface. Small seeded weeds germinating in the allelopathic zone are poisoned, whereas those germinating below the allelopathic zone exhaust seed reserves before reaching the soil surface. In contrast, large seeded crops are planted below the allelopathic zone and have sufficient seed reserves to reach the surface. Small seeded weeds are also physically hindered by the mulch more than large seeded crops are.

vigorously on medium- to fine-textured soils.

The ancestors of several agricultural weeds were species of wetlands or periodically flooded land, such as yellow nutsedge, waterhemp and barnyardgrass. Accordingly, such species tolerate poor soil drainage and are favored over crops in such conditions. At the other extreme are species like Palmer amaranth and common purslane, which are adapted to deserts and droughty soil. They naturally become intensely competitive with crops during drought. Some species such as kochia and Russian-thistle are well adapted to saline soils. When poor irrigation practices allow buildup of salt in the soil, these species thrive, placing the crop under double stress from salt and competition, particularly for water.

Few crops are well adapted to compacted soil. Although no weeds actually grow better in compacted soil than in friable, uncompacted soil, some species, such as giant foxtail, are inhibited much less by compaction than are the crops with which they compete.

In general, for every soil condition stressful to the crop, there will be some weed species that can take advantage of it. Relieving stress conditions through tile drainage, irrigation or better management of soil tilth will not eliminate all weeds, but it may provide leverage in the management of particular problem species.

RESPONSE TO SHADE

Competition for light is a primary factor determining the outcome of crop-weed interactions (see "Growth and Competition for Light"). Although shade from a crop generally reduces weed growth, weed species greatly differ in their capacity to tolerate shade. Weed scientists conduct shading experiments with shade cloth to isolate shading effects from those imposed by crops, where both competition for light above ground and for nutrients and water below ground can occur. Weeds that are relatively short in stature respond variously, with weeds such as common purslane and yellow nutsedge being relatively shade sensitive, while weeds such as plantains and large crabgrass are relatively shade tolerant. Thus, the shade sensitive species are often most troublesome in vegetable crops with incomplete leaf canopies, and the shade tolerant species are found in pastures or lawns with a more complete leaf canopy.

The response of tall growing weeds to shade has been studied more thoroughly. Generally, at intermediate levels of shade, these species (e.g., common lambsquarters, giant foxtail, jimsonweed, pigweed species and velvetleaf) increase in height but decrease in stem thickness and number of branches as they attempt to grow above a competing crop. Responses also can include production of larger, thinner leaves that increase the surface area available for intercepting the light that is available. Common ragweed has a temperature dependent response whereby growth is only reduced by moderate shade at lower temperatures but not at higher temperatures. Common cocklebur has the capacity to maintain function of lower leaves within the crop canopy while increasing light capture by upper leaves when lower leaves are shaded. Thus, it is highly adapted to compete with crops of intermediate height such as soybeans.

At very high levels of shade, however, most species will become uncompetitive, and time to flowering may increase. For example, shade can double the time to flowering for Palmer amaranth. However, many species are still capable of producing seeds under heavy shade, so elimination of uncompetitive plants is still important for reducing long-term seed population levels.

THE TIMING OF REPRODUCTION

Weeds vary in their approach to reproduction. Most perennial weeds do not produce seeds in the first growing season of their life. Instead, they save resources below ground as roots and rhizomes. Most perennials then reproduce more or less continuously over many years beginning in the second year of life. Relatively few agricultural weeds, such as wild carrot, common teasle and common mullein, sometimes referred to as "biennials," produce a massive number of seeds when they reach a minimal size, and then die. Usually this occurs in the second year of life (hence the name) but may require longer if the plant is stressed or damaged. In regularly tilled fields, most "biennials" occur along edges where soil disturbance does not occur every year, or as young plants that never reach maturity. They can, however, be abundant in the second year of a hay seeding and in heavily grazed pastures.

Creeping perennials like quackgrass and field bindweed usually produce relatively few seeds, and as explained in the section on pollination, some populations of creeping perennials produce no viable seeds at all (for example, purple nutsedge).

Annual weeds also vary in their approach to reproduction. Some species flower near the end of their lifespan so that death and reproduction occur as an integrated process. Redroot pigweed and common lambsquarters are examples of species that conserve resources for a "big bang" of seed production late in their lifespan. This type of species is often capable of growing to large size in favorable conditions. Consequently, control early in the season is critical. The massive seed production from even one 3–4-foot plant will ensure a viable population for years to come.

Other species, like hairy galinsoga, common purslane and common chickweed, begin to flower and set seed while still small plants. Under favorable conditions these species will begin releasing seeds when just a few weeks old. They do not make many seeds at a time, but they continue to grow, flower and dribble seeds throughout most of their lifespan. These little plants may hide among larger crop plants, producing thousands of seeds, week after week. Most species that behave as winter annuals reproduce quickly after the weather warms and continue to produce seeds for many weeks. This reproductive behavior allows them to take advantage of periods of favorable weather conditions during a time of year when conditions are unpredictable.

Many weed species are sensitive to day length. "Short day" species like redroot pigweed and sicklepod flower more quickly as days shorten in mid to late summer. Consequently, individuals establishing earlier in spring take longer to flower and generally grow larger than those that emerge later. This behavior has a selective advantage since it allows early emerging individuals to use the full growing season to produce the maximum number of seeds while ensuring that late emerging individuals still produce some seeds. The opposite behavior, in which the plant flowers as days lengthen, is less common among weeds but is exhibited by some winter annuals, like Italian ryegrass.

The "big bang" annuals tend to predominate in full season crops that become difficult to weed as their growth approaches full size. They are often the worst weeds in fields of corn, spring-planted grains and soybeans. The "dribbling" annuals, in contrast, are better adapted to the frequent cultivation that often occurs on vegetable farms. They produce seeds before they are noticed, and often their seed dormancy (see "Season of Weed Emergence") is such that a new generation can immediately sprout to replace plants that have just been killed. Weeds are disturbance-adapted plants, and the dribblers are the most disturbance adapted of the weeds.

Because different types of weeds are adapted to different disturbance regimens, vegetable fields frequently undergo a succession of different weed communities. Often perennials and "big bang" annuals predominate on new vegetable farms as holdovers from previous uses of the land. Frequent cultivation and hand weeding may eventually bring these species under control, though they may persist indefinitely if weeding is less rigorous. Meanwhile, the dribbling annuals slowly increase due to their ability to rapidly produce seeds during the inevitable brief periods of neglect (Mohler et al. 2018). Some people say that the "dribblers" increase because they no longer face competition from the larger "big bang" species, but more likely the shift is just a natural but slow response to the change from one disturbance regimen to another.

POLLINATION

Most annual and stationary perennial weeds self-pollinate. This adapts them well to high-disturbance environments, since one lucky individual that escapes destruction can begin a new population or regenerate a population that has been nearly eliminated. Virtually all annuals and stationary perennials do cross pollinate occasionally by wind or insects. This generates genetic variation and allows the population to adapt to new conditions.

In contrast with the annuals and stationary perennials, most species of creeping perennial weeds must cross pollinate with a genetically distinct individual to produce seeds. Since populations of these species often exist as large clones of genetically identical plants, many populations of creeping perennials never produce seeds. Because their populations are maintained by vegetative reproduction, however, lack of seed production does not inhibit their spread through a field. For creeping perennials, seeds are mostly a means for long distance dispersal to other sites. Production of genetically diverse seeds by cross pollination means that these seeds are suited to a range of conditions, and some may find a new location suitable for growth.

Species with a high degree of self-pollination, and those that propagate vegetatively with only occasional reproduction from seed, sometimes exist as a series of distinct, self-propagating forms referred to as *biotypes*. Herbicide resistant weed biotypes are perhaps the best known, but others have substantial consequences for weed management. For example, wild-proso millet biotypes differ substantially in their seed color, seed dormancy, seed longevity in the soil, growth form and tendency to shatter. Similarly, dandelion, whose seeds form vegetatively without genetic mixing, has populations composed of biotypes that differ in growth form and in seed production, and these biotypes respond differently to management practices (Solbrig and Simpson 1977).

Outcrossing species are ones that form seeds primarily by cross pollination. If such species lack vegetative reproduction, they usually also lack distinct biotypes, but they may form races that differ in habitat preference or geographical distribution. For example, the "common" and "tall" forms of waterhemp are sufficiently distinct that they were formerly considered different species. Both invade fields, but the "common" type has a more western distribution than the "tall" type and is more frequent as a weed in farm fields.

THE MAGNITUDE OF REPRODUCTION

Most annual weed species produce high numbers of seeds if grown in isolation but produce considerably fewer when grown in competition with a crop. We give a range of seed output in the in the individual species chapters, ranging from typical production without competition to typical production with crop competition. These reported figures are very rough, however, because seed production is highly dependent on plant size. For a given annual weed species during a particular growing season, the number of seeds produced is generally proportional to the weight of the plant. The size of the plant is highly dependent on the intensity of competition from the crop and other weeds, as well as on local weather conditions.

Annual weeds generally divert about 15–30% of their resources to seeds and seed producing structures during the course of the season, with the "big bang" reproducers tending toward the upper end of this range, and the "dribblers" tending toward the lower end. Although the pattern of seed production over the life cycle of the species has some effect on seed production, the main plant attribute that determines the number of seeds produced is seed size: On plants of the same size, species with large seeds produce fewer seeds than small seeded species. Large seeded annual weeds like common cocklebur invest a lot in each offspring and are rewarded, on average, by relatively low seedling mortality. In general, the high disturbance rates in agricultural fields have selected for species that package their reproductive output into many very small seeds. Since most will die, only individuals that produce many seeds have their genes represented in the next generation, and those genes, in turn, program for plants that also produce many small seeds.

Because annual and biennial weeds are such prolific producers of seeds, preventing or reducing seed production is an important component of weed management. Reducing seed production generally involves eliminating early flushes of seedlings, since those have the potential for producing the largest plants and the maximum seed numbers. In addition, development of a highly competitive crop canopy is important for depriving later emerging weeds of light needed for establishment and growth. For many weed-crop combinations, significant reductions in seed production can be achieved by cleaning up the field immediately after harvest when weeds can otherwise resume growth with access to full sunlight. Various management approaches are discussed in chapters 3 and 4.

Within a population, plants vary greatly in size and thus in the number of seeds they produce. The distribution of plant size (and seed production) within a population of annual weeds is often a lopsided curve with many small plants and only a few large ones. One practical consequence is that *removing the largest individuals has a disproportionate effect on reducing the seed production of the population.* Thus, hand rogueing large plants out of intensive vegetable production systems can help suppress weed populations. Weed pullers, elevated mowers and electrical discharge weeders have been developed to kill weeds that overtop row crops in land-extensive cropping systems. Such tools are discussed in Chapter 4.

DISPERSAL

Most attentive growers find a new weed species occasionally sprouting up in a field. Understanding how weeds move about in the landscape can help you prevent the arrival of new, difficult-to-manage species. This section reviews the many ways weeds can

disperse, both naturally and due to human activity. See the section "Preventing the Arrival of New Weed Species" in Chapter 3 for guidance on how you can limit the spread of weeds.

Weeds disperse in a variety of ways. Some of these, like the explosive shattering of yellow woodsorrel capsules or the caching behavior of small rodents, move seeds only a few feet. Such processes are unlikely to bring new weeds into a field unless they are present in immediately adjacent habitats, like a hedgerow or a neighboring field. A few agricultural weeds like dandelion and common milkweed disperse by means of hairs attached to the seed that provide buoyancy for travel on wind currents. Most individual seeds of wind-dispersed species travel only a few yards or less and thus remain in the field of origin (Figure 2.10). Occasionally, however, wind-dispersed seeds get caught on updrafts and are carried for long distances. Horseweed seeds are particularly suited to this form of long-distance dispersal. Other weeds break off and roll over the ground as "tumbleweeds," dispersing seeds as they go. Such species are most common in the open grazing lands of the western United States. Russian-thistle and witchgrass behave this way and are well adapted to dispersal across recently tilled ground. A few agricultural weeds like catchweed bedstraw have prickles that catch on clothing and the fur of animals. If the person or animal cleans the seeds off outside, they have dispersed the seeds to a new location. Consequently, paying attention to where you clean the seeds out of your socks can potentially save you a lot of work.

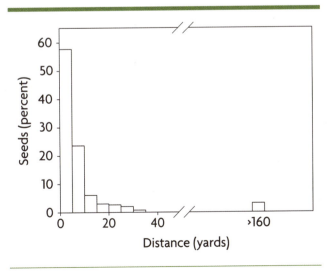

Figure 2.10. Dispersal distances of common milkweed seeds released at a height of 3.3 feet (redrawn from Morris and Schmitt 1985; 1982 data).

Finally, a few weeds, mostly in the nightshade family (for example, horsenettle and the black nightshades), make berries that are consumed by fruit-eating birds. The birds digest the fruit but regurgitate the seeds or pass them in their feces.

Despite these examples of species that possess special adaptations for dispersal, the great majority of agricultural weeds lack obvious dispersal adaptations. Most species of agricultural weeds have unadorned seeds that simply fall to the ground around the parent plant. This leads to the interesting paradox that the world's most widespread species have no apparent adaptation for dispersal! Many of the most abundant and problematic agricultural weeds are included in this category, for example, common lambsquarters, most of the pigweeds, common purslane, the foxtail grasses, bindweeds and morningglories.

Before the advent of human agriculture, seeds of agricultural weeds probably moved around the landscape primarily in the guts of grazers and in soil clinging to their feet and hides. The high persistence of most weed seeds in the soil allows them to reach typical densities of hundreds to thousands per square foot, making it highly probable they are present in soil clinging to animals. Moreover, the tough seed coat that allows most weed seeds to persist in the soil also allows them to pass unharmed through the digestive tracts of grazers. The high palatability of most agricultural weed species may not be accidental. Dispersal on and in large mammals would have placed seeds of these weeds in exactly the locations in which they thrive: ground disturbed by hooves and fertilized with dung.

Today the many agricultural weeds that lack special adaptations for dispersal are moved around the landscape by human activity. They are carried in the soil clinging to shoes, tractor tires and tillage implements. They move longer distances on car tires and wheel wells, and in the root balls of landscaping plants. They are also often abundant in barnyard manure. For farms with livestock, manure is often the means by which weed species spread from an initial invasion point to infest the whole farm. Commercially composted manure is usually free of weed seeds, but as explained in the next chapter, killing weed seeds by composting requires optimum composting conditions (see "Weed Seeds in Compost and Manure"). Since weed seeds usually remain in the guts of livestock for several days, weeds can travel long distances when livestock are moved between locations.

New weed species also arrive in crop seed. Cover crop seed is especially likely to contain weed seeds, since it is rarely certified for quality. Be especially cautious about seed produced far from your farm, since it is more likely to contain species that are new to your area.

Note that when your fields were native forest or grassland, they probably contained no agricultural weeds at all. All of them have arrived there since people began disturbing the soil, and most have arrived in the last 150 years or less. Even the species native to your region were probably not present on your farm prior to European settlement. Instead, most were relatively rare plants that existed in naturally disturbed habitats like the edges of watering holes and the banks of streams. They were brought to your fields by people not so different from you. Some thoughtful prevention can save a lot of future work.

NATURAL ENEMIES

All plant species, including all agricultural weeds, are attacked by a diversity of other organisms, including fungal and bacterial diseases, insects, mites and mammals. Except in cases where the enemy has only recently arrived in the region, these organisms do not devastate the plant population because the plant's defensive mechanisms will have evolved to blunt the enemy's attack. Consequently, examples of a natural enemy consistently controlling a weed species are few. They are mostly limited to cases of classical biological control in which an enemy species, usually an insect, is introduced from another continent to control an introduced weed. Classical biological control has had some notable successes in managing perennial rangeland weeds. However, it has not been successful for management of agricultural weeds. Agricultural weeds mature and die back before the insect population can build up. Also, frequent disturbance in agricultural systems interrupts the growth of insect populations.

Diseases occasionally devastate weed populations, but usually the conditions for rapid spread are absent or short lived and the impact of fungi and bacteria on weed shoots is minor. A few microbial strains have been developed as host-specific bioherbicides. The bacteria or fungal spores are dried for storage and distribution. They are then sprayed on the crop and weed in massive numbers, usually with chemical agents that prolong favorable moisture conditions to ensure high infection rates. Because the pathogen is host specific for one or a few related species, the weed is killed but the crop is not affected. As of the writing of this book, no bioherbicides are commercially available due to limited demand for herbicides that target only one or a few weed species.

These considerations have led many farmers and weed scientists to conclude that natural enemies have little impact on weed populations. The effects of natural enemies can be substantial, however, without being obvious. We measured the mortality of two weed species in the absence of any control measures (Mohler and Callaway 1992). Eighty percent of the first cohort of common lambsquarters emerging after tillage died before maturity in both years of the study. Later-emerging cohorts suffered even greater mortality. Some deaths were due to damping off fungi, but many individuals were just missing and were presumably consumed by insects. Results for redroot pigweed were generally similar. Weed populations are lower following successful management, but the extent to which the observed decline is due to natural enemies acting either independently of the management or in conjunction with the management practice is generally unknown. For example, fungi probably kill many seedlings that emerge in the moist, shady conditions under an organic mulch, but the contribution of fungi to weed suppression by mulch has not been assessed.

The most substantial effects of natural enemies appear to occur in the soil and on the soil surface. The heavy consumption of weed seeds by insects and small mammals is discussed in Chapter 3 (see "Promoting Weed Seed Predation After Seed Dispersal") along with methods for managing the impact of these natural enemies. In addition, earthworms consume many seeds in the soil and digest a substantial proportion of what they ingest. The effects of fungi and other microflora in the soil may be as large as macrofauna, such as insects and worms. Any seed with an even slightly damaged seed coat quickly succumbs to fungal attack. Moreover, seed coats themselves can be attacked by fungi. Fungi also kill many weed seedlings in the white thread growth stage. How to manage microflora for weed control is less clear, but we have demonstrated that incorporation of green plant materials into the soil reduces weed seedling emergence, and that the effect is associated with attack by soil fungi (Mohler et al.

2012).

PALATABILITY

The foliage of most weeds is palatable to livestock, particularly when the plants are immature. The use of weeds by livestock is reflected in common names such as henbit and pigweed. However, a few weeds, such as jimsonweed, are highly toxic (Burrows and Tyrl 2006), and some others, such as common groundsel and kochia, cause health problems when they form a large proportion of the diet for several days or weeks (Burrows and Tyrl 2006). Some other species, notably many of the pigweeds, concentrate nitrogen as nitrate when growing on highly fertile soil, and this can lead to health problems in livestock. Although some species retain palatability as they mature, some, like velvetleaf, become highly fibrous and a few, notably the sandburs, pose a physical hazard to livestock when mature. Despite these issues, many weeds are good forage for livestock (Abaye et al. 2009), and the integration of livestock into weed management plans can be useful.

The general palatability of weeds is related to their evolutionary origins. Most of the world's perennial species have physical and chemical defenses. Perennials are continuously present in the habitat, which means animals that feed on them can converge. Small animals, like insects, can build up a population to exploit the plant. Consequently, most perennial plants have tough leaves with low nutrient content, and they often contain tannins and various toxins that make the plant hard to digest. In contrast, annual plants in natural conditions largely escape their predators by being unpredictable in space and time (Feeny 1976). They sprout up from a buried seed bank following disturbance or are carried in from another location in or on animals. They then may be absent from that location for several years. Consequently, natural selection has favored diversion of resources from defense to reproduction since many seeds die in the soil or are lost on their way to a suitable disturbance. The shift from defense to escape results in high palatability. For some weeds, palatability to grazers may be an adaptation to promote the consumption of seeds, since a high proportion of most species of weed seeds pass through herbivore digestive tracts and are dispersed by this means.

Most creeping perennial weeds originated in prairies and wet meadows. Such places are generally grazed. Fast growing, disturbance-adapted weeds tolerate grazing by rapidly recovering from roots and rhizomes. They may even benefit from the passage of a herd of grazing animals that consumes slower growing competitors. The advent of agriculture provided an opportunity for the most rapidly growing and disturbance-adapted perennial meadow species to explode across landscapes.

Many weeds are suitable for consumption by humans as well as livestock, and in some cases, their desirable properties have led to development of domestic cultivars (Table 2.4). A few, like common chickweed, common lambsquarters and many of the pigweeds can be used as a salad green when young, and these and

Table 2.4. Weed Species Treated in This Book That Have Value as Forage or Human Food[1,2]

Species	Forage Quality[3]	Human Usage[4]
Grasses and sedges		
Annual bluegrass	Good	
Barnyardgrass	Good	Seeds (Japanese millet)
Bermudagrass	Good (cultivated)	
Downy brome	Good when young; mature plants irritate	
Fall panicum	Good when young; can cause photosensitivity	
Foxtail, giant	Good when young; mature plants unpalatable	
Foxtail, green	Good when young; mature plants unpalatable	Seeds (foxtail millet)
Foxtail, yellow	Good when young; mature plants unpalatable	Seeds (kora)
Goosegrass	Good	Seeds (one ancestor of finger millet)
Italian ryegrass	Good (cultivated)	
Johnsongrass	Widely cultivated; stressed plants can be toxic	
Large crabgrass	Good	Seeds
Purple nutsedge	Tubers relished by pigs	
Quackgrass	Good	
Sandbur species	Acceptable when young; mature plants hazardous	
Shattercane	Good for ruminants; can poison horses	Seeds (sorghum)
Wild oat	Good; mill screening used as feed grain	
Wild-proso millet	Good; can poison young sheep and goats	Seeds
Yellow nutsedge	Tubers relished by pigs	Tubers (chufa)
Broadleaves		
Bindweed, field	Good	
Common chickweed	Good; may contain toxic levels of nitrate when grown on fertile soil	Shoots; raw or cooked
Common lambsquarters	Good; use in moderation for sheep and pigs	Seeds; young leaves as salad greens, leaves as cooked greens
Common milkweed	Toxic	Young shoots and buds as cooked greens following several changes of water
Common purslane		Salad or cooked greens (domestic purslane)
Common sunflower	Good	Seeds (culinary and oil seed sunflowers)
Dandelion	Good	Salad or cooked greens
Dock, broadleaf	Poor	Cooked greens
Field pennycress	Poor	Young shoots as salad or cooked greens
Flixweed	Poor	Cooked greens, following changes of water
Galinsoga		Cooked greens

Species	Forage Quality[3]	Human Usage[4]
Broadleaves (cont.)		
Giant ragweed	Good	Seeds
Hemp sesbania	Good	
Henbit	Acceptable in moderation	Salad or cooked greens
Purple deadnettle	Acceptable	Salad or cooked greens
Kochia	Good (cultivated)	
Morningglory, ivyleaf	Good	
Morningglory, tall	Good	
Nightshade, black	Can be toxic but often is acceptable	Berries; leaves as cooked greens (garden huckleberry)
Palmer amaranth	Good	Seeds; cooked greens
Perennial sowthistle	Good for cattle; poor for lambs	Cooked greens
Pigweed, redroot	Good	Salad or cooked greens; seeds
Pigweed, smooth	Good	Salad or cooked greens; seeds (grain amaranth A. cruentus)
Plantains	Good; buckhorn plantain cultivated	Broadleaf plantain as cooked greens
Powell amaranth	Good	Salad or cooked greens; seeds (grain amaranth A. hypochondriacus)
Prickly lettuce	Young, fresh is toxic; mature or dry is acceptable	Extremely bitter but is the ancestor of domestic lettuce
Prickly sida	Acceptable	
Russian-thistle	Good when young and growing on saline soil	
Shepherd's-purse		Cooked greens
Sicklepod	Poor	Young leaves as cooked greens
Velvetleaf	Acceptable for sheep but not other livestock	Seeds
Waterhemp	Poor	
Wild buckwheat		Seeds
Wild mustard	Young plants acceptable; mature plants moderately toxic	
Wild radish	Yong shoots acceptable; mature plants toxic	Leaves as salad or cooked greens

[1]Additional information is given in the "Palatability" section of species chapters.
[2]Domesticated species and cultivars derived from the species are given in parentheses.
[3]A blank in the Forage Quality column indicates a lack of information.
[4]A blank in the Human Usage column indicates that the plant is not consumed by humans.

many other broadleaf weeds can be used as a cooked green. Because most annual weeds concentrate mineral nutrients at levels higher than common crops (see "Nutrient Use"), they can provide an important source of these nutrients in the diet. They also frequently contain high levels of vitamins and antioxidants. Although harvesting weeds for sale is rarely a viable control tactic, weeds can add diversity and customer interest to a vegetable business and can provide nutritious food for a farm family.

SUMMARY
Several important lessons for weed management can be deduced directly from the biology of weeds discussed in this chapter. Implementation of these principles in the field is the subject of the next two chapters.

- The relative strengths of the many processes that affect the death and reproduction of weeds change with weed density. Thus, *the goal of ecological weed management is to manipulate birth and death processes to keep the density of weed populations low most of the time and to reduce them quickly when they rise.*
- Chopping up the roots and rhizomes of perennials with tillage releases dormant buds, thereby increasing the number of shoots present in the field.

However, *the smaller the pieces, the less vigorous are the resulting plants*, and this makes them susceptible to further soil disturbance and other management procedures.

- Early season growth of perennials drains resources out of rhizomes and storage roots and makes them increasingly susceptible to management. They are most susceptible when storage reserves are lowest.

- Most weed species have a characteristic season in which seeds germinate due to mechanisms that keep seeds dormant the rest of the year. They are often less abundant in crops that are planted when the seeds are dormant. Consequently, *rotation between spring, fall and summer planted crops tends to interrupt most weed life cycles and prevent any one suite of species from becoming extremely abundant.*

- Seeds of most weed species respond to environmental cues that indicate soil disturbance, a near-surface environment and the absence of competing plants. *Manipulating these cues provides a means for controlling when weed seeds germinate, so they can be subsequently eliminated or, alternatively, so that germination is reduced in the crop.*

- Seeds of most weed species survive a few to many years in the soil. The rate of mortality varies greatly between species, but *seed mortality for all weed species increases with the frequency of soil disturbance and decreases with the depth of burial in the soil.*

- The much greater size of most crop seeds relative to most weed seeds gives the crop an initial advantage. The same advantage can be produced in many small seeded crops by using transplants. Small seeded weeds usually have a greater relative growth rate, however, and this can allow them to outgrow crops if given a chance. Thus, two critical principles of weed management are *1) attempt to always maintain a positive size difference between the crop and the weeds, and 2) ensure that crops occupy as much of the space available as possible.*

- Weeds are very good at concentrating nutrients in their tissues and saving them for later use. Hence, *nutrient sources that release steadily as the crop grows tend to favor crops, whereas a pulse of nutrients at planting tends to favor weeds.*

- Many weeds continue to respond to additional nutrients even when fertility is high. This can result in increased weed pressure when the crop receives more fertility than it can use. Hence, *avoiding over-fertilization assists weed management.*

- Most annual weed species are prolific seed producers. Consequently, *preventing seed production reduces weed pressure in future crops.* Moreover, because a few large individuals produce most of the seeds, *removing the largest individuals has a disproportionate effect on reducing the seed production of the population.*

- Additional weed species can be expected to disperse onto your farm by a variety of means. *Blocking the avenues for weed dispersal onto the farm, and quick recognition and eradication of newly arrived weed species, can greatly reduce future weed problems.*

REFERENCES

Abaye, A.O., G. Scaglia, C. Teutsch and P. Raines. 2009. The nutritive value of common pasture weeds and their relation to livestock nutrient requirements. Virginia Cooperative Extension, Publication 418–150.

Baskin, J.M. and C.C. Baskin. 2000. Evolutionary considerations of claims for physical dormancy break by microbial action and abrasion by soil particles. *Seed Science Research* 10: 409–413.

Baskin, J.M. and C.C. Baskin. 1985. The annual dormancy cycle in buried weed seeds: a continuum. *BioScience* 35: 492–498.

Blackshaw, R.E., R.N. Brandt, H.H. Janzen and T. Entz. 2004. Weed species response to phosphorus fertilization. *Weed Science* 52: 406–412.

Burrows, G.E. and D.J. Tyrl. 2006. Handbook of Toxic Plants of North America. Blackwell: Ames, IA.

Cousens, R. 1985. A simple model relating yield loss to weed density. *Journal of Applied Biology* 107: 239–252.

De Wet, J.M.J. 1978. Systematics and evolution of sorghum sect. sorghum (Gramineae). *American Journal of Botany* 65: 477–484.

Feeny, P. 1976. Plant apparency and chemical defense. In *Biochemical Interactions between Plants and Insects*, eds. J.W. Wallace and R.L. Mansell, pp. 1–40. Plenum, New York.

Garcia de Leon, D, R.P. Freckleto, M. Lima, L. Navarete, E. Castellanos and J.L. Gonzalez-Andujar. 2014. Identifying the effect of density dependence, agricultural practices and climate variables on the long-term dynamics of weed populations. *Weed Research* 54: 556–564.

Gardarin A., C. Dürr and N. Colbach. 2010. Effects of seed depth and soil aggregates on the emergence of weeds with contrasting seed traits. *Weed Research* 50: 91–101.

Grundy, A.C., A. Mead and S. Burston. 2003. Modelling the emergence response of weed seeds to burial depth: interactions with seed density, weight and shape. *Journal of Applied Ecology* 40: 757–770.

Håkansson, S. 1969. Experiments with *Sonchus arvensis* L.I. Development and growth, and the response to burial and defoliation in different developmental stages. *Lantbrukshögskolans Annaler* 35: 989–1030.

Holm, R.E. 1972. Volatile metabolites controlling germination in buried weed seeds. *Plant Physiology* 50: 293–297.

Hughes, M.A. and M.A. Dunn. 1990. The effect of temperature on plant growth and development. *Biotechnology and Biological Engineering Reviews* 8: 161–188.

Jordan, N.R., J. Zhang and S. Huerd. 2000. Arbuscular-mycorrhizal fungi: potential roles in weed management. *Weed Research* 40: 397–410.

Liebman, M., C.L. Mohler and C.P. Staver. 2001. *Ecological Management of Agricultural Weeds.* Cambridge University Press: New York.

Little, N.G., C.L. Mohler, Q.M. Ketterings and A. DiTommaso. 2015. Effects of organic nutrient amendments on weed and crop growth. *Weed Science* 63: 710–722.

Mohler, C.L., B.A. Caldwell, C.A. Marschner, S. Cordeau, Q. Maqsood, M.R. Ryan and A. DiTommaso. 2018. Weed Seedbank and weed biomass dynamics in a long-term organic vegetable cropping systems experiment. *Weed Science* 66: 611–626.

Mohler, C.L., C. Dykeman, E. Nelson and A. DiTommaso. 2012. Reduction of weed seedling emergence by pathogens following incorporation of green crop residue. *Weed Research* 52: 467–477.

Mohler, C.L. and J.R. Teasdale. 1993. Response of weed emergence to rate of *Vicia villosa* Roth and *Secale cereale* L. residue. *Weed Research* 33: 487–499.

Mohler, C.L. and M.B. Callaway. 1992. Effects of tillage and mulch on the emergence and survival of weeds in sweet corn. *Journal of Applied Ecology* 29: 21–34.

Morris, D.H. and J. Schmitt. 1985. Propagule size, dispersal ability, and seedling performance in *Asclepias syriaca. Oecologia* (Berlin) 67: 373–379.

Roberts, H.A. and P.A. Dawkins. 1967. Effect of cultivation on the number of viable weed seeds in soil. *Weed Research* 7: 290–301.

Seibert, A.C. and R.B. Pearce. 1993. Growth analysis of weed and crop species with reference to seed weight. *Weed Science* 41: 52–56.

Solbrig, O.T. and B.B. Simpson. 1977. A garden experiment on competition between biotypes of the common dandelion (*Taraxacum officinale*). *Journal of Ecology* 65: 427–430.

Taylorson, R.B. and H.A. Borthwick. 1969. Light filtration by foliar canopies: significance for light controlled weed seed germination. *Weed Science* 17: 48–51.

Thompson, K., S.R. Band and J.G. Hodgson. 1993. Seed size and shape predict persistence in soil. *Functional Ecology* 7: 236–241.

Vengris, J., W.G. Colby and M. Drake. 1955. Nutrient competition between weeds and corn. *Agronomy Journal* 47: 213–216.

Villiers, T.A. and D.J. Edgecombe. 1975. On the cause of seed deterioration in dry storage. *Seed Science and Technology* 3: 761–774.

Wesson, G. and P.F. Waring. 1969. The induction of light sensitivity in weed seeds by burial. *Journal of Experimental Botany* 20: 414–425.

Ziska, L.H. 2001. Changes in competitive ability between a C_4 crop and a C_3 weed with elevated carbon dioxide. *Weed Science* 49: 622–627.

CHAPTER 3

Cultural Weed Management

MANY LITTLE HAMMERS

Wise management of weeds in an organic cropping system involves integration of many separate management tactics. Which tactics you use will depend on the weed species present, the crop, the time of year the crop is planted, the type of equipment you have available, other crops in the rotation, and other site- and operation-specific factors. This is why understanding how weeds operate as species is so critical: Only through this understanding can you effectively match your tactics to the weed problem at hand.

Most of the individual tactics discussed in the following sections cannot control weeds by themselves. Instead, they shift the population dynamics of the weeds so that mortality increases by some percentage and consequently fewer individuals grow into the next size class. Matt Liebman and Eric Gallandt have referred to the use of multiple tactics as the "many little hammers" approach to weed management (Liebman and Gallandt 1997). Instead of attacking the weeds with a single, big hammer like an herbicide that is intended to kill off all the weeds at once, many tactics, each of which may be relatively ineffective when used alone, can be used together to accomplish successful management. For example, tactics that each reduce the number or size of weeds present by only one-half can prove highly effective in combination if they are cheap, easy and compatible with the overall cropping plan. In some cases, one tactic may enhance the effectiveness of another (synergism), so discovering multi-tactic, synergistic combinations will prove particularly effective for managing your weeds.

We introduce cultural tactics in this chapter before discussing tillage and cultivation in the following chapter because organic farmers too often place excessive emphasis on cultivation for weed control. Cultivation is an important part of weed management on many farms, but it can damage soil structure and cause crusting and compaction. If used thoughtfully, cultivation will complement cultural practices so that both become more effective.

CROP ROTATION AND WEED MANAGEMENT

Rotation between spring, summer and fall planted crops is an important strategy to reduce overall weed problems becaue it interferes with the life cycles of weed species that have preferred seasons of germination. (see the "Season of emergence" section in each species chapter.) For example, spring germinating weeds will be destroyed during seedbed preparation for summer planted crops, and relatively few individuals of those species will subsequently sprout because the season is not favorable for their germination. During the summer, fall and winter, some of the dormant seeds of the spring germinating species will be eliminated by accidental germination deep in the soil and because earthworms, carabid ground beetles and other soil organisms will consume them. (see the "Seed longevity" section in each species chapter.) Hence, a summer planted crop decreases future pressure from spring germinating weeds.

Similar processes occur when rotating between spring and fall planted crops or between summer and fall planted crops. Fall planted crops like spelt and winter wheat are well established and growing vigorously by the time spring weeds emerge. Thus, spring germinating weed species tend to suffer from severe competition in a fall seeded grain. Furthermore, you will usually harvest a winter grain before the spring weeds set seed. If the field is cleaned up after harvest, for example by light disking and planting a cover crop, seed production by spring germinating weeds can be prevented. Alternatively, if the winter grain was interseeded with forage or a clover cover crop, combining the grain will cut off immature flowering stalks of many spring germinating species, and the interseeded crop will compete heavily with the weeds as they attempt to regrow.

Some crops are hard to keep weeded and others are relatively easy. Also, the best method for weeding varies between crops. Row crops can be intensively cultivated whereas cultivation in grain crops is largely limited to harrowing. Hilling potatoes or mounding soil around the base of corn stalks in a timely fashion can kill most of the weeds present by uprooting those in the inter-row and burying those in the row. In contrast, hilling would ruin lettuce and many other short-statured vegetable crops. Late germinating weeds in corn, cotton or onion rows can be killed by flame weeding, whereas few other crops will tolerate flaming. Some crops like soybeans are naturally highly competitive and effectively suppress weeds after they are well established, while other crops like onions require regular weeding throughout the growing season. Consequently, alternating crops in which you use different weed control tactics varies the types of pressure you can apply to weed populations. This prevents strong dominance by a single hard-to-control species and lowers the abundance of most of the species present.

The length of the crop-growing period is also a critical factor affecting the promotion or control of weeds. Short cycle crops like lettuce, radishes, spinach and mustard greens that are harvested only a few weeks after planting tend to reduce weed populations if you clean up the area after harvest. They are in the ground for so short a time that the weeds do not have a chance to go to seed and thus act in a manner similar to a tilled fallow. In contrast, weeds commonly go to seed in winter squash and field corn, which have long growing seasons and are difficult to weed late in the growth cycle. Consequently, using short cycle "cleaning crops" after long cycle crops in which weeds go to seed helps keep weed populations under control.

Rotating grain and vegetable crops with forage crops can be effective in reducing weed populations. Annual forages like triticale and sorghum-sudangrass cut for silage or hay have been shown to reduce wild oat populations in a subsequent field pea crop, since cutting prevented the wild oat from going to seed. Only annual forages that continued growing the full season, however, were effective for reducing broadleaf annuals. Perennial forages such as alfalfa are particularly effective for suppressing many weed species because the repeated mowing prevents most annual weeds from going to seed and tends to deplete root reserves of many perennials. A survey of Canadian grain fields found that a wide range of weed species were less abundant when grain followed alfalfa than when it followed another grain crop (Ominski et al. 1999). Field pennycress and dandelion, however, were more abundant following alfalfa because field pennycress had time to set seed before the first cutting and because dandelion established easily in alfalfa and tolerated mowing. A study of long-term organic cropping systems in Maryland found that seed banks of smooth pigweed and common lambsquarters sampled before corn were lower following hay than following soybeans or wheat, but that the seed bank of prostrate-growing annual grasses was greater following hay (Teasdale 2018). The weed community on land rotated between perennial hay and annual crops appears to undergo a cyclical change in species composition, with many species decreasing during the hay part of the rotation while others increase. Whether a species increases or decreases depends on its biology but also on whether or not it has an opportunity to set seed during the forage establishment year. Thus, the utility of forage crops for suppressing weeds in a crop rotation will depend on the identity of your problem weeds and on your management practices.

This discussion has highlighted that the diversity of field operations associated with specific crops is as important as the diversity of crops themselves (Teasdale 2018). Each crop has a unique set of tillage, planting patterns, cultivation, fertility, harvest and other operations that suppress those weed species that are not adapted to those operations. Likewise, operations are conducted at different times for each crop, thereby adding to selective suppression of species whose life cycles are unlike that of the crop. Therefore, planning a diverse rotation should also include planning for the diverse set of associated management operations that will target and suppress the most problematic weed species. Carefully planned rotations of crops and operations can be an effective preventive approach that will lower the initial weed populations faced in rotational crops and increase the potential effectiveness of the weed tactics used within each crop.

CROP COMPETITIVENESS

Weed control should not be viewed as simply a set of tactics designed to kill or impair unwanted weedy plants but rather as an integrated set of cultural practices that give the species being grown (the "crop") a competitive advantage over competing species (the "weeds"). A large proportion of weed control comes from the weed having to compete with the crop. For example, a study with field corn found that the weight of velvetleaf and Pennsylvania smartweed was reduced 96% and 99% by the presence of the crop (Jordan 1979). You can easily demonstrate the effect for yourself by leaving two patches of ground unweeded for a few weeks in the middle of the season: one in the crop and one that is left unplanted. Both will become weedy, but the weeds in the patch with the crop will likely be much smaller. Thus, crop competition is the foundational element of weed management, and all practices that favor crop growth, from seedbed preparation and selection of vigorous seed, to density and spacing, fertility and water management should be considered part of a weed management program.

The usefulness of crop competition in managing weeds depends on weed density. If weeds are abundant many will likely escape control methods and, in this case, managing crop competition becomes critical. In contrast, if the density of weed seeds and perennial storage organs in the soil is low, then the competitive ability of the crop is less important for preventing yield loss. However, strong crop competition can still be useful for reducing reproduction of the few weeds present.

Crop Vigor and Uniformity

Producing a competitive crop begins with a vigorous, uniform crop stand. Planters should be kept in good repair and adjustment so that seeds are placed at a precise depth without skips. Remember that any skip in the row provides an opportunity for weeds to grow with reduced competition. Similarly, if seeds are placed too deep or too shallow, they may emerge slowly with reduced vigor, a situation that favors weeds. In addition, planting too shallow or too deep, or at an inconsistent depth will cause variation in crop emergence, leading to challenges in timing cultivation for optimum crop size. Generally, grain drills and planters with double disk openers provide more accurate seed placement than those with single disk openers.

Always use high quality seed. Old, moldy or damaged seed will produce a crop stand that has skips, overall low density and slow-growing plants that compete poorly with the weeds. As discussed in Chapter 4, rapid crop growth is important not only for direct competition of the crop with weeds but also for maintaining a size difference that allows for effective mechanical weed control. If you produce your own seed or save seed from a previous year, you will want to test it for viability and speed of germination before you plant. Germination speed is a good indicator of seed vigor and tends to indicate seedling growth rate. Small flat corn seed has the most rapid emergence, but medium flat seed tends to produce faster-growing plants after emergence. Large round corn seeds can be damaged during harvest and consequently have low vigor. And even if undamaged, they take up water more slowly than flat seeds because of less surface area relative to seed volume.

For vegetable crops, be sure transplants are vigorous. Growing mixes vary greatly in nutrient availability, and finding the right mix for each crop can pay off in improved vigor, weed suppression, yield and quality. Also ensure that the transplants are at the optimum growth stage when the time comes to set them out. If transplanting will be early or delayed, try adjusting the temperature in the greenhouse to ensure that the plants are the right size at transplanting. Ensure that the plugs are wet before you put them in the ground, and give them water during or immediately after transplanting to ensure a good start. Adjust press wheels so that 1) the plug is well covered with loose soil to ensure that the spongy potting soil does not wick away the water, and 2) that soil is firmed in around the plug so that roots are not blocked by air pockets. Good planting practices ensure faster crop growth and better competition with weeds. Rapid establishment of the crop also allows earlier and more effective cultivation (see Chapter 4).

Some evidence indicates that crops tolerate weed competition better when steps are taken to maintain soil quality. For example, in the long-term cropping systems study at the Rodale Institute in Pennsylvania, corn in the organic treatments, which received cover crops and compost, suffered less yield loss from a given level of weed competition than the conventionally managed system that lacked those soil building inputs (Ryan et al. 2010). Similarly, few crops tolerate poor soil drainage, but some weed species like yellow nutsedge and barnyardgrass thrive under such conditions. Hence, practices that improve drainage and water infiltration will tend to shift the competitive balance in favor of your crops.

Dense Planting

High crop density provides early leaf canopy closure and more intense competition against weeds within the row than a sparse planting. Note in this regard that plant density recommendations are always developed in weed free conditions. If your field usually has moderate to high weed density, typical recommendations

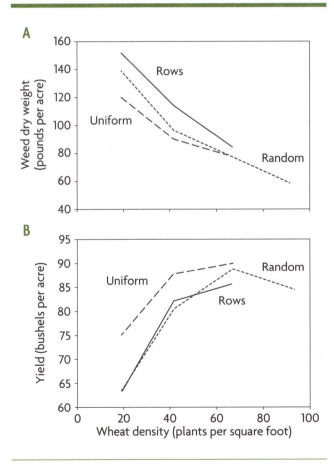

Figure 3.1. Effect of spring wheat density on (A) weed dry weight and (B) wheat yield with three planting patterns. Rows were drilled 5 inches apart. Seeds in the uniform treatment were planted in a square grid using a custom-built machine. The random treatment was created by bouncing seeds from a drill box onto a metal plate and then covering with 1.5 inches of topsoil. Drawn from data of Olsen et al. 2005.

may be inappropriate. In a survey of the many studies on competition between a wide range of crops and weeds at various crop densities, we found that increasing crop density nearly always reduced the growth of weeds regardless of the crop (Mohler 2001 and, for example, Figure 3.1).

Not all crops, however, can tolerate high densities. Crops that produce a single unit (corn, root crops and crops producing heads) may have lower yield quantity and quality at higher than recommended densities. Root crops will tend to make small roots if planted too closely, and this can lead to poor yields in marketable size classes. Similarly, the size of lettuce, cabbage, broccoli and cauliflower heads will shrink with increasing density, and the sprouts of Brussels sprouts will be too small for use if the plants are closely spaced. As the great Dutch agronomist C.T. DeWit once commented, "Brussels sprouts grown at high density are collards." Sweet corn will tolerate small increases in planting density, but ear size shrinks, and the frequency of barren plants increases with increasing density. Also, a high-density planting can encourage disease or cause lodging. Evaluate where your problems lie!

Any crop that makes multiple units of produce on a single plant (for example, wheat, soybeans, squash or tomatoes) and most leafy greens (for example, chard, leaf lettuce or kale) can be planted at higher than recommended rates without yield loss. Each plant will yield less, but the higher density will compensate. For crops that tolerate high density, yields generally continue to increase with planting density when the field is weedy, whereas yields typically plateau in weed free conditions. The extent to which the extra yield compensates for the extra cost of seed depends on the crop, weed pressure, how effectively the crop can be cultivated and other factors. In our experience, a 50% increase over recommended densities is often worthwhile. A North Carolina study in which soybeans were planted in 30-inch rows at 75,000 per acre to 225,000 per acre and in which the weeds were managed with cultivation found the lowest weed cover and highest net returns consistently occurred at the 225,000 per acre planting density (Place et al. 2009). Even if a higher crop density does not improve net income during the current year, the decreased weed seed production will benefit subsequent crops.

Weed suppression by cover crops increases substantially with dense plantings. Sowing

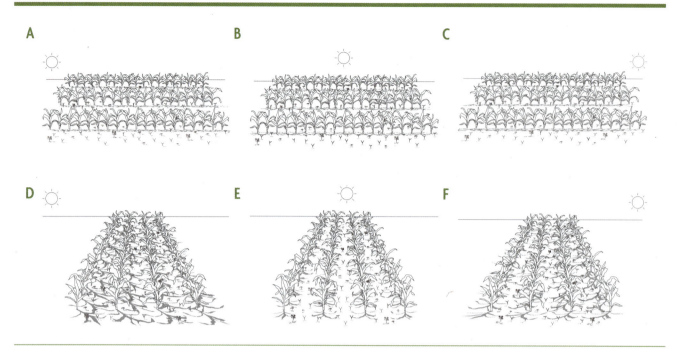

Figure 3.2 Row orientation can affect the vigor and competitiveness of the crop relative to weeds. East-west oriented rows in the morning, noon and afternoon are (A), (B) and (C), respectively; north-south oriented rows in the morning, noon and afternoon are (D), (E) and (F), respectively. Illustration by Vic Kulihin.

recommendations for cover crops are usually derived from use of the same species for forage or grain. When the crop is used for cover, however, higher densities are possible, and these give substantially better weed suppression. The principle upper limit on cover crop density is the cost of the seed. Winter cover crops sown at very high density (e.g., 10 times a normal rate), however, may suffer winter damage due to the weakness of individual plants.

Row Spacing

As a general rule, the closer crop plants come to a square grid arrangement, the more competitive they are against weeds (Figure 3.1). This means that, in principle, narrowly spaced crop rows provide better competition against weeds than do widely spaced rows. In row crops, the need to provide space between the rows for cultivation severely limits options for narrowing row spacing. Work with small grain crops, however, has shown that narrow row spacing often improves both weed control and yield, and that the benefits are magnified further by moderate increases in sowing density (Olsen et al. 2005). Both narrow row spacing and wide row spacing, plus inter-row cultivation, improve weed control in small grains relative to using neither of these practices. One way to halve the

row spacing in a small grain is to link two drills such that the second is offset from the first.

In the northern United States, conventional soybeans are usually planted in narrowly spaced rows with a drill. Providing weeds are well controlled, soybeans in closely spaced rows out yield those in widely spaced rows due to more optimal resource use. In contrast, organic soybeans are typically sown with a planter in rows spaced 2.5 feet (76 centimeters) apart to allow room for inter-row cultivation. However, with an adequate cultivator guidance system and narrow shovels, cultivating more closely spaced rows is possible. Rapid canopy closure would reduce the number of inter-row cultivations to one instead of the usual two and would reduce weed growth in the row by decreasing penetration of light from the side. As a general rule, weed suppression from narrow row spacings will be most effective when combined with increased crop density.

Row Orientation

Several mathematical models of orchard and row crop canopies have shown that during most of the growing season in temperate latitudes, crop plants intercept more light if the rows are oriented in a north-south direction rather than in an east-west direction (Smart

1973). Late season and winter crops, however, capture more light if planted in east-west rows. The reason north-south rows capture more light during the summer growing season is because, in the early morning and late afternoon, sunlight strikes the side of the row rather than shining on the bare ground between rows (Figure 3.2). Moreover, this is the time of day when light is most limiting to crop growth. The effect increases as one moves south from 55° to 25° latitude.

By increasing light capture by the crop, north-south planting increases shade cast by the crop onto weeds, which should decrease their growth. Clover interseeded into oats had lower density in north-south oriented plots, especially in the crop rows. A study of cover crops interseeded into corn showed that, when rows were oriented east-west, a zone of high light on the south side of rows reached to the base of the crop, whereas with north-south orientation shade was most intense in the row. This result is important since weeds are more easily controlled in the inter-row area than in the rows.

In contrast with summer crops, which should be more weed suppressive in north-south oriented rows, winter crops should be more weed suppressive in east-west oriented rows since the sun will be low in the sky through most of the period of active crop growth. As predicted, winter wheat and barley in Western Australia had greater light interception and usually had greater yield and lower weed biomass in east-west rows than when planted in north-south rows (Borger et al. 2010).

The effect of row orientation should increase with the size of the crop species and the distance between rows. For short crops in closely spaced rows, the benefit will probably be small, whereas in larger crops in widely spaced rows, they should be noticeable. For example, a study of apple orchards calculated a 24% increase in light interception with north-south planting (Jackson and Palmer 1972). The potentially small weed control benefits from adjusting row orientation should be considered relative to potential consequences on operational efficiency, cost and soil conservation.

Competitive Varieties

Many factors enter into variety selection. When selecting varieties, most farmers consider market demand, produce quality, yield and disease resistance more important than ability to suppress weeds.

Nevertheless, factors such as vigorous early growth, speed of leaf canopy closure, height and foliage density should be considered as qualities contributing to weed suppression. All crop varieties differ in capacity for weed suppression. If you produce several varieties of a given type of crop, consider planting the more robust, rapidly growing and competitive varieties in weedier fields and planting the less competitive varieties in cleaner fields. For example, the Russet Burbank potato produces more shade and suppresses hairy nightshade better than Russet Norkotah (Hutchinson et al. 2011) and suffers less yield loss in weedy conditions (Colquhoun et al. 2009), so Russet Burbank would be a better choice for the weedier of two fields. Similarly, butternut squash varieties are more competitive than delicata type squash, and Danvers type carrot varieties are more competitive than Nantes type varieties.

Modern cereal grains have been bred for a high harvest index by shifting allocation of plant resources from the stem to the seeds. This increases yield in weed free conditions, but if a short-statured crop is overtopped by weeds, the higher yield potential may not result in greater actual yield. Consequently, some organic growers are experimenting with older, tall-statured grain varieties. Even among short-statured, high yielding varieties, weed competitiveness can vary substantially. Specific recommendations are necessarily constrained by the need for good adaptation to regional climate and soil conditions.

Seed Size

Because seed size affects the rate of growth shortly after emergence and the rate of leaf canopy closure, crop varieties with larger seeds tend to be more competitive than varieties with smaller seeds. Some experiments have shown that when grain and soybean seeds are screened into size classes, planting the large size seeds resulted in a crop that competed more effectively against weeds (Stougaard and Xue 2005). Note that if the seed is larger then more pounds of seed must be planted per acre to achieve the same planting density. Some work, however, has shown that the increased yield from the larger seeds can more than compensate for the increased seed cost (Stougaard and Xue 2005). In principle, a grower could invest in either larger seeds or in more seeds per acre. Since 1) the mechanisms increasing competitive ability may differ between seed size and seed density and 2) the incremental effects of both factors decrease as seed size and density increase, the optimum strategy may be to increase both seed size and density moderately rather than concentrating investment on one tactic or the other. Thus, for example, a study on spring wheat competing with wild oat showed a 12% yield increase with higher seed density, an 18% increase with larger relative to smaller seeds and a 30% yield benefit from using both a high density and larger seeds (Stougaard and Xue 2004).

Use of Transplants for Small Seeded Crops

Most annual weeds have very small seeds and consequently they establish relatively slowly. If the crop is also relatively small seeded (for example, cole crops, lettuce, tomatoes, leeks, etc.) then growing the crop in weed free soil and transplanting after the plants are well established gives the crop a substantial head start over the weeds. Few experienced organic growers direct seed small seeded crops on any substantial scale if the crop tolerates transplanting.

If the bed was prepared before the transplants are ready, the surface soil should be tilled thoroughly to a depth of 1–2 inches with a harrow or rotovator to kill any weed seedlings before transplanting. This should be done even if weed seedlings have not yet emerged: they are on their way up and you always want to give the transplants the maximum head start over the weeds. The depth of tillage depends on the depth that the dominant weeds in the field are likely to emerge from (see the "Emergence depth" section in each species chapter). You want to kill those weeds that have already germinated and are likely to emerge, while minimizing the number of new seeds that are brought to a shallow depth where establishment is more likely.

Planting Date

Every season has weeds that are well adapted to the prevailing weather conditions at that time of year, and every weed has a period of the year when it is most likely to emerge and grow. This optimal emergence period can vary with location and seasonal weather variability (see the "Dormancy" and "Season of emergence" sections in each species chapter). Generally, crop species will be most competitive when planted during a period when the dominant weed species in a field are less likely to emerge and grow vigorously. This can both reduce the number of weeds emerging and the vigor with which emerged weeds will compete with the crop. Several strategies discussed in this book, including rotation, cover cropping and stale seedbed, will disadvantage weeds by manipulating crop planting and tillage dates relative to weed emergence dates.

Many of our crops have their origin in the tropics or subtropics, for example corn, cotton, beans, squash, tomatoes, peppers, etc. Consequently, they grow most vigorously when the soil and air are warm. As a result, pushing warm season crop species to get an exceptionally early or late harvest puts the crop at a disadvantage relative to the weeds. You may find harvesting warm weather crops outside of their usual season is worth the additional effort but anticipate weed problems and take measures to compensate for them. For example, use low or high tunnels, or ridge planting to promote early soil warming.

Corn is a crop that illustrates this point. Conventional field corn growers in the Northeast and the upper Midwest get maximum yields by planting long season hybrids in late April or early May. Growers transitioning to organic practices often encounter difficulties if they continue planting so early. Wet soil often prevents timely cultivation, and slow emergence and growth of the corn makes it relatively uncompetitive with weeds that are favored by cool spring conditions. Moreover, without fungicidal seed treatments the seeds often rot in the cold soil, leaving skips in which weeds grow out of control. Consequently, many organic growers plant corn two to four weeks later than their conventional neighbors. Although the organic corn yields are often lower than the best conventional yields, the higher soil quality on organic fields often compensates for the shorter season varieties and results in yields above the county average. Organic sweet corn growers who want to produce an early crop increasingly transplant their corn. This allows the young sweet corn to establish in optimal conditions and assures a uniform stand.

Intercropping

Another way to increase the competitiveness of crops is to plant them in mixtures. Some mixtures may be less competitive against weeds if the crops compete more with each other than they do with the weeds, so the mixtures must be chosen carefully.

- Clover or alfalfa overseeded into winter grain in the spring or planted simultaneously with a spring grain establishes too slowly to compete with the grain crop but slows down weed growth when the grain matures and light penetrates through the crop canopy. Similarly, red fescue planted in the fall with winter wheat does not reduce wheat production but helps suppress quackgrass (Bergkvist et al. 2010). Conversely, when grain and hay are sown together, the fast, early growth of the grain suppresses weeds that would otherwise compete heavily with the slower-establishing hay species.

- Lettuce is harvested much sooner than tomatoes if they are planted at the same time. Consequently, a row of lettuce next to a row of tomato plants competes with the weeds when the tomato plants are small and is harvested before it is overtopped by the tomatoes. Similarly, kohlrabi can fill the space between young Brussels sprouts and is harvested about the time the Brussels sprouts start to shade the whole bed.

- Light can penetrate the leaf canopy of sweet corn and allow weed growth. Interplanted winter squash or pumpkin vines that run under the corn will compete with the weeds and improve late season weed control. Squash or pumpkin yield will be greatly reduced (for example, by 50–75%) under the corn, so the total planting should be increased accordingly. You need to carefully plan the planting date and maturity of both crops to avoid trampling the vines when you harvest the corn.

- On intensive vegetable farms, when skips occur in a row due to poor establishment, some other crop can be planted to fill in the gap. If you do not have a crop that fits appropriately into the space, you can sow a rapidly growing cover crop like oats or buckwheat (see "Summer Cover Crops"). You may find it worthwhile to keep a supply of cover crop seed on hand for such emergency purposes.

Nutrients and Water

Several studies have shown that weeds are often better equipped for taking up mineral nutrients like nitrogen, phosphorus and potassium than are the crops with which they compete. Not only do many weeds produce root surface area at a faster rate than typical crops, but they also concentrate nutrients in their tissues (see the "Response to fertility" section of each species chapter). Consequently, highly available forms of nutrients like chemical fertilizers and rapidly decomposing organic

fertilizers like blood meal tend to favor weeds relative to crops. In contrast, the slow release of nutrients from green manures and compost tends to favor crops relative to the weeds. Particularly for long season crops, the slow release from organic materials may slow early top growth slightly but encourage a stronger root system and an overall healthier, more productive crop by harvest time.

Most of the mineral nutrition of the crop should come from soil organic matter built up by feeding the soil with green manure and compost. Some crops, like field corn, will often benefit from an additional dose of starter fertilizer banded next to the row, and some heavy feeding crops like broccoli may yield better if given an additional nitrogen source like composted chicken manure after they are well established. Applying such supplements in a band next to the row will avoid feeding inter-row weeds before the crop can reach the fertilizer. Foliar application of a soluble fertilizer like fish emulsion can similarly direct nutrients specifically to the crop, provided weeds are not already established within the crop row.

Similarly, drip irrigation, which applies water next to the crop plants rather than sprinkling it over the whole field, will favor crop plants relative to weeds. This is especially useful for managing weeds along the edges of vegetable beds where they usually receive less shade and root competition from the crop. If the wetting zone is too narrow, however, roots may fail to spread throughout the bed, leading to inefficient use of nutrients released from soil organic matter and incorporated amendments.

COVER CROPS

Sowing a cover crop into or after the final crop of the year can be an effective tactic for managing weed populations. They compete with weeds during periods when cash crops are not grown, thereby suppressing growth and seed production of weeds that otherwise could replenish their seed banks. Soil preparation for planting and terminating cover crops also can destroy weeds that would otherwise have established and produced seeds. In addition, cover crops provide several advantages to the agroecosystem. Benefits for the soil include prevention of erosion, reduced leaching of nutrients and increased favorable biological activity. The dense root systems and the organic material that is incorporated into the soil in the spring also improve soil structure. As explained in the section "Principles of Mechanical Weeding," maintaining good soil properties by using cover crops makes weeding easier.

Winter Cover Crops

Winter cover crops usually are planted and establish in late summer or fall, have the capacity to survive winter conditions and complete their life cycle or are terminated before cash crops are planted in spring. Winter cover crops directly compete with weeds, particularly species that thrive in cool weather, like common chickweed, shepherd's-purse and quackgrass. They may provide little benefit for control of the warm season weeds that will infest the succeeding cash crop, however, when the crop rotation moves to a cool season vegetable crop or winter grain, the reduced input of cool season weed seeds to the soil may prove helpful. Because some cold tolerant weeds like common chickweed are relatively shade tolerant, ensuring a highly competitive cover crop through high planting density and optimal planting date may be required to effectively control seed production.

Various crop species are harvested on different dates, and this will determine to a large extent the type of cover crop that can follow those cash crops. In the northern states, the only readily available annual cover crops that survive the winter well are winter wheat, spelt, grain rye, triticale, annual ryegrass, hairy vetch and Austrian winter peas. In more southern regions, the list of winter covers increases substantially to include crimson, berseem and subterranean clover, winter barley, winter oats, black oats, and brassica species. Although these are useful in many situations, the cold adapted species are among the most popular cover crops in southern regions of the United States as well as in the north.

Hairy vetch, grain rye and annual ryegrass can all become severe weed problems in winter grain crops, though their potential for weediness seems to vary geographically. Caution should be exercised when using these species as cover crops if winter grains are part of the crop rotation. Regardless of the rotation, no cover crop should be allowed to go to seed without having a plan for managing volunteer plants growing from this seed. Hairy vetch usually has a small percentage of hard seed that does not germinate the year of sowing and thus can infest a field in later years even if the cover crop does not set seed.

Summer Cover Crops

Weeds grow rapidly during warm weather and consequently, large seeded cover crops that quickly produce leaf area will suppress weeds best. Common summer cover crops with large seeds include buckwheat, sorghum-sudangrass hybrid (sudex), cowpeas (also called blackeyed peas) and soybeans. If a crop will not be planted until mid-summer and no winter cover crop was planted the previous year, you can use a spring sown cover crop to protect the soil, suppress weeds and add organic matter to the soil until you are ready to plant. Many organic growers use oats, grain rye, field peas or bell beans (a relatively small seeded variety of the normally large seeded broad bean) for this purpose. On grain farms, mustards can be sown in early spring for a weed- and disease-suppressing cover crop before early-summer-sown soybeans or dry beans. We do not recommend the use of mustard cover crops on farms that produce cole crops since frequent planting of members of the mustard family in the crop rotation can promote disease and insect pests.

Several of the cover crops mentioned above have special properties that are worth noting. White clover forms a low growing sod that works well between beds of long season vegetables on plastic. To prevent the white clover from becoming very weedy, sow it with oats (spring planting) or buckwheat (summer planting) and then mow the nurse crop after the clover is well established. Because they have large seeds, a thick sowing of bell beans, cow peas, field peas or soybeans will completely cover the ground within two weeks after emergence and effectively smother annual weeds. Similarly, buckwheat has large horizontal leaves that cast dense shade and is thus especially effective at suppressing annual weeds. It is relatively short, however, and thus can be overtopped by tall or vining perennial weeds that survive on belowground reserves until they reach sunlight. Sorghum-sudangrass hybrid (sudex) is an extraordinarily competitive cover crop. It grows 6 feet tall or more and can thus compete effectively even with most perennial weeds if you sow it thickly. Height, however, has disadvantages. A tall stand of sorghum-sudangrass can shade neighboring crops, and the long stems are difficult to incorporate when you are ready to plant the next crop. Also, vigorous stands of sorghum-sudangrass require relatively high nitrogen fertility, and when N is in short supply, the cover crop is likely to be thin and weedy. Part of the weed suppression by sorghum-sudangrass is due to allelopathy (Weston et al. 2013). Because of its allelopathic potency, direct seeding of crops should be delayed 10–14 days after incorporating the residue (see the section "Cover Crop Management").

Cover Crop Mixtures

Grasses are often more competitive against weeds than are legumes, possibly because grasses develop a more extensive and competitive root system and extract nitrogen from the soil profile (Teasdale 2018). Grass-legume mixtures are sometimes more competitive than the grass or legume component alone; for example, rye-hairy vetch mixtures can be more suppressive than either rye or hairy vetch alone. Multi-species mixtures of up to eight species are being investigated as a means for achieving more consistency in productivity and ecological benefits across a range of climatic conditions. Such mixtures often show lower variability and improved productivity compared to the average performance of individual species, although even these multi-species mixtures often produce no greater biomass than the most productive individual species (Teasdale 2018). Mixtures that pair complementary species, such as nitrogen-fixing legumes with non-legumes or highly winter-hardy with partially winter-hardy species, can potentially achieve multiple benefits with greater resilience over a broader range of conditions. However, if a cover crop is used for a single purpose, such as weed suppression, then the single species that functions best for that purpose will often be the best choice. For example, the legume component of a mixture can dilute the weed suppressive ability of the more competitive grass, causing the mixture to be less competitive than the grass alone (Mohler and Liebman 1987, Brainard et al. 2011). Thus, the grass component may be the better choice for pure weed suppression, but the legume component would be important for achieving a combination of nitrogen fertility and weed suppression goals.

Cover Crop Management

Termination of the cover crop has important consequences for weed suppression in the following crop. Usually, you will want to mow a cover crop before incorporating it. This is particularly important if the cover crop is taller than 8–12 inches or if you are using an implement other than a moldboard plow. If the cover crop is not well incorporated, it may recover and compete severely with the subsequent cash crop. Flail mowers work best because they cut the material into short pieces and leave it uniformly distributed on the ground. Rotary mowers tend to bunch the crop residue into clumps that are difficult to incorporate. If the cover crop has grown tall and you are using a sickle bar or disc mower, cut the top first and then cut closer to the ground. Rye stems longer than 12 inches will make rotary tillage completely impractical. Long rye stems also collect on shanks of chisel plows and field cultivators. Hairy vetch is easier to incorporate, but by late spring the stems will become tough and will wrap on implements. Most experienced growers incorporate their cover crops 10–14 days before they plant. This gives the material time to rot and for allelopathic compounds that can harm sensitive crops to decompose (see "Allelopathy"). Many damping-off and root rot fungi thrive on fresh green organic matter, and allowing the green manure to rot for a week or two allows time for these to be replaced by beneficial fungi (Hoitink et al. 1996). A lag between incorporation and seedbed preparation also gives many weed seeds an opportunity to germinate in response to tillage and be killed by final seedbed preparation (see "Seed Germination: Why Tillage Prompts Germination"). Ultimately, the goal is for rapid establishment of a uniform and competitive cash crop.

Cover crops also can be terminated with the residue left on the soil surface as an organic mulch. This option provides a great deal of flexibility for conventional no-till producers who can kill the cover crop with a burndown herbicide at the optimum time for planting spring crops. In addition to herbicides, cover crops can be terminated by mechanical means with equipment such as a flail mower that shreds vegetation and drops it in place, a sickle bar mower that drops intact residue in place, a light disk set to slice over the vegetation and leave it on the surface, an undercutter that severs roots from stems just below the soil surface and a roller-crimper that crushes vegetation. The success of a mechanical approach to terminating cover crops requires waiting until the cover crop is flowering, otherwise it will usually recover and regrow. The requirement to wait until flowering may require delaying cash crop planting. However, delayed plantings can also be beneficial for weed management purposes (see "Planting Date"). Finally, no-till planters need to be fitted and adjusted to plant through the dense layer of cover crop residue on the soil surface. An alternative is strip-till, which allows the cash crop to be planted in narrow tilled strips but retains the cover crop residue on the untilled area between crop rows. When accomplished successfully, no-till production with surface cover crop residue offers multiple environmental benefits besides suppression of emerging weeds, including protection from soil erosion and nutrient runoff, improved rainfall infiltration, reduced evaporation of soil moisture, and increased soil organic matter.

Additional information on the properties and management of winter and summer cover crops can be found in SARE (2007, www.sare.org/mccp) and at attra.ncat.org.

Rotational No-Till Cropping with Rolled Cover Crops

The capacity of cover crop residue to suppress weeds provides an opportunity for no-till production of cash crops without herbicides. Organic growers are often faced with the tradeoff that multiple pre-plant tillage and post-plant cultivation operations are required to manage weeds, but this can lead to loss of organic matter and destruction of soil structure. In theory, cover crops can provide a solution to this dilemma by utilizing a uniform, dense layer of cover crop residue on the soil surface to suppress weed emergence, while at the same time providing the benefits of no-till, adding organic matter to the soil and releasing nutrients to the crop. Recent research has identified the roller-crimper as the tool of choice for cover crop termination because it is fast and leaves the cover crop tissue intact, thus slowing decomposition and maximizing weed suppression. It is typically designed with blades welded to a drum that simply roll down the cover crop stems in one direction and kill the cover crop by crushing the stems where the blades pass over them. The cash crop is usually no-till planted in the same direction to facilitate seed placement.

The production of a uniform, dense layer of cover crop residue is required to maximize suppression of weed emergence. This means that tillage to prepare a seedbed is usually needed to achieve rapid establishment of a uniform cover crop stand. This explains why this approach is referred to as "rotational no-till," because it relies on tillage within the rotation for maximizing cover crop production, while the cash crop is produced using no-till techniques. Maximum cover crop biomass is then achieved by planting the cover crop early enough in the fall for good establishment before winter and by allowing sufficient growing time in spring to maximize production before termination at flowering. As a rule of thumb, a cover crop residue biomass of at least 7,200 pounds per acre needs to be left after crimping to provide consistently good weed suppression (Teasdale and Mohler 2000). Biomass levels less than this do not completely cover the soil but leave small openings through the dead residue where weeds can receive light and emerge (see discussion of "How Much Mulch" in the "Organic Mulch" section). Achieving such a high production of biomass often requires a cover crop mixture such as a legume and rye.

One drawback to reliance on cover crop residue for weed suppression is that if weeds do escape because of deficiencies in the mulch coverage, the dense mulch and un-tilled soil can hinder post-planting cultivation operations to destroy these weeds. High-residue cultivation techniques can be used, whereby a coulter cuts through residue in front of a flat cultivator shank that moves just below the soil level, severing weed shoots from roots with minimal disturbance of the residue. Generally, cultivation in no-till soils with high residue is not as efficient as cultivation of tilled soils. Thus, cultivation and no-till cover crop mulches are not complementary weed control tactics but instead tend to antagonize each other (Teasdale 2018). On the other hand, growing the cash crop at high densities can be complementary with a cover crop mulch. Research has shown that the cover crop residue reduces and delays weed emergence enough to permit a crop such as soybeans to develop a leaf canopy faster than weeds and enhance the competitiveness of the crop relative to weeds. In addition, reduction of the weed seed population and elimination of perennial weeds in rotational crops prior to no-till crop production can complement cover crop mulching and can greatly enhance the weed-suppressing efficacy of rolled cover crops.

There are several constraints on crop performance in this no-till system, including the difficulty of planting through a dense layer of cover crop residue, the short growing season in northern areas, low soil temperatures under the dense residue, destructive insect populations sheltered by the mulch, and nutrient release that is poorly synchronized with crop needs. The need to plant cover crops early and terminate late to achieve high biomass levels can conflict with the need to harvest and plant cash crops within a recommended time frame. A fall grain crop or a relatively short-season vegetable crop can be planted during the preceding year to allow time for optimal establishment of the cover crop in late summer/early fall. In addition, a sufficiently long growing season is needed after the cover crop flowers and is terminated to grow a productive cash crop. These requirements generally mean that the rotational no-till system works best in the southern or middle latitude states, but usually insufficient time is available to fit a long cover crop growing period into rotations in northern areas. If these constraints can be overcome, this rotational no-till system offers the opportunity for reducing tillage operations in organic crop rotations.

ORGANIC MULCH

In addition to growing a mulch in place by killing a cover crop before planting, organic materials can be brought in from other locations to mulch the soil surface surrounding crop plants. Mulches of organic materials are highly effective for suppressing emergence of small seeded (that is, less than 2 milligrams) annual weeds. Since most agricultural weed species are small seeded annuals, the use of mulches is broadly effective against many species. Mulches are ineffective, however, for controlling perennial weeds because these have sufficient energy stored in the roots or rhizomes to push shoots up through even very thick layers of mulch. For example, we once observed hedge bindweed sprouting up through an 18-inch-thick pile of bark mulch that was waiting to be spread around ornamentals. Large seeded weeds (for example, greater than 5 milligrams) may also emerge through substantial layers of mulch, though a sufficiently thick and dense mat can suppress all but the largest-seeded annual weeds. Grass weeds can usually emerge through more mulch than can broadleaf species with the same size seeds.

How Much Mulch

Two attributes of a mulch mat are critical for weed suppression (Teasdale and Mohler 2000). The first of these is the number of layers of mulch particles, for example leaves or stems, that cover the soil surface. The second is the fraction of space in a mulch mat that is occupied by solid material rather than air. The

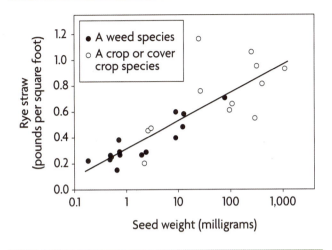

Figure 3.3. Mean amount of loose rye straw through which seedlings emerge from seeds of various weights. The data includes both weeds and crops (Mohler, unpublished data).

importance of the first of these is that more layers of material block more light from reaching the soil surface and a seedling must twist and turn more as it grows up into the mulch. The importance of the second factor is that when the mulch material is more tightly packed, seedlings have difficulty finding gaps in which to grow upward. Also, the more space is occupied by solid material, the less light is reflected from particles and down into the mat.

Since both factors are important, specifying how thick a layer of mulch material needs to be to suppress weeds depends on the type of mulch as well as the species of weeds present. Larger seeded weeds require more mulch for suppression than do smaller seeded species (Mohler and Teasdale 1993, Figure 3.3). A 5-inch layer of loose straw, hay or leaves that subsequently settles to about 2–3 inches is generally effective against most small seeded annual weeds. Because compost is denser, about 2 inches is usually enough to suppress most weed seedlings. Applying compost at such a high rate, however, is likely to create excessive fertility and stimulate weed growth in subsequent years (see "Nutrient Use"). Compacted hay or straw used as slabs from a bale is highly effective at thicknesses of about 1.5–2 inches. Often weeds emerge between the slabs if they do not cover the ground completely. To compensate for this, thin slabs of mulch (say 1 inch) should be placed over the joints between main slabs to obtain full coverage.

Although dense material like compost and slabs of hay or straw are very effective at blocking growth of weeds up from the soil, they also hold water well and provide a good medium for germination of windblown seeds like dandelion or seeds that are present in the mulch itself. Hay and straw become progressively more prone to support windblown or resident seeds as these mulches rot.

When to Mulch

The optimal time to mulch a crop depends on the crop and the season of planting. Large seeded and transplanted crops can be mulched almost immediately after planting. For summer planted vegetable crops, a heavy cover crop can be mowed or rolled, or new mulch material can be laid and the vegetable crop transplanted through the mulch. For spring plantings, however, you may want to delay application until the soil warms. Also, removing the first flush of weeds prior to laying organic mulch is often helpful. Although a thick mulch eliminates most light at the soil surface, even a homogenous-appearing mulch layer has partial "windows" through which some light penetrates. If weeds have already emerged before laying mulch, more weeds will be positioned to exploit such windows and emerge through the mulch. In contrast, if the first flush of weeds has been removed then fewer weeds will successfully penetrate the mulch.

Weed Seeds in Mulch

The most commonly used mulch materials are straw, hay, compost, tree leaves and bark chips. Regardless of the material used, you should thoroughly check it for weed seeds. Hay, and particularly late cuttings of grass hay, often contain mature weeds and perennial grasses that you do not want in your fields. Straw is generally free of weed seeds but may have thistle seed heads and some grain seed heads that did not get picked up by the combine. Tree leaves are generally free of weed seeds but may contain acorns and other tree seeds that subsequently germinate and compete with crops.

Applying Mulch

Applying straw or hay mulch is often unpleasant. Doing so by hand is backbreaking, and it is usually scratchy. It is also usually dusty enough that you should wear a dust mask. Bale choppers and blowers are available, however, which can automate the process. The gun type blowers that are used to apply mulch in landscape seedings need to be modified for

most work in vegetables. They are useful for blowing mulch between plastic covered beds, provided the beds are swept off afterwards (either by hand or with a rotary brush). Adding a flexible tube allows mulch to be blown in under established crop plants, but you may have to modify the fan speed. Wetting loose organic mulch with sprinkler irrigation before it is subjected to high winds will help keep it in place.

Sources of Mulch

In grain growing regions, grain straw is often plentiful and cheap. The source of straw should be checked to ensure the grain crop was not grown with phenoxy herbicides, which can later volatilize off the straw and injure sensitive crops like tomatoes. In areas with dairy farms, straw will be in demand for bedding, but spoiled hay will likely be cheap and readily available, at least in some years. Growers should assess available sources of organic materials in their region for creative ways to mulch soils while also improving soil quality (Box 3.1).

In hilly regions, many vegetable farms use the best bottom land soils for vegetables but have upland parts of the farm that are underutilized. These upland areas can produce a steady supply of mulch for weed control and nutrient supply for the vegetable fields, provided you replace mineral nutrients (especially P, K and Ca) exported with the mulch. These minerals will end up in the vegetable production soils after the mulch rots or is incorporated, and the soil should be monitored to avoid nutrient imbalances. In particular, build up of K to excessive levels can be a problem when organic mulches are used to the exclusion of other weed management methods.

Notes on Particular Mulches

Some mulches pose special problems and advantages. Bark and wood chips can pose problems in vegetable cropping systems because their high C:N ratio encourages decomposer microbes to take nitrogen from the soil, thereby starving the crops. The problem is not during the year the mulch is applied, since the microbes only have access to the mulch at the mulch-soil interface. But when the soil is eventually tilled, N will be tied up by the decomposing wood.

Unlike hay and straw, tree leaves do not tangle into a mat and thus sometimes blow about and smother small crops. Applying the leaves after the crop is established helps hold them in place. Chopping the leaves also helps prevent movement.

Rye straw commonly releases allelopathic compounds (see "Allelopathy") that are toxic to other plants. Since small seeded weeds are more susceptible than large seeded or transplanted crops, this can be advantageous. However, in some circumstances, the toxins may also slow crop growth. For example, early growth of corn can be slowed by a rye mulch, possibly due to an interaction between soil cooling, N immobilization and allelopathy. However, application of rye straw after sweet corn is well established seems to be a safe practice.

CONTINUOUS NO-TILL VEGETABLE PRODUCTION USING ORGANIC MULCHES

Although continuous no-till cropping is generally impractical without the use of herbicides, a few small-scale intensive vegetable farms have developed continuous no-till systems that rely on mulch and hand weeding. Mulch is used to prevent emergence of weed seedlings, and the few that do emerge are hand pulled to prevent reproduction and competition with the crop. After several years, the near-surface seed bank is depleted, and the amount of mulch and hand weeding required can be reduced.

The approach is illustrated by Jay and Polly Armour's Four Winds Farm in Gardiner, N.Y. (www.fourwindsfarmny.com). This 24-acre farm has four acres in permanent raised vegetable beds, with the rest of the land used to produce grass-fed beef cattle. The cattle manure that accumulates in the barn over the winter is composted with horse manure from a nearby farm to make compost. The temperature in the pile is monitored during composting to ensure that weed seeds in the manure are killed. The compost forms the principal mulch material used on the vegetable beds. When beds are first brought into production, compost

---

Box 3.1. Mulched Beds for Onion Seedling Production

Relinada Walker farms 67 acres of organic vegetables and grains near Sylvania, Ga. One of her specialties is organic onion transplants, which she ships to various locations in the South and Northeast. Mulch is a key component of her onion transplant production system.

The first step is to create a clean seedbed. When growing transplants for the late fall market in the Southeast, she starts by plowing under a cover crop of brown-top millet around September 1. She lets this break down for 2–3 weeks before rotovating and creating beds with a bed shaper. This break between operations not only gets the soil ready for making beds but also allows time for a flush of fall weeds to come up and be killed during seedbed preparation. She then ensures that the beds are moist, if necessary by irrigation, and spreads 1 inch of compost as a mulch over the beds. The compost

is made of some mixture of vegetable waste, cotton gin trash and peanut hulls, depending on what is available. She adds bark fines to the finished compost to create a fine textured but fibrous material that spreads uniformly over the beds. Although the mulch layer is relatively thin, the pores between mulch particles are small so that no light reaches the soil surface. She sprinkle irrigates the mulch to wet it down and then lets it set for one day, or ideally two. This allows her to be sure that the compost mulch will not reheat. The onion seeds are planted into the mulch with a gang of four EarthWay seeders pulled by a tractor. The mulch holds moisture to facilitate good seedling establishment while suppressing the weeds. Of course, eventually some weeds do come through the mulch, but the mulch keeps the soil loose so that these are easier to hand weed. With this system the onions are ready for transplanting around Thanksgiving.

She uses a similar system for more northerly markets. For shipping to the mid-Atlantic states, she plants in October and pulls the seedlings from late January through early March. Before plantings in January and early February she has used clover and rye as cover crops but is concerned about the possible allelopathic effect of rye on tiny onion seedlings. Consequently, she is still experimenting to find the best cover crop to use with these mid-winter plantings. January plantings are ready to pull in April for buyers in Pennsylvania.

Walker grows one to two acres of onion transplants each year, depending on the volume of advance sales. Using mulch allows her to keep hand weeding costs down and thereby grow a product that other organic farmers can afford. The increased organic matter from the mulch and the lack of weed seed shed during the production of onion seedlings also create a good location for direct seeded rotation crops like carrots.

---

is spread 2 inches thick and seeds or transplants are planted directly into the compost. The thick layer of compost effectively isolates the soil surface from the cues that prompt weed seed germination (see "Why Tillage Prompts Seed Germination"). The few weeds that arrive by wind-blown seeds, particularly dandelion, are hand pulled. After nearly 20 years without wheel and foot traffic, tillage or weed seed production, soil on the beds has a high tilth and a very low seed bank near the surface. These factors allow beds to be planted without tillage and with compost applied only to meet the nutrient demands of the crop.

Although repeatedly applying compost at high rates can be expected to create excessive phosphorus concentrations in the soil, if weed seed production is prevented in the early years, the high P may not exacerbate weed problems. In principle, other mulch materials could be used, at least for some crops, once soil tilth is sufficiently good to allow planting without tillage for seedbed preparation.

SYNTHETIC MULCH

Many synthetic mulch materials are marketed for use in vegetable production, including plastic films of various colors, spun and woven cloth that is permeable to water, and plain and oiled paper. All of these can control weeds, but they all pose problems as well.

Plastic Mulches

Plastic films come in various colors. Clear plastic warms the soil better than black plastic, but it allows weeds to grow. Normally it is used in conjunction with residual herbicides. Infrared transmitting films warm the soil about as well as clear plastic but suppress weeds by blocking light that triggers weed germination. Plastic films warm the soil for early production and can be highly effective for suppressing weeds in the crop row, where they are difficult to control with cultivation. Plastic films are especially useful for weed control in onions, which grow slowly, and in full season vegetable crops like peppers and winter squash that tend to get weedy late in the season (Figure 3.4). They are also particularly useful for vegetables like trellised tomatoes, where the supports interfere with cultivation and even with hand hoeing.

Weeds frequently grow in the planting holes and may be sufficiently competitive to require laborious hand weeding. In particular, perennial weeds that

Figure 3.4. Peppers growing on black plastic mulch with a cover crop of annual ryegrass and Dutch white clover between the rows.

sprawl or vine will grow toward light entering at planting holes and will thereby be directed onto the crop.

Landscape Fabrics

Spun cloth ground covers are similar to floating row covers but are colored brown or black to block light from weeds. These are reasonably effective at preventing the growth of annual weeds. Many perennials, however, can penetrate these materials. Pulling these weeds pulls on the cloth, and that may disturb crops planted in holes in the material. Moreover, great masses of quackgrass and other perennials will cling to these ground covers when they are collected, thereby increasing the expense of disposal.

Woven landscape fabrics are generally much heavier than the spun fabrics (5–7 ounces per yard as compared to 2 ounces per yard for the spun materials). Woven fabrics effectively block growth of perennial weeds for several years. They are much more resistant to tearing than plastic film or spun fabrics and are suitable for long-term installation around grape vines and fruit trees. In this application, the fabric should be covered with bark mulch or rounded pebbles to prevent deterioration by ultraviolet light. Avoid crushed stone, which has sharp edges that can puncture the fabric. After a few years, weed seeds blown in from adjacent areas will germinate in the organic mulch or in soil that naturally accumulates on the fabric. Whereas shoots cannot push up through the fabric, fine roots can penetrate it, allowing establishment of the weeds. In addition, weeds that spread by runners may also begin growing on the fabric. Consequently, the period of effective weed control is often substantially shorter than the lifetime of the fabric.

Organic standards currently require that synthetic mulches be removed annually. Woven fabrics can be reused repeatedly for growing annual crops, but removing debris from the fabric can be laborious. While these durable fabrics are initially substantially more expensive than other synthetic mulches, they can be used for multiple years.

Biodegradable Mulches

Brown kraft paper can be used as a biodegradable mulch, but it presents several challenges. The rolls of paper mulch are heavy, bulky and difficult to handle. Paper alone is a poor mulch material because it tears easily during installation, and this problem is made worse by wind. Usually, paper mulches will not endure long enough to provide weed control during the entirety of a long season. Kraft paper treated with vegetable oil or other stabilizers is somewhat more durable, but still compares unfavorably with plastic film. Organic certifiers generally allow paper mulch, including non-colored newsprint, to be incorporated into the soil at the end of the growing season.

Although paper is relatively ineffective as a mulch, paper covered with an organic mulch material can be more effective than using either the paper or the organic material by itself. The paper provides a thin but dense layer that blocks weed growth while the organic material on top helps hold the paper down and intercepts light that would pass through the paper.

Recently, high-performing, biodegradable, starch-based plastic films have become available as substitutes for polyethylene plastic mulches. These require careful field application to optimize their performance. The thickness and polymers in the mulch affect biodegradation rates, and weather conditions affect the speed of degradation. The initial area of mulch breakdown is usually along the edge where the mulch is covered with soil. Some starch-based, biodegradable mulches are not approved for soil incorporation on certified organic farms.

Weeds Along Edges of Synthetic Mulch

All synthetic mulches must be anchored along all edges to prevent the material from blowing in the wind. Normally, this is accomplished by pressing soil along the edge of the mulch material, usually with hilling disks attached to the mulch layer. This soil is, of course, above the mulch and so tends to become weed infested. Although the weed roots cannot grow directly down through the mulch, they can grow around the edge and into deep soil. Also, getting cultivating tools close to the mulch without snagging and cutting it is a problem. One way to cope with weeds along the edges of the mulch is to use vegetable knives that are set deep enough to reach in under the mulch without catching it. Alternatively, a rolling spider gang cultivator with the gangs tilted upward toward the bed can be run along the edges. Another approach is to mulch the inter-bed areas, including the anchoring soil along the edges of the plastic, with an organic mulch such as straw. In addition to suppressing weeds, the organic mulch will protect the soil and makes a mud-free path during harvest. A fourth alternative is to apply a band of a natural product burn-down herbicide along the edge of the plastic to kill young weeds (see "Natural Product Herbicides"). Multiple applications may be required for a long-season crop. The effect of these herbicides on various mulch materials has not been well studied, and some may speed deterioration of certain synthetic mulches. We suggest you check with the manufacturer of the mulch or apply a normal application rate to a small area of stretched mulch for a season before using any herbicide extensively on mulched beds.

Tarping

A different approach to the use of synthetic soil covers for weed control involves covering the soil with a large, opaque tarp for several weeks and then removing it prior to planting. Unlike solarization (see the next section), in which the objective is to achieve soil temperatures sufficiently high to kill weed seeds, the objective of tarping is to smother emerged weeds to create a stale seedbed and to shift soil properties in ways that are detrimental to seed persistence as described below. Tarps are typically 6 mil black polyethylene plastic, which is tough enough to be rolled up and reused repeatedly. Tarps can cover multiple beds, but tarps larger than 30 by 100 feet can be difficult to handle (Maher and Caldwell 2018). Often, beds are prepared prior to tarping, in which case the tarp is used to create a stale seedbed (see "Stale Seedbeds" in Chapter 4). Alternatively, tarps can be used to smother existing weeds, harvested crops or cover crops prior to no-till planting. In the latter application, any living plant biomass should be mowed prior to tarping to help manage residues. Tarping durations are typically 3–10 weeks depending on the cropping plan, though tarps can be applied in the fall as a way to preserve beds and control soil moisture prior to early spring planting. They can also reduce fluctuations in soil moisture and conserve moisture during dry periods. Tarping periods less than three weeks are unlikely to completely kill emerged weeds. Tarps can add crop management flexibility by suppressing weeds and holding beds idle when time, equipment or field constraints limit other types of bed preparation.

Tarping is a relatively new practice and consequently has received little systematic study. Weeds vary in susceptibility to tarping. As might be expected, perennial weeds are generally resistant to tarping but can be stressed by extended tarping periods. Annual weed species also vary in their susceptibility. The mechanisms whereby tarps affect weed seeds in the soil is unknown. The primary mechanism may be the stimulation of germination that results in death due to lack of light. The magnitude of this effect could depend on soil conditions prior to and during tarping, including the intensity of pre-tarp soil disturbance, soil moisture and the time of year. The covered soil is, on average, usually a few degrees warmer than adjacent uncovered plots, even during cool spring weather in the northeastern United States (Maher and Caldwell 2018). Also, since growing plant roots are not present to take up nitrate and water cannot percolate through the soil to wash away nitrate, nitrate levels rise with the period of tarping (Ryberger et al. 2018). Both elevated temperatures and the presence of nitrate in the soil are known to stimulate the germination of many weed species, and the resulting seedlings would then die in the soil or after emergence under the tarp. Elevated soil nitrate after tarp removal could also favor some weeds in the following crop. Biological activity that is detrimental to weed seeds may also occur under tarps, but such effects have not yet been demonstrated.

Disposal

Most synthetic mulches pose significant end-of-season disposal problems. The labor and cost of disposing of large amounts of dirty, and probably wet, material should be considered when contemplating the use of these materials. If biodegradable mulches are not gathered and composted, they can be difficult to fully incorporate and fragments can blow about, leaving the farm an unsightly mess. Biodegradable mulches vary in how quickly they decompose. Combining paper mulch with an organic mulch material like straw or compost improves decomposition of the paper and helps hold it in place during the growing season (see "Biodegradable Mulches"). Organic standards currently require complete annual removal of most synthetic mulches other than paper, even if they are biodegradable.

WEED MANAGEMENT DURING TRANSITION TO ORGANIC PRODUCTION

Vegetable Farms

Organic vegetable farms are often established on old hayfields or pastures. This shortens the time until certification since semi-abandoned land often receives no chemical fertilizers or pesticides. Such fields, however, often have severe infestations of perennial weeds and dense seed banks of annuals.

Avoid planting vegetables the first year when you are starting vegetable production on very weedy ground. Instead, till, plant a cover crop, till in the cover crop before perennial weeds get large or annuals go to seed, and repeat this at four- to six-week intervals throughout the summer. This will reduce the weed seed bank and exhaust the storage organs of perennial weeds, while simultaneously building up soil organic matter and soil tilth (see "Tilled Fallow" and "Soil Tilth and Cultivation"). Oats or barley planted early in the spring at a high seeding rate make suppressive cover crops for the beginning of the season. Buckwheat and sorghum-sudangrass are fast growing, competitive summer cover crops. Near the end of hot weather, plant oats or an oat-field pea mixture on land that will be planted early the next year. This cover crop will compete with weeds in the fall, but frost kill in most regions, leaving the field ready for early planting. On fields that will be planted to vegetables after the last frost in the spring, plant hairy vetch, rye or a mixture in early fall (see "Winter Cover Crops"). To compete effectively with weeds, sow them all at high density (see "Crop Competitiveness").

Although there are costs to keeping fields in cover crops/fallow for a year before planting, these costs are often less than the labor required to control weeds in relatively uncompetitive crops. Since most growers bring fields into production on a staggered basis as their operation grows, usually opportunities exist for getting the weeds at least partially subdued before planting cash crops.

You have several good options if weed problems seem likely following the fallow/cover crop year. One is to plant short cycle transplanted crops like lettuce and kohlrabi that can be cultivated throughout their growth cycle and are harvested before most weeds get a chance to shed seeds. A second option is to plant sweet corn or potatoes. They can be cultivated with a tine weeder before and after emergence to reduce the density of annuals. They can also be cultivated aggressively between the rows and will tolerate having a lot of soil thrown into the rows to bury small weeds. Unless you can hoe or flame weed the corn, some weeds are likely to go to seed. Further weed suppression in potatoes can be obtained by mulching with straw, and in warmer climates this will keep soil temperatures in a favorable range for tuber production. Both the short cycle crops and intensively cultivated sweet corn or potatoes can continue the cleaning of the soil you began the previous year if they are managed well. A third option is to plant a highly competitive crop like winter squash. The squash can be cultivated until it starts to run, and it will tolerate the weeds that come up later. Unless you mulch heavily, you will likely have substantial weed seed production with this option, but at least you can get a crop off the field. Finally, you can grow a crop that you would produce on plastic for cultural reasons. You will probably have to hand weed the planting holes and possibly the edges of the plastic (see "Synthetic Mulch"), but with only moderate effort you can get a crop and still reduce the weed problem. Laying straw mulch between the plastic strips is initially laborious but saves labor later. If at all possible, avoid planting a slow growing, poorly competitive crop like onions, carrots or parsnips on the field until you have the weeds under control. Onions, however, can be transplanted into plastic mulch, which makes their production in high weed pressure conditions more practical.

Grain and Mixed Grain and Livestock Farms

Grain fields transitioning from conventional agriculture often have two weed management problems. First, the soil may have relatively low tilth, and this can make cultivation relatively ineffective. To understand why low tilth interferes with cultivation and how to improve your soil, see "Soil Tilth and Cultivation" in Chapter 4. The other problem is that you may have high densities of a few weed species that the previous herbicide regimen did not control well. Study the sections on those particular weeds and expand on the general strategy below accordingly.

If you have livestock, the best crop for transition is alfalfa or a grass-alfalfa mix. Repeated mowing will bring many perennials under control, and natural seed mortality will destroy part of the seed bank of annual weeds. Grass or clover will have similar benefits for reducing populations of annual weeds, but the less frequent mowing may not effectively suppress some perennials. If you do not have livestock, perennial sod crops have the same advantages for weed management, but you will export nutrients in the hay or haylage you sell. If your P or K are excessive, that may be desirable since excess nutrients can promote certain weeds. If P and K are very low to medium, however, exporting them in forage could set back your overall transition.

From a weed management perspective, soybeans are also a good crop to begin transition since 1) it is a competitive crop and 2) it can be cultivated aggressively early in the season with a tine weeder or rotary hoe and then later with a row-crop cultivator. From a nutrition perspective, soybeans are also a good crop to start transition since it is a nitrogen fixer and therefore does not require external nitrogen inputs. If the farm has been largely cropped in the past with summer row crops and spring grains, a winter grain crop early in the transition will help reduce weed populations. Due to the history of spring crops, the field will have few fall germinating species, and harvest and post-harvest operations will kill spring germinating species before they set seed (see "Crop Rotation and Weed Management"). For most farms a spring grain crop is a poor crop to start transition because the options for weeding are limited to early harrowing and many weeds will certainly go to seed before harvest. The Farming System Trial at Rodale showed that transitions beginning with corn tended to be weedier than a transition beginning with soybeans or winter wheat (Liebhardt et al. 1989), probably because corn grain is harvested late in the season and because the relatively open canopy allows light to penetrate to late maturing weeds. The low nitrogen status of most soils at the beginning of transition will also reduce the growth potential of high-N-requiring crops like corn, further reducing their competitiveness with weeds. Generally, corn harvested for grain causes more weed problems than silage corn or sweet corn because many more weed seeds will mature before grain has dried sufficiently to harvest.

SOLARIZATION

Soil solarization is an approach for killing weed seeds located near the soil surface. It is suitable for regions with warm climates and intense sunlight. Typically, the soil is tilled, firmed to create good soil-seed contact, irrigated and then covered with a clear polyethylene tarp for several weeks. To maximize heating, the tarp is laid close to the soil surface and the edges are covered with soil. The plastic transmits and traps sunlight, thereby heating the surface soil. Clear plastic is more effective for heating the soil than black plastic (Standifer et al. 1984). The soil is irrigated prior to covering because 1) moist, biologically active seeds are more susceptible to heat damage than are dry seeds, 2) moist soil conducts heat better than dry soil and 3) moistening the soil increases biological activity of microorganisms that attack seeds. Although solarization will kill many dormant seeds, much of the action appears to be against seedlings that result when seeds are prompted to germinate by warm, moist conditions under the tarp. Generally, seedlings are more susceptible to heat stress than are the seeds they come from.

Most weed species adapted to the warm climates where solarization is practical can tolerate temperatures up to 122°F, but few tolerate prolonged or repeated exposure to temperatures higher than that. Many studies have shown the effectiveness of solarization for achieving lethally high surface soil temperatures in warm climates. In an experiment in India, soil temperatures at 2 inches deep under clear plastic exceeded 132°F on 23 out of 32 days and exceeded 140°F for 7 days, whereas temperatures at the same depth in uncovered soil never exceeded 122°F. Similarly, in Mississippi, soil temperatures at half an inch reached 149–156°F under clear plastic, but only 103–122°F in the uncovered soil (Egley 1983). Daily maximum temperatures at 2 inches averaged 18°F higher under the plastic than in bare soil.

The many studies on soil solarization indicate several important points. First, because solar heating

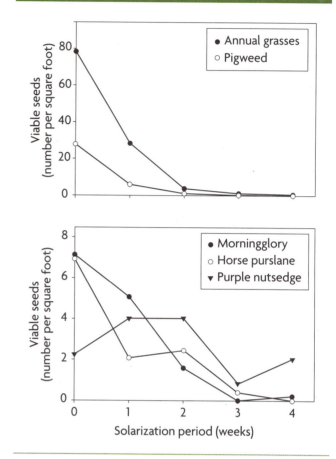

Figure 3.5. Emergence of weed seedlings during a three-month period (June 10 to September 10) following one, two, three or four weeks of solarization in Mississippi. Data from Egley, 1983.

of the soil acts primarily near the soil surface, usually the seed bank is substantially reduced only in the top few inches. For example, a study in Louisiana found that annual bluegrass and barnyardgrass seeds were completely eliminated from the top 1.2 inch and substantially controlled down to 2.4 inches, although some reduction in the barnyardgrass seed bank occurred down to 6 inches (Standifer et al. 1984). Since most weed seeds need to be close to the soil surface to produce seedlings, the minimal damage to deep seed banks is not critical for weed control in the crop that immediately follows solarization. However, since viable seeds may remain just below the cleaned soil layer, minimizing soil disturbance before, during and after planting is critical for the success of this method.

Second, the soil must remain covered for several weeks to effectively kill weed seeds (Figure 3.5). This factor largely precludes use of solarization in regions with relatively short growing seasons, even if mid-summer is hot and sunny. Similarly, the long periods of hot, sunny weather required for effective solarization largely limits use of the procedure to late-summer and fall planted crops.

Finally, species differ substantially in how well they are controlled by solarization. For example, pigweed and morningglory species were controlled by three weeks of solarization whereas grasses and horse purslane required longer periods of treatment (Figure 3.5). Purple nutsedge, a perennial emerging from tubers, was not significantly affected by solarization at all. Generally, perennials are not effectively controlled by solarization because they can emerge from large storage organs deep in the soil where the killing effect of the solar heat does not penetrate. Similarly, large seeded annual species can sometimes emerge from below the depth of the well-cooked soil.

In addition to killing weeds, solarization can kill soilborne plant pathogens, mobilize nutrients and increase crop yields. However, the high cost of the approach, the long amount of time the field has to remain covered and the ineffectiveness of the method against some weed species makes solarization only appropriate for selected regions, fields and cropping systems.

NATURAL PRODUCT HERBICIDES

Several natural product herbicides are approved for use on organic farms. Except for corn gluten, all registered materials are "burn-down" type herbicides that kill or damage only the green tissues they contact. Destruction of the roots of well-established plants requires multiple applications, which eventually exhaust the plant's belowground reserves. Note that these burn-down products are non-selective, meaning that they will kill green crop tissue as easily as weed tissue. Therefore, direct applications away from crop plants.

Acetic acid, the active ingredient in vinegar, is the best-known natural product herbicide. Note that organic standards require that the acetic acid be derived from natural fermentation. Generally, it is most effective against small annual broadleaf weeds that do not have a waxy or densely hairy leaf surface. Apparently, acetic acid does not stick well to waxy surfaces, and the dense leaf hairs on plants like velvetleaf prevent the acid from reaching vital tissues before it evaporates. Tests in several states have shown that acetic acid concentrations of 5–10% are relatively ineffective against even small annual broadleaf weeds, and that effectiveness in the 15–30% range increases with

concentration (Brainard et al. 2013). High application volumes whereby weeds become visibly wet are required for consistent control. Grasses tend to resprout after getting burned back. Control is improved when temperature and relative humidity are higher at the time of application (Brainard et al. 2013).

Several other burn-down products based on clove oil, lemongrass oil, citrus oil and capric plus caprylic acid are also available. Trials with these products indicate that concentrations and application rates similar to those for acetic acid are required to achieve reasonable control, but they may be more or less effective against specific weeds or growth stages. OMRI-approved herbicidal soaps are available, but their use is not allowed on organically certified crop land. They do provide a way to burn down weeds along fences and around building foundations where mechanical management can be difficult.

Because natural product herbicides require several to many gallons of concentrate per acre, they are an expensive approach to weed management. Consequently, they are most cost effective when applied to intermittent patches of weeds, such as along the edges of mulch, or in a band, such as to the row just before emergence of slowly emerging crops like carrots. In these applications, inter-row cultivation can be used to inexpensively remove weeds between the crop rows. Natural product herbicides may also find some use in preparation of stale seedbeds for high value crops. In such applications, however, high rates of the active ingredient will be necessary to avoid escapes that then have a substantial head start on the crop.

A corn gluten meal product is currently the only OMRI-approved pre-emergence herbicide. Corn gluten that is not specifically OMRI approved is likely to have been made from genetically modified corn. The active principles in corn gluten are very short chain proteins. The material thus also doubles as a nitrogen fertilizer (10% N). Corn gluten kills a substantial percentage of most weed species during germination. It inhibits root growth, and the seedlings die of drought stress. It is not effective against perennial weeds. A drawback to the material is that it is toxic to most crops when they are germinating and, consequently, the manufacturer recommends waiting 4–6 weeks before planting seeds. Whether corn gluten is more effective for weed control than a 4–6-week tilled fallow seems doubtful. The material can be used safely, however, with transplanted crops. Several crops, including broccoli, cauliflower and strawberries, have been transplanted into soil recently treated with corn gluten without harm, and it is probably safe for many other transplanted crops as well. Since it is a legitimate fertilizer, the material can be used in the many states where it is not registered as an herbicide. The recommended application rate is 446 pounds per acre, which makes it a bulky and expensive way to manage weeds. Other seed meal byproducts from processed mustard and soybean oil crops have shown potential for controlling weeds, but their bulk and cost represent major obstacles to adoption.

LIVESTOCK FOR WEED MANAGEMENT

Farmers have used cattle, sheep and goats to control weeds for many centuries. Most common agricultural weeds make nutritious forage. Of course, some are toxic (Burrrows and Tyrl 2006), and others become unpalatable when they mature (see "Palatability" in Chapter 2 and the "Palatability" section of each species chapter for details). The most common way growers use large livestock is to clean up fields after harvest. A brief period of intensive grazing is most effective for this purpose since the animals then trample any weeds that they do not eat. Intensive grazing will kill most annuals and set back perennials, but most perennials will resprout after the livestock are removed.

Cattle feed almost exclusively on grasses and herbs and can therefore be used to manage ground vegetation in orchards, though care must be taken to avoid damage to tree roots. You can run sheep in orchards too if the trees are tall, but they will nip buds and young branches from dwarf trees. They will also strip bark if preferred forage is not available. Goats prefer woody browse over most herbaceous plants, so they are inappropriate for use in orchards. For the same reason, however, goats can very effectively reclaim brushy pastures. We once watched a flock of goats attack a patch of the introduced Himalayan blackberry in Oregon. They ate it with relish despite the large thorns that covered the tall, tough canes. Sheep do not eat as wide a range of woody browse as goats do, and they cannot reach as high, but sheep can effectively control many nuisance shrubs in cattle pastures, including red cedar, multiflora rose and speckled alder.

Pigs can root out perennial weeds that are otherwise difficult to control. They are especially effective against species like quackgrass, perennial sowthistle and the nutsedges, where most of the rhizomes, storage roots or tubers are in the plow layer. Breaking the soil, for example with a chisel plow, will allow the pigs better access to the weed storage organs. To minimize damage to soil structure, avoid both prolonged grazing by pigs and grazing when the soil is wet.

A flock of chickens makes an excellent adjunct to a vegetable operation. Since they essentially spread raw manure as they graze, they should not be put into the field before a short season crop. The National Organic Program requires 90 days between manure application and harvest of crops in which the edible portion does not touch the soil and 120 days for crops where the edible produce does touch the soil. Following this standard protects public health regardless of whether or not your farm is certified organic. Chickens can be released into the field after harvest, however, and will pick out weed seeds, clean up perennial weeds and also eat slugs and insect pests. Chickens relish dandelion, quackgrass and most other weeds but will reject some members of the mint and parsley family. They can also help weaken a cover crop in preparation for incorporating it or speed the decomposition of a cover crop after it is mowed prior to incorporation. A low, temporary fence coupled with clipping the wing feathers keeps them confined to the area where you want them. Chickens should not be left in the same area for long periods, however, because their constant scratching will ruin soil structure. Their potential for damaging the soil is particularly great if the soil is wet and the ground is mostly bare.

Geese can also be useful for weed management. They are true grazing animals. They selectively eat grass and a few other weeds (chickweed, horsetail) but avoid most broadleaf species, including most fruit and vegetable crops. Goslings can be trained to eat a wider range of weeds by feeding them the weeds with little choice of other forage when they are young. Although geese have been used successfully in vegetables, particularly potatoes, their droppings constitute a potential health risk that most growers will want to avoid. Geese were widely used in cotton, however, prior to the development of herbicides. They can be helpful in controlling weeds in nurseries and in other non-food crops like cut flowers. They have also been used successfully in perennial crops including orchards, grapes, strawberries, brambles, blueberries and asparagus. These crops often have severe problems with perennial grass weeds, and geese can help manage these weeds. To avoid health risks, the geese should be removed four months before harvest, which in most crops restricts their use to postharvest weeding. Geese should be fenced, and electrified mesh is probably the best choice for simultaneously confining the geese while keeping out predators. Goslings are generally preferred over adult geese for weeding since they are more active and less likely to trample crops. Two to six goslings per acre are sufficient to keep weeds under control if they are released against the weeds early in the season. The optimal number depends on weed density and whether inter-row areas are cultivated. Placing water and supplemental food at opposite ends of the pen encourages the geese to walk through the whole area, weeding as they go (Geiger and Biellier 1993).

Very few farmers will keep livestock just for weed control. Sustainable farming, however, is based on ecological integration. Using your livestock to help control weeds is one more way to achieve that integration.

PREVENTIVE WEED MANAGEMENT

Minimal Weed Competition Versus Preventive Management

Farmers take one of two approaches to weed management. Most seek to keep weed populations sufficiently low that they can obtain good yields but otherwise do not worry if their fields have some weeds present. Ecological theory predicts that even minor weed infestations reduce crop yield, and that the impact per weed is actually greatest when weed density is low (Cousins 1985). Variation in yield from year to year and field to field is so large, however, that weed abundance usually has to be surprisingly high before yield loss can be detected statistically. The bottom line is that although weeds usually hurt yield, the impact on profits from a moderate stand of weeds may be too low to matter most of the time. But, given their great powers to reproduce, even maintaining weeds at moderate levels usually requires substantial effort and diligence. Moreover, a moderate density of weeds can hurt your crop in a year when growing conditions prevent timely cultivation or if the crop is stressed.

The other approach to weed management is to consistently minimize weed reproduction through a program of preventive management coupled with attack on soil seed banks and reserves of perennial

roots and rhizomes. Farmers who follow this strategy may eventually have fields with low weed populations that allow them to greatly reduce weed management efforts. Their small seed banks and low populations of perennial weeds also buffer them against yield loss to weeds in years when weather makes timely cultivation difficult.

Preventive management often requires more precise cultivation with a wider array of implements. It also usually involves extra hand pulling, flame weeding and hoeing of weeds to prevent seed set, even though the impact of the weeds on the immediate crop may be negligible. Although hand weeding is usually

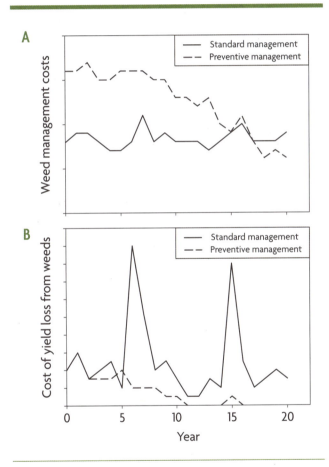

Figure 3.6. A conceptual illustration of the effects of standard management and preventive management on (a) the costs of weed management and (b) the costs of yield loss due to weeds. Standard management aims to keep crop yield losses due to weeds low in most years (in this case, nine out of 10 years). Preventive management aims to prevent all weed reproduction. Preventive management has a relatively high investment in weed management in early years, but low risk. Standard management has an approximately constant investment in weed management over years, but moderate risk. The example is hypothetical. The time course of costs for each strategy on an actual farm will depend on the weed species present, the resources of the farm and the management skills of the farmer.

associated with vegetable production, see profiles of Carl Pepper and Paul Mugge in Chapter 5 for examples of applying this approach cost effectively in field crop production. Many growers who follow a preventive management approach use tilled fallows to flush seeds from the soil (see "Tilled Fallow" in Chapter 4) and intensive cover cropping to prevent seed production when a cash crop is not present. If nothing else, seedbed preparation for a cover crop after harvest kills weeds before they can make additional seeds. Preventive weed management sometimes even involves sacrificing a crop by tilling it under rather than allowing heavy weed seed production. The loss of the crop has an immediate cost, but it is a cost that can be partially offset by planting another (often different) crop. Farmers following a preventive management strategy sustain a cost they understand and can manage compared to the unknown costs of future weed problems.

Neither strategy is "correct" (Brown and Gallandt 2018). Farmers following each of these strategies have been highly successful for many years. With the more common strategy of weeding enough to get a good yield in most years, weed management costs vary moderately from year to year, with occasional spikes to bring weed problems under control after bad years (Figure 3.6a). On average, however, the cost of weed management remains about the same over the long run. With this strategy, costs due to yield losses from weeds are generally low but occasionally become substantial in bad years.

With a preventive management strategy, costs of weed management are relatively high in the early years but drop with time along with weed populations (Figure 3.6b). If the preventive strategy is highly successful, the farmer may eventually be able to relax management in later years and actually have lower management costs than the grower following the standard "weed enough to get a good yield" strategy. Greatly reduced weeding may never prove possible, however, if any of a variety of factors prevent perfect control in every year. The great benefit of a preventive strategy is that yield losses due to weeds decline essentially to zero after a few years. Most importantly, the large yield losses due to weeds in occasional bad years disappear.

Cleaning up After Harvest

Even if you do not follow a strict program of preventive management, steps to reduce seed production will help keep weeds under control. Probably the most important of these is to clean up the field as quickly as possible after harvest. Once the crop is removed, it no longer competes with the weeds and they will then grow and mature rapidly. Depending on the crop, the weeds and the season, delaying cleanup by a week may increase seed shed by several fold; delaying cleanup by a month may increase seed production by a hundred-fold. Cover crops are another important tactic for reducing seed production. Because they reduce weed density and slow growth and maturation of weeds, a competitive cover crop can mean the difference between no seed production and substantial seed production prior to the next crop. Tillage associated with cover crop establishment also can help eliminate weeds that would otherwise mature after harvest. Experienced growers recommend keeping a supply of frequently used cover crop seeds on hand so that cover crops can be planted without delay after crop harvest.

Reducing Seed Shedding During Combining

Combine harvesters normally spread weed seeds throughout a field. Most weed seeds are dispersed with the chaff rather than with the straw. Consequently, several companies have developed systems for collecting the chaff, usually in a wagon pulled behind the combine. In a Canadian study on spring wheat, chaff collection reduced the number of wild oat seeds dispersed by the combine by 77% (Shirtliff and Entz 2005). Although 50–60% of the seeds had already fallen from the weeds by harvest, chaff collection still captured roughly one third of the seeds produced. Chaff collection also greatly reduced the average distance weed seeds were spread and would thereby tend to keep weed problems more localized. As an alternative to chaff collection, a Harrington Seed Destructor has been developed in Australia that mechanically kills seeds in the chaff before the chaff is released onto the field (Walsh et al. 2012). Mechanically killing the weed seeds avoids the need to handle large volumes of chaff.

A related approach is to direct the straw and chaff into narrow windrows that are then burned. This has the disadvantage of destroying crop residues that would otherwise support soil health, and it is not allowed in some areas due to air quality regulations. Nevertheless, the practice is widespread in some regions.

Depending on the weed and the crop, the combine can be set to collect weed seeds along with the grain.

The weed seeds can then be separated by subsequent cleaning, thereby removing them from the field. We know a grower who regularly removes wild mustard seed from small grain fields in this way. He sells the seeds to an artisanal mustard producer. Computer models show that removing a substantial portion of weed seeds during harvest can tip the balance between an increasing weed problem and a decreasing one.

The benefits of removing weed seeds during harvest depend on the crop, your particular weed problems and other factors. Most weed seeds have already dispersed by the time corn or soybeans are harvested, though a substantial proportion of seeds of some species like giant ragweed, waterhemp and common lambsquarters remain on the plant well into the fall. In contrast, collecting weed seeds during the harvest of any spring grain will substantially reduce seed dispersal by many weed species. In winter grains, it can often reduce the number of weed seeds reaching the soil from a moderate number to almost none, but the effects will depend on how fast the grain and the weeds mature. The captured weed seed and chaff can be fed directly to poultry, ground with a hammer mill to destroy viability and then fed to other livestock or used for biofuel.

Promoting Weed Seed Predation After Seed Dispersal

Predation of weed seeds can substantially reduce seed populations. A summary of 10 experiments estimated that short-term weed seed predation rates averaged 52%, with the potential for considerably higher losses (Davis et al. 2011). Generally, loss rates from seed predation at the soil surface are higher than losses would be from aging and decay if seeds were buried in the soil. In one study, seed removal by predators was responsible for a 38% reduction in seedling emergence and an 81% reduction in weed biomass in the subsequent season (Blubaugh and Kaplan 2016). However, seed predators do not affect the abundance of seeds already buried in the soil, suggesting that additional tactics would still be needed for long-term seed bank management (Blubaugh and Kaplan 2016).

Both invertebrates and vertebrates contribute to weed seed predation. The most prominent invertebrate predators include carabid beetles and crickets. Invertebrates are generally most active during warmer months in mid to late summer and have a foraging range of tens of yards. The most prominent vertebrate predators are rodents and birds, which can be active throughout the year and have a foraging range of hundreds of yards or more. Generally, the larger the predator, the larger their weed seed size preference. Thus, invertebrates tend to consume smaller seeded weeds and vertebrates prefer larger seeded species. From a broad ecological perspective, seed predators should be considered as part of a larger food web, so their behavior not only reflects search strategies for food sources but also avoidance of predators that can consume them. The preference of most grain-eating beetles and rodents for denser vegetation, and grain-eating birds and ants for open patches, is probably related to their relative need for predator avoidance.

Maintaining fields in an untilled condition is the most important management practice for encouraging weed seed predation because it keeps weed seeds on the soil surface where predators can more easily find them. This practice is most advantageous for reducing a high density of seeds in a year with poor weed control; once seeds are incorporated into the soil seed bank, the opportunities for predation are limited. Leaving seeds on the soil surface is most beneficial when seed shed occurs at or before crop harvest. For example, peak activity by seed predators often occurs in late summer, which coincides with weed seed production in full season crops but is not well synchronized with crops harvested in mid-summer. Caution is required, however, to ensure that leaving fields untilled after harvest does not permit weeds to grow and produce additional seeds, thereby undoing the benefits of predation.

A second practice that usually enhances weed seed predation is maintenance of high vegetation cover (Meiss et al. 2010). Including forage crops, cover crops or dense plantings of row crops in rotations can increase vegetative ground cover and the potential activity of seed predators in fields. Although an abundance of living vegetative cover enhances seed predatory activity, dead crop residue on the field has little influence. In addition, some research has indicated that seed predation is higher near the edge of fields and that increasing the field-edge-to-area ratio may enhance seed predator activity, but other research shows high local variability and the lack of a definitive response to border management.

Generally, seed predation is controlled by a complex interplay of environmental and biological interactions that occur locally within individual fields and can, therefore, be a very dynamic and ephemeral process. Although weed seed predation can provide substantial weed management benefits, weed seed predation should not be a primary factor driving crop management decisions.

Maintaining Clean Field Margins

A final step to help reduce weed reproduction is good management of field edges and driveways. Weeds often survive in a field primarily by seeds entering from un-tilled land adjacent to the field. Although some individuals within the tilled field may set seed in some years, that reproduction may not be sufficient to maintain the population without supplementary seeds from the field margins. For example, we believe that curly dock and broadleaf dock populations are often maintained in this way in regularly tilled fields. Similarly, field edges may be critical for maintaining populations of creeping perennial weeds and those with wind-dispersed seeds. Your management of the field may be continuously eliminating the weed, but roots or rhizomes grow into the field each year and are then spread about on tillage implements the next season. If a species is more abundant along the long edge of a field than it is toward the middle, this indicates that increased management of the field margin may prove worthwhile. Weeds are often more abundant on headlands due to soil compaction and gaps in the stand as well as seed input from the field margins.

To avoid spread of weeds from the field edges, keep grassy margins and driveways mowed frequently enough to prevent seed production. Regular mowing will also force perennials to put energy into shoots rather than into roots or rhizomes that spread into the field. Where a grass way abuts to a hedgerow or woods, keep the brush trimmed back so that you can mow consistently along the edge. When using a rotary mower along field edges, drive around the field in the direction that throws the cuttings, which contain weed seeds, away from the field rather than into it. Eliminate irregular edges along tilled fields so that the edge is straight or smooth. Old rock piles or encroaching brush that prevent mowing or tillage also create pockets where weeds thrive and then spread.

PREVENTING THE ARRIVAL OF NEW WEED SPECIES

Most farmers who know their weeds eventually observe a new species that has not been on the farm previously. Invasive species from other continents and regions are spreading through the United States and Canada. Other long-naturalized species are shifting their ranges as the climate changes. When a new species shows up on the farm, prudence dictates that you attempt to eradicate it during the year it arrives. Once it has multiplied and spread, eradication becomes much more difficult, and you will likely have to learn to live with the new weed species. Consequently, learn to recognize at least the most problematic species found in your region even if they are not yet common, and keep an eye out for new species as you work. Many resources are available for helping you identify weeds (see Box 3.2).

New weed species can arrive on the farm by any of the means discussed in the "Dispersal" section of Chapter 2. You can protect yourself against most of the ways new species arrive.

Weed Seeds in Forage, Cover Crop and Grain Seed

Corn, soybean and small grain seed produced off the farm should be confirmed free of weed seeds before it is sown, even if you are just using it for a cover crop. Because they have very large seeds, corn and soybeans can usually be cleaned completely. Small grain seed may still have a low level of contamination that is sufficient to start a weed infestation, even after cleaning. Blue tag certified grain seed is usually safe. To give perspective, for grain sown at 110 pounds per acre, a 0.01% contamination translates to about 5,000 1-milligram seeds per acre, though often the contamination level is much lower than the maximum indicated on a bag of certified seed. Always visually inspect non-certified seed for contamination. We have frequently found grain rye being sold for cover crop seed that was severely infested with noxious weeds.

Forage seed often contains weed seeds because many weed species have seeds that are similar in size and density to various forage species. Always inspect forage seed for weed contamination. Be especially cautious of seed produced in other regions, since it is more likely to contain new weed species you want to avoid.

If contaminated seed has weeds of a species that is already common on your farm, calculate how the level of contamination you are observing translates into the number of weed seeds you are sowing. Remember that your soil probably already contains at least several

Box 3.2. Resources for Weed Identification

Weed identification guides are available for most regions of the United States and Canada.

Alex, J.F. 1992. Ontario Weeds: Description, Illustrations and Keys to their Identification. Consumer Information Centre, Ontario Ministry of Agriculture and Food: Toronto, Ontario.

Barkley, T.M. 1983. Field Guide to the Common Weeds of Kansas. University Press of Kansas: Lawrence, Kansas.

Bouchard, C.J., R. Néron and L. Guay. 1999. Identification Guide to the Weeds of Quebec. Centere ARICO Direction des services technologiques MAPAQ: Quebec, Quebec.

DiTomaso, J. 2007. Weeds of California and Other Western States (2 volumes). University of California Dept. of Agriculture and Natural Resources: California.

DiTommaso, A. and A.K. Watson. 2003. Weed Identification, Biology and Management. Two CD set. HRC Photo.

Gains, X.M. and D.G. Swan. 1972. Weeds of Eastern Washington and Adjacent Areas. Camp-Na-Bor-Lee Association, Inc.: Davenport, Washington.

Georgia Cooperative Extension Service. 1987. Weeds of the Southern United States. Cooperative Extension Service, College of Agriculture, University of Georgia: Athens, Georgia.

Hall, D.W. 1994. Weeds of Florida. University of Florida, Cooperative Extension Service, Institute of Food and Agricultural Sciences: Gainesville, Florida.

Haragan, P.D. 1991. Weeds of Kentucky and Adjacent States: A Field Guide. The University Press of Kentucky: Lexington, Kentucky.

Harrington, H. and R. Zimdahl. 1974. Weeds of Colorado. Colorado State University Extension: Fort Collins, Colorado.

Kinch, S.S. 1975. South Dakota Weeds. South Dakota State Weed Control Commission: South Dakota.

Muenscher, W.C. 1955. Weeds, Second Edition. Comstock: Ithaca, New York.

Stearmar, W.A. 1941. Weeds of Alberta. A. Shnitka, King's Printer: Edmonton, Alberta.

Stucky, J.M., T.J. Monaco and A.D. Worsham. 1981. Identifying Seedling and Mature Weeds Common in the Southeastern United States. North Carolina Agricultural Research Service Bulletin No. 461: Raleigh, North Carolina.

University of California Statewide Integrated Pest Management Program. Weed photo gallery. http://ipm.ucanr.edu/PMG/weeds_intro.html

University of Missouri, Weed ID Guide. University of Missouri, Division of Plant Sciences. https://weedid.missouri.edu/

Uva, R.H., J.C. Neal and J.M. DiTomaso. 1997. Weeds of the Northeast. Cornell University Press: Ithaca, New York.

XID Services. 2012. 1200 Weeds of the 48 States and Adjacent Canada. Interactive CD. XID Services: Pullman, Washington.

XID Services. 2012. 1000 Broadleaf Weeds of North America for Android. Android App. XID Services: Pullman, Washington.

Whitson, T.D. 2006. Weeds of the West. Western Society of Weed Science in cooperation with the Western United States Land Grant Universities Cooperative Extension Services: Laramie, Wyoming.

Color photographs of many weed species can be found at the Weed Science Society of America (WSSA) website http://wssa.net/wssa/weed/weed-identification/. Additional photographs and distribution information can be found at the U.S.D.A. Agricultural Research Service's PLANTS database, http://plants.usda.gov/.

hundred weed seeds per square foot, so it is only the unusual weeds you have to worry about. Resources for visual identification of weed seeds include Davis (1993) and Ohio State University's OARDC Seed ID Workshop (www.oardc.ohio-state.edu/seedid/), although the latter requires you to guess at the seed's identity. Alternatively, grow some of the weed seeds out in flowerpots to see what comes up: If you cannot identify the weeds as species already on your farm, then you probably have a potential problem.

Weed Seeds in Feed and Forages

You can introduce new weeds onto your farm by purchasing weedy feed grains and forages. Since seeds of most weed species readily pass through the guts of livestock alive, weed seeds can spread with manure throughout the farm in a single season. Note that even a few dozen seeds per ton may be enough to start a serious weed infestation. For example, velvetleaf-infested corn from the Midwest probably caused the rapid spread of this weed across the dairy farms of New York and New England during the 1970s and 1980s.

Discussing potential weed problems with the farmer from whom you are buying feed is probably the best insurance against accidental introduction of a new weed species in feed. The alternative is to inspect the feed for weed contamination. If you are buying grain, attempt to check samples from the bottom of the bin.

Small weed seeds tend to sift downward as the bin is filled so the bottom material will likely show them. Carry along a screen and sift off the grain so you can examine the material that is left. If you can identify the seeds of the weeds you already have on the farm, you will recognize any that are unfamiliar. Some Extension agents can also help identify weed seeds. With forages, the easiest way to check for weeds is to walk the field before it is chopped or hayed. Identifying weeds in green chop is usually hopeless. Buying a few sample bales from different parts of the stack and pulling them apart to look for seed stalks, however, can help you avoid buying half a barnful of contaminated hay. The same principle applies to bedding straw, since it will eventually end up in the field as well. We have frequently observed Canada thistle seeds in straw, and we are aware of an infestation of mayweed chamomile that was initiated by infested straw used as mulch.

Weed Seeds in Compost and Manure

Composting can effectively kill weed seeds if it is done properly. This is not easy, however, and if you make your own compost it is safer to use seed-free material rather than relying on the composting process to protect you. The problem with adding weedy compost or manure to your fields is not that you will immediately be overwhelmed with weeds: Usually the weed seed density of field soil is higher than that of manure or compost. Rather, the problem is that you may introduce a new pernicious weed species that will cause management problems for years to come. By the same principle, however, if you only use materials generated on the farm to make compost (including the feed that made the manure), you probably have nothing to fear from weed seeds in that compost.

Many materials used for making compost are commonly contaminated with weed seeds. Late cut hay will certainly contain weed seeds. Straw can be examined for fruiting stalks of weeds, but it is often clean. Leaves from street trees usually have very few weed seeds but may contain traces of toxic materials like motor oil and rubber dust. Cotton gin waste is usually heavily contaminated with weed seeds, and compost made from gin waste is frequently contaminated as well (Norsworthy et al. 2009). You should consider all manure from off the farm to be contaminated unless you have tested it. Horse manure and manure from livestock that have access to weedy pastures or pastures along roadsides are most likely to contain a high density and diversity of weed seeds.

You can test manure and compost for weed seeds by mixing several quarts taken from various parts of the pile with potting mix in a 1:1 ratio and spreading it in flats. Keep the flats warm during the day and cool, but not cold, at night. For example, run the test inside on a windowsill in winter, outside in the summer, and in a cold frame or unheated greenhouse during the spring or fall. Water the flats regularly and observe any weed seedlings that emerge over the following two to three weeks. This test will usually show if weed seeds are present, but it may not accurately predict their density or which species are present, since some seeds may be dormant.

To kill weed seeds during the composting process, the pile should reach at least 140°F for at least two weeks. Some of the more resistant species may not be killed even by this treatment. For a small compost pile, achieving a high sustained temperature will prove difficult. Thoroughly mixing the pile several times is necessary to ensure all the materials are aerated and attain the required temperature for a sustained period. Mixing with a front-end loader is less likely to provide the sustained high temperatures needed to kill weed seeds than using a compost turner. Whenever you use compost that is potentially contaminated, watch your weed flora carefully for the next year and vigorously attack any new species that appear.

Weed Seeds on and in New Livestock

Weed seeds commonly travel on the fur and soil clinging to the feet of livestock. Clean off any newly purchased animals and dispose of the waste in a way that will avoid introducing weeds into the field. Most species of weed seeds pass largely unharmed through the guts of horses and all ruminants. Isolate newly purchased animals for a few days indoors or in a small exercise yard until seeds have passed before releasing them onto pastures. Dispose of any manure collected during those few days. Monitor the exercise yard and promptly eliminate any weed species you don't recognize.

Weed Seeds on Machinery

Tillage and cultivation implements commonly transport weed seeds and storage organs of perennial weeds between fields. The problem arises primarily in two situations: 1) when you are trying to contain a problem weed to one part of the farm and 2) when you are doing custom operations on another farm or having them done on your farm. The more soil that clings to an implement, the more likely you are to transport weed seeds. Plows and disks can transport a lot of soil. For example, we once cleaned 46 pounds of soil off of a 6-bottom plow after it had traveled 1.5 miles back from the field. Cultivators usually carry less soil, but the hooked arrangement of shovels is effective at transporting perennial roots and rhizomes. Rotary tillers are also good at transporting perennials because shoots wrap around the tine shaft, and they drag roots and rhizomes along with them. Tractor tires can also carry a substantial amount of soil.

Some idea of the potential of weed seed transport in soil can be judged from the density of seeds in the soil. Weed seed densities often reach several hundred seeds per square foot of soil. A density of 100 seeds per square foot in a plow layer 8 inches deep translates to about one seed per half pound of soil. Thus, the risk of transporting seeds in soil is relatively low until the weed becomes abundant in the field. The chance of picking up seeds on equipment increases, however, if conservation tillage leaves the seeds concentrated near the soil surface. Also, species are often more abundant on headlands, which tend to be tilled or cultivated last—just before the implement leaves the field. Finally, seeds of weeds along field edges can easily get stuck in the soil clinging to tractor tires. Thus, watching for unwanted species on headlands and field edges when doing custom work can help prevent weed spread.

If transport of an unwanted species seems likely, wash your implement either before or after transporting it, depending on which is appropriate. A power washer is more effective than a hose for cleaning caked soil out of crevices in implements. The best place to wash is in the field before you leave, but that is rarely practical. A thick, healthy lawn is a good place to clean implements since few agricultural weeds can establish in a frequently mowed lawn. Avoid washing on weedy areas where you park implements, since establishing a population there will lead to seeds subsequently getting carried into the fields on tires. Also, avoid washing the soil into road ditches. Weed populations often establish and spread along ditches and embankments first before subsequently spreading into fields.

Combine harvesters are probably even more effective than tillage implements for transporting weed seeds between fields. They have a variety of locations where seeds can lodge and then later become dislodged. One study found that more weed seeds accumulated in the central divider assembly of a corn head than elsewhere, and this can be easily cleaned with a vacuum. Forage harvesters have been shown to move seeds long distances within fields when chopping corn (Heijting et al. 2009), and this indicates their potential for dispersing seeds between fields as well.

Weed Seeds in Irrigation and Flood Water

Most common agricultural weeds will survive for several months completely immersed in fresh water, and many can survive immersion for more than a year (Comes et al. 1978). Moreover, flower and fruit parts cling to the seeds of many species, trapping air and providing buoyancy. Consequently, seeds can be carried long distances by flood waters and deposited in low lying fields when the flood recedes. Therefore, watching for and eliminating new species that show up after floods can prevent weed problems later.

Many weed species are also dispersed in surface irrigation water (Kelley and Bruns 1975). For example, 164 different weed species were found in water from the Columbia River and irrigation canals supplied from the river. The density and diversity of weed seeds in irrigation water can be reduced by managing vegetation along canal banks and drainage ways feeding into the canals, for example, by grazing or tillage (Kelley and Bruns 1975). Ultimately, however, the most effective way to prevent new weed species from entering a field in irrigation water is to screen the water before it is distributed. Since the density of most species in irrigation water is low, growers should not assume that just because a field has been irrigated for many years, all potential species have already arrived. Moreover, water courses of all types can act as effective routes for the dispersal of invasive species new to the region, as is demonstrated by the close correlation of giant hogweed sites in the United Kingdom with stream courses.

Awareness of Emerging Weed Problems

Knowledge and vigilance are the keys to successfully preventing outbreaks of new weed species and managing existing problems. To repel new invaders, you need to know that they are new to your farm. Moreover, to use the species-specific strategies outlined in later chapters of this book, you need to know which species you are trying to control. Although distinguishing between a few closely related species can be challenging, for the most part weed identification is not difficult. Learning to identify the weeds on your farm plus a few likely potential invaders will pay dividends for you and for future generations.

SUMMARY

- The core principle of ecological weed management is the integration of multiple tactics that attack weed populations at multiple points in their life cycles.
- Sound ecological weed management begins with a diverse crop rotation that allows implementation of the preceding principle.
- Rotation between crops with different planting dates, growth periods and harvest times tends to disrupt weed reproduction. Rotation between different types of crops allows for the use of diverse tactics during the overall crop rotation.
- Enhancing crop competition is a critical component of weed management and becomes more important as weed pressure increases. Increasing crop density always improves competitiveness of the crop, and narrowing row spacing often improves crop competitiveness as well. For summer crops, north-south oriented rows are most competitive, whereas for cool season crops east-west oriented rows are most competitive. Use of crop cultivars that are good competitors can facilitate weed management. Use of transplants increases the competitiveness of small seeded vegetable crops. Sometimes, additional crop competitiveness can be achieved by growing two crops together, particularly when one matures more quickly than the other.
- Nutrient amendments like green manures and composts with high organic matter content tend to favor long season crops like corn relative to weeds, whereas amendments that rapidly release nutrients like poultry manure and Chilean nitrate will tend to favor weeds. Banding rapidly released amendments next to the crop row or side dressing them after crop establishment will help direct nutrients to the crop rather than to the weeds.
- Cover crops can contribute to the diversity of crop rotations and provide opportunities to suppress weed growth and seed production during periods between cash crops, in addition to their many benefits for improving soil quality. To achieve good weed suppression, attention should be given to planting cover crops to ensure a dense, uniform stand.
- Cover crop residue can provide weed suppression for no-till crop production. When cover crops flower, they can be killed with a roller-crimper and the next cash crop can be planted directly into the residue. If the cover crop has been sufficiently productive, the residue will suppress the emergence of most annual weeds. This procedure allows the soil conservation benefits of no-till planting for some crops within the crop rotation.
- Organic mulch materials can also be brought to a field or bed and spread to suppress emergence of annual weeds. Weed species vary in their susceptibility to suppression by mulch, with smaller seeded species more susceptible than larger seeded species.
- A variety of synthetic mulch materials are available to provide a physical barrier to prevent weed emergence. These vary greatly in durability and difficulty of disposal after harvest.
- When transitioning an old hay field to organic vegetable production, reducing perennial weeds and the weed seed bank prior to planting the first vegetable crop is often less costly than expending labor to manage extreme weed pressure. When transitioning a field from conventional to organic field crops, perennial forage crops or soybeans are likely to have fewer weed problems and lower yield loss than corn.
- In warm, sunny climates, covering the soil with clear polyethylene tarps for several weeks can heat-kill most weed seeds in the top few inches of soil (solarization), thereby creating a weed-free seedbed for high value crops.
- Natural product herbicides like acetic acid and essential oils can effectively burn down weeds if used at high concentrations. They are, however, expensive and will damage crops, so their use is restricted to localized directed sprays or pre-emergence applications (e.g., for creating a stale seedbed).
- Livestock can provide substantial weed control. Goats and sheep can clear brushy pastures. Pigs can root out storage organs of difficult-to-control perennial weeds. Chickens can clean up weeds and weed seeds after harvest. Geese can remove grassy weeds from cotton, fruit and perennial vegetable crops. Maintenance of food safety requires thoughtful timing in the use of livestock for weed management.

- Strategies focused on preventing weed reproduction can help reduce weed problems and the expense of weed management in future years. Some approaches to prevention include hand rogueing, promptly cleaning fields after harvest, capturing or destroying weed seeds during combine harvesting, and maintaining clean field margins.

- Simple steps can block the arrival of new weed species onto the farm and thereby prevent future problems. These steps include careful inspection of forage, cover crop and grain seed, feed grain, and purchased hay and straw; knowledge of weed problems on farms supplying your forage, compost or manure; precautions to prevent arrival of weed species on or in the guts of newly purchased livestock; cleaning machinery that has been on other farms before using it in your fields; and filtration of surface irrigation water. Awareness of new weeds moving into your region and an understanding of weed problems on other farms in the neighborhood can provide early warning of problem weeds that may show up on your farm.

REFERENCES

Bergkvist, G., A. Adler, M. Hansson and M. Weih. 2010. Red fescue undersown in winter wheat suppresses *Elytrigia repens*. *Weed Research* 50: 447–455.

Blubaugh, C.K. and I. Kaplan. 2016. Invertebrate seed predators reduce weed emergence following seed rain. *Weed Science* 64: 80–86.

Borger, C.P.D., A. Hashem and S. Pathan. 2010. Manipulating crop row orientation to suppress weeds and increase crop yield. *Weed Science* 58: 174–178.

Brainard, D.C., W.S. Curran, R.R. Bellinder, M. Ngouajio, M.J. VanGessel, M.J. Haar, W.T. Lanini and J.B. Masiunas. 2013. Temperature and relative humidity affect weed response to vinegar and clove oil. *Weed Technology* 27: 156–164.

Brainard, D.C., R.R. Bellinder and V. Kumar. 2011. Grass–legume mixtures and soil fertility affect cover crop performance and weed seed production. *Weed Technology* 25: 473–479.

Brown, B. and E.R. Gallandt. 2018. A systems comparison of contrasting organic weed management strategies. *Weed Science* 66: 109–120.

Burrows, G.E. and R.J. Tyrl. 2006. Handbook of Toxic Plants of North America. Blackwell: Ames, IA.

Colquhoun, J.B., C.M. Konieczka and R.A. Rittmeyer. 2009. Ability of potato cultivars to tolerate and suppress weeds. *Weed Technology* 23: 287–291.

Comes, R.D., V.F. Bruns and A.D. Kelley. 1978. Longevity of certain weed and crop seeds in fresh water. Weed Science 26: 336–344.

Cousins, R. 1985. A simple model relating yield loss to weed density. *Annals of Applied Biology* 107: 239–252.

Davis, A.S., D. Daedlow, B.J. Schutte and P.R. Westerman. 2011. Temporal scaling of episodic point estimates of seed predation to long-term predation rates. *Methods in Ecology and Evolution* 2: 682–690.

Davis, L.W. 1993. Weed Seeds of the Great Plains, A Handbook for Identification. University Press of Kansas: Lawrence, KS.

Egley, G.H. 1983. Weed seed and seedling reduction by soil solarization with transparent polyethylene sheets. *Weed Science* 31: 404–409.

Geiger, G. and H. Biellier. 1993. Weeding with geese. University of Missouri Extension G8922.

Heijting, S., W. Van Der Were and M.J. Kropff. 2009. Seed dispersal by forage harvester and rigid-tine cultivator in maize. *Weed Research* 49: 153–163.

Hoitink, H.A.J., L.V. Madden and M.J. Boehm. 1996. Relationships among organic matter decomposition level, microbial species diversity, and soil borne disease severity. In *Principles and Practices of Managing Soilborne Plant Pathogens*, ed. R. Hall. pp. 237–249. APS Press: Saint Paul, MN.

Hutchinson, P.J.S., B.R. Beutler and J. Farr. 2011. Hairy nightshade (*Solanum sarrachoides*) competition with two potato varieties. *Weed Science* 59: 37–42.

Jackson, J.E. and J.W. Palmer. 1972. Interception of light by model hedgerow orchards in relation to latitude, time of year and hedgerow configuration and orientation. *Journal of Applied Ecology* 9: 341–357.

Jordan, J.L. 1979. The growth habit of Pennsylvania smartweed and velvetleaf. Proceedings of the North Central Weed Control Conference 34: 48.

Kelly, A.D. and V.F. Bruns. 1975. Dissemination of weed seeds by irrigation water. *Weed Science* 23: 486–493.

Liebhardt, W.C., R.W. Andrews, M.N. Culik, R.R. Harwood, R.R. Janke, J.K. Radke and S.L. Rigger-Schwartz. 1989. Crop production during conversion from conventional to low-input methods. *Agronomy Journal* 81: 150–159.

Liebman, M. and E.R. Gallandt. 1997. Many little hammers: ecological management of crop–weed interactions. In *Ecology in Agriculture*, ed. L. E. Jackson. pp 291–343. Academic: San Diego, CA.

Maher, R. and B. Caldwell. 2018. Take me out to a tarped field: learning a small-scale organic method to reduce tillage with less weeds. Cornell Small Farms Program.

Meiss, H., L. Le Lagadec, N. Munier-Jolain, R. Waldhardt and S. Petit. 2010. Weed seed predation increases with vegetation cover in perennial forage crops. *Agriculture, Ecosystems and Environment* 138: 10–16.

Mohler, C.L. and M. Liebman. 1987. Weed productivity in sole crops and intercrops of barley and field pea. *Journal of Applied Ecology* 24: 685–699.

Mohler, C.L. and J.R. Teasdale. 1993. Response of weed emergence to rate of *Vicia villosa* Roth and *Secale cereale* L. residue. *Weed Research* 33: 487–499.

Mohler, C.L. 2001. Enhancing the competitive ability of crops. In *Ecological Management of Agricultural Weeds*, ed. Liebman, M., C.L. Mohler and C.P. Staver. pp. 269–321. Cambridge University Press: NY.

Norsworthy, J.K., K.L. Smith, L.E. Steckel, and C.H. Koger. 2009. Weed seed contamination of cotton gin trash. *Weed Technology* 23: 574–580.

Olsen, J., L. Kristensen, J. Weiner and H.W. Griepentrog. 2005. Increased density and spatial uniformity increase weed suppression by spring wheat. *Weed Research* 45: 316–321.

Ominski, P.D., M.H. Entz and N. Kenkel. 1999. Weed suppression by *Medicago sativa* in subsequent cereal crops: a comparative survey. *Weed Science* 47: 282–290.

Place, G.T., S.C. Reberg-Horton, J.E. Dumphy and A.N. Smith. 2009. Seed rate effects on weed control and yield for organic soybean production. *Weed Technology* 23: 497–502.

Ryan, M.R., D.A. Mortensen, L. Bastiaans, J.R. Teasdale, S.B. Mirsky, W.S. Curran, R. Seidel, D.O. Wilson and P.R. Hepperly. 2010. Elucidating the apparent maize tolerance to weed competition in long-term organically managed systems. *Weed Research* 50: 25–36.

Rylander, H., A. Rangarajan, R. Maher, B. Caldwell, M.G. Hutton and N.W. Rowley. 2018. Reusable plastic tarps suppress weeds and make organic reduced tillage more viable. Abstract and Video recording. American Society for Horticultural Science 2018 Annual Conference. https://ashs.confex.com/ ashs/2018/meetingapp.cgi/Paper/28745.

SARE. 2007. Managing Cover Crops Profitably, 3rd Edition. Sustainable Agriculture Research and Education Program: College Park, MD.

Seed ID Workshop. Department of Horticulture and Crop Science, The Ohio State University.

Shirtliff, S.J. and M.H. Entz. 2005. Chaff collection reduces seed dispersal of wild oat (*Avena fatua*) by combine harvester. *Weed Science* 53: 465–470.

Smart, R. 1973. Sunlight interception by vineyards. *American Journal of Enology and Viticulture* 36: 230–239.

Standifer, L.C., P.W. Wilson and R. Porche-Sorret. 1984. Effects of solarization on soil weed seed populations. *Weed Science* 32: 569–573.

Stougaard, R. N. and Q. Xue. 2004. Spring wheat seed size and seeding rate effects on yield loss due to wild oat (*Avena fatua*) interference. *Weed Science* 52: 133–141.

Stougaard, R.N. and Q. Xue. 2005. Quality versus quantity: spring wheat seed size and seeding rate effects on *Avena fatua* interference, economic returns and economic thresholds. *Weed Research* 45: 351–360.

Teasdale, J.R. 2018. The use of rotations and cover crops to manage weeds. In *Integrated Weed Management for Sustainable Agriculture*, ed. Zimdahl, R.L. pp. 227–260. Burleigh Dodds Science Publishing: Cambridge, UK.

Teasdale, J.R. and C.L. Mohler. 2000. The quantitative relationship between weed emergence and the physical properties of mulches. *Weed Science* 48: 385–392.

Walsh, M., R.B. Harrington and S.B. Powles. 2012. Harrington Seed Destructor: a new nonchemical weed control tool for global grain crops. *Crop Science* 52: 342–347.

Weston, L.A., I.S. Alsadawi and S.R. Bearson. 2013. Sorghum allelopathy—from ecosystem to molecule. *Journal of Chemical Ecology* 39: 142–153.

CHAPTER 4

Mechanical and Other Physical Weed Management Methods

ESSENTIAL CONCEPTS OF MECHANICAL WEED MANAGEMENT

Soil disturbance is the most powerful lever a farmer has for manipulating the ecological processes of a field. Soil disturbance changes the physical environment of all species present, including soil microorganisms, insects, diseases and, most importantly for the purposes of this chapter, weeds and crops. Tillage and cultivation directly affect weeds by burying shoots, by cutting up plants and by uprooting plants so that they desiccate. As discussed in Chapter 2, soil disturbance stimulates the germination of many weed species (see "Seed Germination: Why Tillage Prompts Germination"). Finally, tillage changes the vertical distribution of weed seeds in the soil, and this affects their probability of survival, germination and emergence.

A central principle of this chapter is that *the effect of tillage or cultivation on a weed population depends on the interaction between the nature of the soil disturbance and the ecological characteristics of the weed.* The size, shape, position and physiology of shoots and underground organs, and particularly whether the species is an annual or a perennial, greatly affect whether a weed survives a particular tillage or cultivation practice. Similarly, the size, longevity and germination characteristics of seeds largely determine how they respond to a tillage or cultivation event. Thus, to effectively control weeds with tillage and cultivation, you need to think about the properties of the particular species and size of weeds in a field and choose tillage and cultivation practices that prey on the weaknesses of those weeds. Moreover, your best choices will vary depending not only on the weeds but

also on the crop stage, the weather and soil conditions. Understanding what you are specifically trying to do to the weeds, and balancing this with the machinery options available, is the essence of good mechanical weed management.

A second general principle of mechanical weed management is that *timing determines how successfully tillage and cultivation control weeds.* First, season, weather and soil conditions greatly affect how the operation will affect the soil and the weeds. Second, the effectiveness of tillage and cultivation methods depend on the size of the weed. For example, tine weeding only works on very small weeds; controlling very large weeds often requires heavy tillage implements. Thus, the choice of implement and how it is set up and used depends on the size of the weeds. Finally, the stage of crop development and soil conditions determine the types of cultivation that the crop can tolerate.

A third general principle is that *mechanical weed management is most effective when you use several operations in a planned sequence.* Each operation should target particular types and sizes of weeds in accord with the previous two principles. For example, corn production usually begins with tillage to bury early emerging annuals and disrupt the growth of perennials. After planting, a grower will usually cultivate two or three times with a rotary hoe or tine weeder to remove weeds in the white thread and early cotyledon stage from the crop rows. Typically, these operations will be followed by inter-row cultivation with a row-crop cultivator. During inter-row cultivation soil is usually thrown into the row to bury weed seedlings. Burial only works, however, if the weeds are small. Hence, the success of that operation depends on removal of the first flushes of seedlings during the earlier tine weedings or rotary hoeings. The sequence and types of practices you use will depend on the crop, but in general, your success at eliminating weeds will depend on a well-planned program of multiple operations with a variety of tools.

TYPES OF TILLAGE AND THEIR EFFECTS ON WEEDS

Different tillage implements move soil in different ways, and consequently they differ in their effects on weeds (Table 4.1). Moldboard plows invert the soil and are thus very effective at burying weeds. Since large blocks of soil within the furrow remain relatively undisturbed, a moldboard plow is relatively poor at fragmenting weeds or dislodging soil from weed roots. Chisel plows crack the soil laterally and send a wake of soil rolling away from both sides of the shank. A chisel plow uproots weeds that are directly in the track of a shank, but these are a small proportion of the total. Many weeds are buried, however, by the soil thrown from the shank. Field cultivator blades cut deeper roots of weeds, and their relatively steep pitch churns the soil, thus burying small to moderate sized weeds. In contrast, the lower pitch of a sweep plow blade primarily lifts and loosens the soil. The roots of large weeds are sliced off the shoots by a blade plow, and seedling weeds may subsequently dry out and die due to loss of soil-root contact. Disks and rotary tillers are very effective at chopping weeds into fragments. They also separate weed roots from soil by fragmenting soil masses, and the rotary action tends to bury the fragments. Plowing with either a moldboard or chisel plow is usually followed by disking or harrowing to create a level seedbed of uniform consistency. In the process, weeds are further fragmented, and roots are exposed to additional drying.

Annual weeds that germinated earlier in the same season are usually completely killed by any kind of tillage. Annuals that germinated the previous season are well established by the time of tillage and may respond to tillage more like perennials.

The growth habit of perennial weeds indicates to what extent they are susceptible to uprooting, breakage or burial (Table 4.2). The effect of tillage on a creeping perennial species depends on the depth of the storage organs. Species like field bindweed and common milkweed, in which most of the storage organs are below the normal depth of tillage, are not susceptible to uprooting. Moreover, if the shoot is buried, it will usually quickly resprout from dormant buds on the vertical rhizome just below the soil surface. Thus, the damage these species sustain from tillage comes primarily from severing the shoot from the deep roots or rhizomes. Several cycles of tillage and resprouting are usually required to deplete deep storage organs (see "Exhaust Perennial Storage Organs").

Creeping perennials with shallow storage organs (for example quackgrass or perennial sowthistle) are easier to exhaust because the storage roots or rhizomes can be broken into small pieces, each of which has less stored energy (see "Vegetative Propagation of Perennial Weeds"). Moreover, creeping perennials that have storage roots or rhizomes in the plow layer are more easily controlled by tillage because they are susceptible to uprooting and desiccation (see "Dry Out Perennial Storage Organs"), and may be susceptible to physical removal from the field.

A few perennials spread by rhizomes but overwinter by means of a bulb or tuber (for example, yellow nutsedge, purple nutsedge or wild garlic). Usually, most of the tubers are in the plow layer, though a few may form deeper in the soil. Tillage usually has little effect on dormant tubers. It may redistribute them vertically, but energy stored in the tuber is generally

Table 4.1. Effectiveness of Tillage Implements for Uprooting, Breaking and Burying Weeds (Reproduced from Mohler 2001)

Implement	Uprooting	Breakage	Burial
Moldboard plow	Good	Poor	Good
Chisel plow	Moderate	Poor	Moderate
Field cultivator	Moderate to good	Moderate	Moderate
Sweep plow	Poor	Moderate	Poor
Disks	Moderate	Good	Moderate
Rotary tiller	Moderate	Good	Moderate

Table 4.2. Susceptibility of Perennial Weeds with Different Growth Habits to Uprooting, Breakage and Burial by Tillage Implements

Growth Form	Uproot	Sever Shoot From Root	Fragment Storage Organ	Burial	Examples
Creeping perennials					
Storage organs are below tillage depth	Very low	Moderate	Very low	Moderate	Field bindweed, common milkweed
Storage organs are in the plow layer	Moderate	Moderate	Moderate, propagates	Moderate	Quackgrass, perennial sowthistle
Stationary perennials					
With a taproot	Low	Low	Moderate[1]	High	Dandelion, curly dock
With fibrous roots	Moderate	High	High	High	Broadleaf plantain

[1]Fragmentation can potentially propagate taprooted perennials, but this rarely poses a problem in practice.

enough to allow sprouts to emerge from anywhere in the tilled layer. Once the tubers sprout, however, tillage that breaks or cuts the shoots from the tuber will force resprouting and weaken the plant. Both dormant and sprouting tubers can also be killed by drying. See individual species accounts for details on using these management strategies against particular weeds.

Fixed perennials, including biennials, have to reproduce by seeds, and when they are young, they are susceptible to the same methods that control annual weeds. The response to tillage of fixed perennials that have developed substantial taproots is similar to that of creeping perennials with storage organs in the plow layer. Usually, species with taproots cannot be killed by uprooting unless the taproot is brought to the soil surface and thoroughly dried. Just cutting the shoot off with shallow tillage does not usually control a taprooted weed because it quickly grows new leaves using the large reserves of energy in the root. Using tillage to chop the roots into small pieces and burying them deeply is a good first step toward exhausting the plants, but usually will not be sufficient to control the weed without additional measures. Biennials like bull thistle and common burdock are susceptible to tillage, however, after they have bolted and transferred energy to the shoot.

Tillage easily kills most fixed perennials without taproots (for example, broadleaf plantain). These species cannot easily grow back to the surface after burial, and they do not tolerate getting chopped up. Consequently, these species are rarely a problem in tilled fields but are instead primarily weeds of no-till fields and perennial crops, including fruit, mowed forage crops and pastures. Overwintering populations of annual bluegrass are an exception that proves the rule. They often do survive tillage, but not because they tolerate burial or dismemberment. Rather, the tight little clumps of roots embedded with soil resist breakage and tend to float to the surface if harrowing is used to create the seedbed.

USING TILLAGE AGAINST PERENNIAL WEEDS

Exhaust Perennial Storage Organs

Many perennial weeds have deep storage roots or rhizomes that resprout after the tops are cut (for example field bindweed or common milkweed). Since the storage roots or rhizomes are too deep in the soil to damage with normal tillage, your best hope for controlling these weeds without chemicals is to exhaust the storage organs by repeatedly removing the top growth. Generally, the net flow of carbohydrates is from the storage organ until at least three or four leaves have formed (see "Vegetative Propagation of Perennial Weeds"), but it may be later in some species. Timing your operations to correspond to the target species' minimum food reserves will reduce the population with the least amount of labor. Typically, eradication of deep-rooted perennials requires about six to eight well-timed tillage events the first year followed by three to five the second year. A field cultivator, spring tooth harrow or sweep plow run shallowly are good implements for this job. However, many repeated diskings can cause soil compaction. If previous activities have removed most storage organs from the plow layer, deep cutting at 16–18 inches can greatly slow re-emergence and reduce the number of operations needed to control the weed.

Few growers can afford to remove land from production long enough to completely eradicate perennials by exhausting the storage organs. If the population is light to moderate, one to three cycles of sprouting followed by tillage may, however, be sufficient to keep the weed at tolerable levels.

Species that have their storage organs primarily in the tilled layer of the soil (quackgrass, perennial sowthistle, dandelion) are easier to exhaust than deep rooted perennials. Use a tillage method that breaks up the storage organs into small pieces, but if the problem is localized, avoid dragging the pieces all over the field. Breaking up the storage organs will increase the number of plants but will make each one substantially weaker (see "Vegetative Propagation of Perennial Weeds"). The little sprouts can then be killed by shallow tillage or, in some cases, suppressed by a dense, competitive cover crop. Since the point is to induce the fragments to sprout and then be killed, tilling at a time of year when the weed is actively sprouting is usually best, although breaking up dormant storage organs may be useful as well (Anbari et al. 2011).

If the perennial weed is posing severe problems, try to bury the storage organs as deeply as possible. This will force the weed to put more energy into reaching the soil surface and thereby weaken the plant. For example, a tillage sequence for weakening a perennial with storage organs in the tilled layer might begin with two passes with a disk or one slow pass with a rotary tiller. Note that the tines of a rotary tiller cut off pieces of the well anchored root or rhizome on the first pass but may just throw the pieces around on the second. Thus, the slower the first pass, the smaller the pieces will be. Follow the initial tillage with deep moldboard plowing to bury the fragments and finish with shallow harrowing to create a seedbed. Follow the tillage treatment either with extraordinarily intense competition, as from a densely planted cover crop, or with regular shallow tillage before the plants begin to restore root or rhizome reserves (see species chapters). Either approach should reduce the population to manageable levels. Clearly, such an intensive tillage regimen, and particularly the deep moldboard plowing, will be hard on your soil and should be used only when less stringent measures fail. Use cover crops or heavy applications of high organic matter compost to restore the soil structure and biological activity after this technique.

Remove Perennial Storage Organs

Perennial weeds depend on resources stored in roots,

rhizomes and tubers to establish new shoots the following year. You can reduce the population of creeping perennials with tough rhizomes like quackgrass and johnsongrass by physically removing the rhizomes. Moldboard or chisel plow the soil, and then work the rhizomes to the soil surface with a spring tooth harrow. Rake them off the field or into piles with a bent tooth tine weeder. Piles can then be forked onto a wagon. You can feed them in moderation to livestock or chickens, or dump them in woods or a hedgerow where they cannot grow due to shade. The practicality of this approach depends on the size and configuration of the area to be cleaned. Removing rhizomes from a large field may not be practical, but often only restricted areas of a large field will be infested. If a mowed field margin or grass waterway is nearby, the rhizomes can be raked onto that. The thickened storage roots of most perennial broadleaf weeds are too fragile to allow this technique to work.

Dry Out Perennial Storage Organs

You can also control creeping perennials that have most of their rhizomes or tubers in the plow layer by drying. Start by watching the long-term weather forecast. When the weather will be dry for at least a week, work the perennial storage organs to the surface with a spring tooth harrow as you would to remove them (see "Remove Perennial Storage Organs"). Stir the soil with a spring tooth harrow every couple of days to ensure that the surface soil dries completely, and that all storage organs are fully dried. A peanut digger has been used to bring nutsedge tubers to the soil surface for drying. Drying storage organs is much quicker than exhausting the storage organs, but unless the dry weather period is long and you till deeply, some fragments will likely escape. Nevertheless, the treatment can reduce a severe infestation to a minor one.

As with all procedures for controlling perennials, its long-term success will depend on follow-up measures. If you keep pressure on the population with additional fallow cultivation or strong competition with a vigorous cover crop, the population will continue to decline rather than begin to recover.

European farmers in the early 20th century often had severe infestations of perennial grasses due to the high frequency of cereal grains in their crop rotation. One way in which they dealt with the problem was to plow the soil when wet to create clods. During dry weather, they then stirred the clods with a field cultivator to thoroughly dry them out and kill the grass rhizomes. This is obviously an extreme measure that should only be used in desperate situations. It might be preferable, however, to frequent, long periods of tilled fallow.

In general, all tillage measures to control perennial weeds tend to have a negative effect on soil tilth, and some also expose the soil to erosion. Consequently, when you use these methods you need to compensate for them with actions that build soil quality, both to maintain good crop yields and to cultivate effectively (see "Soil Tilth and Cultivation").

TILLAGE EFFECTS ON WEED SEEDLING DENSITY

Timing of Tillage Affects Weed Density

Tillage has profound effects on weed seeds and seedlings as well as on perennial weeds. Since most agricultural weed species have particular times of the year in which they germinate most profusely, the timing of weed emergence relative to the timing of tillage and planting can either help control a weed species or ensure that it becomes a problem. You can reduce weed problems by timing tillage relative to weed germination in two ways, either 1) by destroying a large proportion of the weeds that will emerge that year with your seedbed preparation, or 2) by planting well before the weeds emerge to increase crop competition and the intensity of cultivation. Which option you use will depend on the crop.

Many spring germinating weeds reach peak emergence in mid-spring, with emergence decreasing later. If you wait until many seeds have already produced seedlings before tilling, then the tillage operation will eliminate a large fraction of the weeds that will appear that year. For example, a study in Wisconsin showed that delaying corn planting from April 25 to May 15 reduced in-row weed density after rotary hoeing by 55%. In organic systems, delayed corn planting has other advantages as well (see "Planting Date" in the "Crop Competitiveness" section).

Since many of the weeds that infest winter grain crops germinate best in late summer or early fall, delaying seedbed preparation and planting until mid to late fall also potentially reduces weed density. Delayed planting can, however, reduce the crop's competitiveness in the spring and may decrease yield.

Delaying tillage and planting for summer planted vegetable crops like tomatoes and peppers tends to be less effective than for spring or fall planted crops. The weeds that dominate following tillage in hot weather tend to be species like hairy galinsoga and common purslane. Seeds of such species do not go dormant in hot weather. Consequently, you cannot out-wait them.

Note that to use tillage and planting date to eliminate early emerging weeds, you need to manage your tillage in a particular way. Specifically, primary tillage must occur long in advance of final seedbed preparation and planting. The reason is that if you till deeply just before planting, you will eliminate the current weed seedlings, but you will also bring many new seeds to the surface. Many of these will be released from dormancy by the usual cues associated with tillage and near-surface conditions (see "Seed Germination: Why Tillage Prompts Germination"). Your crop may then end up as weedy as it would have been if you had not delayed planting. Consequently, a shallow final seedbed preparation will result in less weed emergence than deep preparation.

Spring grains germinate and establish well at low temperatures. If planted in early spring they will emerge before most weeds. In particular, most annual grass weeds do not emerge in large numbers until mid or late spring. Moreover, after the soil settles due to rain and the crop begins to cast shade, the intensity of tillage-related cues that prompt germination will be reduced and consequently, so will weed emergence. The head start the grain gets from early planting also makes the crop more competitive against weeds. The larger difference in size between crop and weeds allows you to tine weed more aggressively to remove weeds that do emerge. In principle, early spring planting of vegetable crops that thrive in cool weather should also reduce weed density and favor the crop relative to the weed. The effect of tillage and planting date on weed-crop interactions in vegetables has not, however, been well studied.

Type of Tillage and Weed Seedling Density

Tillage redistributes weed seeds vertically in the soil column. This has three main effects. First, it changes the proportion of seeds that germinate. Seeds that move downward toward constant soil temperatures and away from light and gas exchange tend to become dormant, whereas those that move up into the light, warmth, fluctuating temperatures and rapid gas exchange of the surface soil tend to lose dormancy and germinate (see "Seed Germination: Why Tillage Prompts Germination"). Second, vertically redistributing the seed bank changes the fraction of germinating seeds that successfully grow to the surface and emerge. Finally, vertically repositioning seeds in the soil will change the proportion of weed seeds that are killed by the several factors that influence seed mortality.

Various tillage implements redistribute seeds differently. A moldboard plow distributes surface sown seeds in a bell-shaped curve, with the peak density usually 5–7 inches deep. The location of the peak depends on how deeply you plow and how well the soil is inverted (Figure 4.1). Regardless, a moldboard plow usually places seeds that were formerly on the surface too deep for emergence. Other implements keep most seeds relatively close to the soil surface, with ever fewer seeds present as depth increases. However, when you use any implement a second time, particularly a moldboard plow, the seed bank becomes more vertically even (Figure 4.1).

With time, seeds become mixed throughout the soil profile. Their exact distribution depends on how fast they germinate or die at various depths and how the tillage implement redistributes the seeds that are already buried. We studied the redistribution of buried seeds by placing millions of colored seed-like beads at various depths in the soil and then tilling with various implements (Mohler et al. 2006). By analyzing the location of the beads after tillage we developed tables showing the probability of beads (or seeds) moving from one position in the soil to another.

As expected, moldboard plowing followed by disking and shallow harrowing tended to move shallow beads downward and deep beads upward. Chisel plowing followed by disking and shallow harrowing tended to move seeds in the top 4 inches downward, but mostly left deeper beads in place. Remarkably, a single pass with a rotary tiller or disk moved beads in a qualitatively similar way to the chisel plow followed by disks.

The utility of this information for weed management comes through understanding the relation of the vertical position of seeds in the soil to survival and emergence (Mohler 1993). As discussed in Chapter 2, weed seedlings emerge best from near the soil surface and germination deep in the soil will produce seedlings that fail to emerge. Consequently, seeds of most weed species have been selected to remain dormant when deep in the soil. Some fatal germination occurs below the surface zone of optimal emergence, particularly if the seeds have been exposed to favorable conditions for germination before they are buried by tillage. Nevertheless, deep burial usually reduces mortality of dormant seeds whereas seeds near the soil surface age rapidly if they do not germinate. Consequently, burying weed seeds with tillage tends to preserve them, both by reducing germination and by slowing the aging process. However, species differ greatly in how fast they die off in the soil.

If effective herbicide regimens are available in a conventional agricultural system, then a strict no-till program provides an opportunity to quickly deplete the near-surface seed bank and may result in fields with very low weed pressure within a few years. In addition to maximizing seed mortality, a no-till system provides good conditions for germination of seeds that fell the previous year, thus making them susceptible to both pre-emergence and post-emergence herbicides. A few intensive organic vegetable growers have produced a similar effect by combining no-till with mulch and intensive hand weeding (see "Continuous No-Till Vegetable Production Using Mulch").

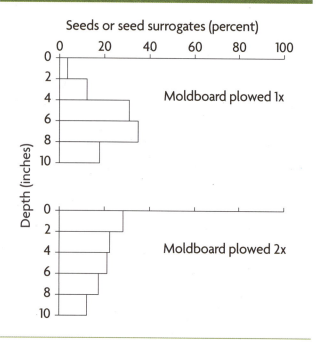

Figure 4.1. Vertical distribution of surface sown weed seeds or soil markers after one (1x) or two (2x) passes with a moldboard plow.

Information on seed movement, seed survival and emergence depth can also be combined to predict the optimal type of tillage, if tillage is a regular practice. First, if the seeds of a species are short lived and the seed must be near the soil surface to emerge successfully, then inversion tillage will help control the weed. Consider hairy galinsoga, whose seeds rarely live more than a year or two and which must be within 0.25 inch of the soil surface to successfully emerge. If the seeds are plowed under, the probability of a seed returning to the top 0.25 inch of soil in any future year is very low. Since seed survival is low, even when well buried, the chances are good that any given seed will die before it is plowed back up again.

Generally, for spring and summer germinating species, plowing is best left until the spring to allow seeds to die off on the soil surface through the winter. Some tillage may be necessary to establish a winter grain, cover crop or late vegetable crop, but if seeds were shed during the summer, then minimizing tillage in the fall will increase winter mortality. If the field must be plowed in the fall, then till shallowly in the spring to avoid plowing up seeds before they have had time to die.

In contrast with shallow emerging species with short lived seeds, a species like ivyleaf morningglory that survives very well when deeply buried and emerges from a depth of 2 inches or more has a high probability of eventually returning alive to its emergence range if you regularly moldboard plow. If the seeds are left on the soil surface over the winter, however, many will die. If you follow this treatment with shallow tillage that prompts germination, you can then kill off many of the remaining individuals with a shallow tillage pass. In neither the galinsoga case nor the ivyleaf mornigglory case will these methods completely control your weed populations. Rather, they provide one component of an overall ecological weed management program.

When is deeply incorporating weed seeds a preferable strategy to keeping them on the soil surface? Although we are working on that question, we do not yet know the answer. However, evidence indicates that cumulative velvetleaf density over several years is lower when seeds are plowed under relative to shallow tillage or no tillage. This occurs despite velvetleaf's good persistence when deeply buried (Figure 2.7) and its high rate of emergence from anywhere in the top

2 inches of soil. This indicates that most species that have regular seed input to the soil probably have lower average density with annual moldboard plowing than with any reduced tillage regimen or rotary tillage, which produces vertical seed distributions similar to typical reduced tillage practices (Mohler et al. 2006). The lower the seed survival rate and the shallower the depth from which seedlings can emerge, the greater will be the benefit of moldboard plowing. Reduced tillage has many benefits for soil health, and we do not advocate continuous moldboard plowing. However, understanding the effects of different types of tillage on weed populations will help in making informed choices about how you manage your cropping systems.

Most fields have a diversity of weed species but, generally, only one or two cause most of the problems. Choosing a tillage strategy that reduces the population of these problem species will allow other management tactics to act more effectively (see "Weed Density Affects Death and Reproduction" in Chapter 2). A shift in tillage regimen may favor other species, even as it is helping reduce the problem species. Our experiments (Mohler unpublished) indicate that sometimes a rotation of tillage regimens is more effective for weed management than following a single practice consistently. For example, the many seeds produced during a weedy year can be plowed under, and then shallow tillage can be used for a few years while the deeply buried seeds die off. (see the Nordell case study in Chapter 5).

RIDGE TILLAGE

Ridge tillage is a highly reduced tillage system in which tillage is done with just the planter and a row-crop cultivator. An attachment on the planter scrapes the top off the ridge, pushing the top few inches of soil, crop residue and weed seeds produced the previous year into the inter-row area. Meanwhile, the planter units place crop seeds on the scraped down ridges. The cleaned area on the ridge can be cultivated with a rotary hoe if needed, and weeds emerging in the inter-row area can be eliminated with a row crop cultivator. At the final row-crop cultivation, soil is piled up around the base of the crop, which buries small weeds and recreates the ridge for next year (Figure 4.2).

Studies have shown that the density of weed seeds near the row immediately after planting is only about one third as high in a ridge till system as with conventional tillage, whereas seed density in the inter-row

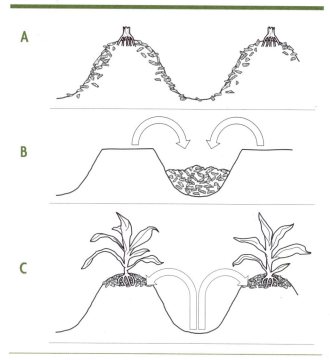

Figure 4.2. Annual cycle of soil movement in a ridge tillage system. (A) Ridges before planting. (B) Crop residue, weeds and soil on the top of the ridge are scraped into the inter-row furrow by an attachment on the planter that runs ahead of the seed openers. (C) After the crop is well established the ridges are rebuilt with a cultivator that throws soil into the row around the base of the crop plants. Illustration by Vic Kulihin.

area is about two times higher. Thus, most of the seeds get moved to the inter-row area where they are easier to control with cultivation after they germinate. Moreover, the soil remaining in the ridge is only slightly disturbed, so fewer seeds there are prompted to germinate than with conventional tillage. Since the soil in and below the base of the ridge is not disturbed by implements or packed by wheel traffic, it tends to develop good crumb structure that promotes healthy plant growth. The permanent, firm ridge bases also allow use of guide cones on the cultivator to ensure accurate and relatively stress-free cultivation.

A comparison of ridge tillage plus rotary hoeing versus conventional tillage plus herbicide in 51 on-farm trials in Iowa showed no difference in corn yield between the two systems and only slightly higher weed densities in the ridge till system. Extensive farmer experience with the system has led to several rules of thumb for ridge tillage:

- The ridge should be wide and flat topped, about 12 inches wide for crop rows spaced 30 inches apart. The wide ridge allows the planter ridge cleaning attachment to remove last year's crop of weed seeds from a strip several inches wide on both sides of the row. It also allows a wide strip of residue-free soil to speed soil warming and thereby encourage rapid crop emergence.
- Usually, you want to remove about 2 inches of soil from the top of the ridge at planting. This is enough to move most of last year's weed seeds out of the row and clear off residue from the previous crop, but to still leave an elevated ridge base for rapid soil warming and cultivator guidance.
- Ridge till planting should occur early enough in the spring that weeds have not yet grown too large to eliminate with the planter attachment. A winter cover crop can help slow spring weed growth so that a good weed kill on the ridges is more likely.
- Ridge building should occur early enough in the season that you can move a lot of soil from the inter-row to the row without pruning crop roots. You probably want to minimize root pruning when cultivating anyway, but root pruning when ridging is especially likely since the cultivating tools dig deeper relative to the base of the crop plant than when cultivating a flat tilled field. We find that a sweep with hilling wings builds a more uniform ridge than disk hillers, and since it operates further from the row, it is less likely to cause root pruning. We have also built excellent, wide, flat topped ridges with potato hillers.

As with other reduced tillage methods, perennial weeds can increase in ridge tilled fields. Certainly, you will not want to use ridge tillage on a field unless it is nearly free from perennial weeds at the outset.

Compost or manure applied prior to planting will become concentrated in the inter-row areas after scraping and planting, as will legume cover crop residue. Consequently, nutrient availability during early crop growth can be an issue, and banding starter fertilizer next to the row may be useful for some crops.

Ridge tillage also poses some problems for crop rotation. With crops like cabbages, beets and lettuce, the ridges have to be built in a separate operation after harvest since ridging at inter-row cultivation will bury the crop. For small grains, some grain drills will not plant well on both the ridges and valleys, and the grain in the valleys may grow poorly in any case. For hay crops, the field has to be tilled flat before planting since ridges will interfere with hay harvest. This limits the soil structure benefits of the undisturbed ridge bases.

Most growers who avoid using herbicides rotate ridge tillage with other tillage systems (see the profile of Paul Mugge). For example, Dick and Sharon Thompson in Iowa used a five-year rotation of hay–corn–soybeans–corn–oats underseeded with hay. The field is moldboard plowed after the hay, ridges are created by cultivation in the first corn crop, the soybean and second corn crops are ridge-till planted, and the field is disked flat for planting the oats and hay. A study of this system showed an increase in the weed seed bank, mostly waterhemp and foxtail, following the oats and hay due to seed production during the oat/hay establishment year (Buhler et al. 2001). Waterhemp declined steadily during the years with corn and soybeans. Foxtail dropped greatly following moldboard plowing and the first corn crop and it was then held low through the following soybean and corn crops. This demonstrates the utility of ridge tillage for weed management in a tillage rotation/crop rotation.

TILLED FALLOW

A tilled fallow is a period during which no crop is planted and the soil is regularly tilled to eliminate weeds. The bare soil and regular stirring of the soil provide cues that prompt seed germination (see "Seed Germination: Why Tillage Prompts Germination") and the young weeds are then destroyed. If many weeds went to seed the previous year, avoid deep tillage that will place many seeds deep in the soil where they will remain dormant and have a higher rate of survival. In addition, if you clean the surface soil with a shallow tilled fallow, avoid using deep tillage when you prepare the seedbed for the next cash crop, because this will bring up more seeds. Tilled fallows are also an effective way to decrease perennial weeds (see "Exhaust Perennial Storage Organs").

A period of tilled fallow prior to planting is sometimes referred to as a "stale seedbed," though properly, this term only applies if the weeds are killed without disturbing the soil surface, for example with a propane flame or an herbicide (see the next section, "Stale Seedbed"). Other opportunities for tilled fallows exist between harvest of early crops (for example spinach or lettuce) and fall crops (for example broccoli), or after the harvest of cereal grains. The optimal time to use a tilled fallow for weed management depends on the seasonality of the weeds that are causing your most severe problems.

How often should you weed during a tilled fallow? The answer depends on the weed species you are trying to control. Most annuals require at least five weeks to set seeds, though common purslane in warm weather can set seed in four weeks. Certainly, you should till before any weeds go to seed! Usually, you will want to till sooner, while the weeds are small and easy to kill. Also, if you are targeting perennials, you will want to kill them before root reserves are replenished (see "Exhaust Perennial Storage Organs"). Once every three weeks is often the right interval.

When cultivating in a crop, part of the objective is to create a loose layer of surface soil (dust mulch) in which weed seeds germinate poorly. In a tilled fallow, however, you want the seeds to germinate so you can kill them, and for this you need good soil-seed contact. Thus, either use disks, which leave a firm seedbed, or firm the seedbed with a roller or cultipacker. Rolling the seedbed will usually be more effective, since it firms the surface layer where the seeds you are trying to flush out are most likely to germinate (see "Seed Germination: Why Tillage Prompts Germination"). Rolling is also easier on soil tilth since the compaction will be localized to the surface inch or two. Waiting a day or two between tillage and firming the soil is often beneficial to allow uprooted weeds a chance to die. Otherwise, you may be firming the soil back around the roots and encouraging re-growth. As with all tillage and cultivation, tillage for a fallow is most effective when the surface soil is drying and the weather is warm and sunny. However, rainfall or irrigation may be needed to bring on the next flush of weed seedling emergence.

Tilled fallows tend to harm soil structure. If your intent is to clean out seeds from the surface soil prior to planting a crop, then only shallow tillage is necessary to kill the weeds, and this reduces the problem since less of the soil column is disturbed and lighter tractors can be used to reduce wheel compaction. Nevertheless, soil-building practices discussed in the section "Soil Tilth and Cultivation" should be used to counteract the damaging effects of tilled fallows. A modification of the tilled fallow is to grow short season cover crops between tillage operations. The cover crops protect the soil from sun, wind and rain, and the incorporation of the cover crops feeds the organisms that build soil structure. With the cover crops slowing weed growth rates due to competition, fewer tillage operations may be needed to prevent weeds from producing seeds. Parallel experiments in Pennsylvania and Maine showed that a sequence of yellow mustard, buckwheat and winter canola with tillage before each cover crop was as effective as a season long bare fallow with four tillage operations (Mirsky et al. 2010). Both treatments nearly eliminated the seed bank of foxtail and reduced common lambsquarters and velvetleaf seeds by over 80%. Systems with less soil disturbance were not as effective, though a snap bean cash crop cultivated between the rows followed by a rye/hairy vetch cover crop also substantially depleted the seed banks of these long season summer annuals.

Another strategy for preserving soil health during a tilled fallow is to chop up the cover crop and then incorporate it only shallowly. The soil then acts like a sponge to absorb rainfall with minimal runoff and erosion, while nitrate released from the decaying cover crop can stimulate emergence of some weed species.

STALE SEEDBED

The stale seedbed technique is a special variation on the tilled fallow. For a stale seedbed, you till the soil and prepare a firmed seedbed. Then after 2–3 weeks, when the weeds have sprouted, you kill them without soil disturbance. In organic systems this can be done with a flame weeder or natural product herbicide like acetic acid. You then immediately plant into the stale seedbed with as little soil disturbance as possible. Since you have depleted the near-surface seed bank and have not provided the disturbance related cues that would prompt another flush of germination (see "Seed Germination: Why Tillage Prompts Germination"), relatively few new seedlings will emerge. The procedure is even more effective if the flame weeding or herbicide treatment is repeated again just before crop emergence.

The stale seedbed technique is commonly confused with tilled fallow. If even a minimal amount of secondary tillage is used prior to planting, for example, by taking a light harrow or basket weeder over the field, then the surface soil is disturbed, not "stale," and additional weed seeds will be prompted to germinate (see "Tilled Fallow").

The stale seedbed technique is an exacting and potentially expensive procedure. Thus, it is usually reserved for high value but hard to weed or slow growing crops like parsnips, carrots or onions. One way to reduce costs for less valuable crops is to stale seedbed only a strip centered on where you intend to plant the row, and then cultivate inter-row areas using shields to keep soil (and therefore weed seeds) off of the cleaned area. If you plan to attempt this, we suggest you set up your cultivator, planter and flame weeder or band sprayer with a good guidance system (see "Cultivator Guidance Systems").

PRINCIPLES OF MECHANICAL WEEDING

Cultivators act by cutting, uprooting or burying weeds with tools that disturb the soil. In addition to cultivators, other types of weeders can damage weed tissues with heat, cold or electricity. The most common of these are the various types of flame weeders that disrupt weed tissues with a propane flame. Mowers are also used to control weeds between rows or beds, especially when sod strips are sown between beds. Weed pullers can pull tall weeds out of crop rows. All of these implements may be classified according to where they work relative to the crop row. Inter-row cultivators remove weeds between crop rows. They are what most growers think of when the word "cultivator" is used. In contrast, in-row cultivators and weeders specifically attack weeds within the crop row. Near-row cultivators and weeders may or may not kill weeds between rows, but they can harm weeds closer to the crop row than most inter-row cultivators. Finally, some machines act similarly on both the in-row and inter-row areas, and we will refer to these as full-field cultivators. These are often used for blind cultivation: that is, cultivation before crop emergence. The most difficult weeds to remove with cultivators are the ones that establish in the crop row, and much of our discussion will focus on implements that control these weeds.

The use of cultivators and other mechanical weeders is guided by several simple principles.

Principle 1: *Cultivators other than full-field implements should work the same number of rows as the planter, or a simple fraction of this number (for example, one half).*

If the planter and cultivator have incompatible sizes, inexact spacing between rows in adjacent planter passes may place cultivator tools too close to some rows and too far from others. The result will be damage to the crop where the rows are too close together and poor control of weeds in the inter-row where the rows are too far apart. Even if the planter passes are only off by 2–3 inches, root pruning and suboptimal weed control will be likely if you use, say, a 4-row cultivator with a 6-row planter or vice versa.

Principle 2: *The action of the cultivator must be appropriate for the growth stages of the weeds and the crop.*

The timing and number of cultivations depends on how fast the weeds are growing and the size range over which the weeds are susceptible to the implement. Based on many years of farming experience, Jim Bender suggests that staggering planting times helps improve the timeliness of cultivation when only short time periods are available between rainfall events.

The degree to which precise timing is critical depends on how close to the crop row the implement operates. Tine weeders, rotary hoes and in-row weeding tools will damage the crop if they dig deeply. Consequently, these machines have to catch the weeds after they have germinated but before they become well rooted. Delay of even a single day may allow many weeds to escape once they begin to appear above ground. Implements that work close to, but not in the row, like spyders and basket weeders, have a larger window of opportunity for killing weeds but still require attention to timing. In contrast, timing is less critical for inter-row cultivation of many crops. For example, experiments in New York showed that timing of cultivation with an S-tine cultivator with goosefoot shovels had little effect on either the dry weight of between-row weeds or on corn yield (Mt. Pleasant and Burt 1994). Of course, if an inter-row cultivator is used to bury weeds in the row, then timing becomes critical.

Principle 3: *Control of weeds in the crop will be more effective if you create and maintain a size difference between the crop and the weeds.*

The most sophisticated mechanical weed management programs begin with a stale seedbed or blind cultivation to ensure that the crop emerges before the weeds. Many vegetable crops can be transplanted to give the crop a head start over the weeds. Regardless of how the initial size difference is created, full-field, in-row and near-row cultivation can then increase in depth and degree of soil movement as the crop grows larger. For many row crops (for example, corn, sorghum, soybeans or potatoes), once the crop is well established, you can move large amounts of soil around the base of the shoots to cover small weeds. Because

most agricultural weeds have a high relative growth rate, however, burying weeds in the row will be most effective if you have already killed the first flushes of weeds that emerge after crop planting. Ultimately, the greater the size differential established, the greater the competitive advantage the crop will have over the weeds.

Principle 4: *The effectiveness of cultivation declines when weed density becomes very high.*

Over a wide range of weed densities, cultivation will kill the same proportion of weeds. However, when weeds become very dense, cultivation can become ineffective for several reasons. First, some weeds in the crop row will escape even a well-timed and well managed cultivation program. If the density is high, enough weeds may escape to cause substantial yield loss. In contrast, if the density before cultivation is low, the few weeds that survive will cause little harm to a competitive crop species and can be cost-effectively hoed or hand-weeded out of a high value crop. Second, soil clings better to a mat of roots from a dense stand of weeds than it will to the roots of an individual weed. Consequently, more weeds will reroot when weeds are dense. Finally, some shallow-working implements will not penetrate well if the soil is tightly bound by roots and the soil surface is lubricated by plants smashed by the cultivation tool. For these reasons, if the field is badly infested with perennial weeds or the weed seed bank is unusually high, consider reducing weed pressure by using a tilled fallow period before planting. Also consider rotating to a vigorous crop that tolerates early and frequently repeated cultivation to keep emerged weeds from accumulating to a high density. Or, rotate to a crop that grows during a different season than the problem weed species.

Principle 5: *Planting density should be increased when using implements that randomly kill a percentage of the crop.*

Rotary hoes, tine weeders and some in-row tools often reduce crop density by several percent. This stand loss associated with tools that attack weeds in the crop row usually takes the form of random missing individuals rather than as blighted row sections. Consequently, you may find planting at a rate near the upper limit of the acceptable range improves yield and competitive pressure on surviving weeds. Good understanding of how your machinery interacts with your particular soils and crops is necessary to balance the risks of yield loss from stand reduction with the yield loss from weed competition. Attempting to preserve every crop plant will usually result in an ineffective in-row weeding operation. As a rule of thumb, we aim for a stand loss of around 2% when using full field and in-row tools, but the optimal balance between tolerable stand loss and the intensity of weeding depends on the density of the weeds.

Principle 6: *Good soil drainage and careful timing relative to changing weather and soil conditions can improve the effectiveness of cultivation.*

Rotary hoes just poke holes in the ground if the soil is too wet. Flame weeders work best when leaf surfaces are dry. Most cultivators kill more weeds during hot, dry weather since uprooted weeds dry out quickly without rerooting. Thus, planning cultivation with the weather forecast in mind will frequently improve results.

Because timeliness is critical to the success of most in-row, near-row and full-field cultivation, adequate soil drainage may make the difference between successful weed management and substantial crop loss. When storm events are following in close succession with short rain-free periods in between, then adequate tile or swale drainage may allow cultivation on fields where it would otherwise be impossible. Similarly, improving surface drainage of poorly structured and compacted soils with deep rooted cover crops and crops can increase the possibility of timely cultivation.

Principle 7: *Effective cultivation requires good tilth and careful seedbed preparation.*

Good tilth is critical for effective cultivation. Good tilth facilitates shaking soil from weed roots. It also decreases the chances of knocking over crop plants with clods when soil is thrown into the row. Moreover, tools that work shallowly in or near the crop row are relatively ineffective in cloddy soil for three reasons. First, seedlings of some species emerge from greater depth in cloddy soil (Cussans et al. 1996), and the tools cannot reach them without damaging the crop. Second, when clods are moved, seedlings emerge that otherwise could not reach the soil surface. Finally, seedlings that sprout in clods may eventually take root if rain or irrigation subsequently allows the clods to

merge into the soil matrix. All of these factors argue for practices that improve soil structure, including cover crops, compost, rotating with sod crops and controlling wheel traffic. They also argue for delaying tillage until soil moisture conditions are appropriate, even if this requires a delay in planting.

Depending on the way the seedbed is prepared, clods can form even in well-structured soil. A coarse seedbed does not inhibit establishment of many large seeded crops and may help reduce erosion. However, for the reasons mentioned above, a coarse seedbed is never an advantage during cultivation.

For shallowly working implements, a level seedbed facilitates depth control. For some implements like basket weeders, a level seedbed is essential for adequate performance.

Principle 8: *Create a dust mulch.*

In regions where the risk of wind erosion is slight, one objective of cultivation should be the creation of a dust mulch. This is a surface layer of loose soil crumbs, typically 0.5–1.5 inches deep. The term "dust mulch" is really a misnomer but is in wide use. Ideally, the loose layer consists not of powder but of small aggregates (0.1–0.4-inch diameter) that allow good air circulation to dry out the surface soil. This loose surface layer can be achieved with most tools that work the soil shallowly. Since most annual weed species emerge from the top inch of soil, maintenance of a loose, dry surface layer of soil greatly decreases weed establishment. At the same time, this loose soil slows upward movement of moisture from deeper in the soil and can facilitate emergence of crops planted deeply into moist soil (1.5–2 inches) while restricting weed establishment in the dry surface soil. Obviously, you cannot maintain a dust mulch during wet weather, but a dust mulch is a highly effective weed management technique when weather allows. It is also an effective way to conserve soil moisture during dry periods.

Principle 9: *Weed early, shallow and often.*

Most annual weed seeds are tiny—often about the size of the head of a pin or smaller (see "Seed Weight"). Because their food stores are small, they cannot emerge from deep in the soil, and they are very thin and fragile shortly after germination. Hence, very shallow disturbance of the soil (less than 2 inches) can effectively eliminate a large percentage of these weeds.

Deeper soil disturbance brings additional seeds to the surface where they will germinate.

If planting has been delayed since the seedbed was prepared, work the surface soil thoroughly with a spring tooth harrow, field cultivator or similar implement before planting to kill any weeds that have germinated. Most tine weeders are not aggressive enough for this job. You may not see the weeds yet, but if the soil is moist enough for germination you can be sure that they are getting a head start on your crop.

Note that you may not be able to see newly emerged seedlings unless you get down on hands and knees. For large seeded crops that are planted more than 1 inch deep, such as corn or snap beans, you can cultivate right over the row with a tine weeder or rotary hoe before crop emergence until the crop is several inches tall. These "blind cultivations" as they are sometimes called are the best technique you have for controlling weeds in the row for crops that will tolerate such treatment. Note, however, that blind cultivation is pointless if the surface soil is too dry for seed germination. So, look for tiny weed seedlings and check the soil for the white threads of germinating weeds before you cultivate.

As weeds get larger, rerooting following cultivation becomes more probable. Hence, inter-row cultivation should occur before the weeds are larger than about 2 inches in height or lateral spread.

Principle 10: Plant carefully and adjust the cultivator.
Weed control will be improved and stand loss minimized if the rows are straight and the space between planter passes is identical to the spacing between rows within a pass. Computer guidance can help ensure planting uniformity (see "Cultivator Guidance Systems").

Adjust tool depth and position relative to the row to ensure safety of the crop and effective killing of the weeds. Be sure the cultivator frame is level front to back and side to side. This will ensure that all tools are working to their maximum effectiveness. For some implements, notably tine weeders, the adjustments must be made after the machine is in the soil, since forward motion in the soil will pull the nose down.

Unless you are using a guidance system to position the cultivator, put on the tractor's sway bars so that the cultivator is held rigidly to the tractor. This will prevent the cultivator from creeping from side to side,

blighting rows and missing weeds.

Expect to readjust the machine, as soil moisture varies across the field. A hydraulic top link can save much starting and stopping and will quickly pay for itself in time saved, decreased crop damage and improved weed control. However, changes in the top link can shift the cultivator out of level and may not compensate for other needed adjustments.

CULTIVATORS AND CULTIVATING TOOLS

A cultivator consists of a frame and one or more types of tools that engage with the soil and weeds. Most commonly, cultivators are belly mounted under the tractor or carried on a rear three-point hitch. Some implements can be front mounted. Belly or front mounted implements are easier to guide and less prone to damage the crop because the tractor operator can see the position of at least one set of tools relative to the row. However, wider, longer (to accommodate more tools) and higher clearance implements are generally more feasible with a rear mount. Belly mounted implements are stabilized by the tractor, but most three-point hitch mounted cultivators require additional stabilization to prevent sideways drift. Usually this is accomplished by large diameter coulters mounted near both ends of the toolbar. Wheels are also available that track the sides of raised beds, furrows made by the planter or the ridges in ridge-till systems. Some rear mounted cultivators have a seat and steering mechanism to allow working very close to the row.

Frames are of two basic types: rigid toolbars or independent parallel gangs. Usually, rigid cultivators are constructed with several parallel toolbars so that tools can be staggered. Staggering tools reduces opportunities for jamming with weeds, crop residue and stones. Rigid toolbars are adequate for flat land and smooth seedbeds. Parallel gang cultivators are more suitable in areas with swales and rocks, and in crops where back and dead furrows are unlikely to be worked flat during secondary tillage. Parallel gang cultivators work better under such conditions because the gangs for each inter-row track the soil topography independently by means of gauge wheels. Usually, the gangs are connected to the toolbar by a parallelogram-shaped linkage that causes the whole length of the gang to move up or down as a unit. This prevents minor variation in the depth of the forward tools or the gauge wheel from creating larger changes in the depth of the rear tools. Usually, parallel gang cultivators are three-point hitch mounted.

A variety of tools are available for mechanical weed management (Table 4.3). The amount and type of information available on these implements varies greatly. For most equipment, comparative data are meager, and some devices have received no scientific study at all. Consequently, much of the information compiled in Table 4.3 and discussed below is based on our personal experience, derived from discussion with farmers and colleagues who have used the implements, or taken from manufacturers' promotional materials. Two excellent videos are available that show many of these tools in action (Grubinger and Else no date, OSU 2005).

Shovels, Sweeps and Knives

Sweeps and shovels are the most commonly used cultivation tools. They are simple and durable. They vary greatly in width, shape and pitch (Figure 4.3). Generally, soil movement away from the shank increases with width and pitch of the sweep or shovel and decreases

Table 4.3: Operating Parameters, Uses and Limitations of Various Types of Mechanical Weeding Tools and Implements

Implement/ Tool[1]	Position of Action	Operating Depth[2]	Speed[3]	Weed Size[4]	Crop Size[5]	Soil Movement	Crops	Soil Limitations
		inches	mph	inches	inches			
Shovels and sweeps (hoes)	Inter-row	1.5–3 (4)	1.5–5 (6)	To large size	Limit set by clearance	Toward row	All row crops	Few soil limitations; high residue models available
Rolling cultivator — spider gangs	Inter-row, near row, sides of beds	1–3	1.5–5	To 12+	Limit set by clearance	Operator's choice	All row crops	Tolerates moderate rockiness; poor in residue
Rolling cultivator — disk gangs	Inter-row	2–3 (1)	1.5–5	To large size	Limit set by clearance	Operator's choice	All row crops	Tolerates moderate rockiness and high residue
Horizontal disk cultivator	Inter-row	1–2.5	4–8	To 16	Limit set by clearance	Toward row	Most row crops	Tolerates moderate residue
Rotary tiller (power hoe)	Inter-row	1–3	1.5–5	To large size	Limit set by clearance	Random	Most row crops[6]	Reduces soil structure
Mower, inter-row	Inter-row	None	2.5–6	To large size	Limit set by clearance	None	Most row crops	Problem with surface rocks
Disk hillers (cutaway disks)	Near row	1–3	1.6–5	To large size	Limit set by clearance	Operator's choice	All row crops	Few limitations
Spyders	Near row	1.5–3.7	1.5–5	To 12+	Limit set by clearance	Operator's choice	All row crops	High residue model available
Basket weeder	Very near row to inter-row	1.5–2.5	4–6	To 2	1–10	Parallel to row	Most row crops[7], tree seedlings	Intolerant of rocks; best with flat seedbed
Brush hoe— horizontal axis	Very near row to inter-row	1–2	1–3	Uproots seedlings, strips larger weeds	To 8 (11)	Parallel to row	Most row crops[6], tree seedlings, cereals	Tolerates wet soil; rocks may jam shields; best with flat seedbed
Brush weeder— vertical axis	Near row, in row	0.5–1.5	0.3–2.5	To 4	Limit set by clearance	Operator's choice	Many row crops	Tolerates wet soil
Vertical axis tine weeder	Near row, in row	1–3	2–4	To 4	Limit set by clearance	Away from row	Many row crops	Tolerates residue
Torsion weeders, spring hoes[7]	In row	1–1.5	1.5–5	Thread to cotyledon	Limit set by clearance	Slightly toward row	Most row crops[6], heavy model for tree and vine crops	Tolerate minor rockiness and residue
Spinners[7]	In row	1–2	1.5–5	Thread to cotyledon (5)	To 4 (8)	Minimal	Many row crops[7]	Tolerate moderate rockiness

Implement/ Tool[1]	Position of Action	Operating Depth[2]	Speed[3]	Weed Size[4]	Crop Size[5]	Soil Movement	Crops	Soil Limitations
		inches	mph	inches	inches			
Rubber finger weeder[7]	In row	1	1.5–5	Thread to cotyledon	To 10 (16)	Minimal, or away from row	High value row crops, nursery stock	Poor in crusted soil or large pieces of residue
Rubber star wheels[7]	In row	1	1–5	Thread to cotyledon	Limit set by clearance	Usually toward row	Most row crops	Poor in crusted soil
Rotary hoe	Full field	1–1.5	7–13	Thread to cotyledon	To 16	Random	Large seeded crops, cereals	Tolerates moderate rockiness; poor in wet soil; high residue models available
Tine weeder (weeding harrow)	Full field	0.7–2	2–5 (6)	Thread to cotyledon	To 6 (8)	Random	Large seeded crops, cereals, transplants	Poor in residue or crusted soil
Spike harrow, chain harrow	Full field	0.7–2	2–5 (7)	Thread to cotyledon	To 6	Random	Large seeded crops, cereals	Poor in residue
Rod weeder	Full field	1.5–2.5	8–12	To large size	Fallow, post-harvest	Minimal	Primarily dryland fallow	Tolerates residue
Flame weeder	Full field or in row	None	0.6–4 (8)	To 2 (8)	Mostly pre-emergence; to large size in a few crops	None	Pre-emergence in most crops; post-emergence in crops with protected buds	Fire hazard in residue
Hot water weeder	Inter-row, full field	None	1–4	To large size	Tall, woody crops	None	Tree and vine crops	Potential for compaction
Weed puller	In row	None	3–5	>8 taller than crop	Mid-season	None	Short row crops	Best in slightly moist soil
Mower, weed topper	In row	None	4–6	>3 taller than crop	Mid-season	None	Short row crops	Not affected by soil conditions
Electric discharge weeder	Full field	None	2–5	>3 taller than crop	To 40	None	Short row crops	Best with dry soil

[1]Implements are in singular form; tools that attach to another implement are given in plural. Synonyms are given in parentheses.

[2]Unusual operating depths that are used in some circumstances are given in parentheses.

[3]Unusual operating speeds that are used in some circumstances are given in parentheses. Speeds listed assume normal operating conditions and no special guidance system. Variation is largely due to variation in crop size and skill of the operator.

[4]"To large size" indicates that the implement is effective against even large weeds; how big a weed the implement can destroy will depend on the weed species and operating conditions, but generally, the effectiveness of the implement is not affected by the size of the weeds. A number followed by a plus indicates that normally the machine is effective to weeds of that size, but that in some circumstances it may be effective against larger weeds.

[5]"Limit set by clearance" indicates that the implement can be used until the crop has spread laterally so much that it is crushed by tractor tires, or is so tall that it will no longer pass under the tractor axle or implement toolbar. "Mid-season" indicates that the implement will generally be used after the crop is well developed but before it is maturing, regardless of crop stature.

[6]The implement is effective in most row crops but is largely limited to high value crops because of the need for time consuming adjustments or for a flat seedbed, or for its slow operating speed, etc.

[7]These are high precision tools that must track the row precisely. This requires either an effective guidance system or a belly mounted cultivator.

with increasing angle between the sides. "Goosefoot" style shovels (Figure 4.3) move less soil than sweeps. Typical cultivators carry three to seven sweeps or shovels per inter-row, each mounted on a separate spring steel shank. S-shaped (Danish) shanks allow greater vibration, which helps bring weeds to the surface and shake soil loose from the roots. They bend to allow the tools to slide by stones. They may also, however, deflect around a well rooted weed, like a large dandelion, rather than digging it out or cutting it off. They are less robust than C-shaped shanks and will do a poor job if

they get bent. If you use S-shanks in rocky soil, keep a spare or two on hand and replace bent shanks promptly to avoid crop damage. C-shanks are usually heavier and are protected by springs or break pins, so they are therefore better suited to cultivating untilled or very stony ground.

Multiple shanks provide flexibility. For example, in one experiment we used 1-inch spikes nearest the row when corn was young to reduce soil movement toward the row, but we changed to 4-inch sweeps to throw more soil into the row at the second cultivation.

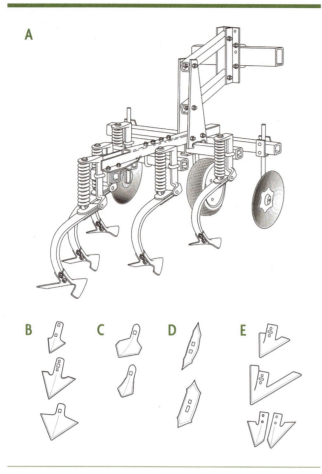

Figure 4.3. (A) One gang of a parallel gang row-crop cultivator showing C-shanks, sweeps, a gauge wheel, hilling disks and parallel linkage. (B) Small cultivating sweeps. (C) Goosefoot shovels. (D) Cultivating spikes. (E) Large V-sweeps and a pair of half-sweeps. Illustration by Vic Kulihin.

Minimum tillage cultivators designed to operate in high crop residue, however, usually have a single shank with one broad sweep per inter-row. This design presents less metal at the ground level to snag debris and uses a coulter in front of each shank to cut residue so it can flow past the shank. Minimum tillage machines operate with sweeps just below the soil surface to sever weed stems from roots with minimal soil disturbance and are best suited for killing larger weeds. However, hilling can be accomplished with wing attachments that increase the lateral displacement of soil or with disk hillers (see below).

Vegetable knives (beet knives) are a special type of half sweep that is useful for cultivating close to the crop row. They have a shallow pitch so that lateral soil movement is relatively small. When the crop is young, most growers run them with the knife tip pointing toward the inter-row so that no soil is thrown onto the young plants. When the crop gets large, you can turn them around so that the tips of the knives reach in under sprawling leaves of crops like cabbage and sugar beets. Whereas most other types of shovels and sweeps are inter-row tools, vegetable knives are effective near-row tools.

Horizontal Disk Cultivators

These machines are built like a typical parallel gang row-crop cultivator, except that instead of sweeps or shovels, they are mounted with rotating disks that travel roughly parallel with the soil surface. The disks are designed to be self-sharpening. They are tipped slightly so that they rotate as they travel through the soil. The disks cut through weed roots and lift the soil, causing it to shatter and exposing near surface roots to drying. The disks, typically three per inter-row, overlap to ensure complete cutting of the weeds. An advantage of the machine is that the soil experiences little inversion and thus residue is retained on the soil surface, though these benefits are not necessarily superior to what can be achieved with flat sweeps. The spindles that carry the disks are spring loaded to deflect on impact, but high-speed encounters with rocks can damage the disks, making the implement inappropriate for rocky soil.

Rolling Cultivators

This implement usually consists of gangs of "spider" wheels (not to be confused with spyders, spelled with a "y"—see below). Each gang is mounted on a separate tube (Figure 4.4). Two gangs work each inter-row. The spiders are ground driven and cut and dig out weeds as they roll. Aggressiveness and amount of soil movement are controlled by adjusting the angle relative to the direction of travel (Figure 4.4). Depending on the setting of the gangs, soil flow is strictly toward or away from the row. The gangs can also be tilted, which makes this the implement of choice for cultivating the sides of raised beds. In European sources these implements are sometimes referred to as "rotary hoes," but they are constructed and used very differently from the American rotary hoe discussed below.

Rolling cultivators are less able to dig out large weeds than are shovel type cultivators, but they work the soil more thoroughly for shallow cultivation. A sweep can be mounted on alternate tubes to work the center of the inter-row, and this is helpful for digging out weeds that slide between the gangs. The sweeps are also useful for ripping up tire tracks. Because soil flow

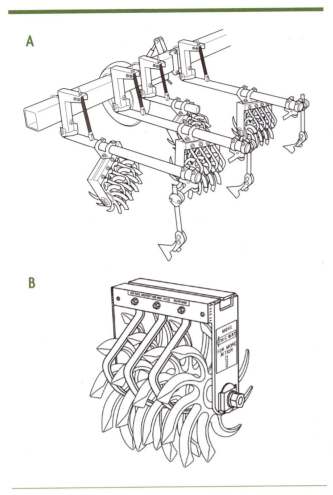

Figure 4.4. (A) Rolling cultivator. Illustration by Vic Kulihin. (B) Detail of one spider gang.

is strictly in one direction, and because the gangs can be tilted to work very shallowly next to the crop row, rolling cultivators can safely cultivate closer to the crop row than can shovel cultivators.

The rolling cultivator can be outfitted with disk gangs for work in high residue. These chop through dense residue without jamming, but the curvature of the disks prevents the tools from working as close to the row as is feasible with spider gangs.

One drawback of rolling cultivators is that the tubes are connected to the toolbar by a hinge linkage rather than by a parallel linkage. This does not affect the performance of the rolling gangs much, but it means that if the rolling gang hits a rock or hard spot, the end of the tube rises and the sweep comes out of the ground. Also, the hinge linkages are not heavy enough for using the disk gangs in previously unbroken ground (for example, in a ridge tilled system) unless the soil has exceptionally good tilth.

Rotary Tillers (Multivators)

These implements consist of gangs of PTO driven, rotating tines that chop up weeds and mix them into the soil (Figure 4.5). The tines are either curved or L-shaped. In most models you can adjust the position and width of the tilled strip. They are an alternative to zone tillage tools for tilling strips into cover crops if deep ripping is not desired. They are also currently the best tools available for strip tillage into perennial living mulches prior to planting. Their principal advantages in cultivation are that they completely incorporate all aboveground weed tissues, and they chop near-surface roots and rhizomes to smaller fragments than do most other implements. Disadvantages include relatively slow speed compared with shovel type cultivators for similar operations and deterioration of soil structure that results from repeated pulverization.

Disk Hillers and Spyders

Disk hillers consist of curved disks mounted on a vertical shank. Usually, they are placed in the front-most position and next to the row on shovel type cultivators (Figure 4.3). Early in crop growth they are set to cut soil and weeds away from the row; later they may be used to hill up soil around the base of the crop. They are aggressive tools that can dig out large annual weeds and cut the stems of rank perennials. This also allows them to perform well in heavy crop residue. Since they are used relatively close to the row, understanding the rooting habit of the crop is required to obtain optimal use of disk hillers without pruning roots.

Spyders are star shaped wheels that are used in much the same way as disk hillers (Figure 4.6). They

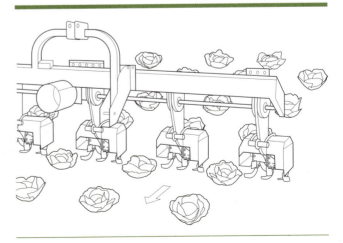

Figure 4.5. Rotary cultivator with shields. The arrow indicates the direction of travel. Illustration by Vic Kulihin.

Figure 4.6. Spyders. Arrows indicate direction of travel. Illustration by Vic Kulihin.

are somewhat similar to the spiders (spelled with an "i") of a rolling cultivator, but, in addition to important details of design, they are run singly rather than in gangs. They are smaller diameter (13 inches) than most disk hillers, which allows you to get closer to some crops. Also, when cutting soil away from the row, they leave a loose soil layer next to the row rather than a bare shoulder, and this probably reduces drying in the row. A disadvantage of spyders is that unlike a disk

hiller that cuts off crop leaves that lie in its path, the thicker blade of the spyder grabs the leaves and drives them into the soil. This jerks the crop plant sideways, causing greater damage than just the loss of the leaf tip that is actually hit. Thus, you can run spyders closer to the row in upright crops, like leeks, than you can in crops where leaves on the young plant spread out, like corn or cabbage. We experimented with shields attached to the spyder shank to protect crop plants from getting snagged. They had to be set close to the ground to push leaves of young plants aside, and consequently they jammed with stones. Shields would probably work well in a stone free soil.

For maximum effectiveness, you should adjust both disk hillers and spyders for angle, distance from crop and depth. On an implement for cultivating six or more rows, this is time consuming. We do not know of any studies directly comparing effects of disk hillers and spyders on different weed and crop species or stages of weed growth.

Basket Weeders

These cultivators consist of two sets of rotating wire cages (Figure 4.7). The forward cages are ground driven; the rear cages are driven by a chain connected to the forward cages, and they turn twice as fast in the opposite direction. Penetration is very shallow (Table 4.3), but they work the soil thoroughly. Consequently,

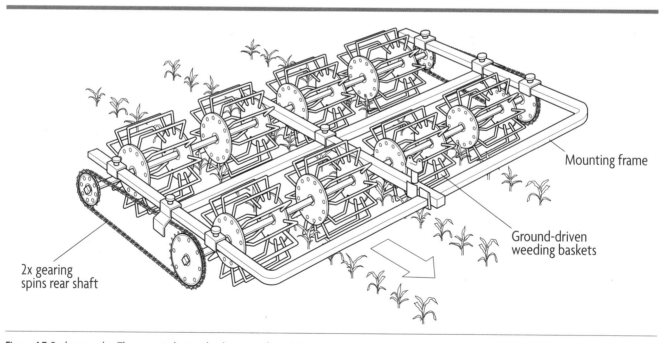

Figure 4.7. Basket weeder. The arrow indicates the direction of travel. Illustration by John Gist.

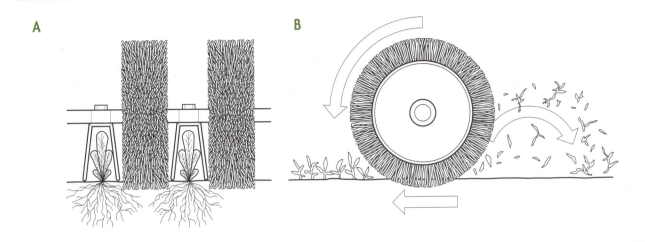

Figure 4.8. Brush hoe. (A) rear view; (B) side view. Illustration by Vic Kulihin.

few small weeds escape substantial damage even if they are not completely uprooted. Since the flow of soil is strictly parallel to the crop row, the implement can get within about 2.5 inches of the crop. Thus, it is very useful for cultivating close to the row in young crops. The manufacturer makes specially shaped baskets for cultivating the sides of beds. Because the baskets must be sized appropriately to the dimensions of the inter-row, you need to have separate machines for each row spacing. Basket weeders are not suitable for stony ground because rocks bend the baskets out of shape and can become caught between adjacent wires.

Brush Hoes

The standard brush hoe consists of PTO driven polypropylene brushes working parallel to the crop row (Figure 4.8). These cultivators uproot small weeds and shear off larger ones. The soil flow is primarily parallel to the row, which, in conjunction with narrow tunnel shields (2.5–8 inches wide) allows you to cultivate very close to small crop plants. You need to switch to another type of cultivator once the crop plants grow too large to move easily through the shields.

The brush hoe resists clogging with large weeds and debris, and works well in wet soils. An additional advantage is that, like the basket weeder, it leaves a loose, uniform soil surface that slows weed germination. However, because the brushes are mounted on a common axle, the implement requires a flat seedbed for consistent depth of operation. Working depth declines with increasing tractor speed, and consequently operating the implement at a high speed

is impractical. Adjusting the row spacing requires substantial disassembly.

Because brush hoes work very close to the crop row, they come equipped with a driver's seat and steering mechanism. They are thus a labor-intensive implement. Moreover, when run on dry soil, they create much dust, which makes steering the implement unpleasant.

Torsion Weeders and Spring Hoes

Torsion weeders consist of spring steel rods that reach within a few inches of the crop row and travel about 1 inch below the soil surface (Figure 4.9a). The compressive action of the springs causes the soil in the row to boil up, thereby heaving out weeds that are not yet fully rooted. For weeds in the row, torsion weeders only work against weeds in the white thread stage, or ones that have just emerged. The tool shears off most of the larger weeds next to the crop row. Spring hoes work in a similar manner to torsion weeders but are more robust and aggressive (Figure 4.9b). A smooth, flat seedbed improves the consistency of weed control.

The inventor of these tools recommended torsion weeders for cultivating small crop plants and spring hoes for larger plants. We have not observed much difference in the action of the two tools in a silt loam soil, but action may vary with soil type. Both torsion weeders and spring hoes usually mount on a shovel cultivator in front of the forward shovels. These tools may perform better if spyders are run ahead of them to loosen the soil, but we know farmers who use the tools successfully without spyders. Several studies have

A

B

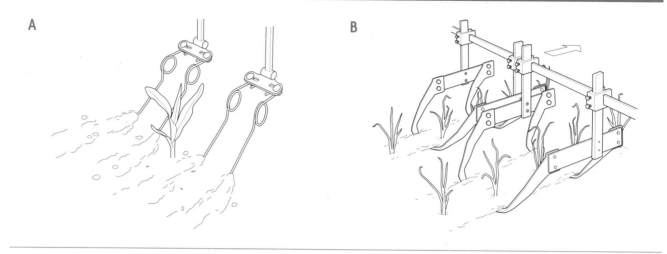

Figure 4.9. (A) Torsion weeders and (B) spring hoes. Illustration by Vic Kulihin.

demonstrated improved weed control and crop yield with these tools plus spyders relative to shovel cultivators without them.

Torsion weeders and spring hoes are precision tools that must be carefully set for depth and distance from the row to achieve good weed control without damaging crop plants. Consequently, they work best when belly mounted or front mounted. With a rear mounted cultivator, you need a guidance system. Otherwise, you will have to set the tools so far apart that the soil in the row will not boil properly. Most gang type, rear mount cultivators do not have enough room to mount these tools plus spyders. Since these tools and the spyders are close to the row you still need all the shovels. If too many tools (and the gauge wheel) are crammed close together, the implement will jam with weeds, crop residue or stones. One option is to lengthen the gang bars. A more elegant solution is to belly mount the torsion weeders (or spring hoes) and spyders, and then use a rear mounted row-crop cultivator to clean out the inter-row area.

Stiff, heavy duty spring hoes are available for work in orchards and vineyards. These scrape the soil surface free of weeds in and near the row. A castor at the tip allows the tool to bend past trunks without scraping the bark.

Vertical Axis Tine and Brush Weeders

These machines are designed to work very close to, or sometimes in, the row. They are rear mounted and have a seat for an operator, though the tine type can be locked for inter-row cultivation without an operator.

The operator controls the position of a pair of rotating wire baskets or polypropylene brushes that stir the soil next to the crop. Handles allow the operator to accurately control the position of the tools. For crops that are well spaced within the row, the operator can bring the tools fully into the crop row between plants, thereby eliminating most in-row weeds. These are precision implements that require operator skill for maximum effectiveness while minimizing crop damage. The speed of use depends on the skill of the operator.

The tine type machines have inverted baskets of nearly vertical spring tines that aggressively stir the soil and can uproot and break weeds of moderate size. These machines are particularly useful in strawberry production because, in addition to controlling weeds, they brush runners into the crop row with little damage. Because the operator is controlling the position of each of the rotating baskets independently of the other, the standard machine cultivates a single row. However, a machine with similar operating principles but controlled by a machine vision system can cultivate multiple rows.

The brush type machines have inverted cones of bristles that brush the soil surface with minimal disturbance of the soil profile. Although they can uproot very small weeds, much of their action involves shearing off the weeds at the soil surface. Because the bristles flex, they can brush against the stems of some crops with little crop damage, which allows the brushes to work in the crop row to control small weeds.

Spinners

These ground driven in-row weeders consist of a

basket-like arrangement of spring steel wires that scratch laterally across the crop row (Figure 4.10). Spinners are normally used in pairs, with the two tools working in opposite directions across the row. This increases the proportion of the row area that gets worked. Usually, the depth is set so that the deepest penetration is a little above the planting depth and squarely in the row. Alternatively, you can set the tool so that it penetrates deepest next to the row and scratches across the row with an ascending or descending stroke. Either way, deeper planting allows more aggressive weeding. Cultivation late on a sunny day, when crop stems are less stiff, reduces mortality in corn. Spinners are primarily effective against weeds in the white thread to cotyledon stages and are only safe while the crop is small. Consequently, the window of usefulness for these tools overlaps greatly with that of rotary hoes and tine weeders. Those full-field cultivators are easier to use but are less versatile than spinners.

A direct strike by the tip of a wire can cut off a crop plant. Consequently, spinners tend to thin crops like soybeans or peas that are planted closely within the row, but mortality from properly set tools is usually no more than a few percent in crops spaced more than 6 inches apart in the row. Spinners should not be used in transplanted crops until the plug is well rooted. If set deep, spinners can be used to deliberately thin an overly dense crop stand. Since the probability of a fatal strike is higher where crop plants are dense, thinning a spotty but overly dense stand with spinners results in a more uniform spacing of crop plants.

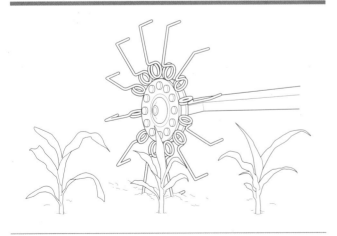

Figure 4.10. One spinner. Normally, these are used in a staggered pair with the two spinners scratching across the row in opposite directions. Illustration by Vic Kulihin.

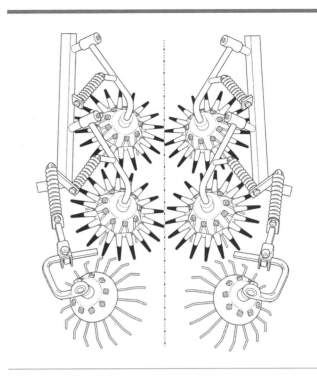

Figure 4.11. Rubber finger weeder.

Rubber Finger Weeders and Rubber Star Wheels

Rubber finger weeders consist of two pairs of ground driven wheels equipped with rubber-tipped fingers that stir the surface soil in the row but bend around well rooted crop plants. These are followed by wire baskets that aggressively stir the area adjacent to the row (Figure 4.11). The implement is usually belly mounted. Single row machines can work with row spacings as narrow as 20 inches, but cultivation of multiple rows requires a minimum row spacing of 34 inches. This essentially limits use of the machine to high value crops that are worth cultivating one row at a time or to crops with wide row spacing. Although the rubber fingers have some ability to flex, they are sufficiently stiff to damage most young crops. Unlike some other implements that attack weeds in the crop row, finger weeders can be used on large, upright crop plants. The upper limit on height is set by clearance under the tractor. This makes them useful in nurseries.

Rubber star wheels (Figure 4.12) are similar in concept to rubber finger weeders but are designed to mount on a row-crop cultivator to provide in-row and near-row weeding that cannot be achieved by sweeps and shovels alone. The rubber stars are available with multiple grades of stiffness, and the more flexible versions can be used in young and relatively delicate

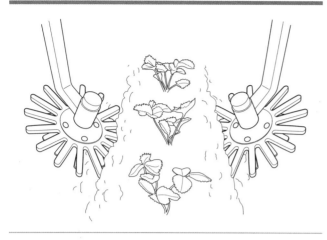

Figure 4.12. Rubber star wheels. Illustration by Vic Kulihin.

crops. The more flexible the rubber is, the more the soil in the row needs to be loose for the tools to work well. As with torsion weeders and spring hoes, finding a place to mount these tools on a cultivator gang can be challenging. As with all in-row weeding tools, exact guidance is required to prevent crop damage and achieve good in-row weeding.

Pneumatic Weeders

These consist of a PTO driven air compressor leading to a pair of small air-blast nozzles in each row. The nozzles point toward the row but somewhat backward, and they travel just under the soil surface. The blast of air blows shallowly rooted weed seedlings out of the ground, without damaging more deeply rooted crop plants. Effective action requires a loose, friable soil surface.

Rotary Hoes

A rotary hoe consists of one or two ranks of wheels each bearing 16 spoon-like projections (Figure 4.13). The wheels are attached to the toolbar by spring loaded arms to allow movement over obstacles. The ground driven wheels typically penetrate to a depth of 0.8–1.5 inches and flick up soil and small weeds as they turn. To disturb the soil effectively, you must operate the machine at high speed (Table 4.3). This allows rapid treatment of large acreage. Shallow penetration allows the wheels to contact young crop plants with little damage.

Most rotary hoes less than 20 feet wide do not come equipped with depth gauge wheels. If your rotary hoe does not have gauge wheels, you should get some.

Although soil conditions that lead to excessive penetration only occur occasionally, running a rotary hoe deeper than planting depth will badly damage the crop. Carrying the implement on the three-point hitch is a poor alternative to gauge wheels because at the high speed necessary for effective operation, the tractor will bounce, and this causes inconsistent penetration.

Rotary hoes mostly kill weeds in the white thread stage; by one or two days after emergence, most weed species will be sufficiently well rooted to survive. Hence, timeliness is critical for success. When soil conditions are suitable for weed germination, typically the implement is used five days after planting and at five- to seven-day intervals after that. Two or three passes are usually all you can fit in before the crop will grow large enough to be damaged. One study showed that whereas three timely rotary hoeings reduced weed dry weight by an average of 72% in drilled soybeans, two untimely rotary hoeings only reduced dry weight by 38%. However, if the timing is good, most of the weed control results from the first rotary hoeing.

Wet soil conditions reduce the effectiveness of rotary hoeing. If the soil is too wet, the spoons just punch holes in the ground rather than throwing the soil and seedlings into the air. In addition, rainfall or irrigation soon after rotary hoeing reduces the percentage of weeds that dry out and die.

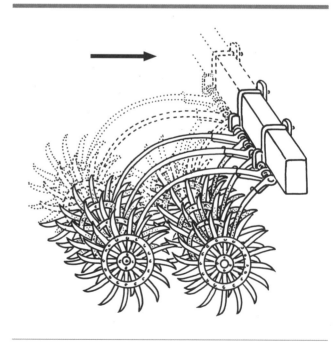

Figure 4.13. Rotary hoe. The arrow indicates the direction of travel. Illustration by John Gist.

Unlike many cultivating implements that have a long lifespan, rotary hoes wear out within a few years. The spoon tips wear off and either tips or the whole wheels have to be replaced. Worn tips will not throw much soil and thus make the implement ineffective.

Tine Weeders (Weeding Harrows)

These implements vary greatly in design, but all consist of a frame with many downward pointing, small diameter tines (Figure 4.14). These machines are a key element in weed management on many organic grain and vegetable farms. Although some growers still use chain harrows and spike tooth or peg tooth harrows for weed control, these more traditional designs are rapidly being replaced by tine weeders. The tines on tine weeders are usually flexible spring steel wires, typically 0.19–0.38 inch in diameter. The popularity of this implement comes from the ability to adjust the tine angle and the down-pressure, and hence aggressiveness, coupled with the ability of the tines to spring around well rooted crop plants. Also, the springiness of the tines causes them to vibrate from side to side, which helps shake the weeds free from the soil. Unlike more traditional harrows with rigid tines, they do not throw rocks onto crop plants. Tine weeders are also sometimes called "finger weeders" but should not be confused with rubber finger weeders.

Makes and models vary in design. Machines with tine carriers rigidly mounted to the frame can transfer weight from the tractor to force tines into the soil. They can be used more aggressively but require careful adjustment to avoid crop damage. Machines in which the tine carriers are suspended from chains float on the soil surface and are less likely to damage the crop. Some designs lift the tine carriers on chains so they can float but also provide means for transferring tractor weight when necessary. The springiness of the tines increases with their length and decreases with their diameter but also depends on the type of steel and the design of the coil. A highly flexible tine will easily deflect around crop plants but may also leave many weeds undisturbed. Tines may be straight or bent at an angle from 45 to 85 degrees. A 45-degree tine will work well in most situations. An 85-degree tine can aggressively pull weeds out of taprooted crops like green beans and soybeans but easily damages corn, transplanted vegetables and other crops with a diffuse root system by hooking under the roots. Straight tines are

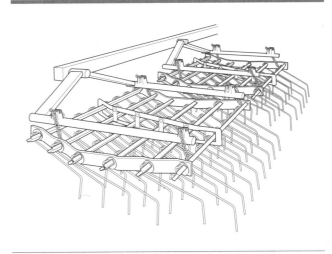

Figure 4.14. A tine weeder. Illustration by Vic Kulihin.

generally less aggressive than bent tines of the same flexibility. A stiff straight tine can bury weeds in soil that is too damp for effective weeding with a springier bent tine.

Generally, corn and soybeans can be tine weeded from pre-emergence to about 7 inches, but avoid weeding soybeans or other beans from ground crack until the seed leaves are horizontal. Tine weeding is most effective against weeds that are just emerging or are in the white thread stage. Thus, tine weeding should be done before the weeds are readily visible. However, if the soil is too dry for weed seed germination, then tine weeding may be pointless. The soil disturbance from tine weeding will often prompt germination of a new flush of weeds. If a pre-emergence tine weeding is done too soon, then these weeds may emerge when the crop is susceptible to damage. Consequently, delaying the first tine weeding until the very first weeds emerge is often optimal. However, effectiveness of the operation declines rapidly once the weeds emerge, so tine weeding early by one day is usually preferable to weeding later by one day. Basically, tine weeding involves balancing timing of weed emergence, crop development, and weather and soil conditions.

We have successfully tine weeded soybeans until the second leaves with three leaflets were beginning to emerge. Tine weeding in soybeans can be aggressive once the soybeans have developed deep taproots. The aggressive tine weeding will flatten many of the weeds that are not uprooted, and these can then be buried by immediately throwing 1 inch of soil into the crop row with a row-crop cultivator. When corn is between emergence and about two leaves, use tine weeders at a low aggression setting to avoid killing the crop.

Regardless of the crop or the weeder, the machine must be carefully adjusted to kill the weeds while preserving the crop. The tine carriers should be level when they are in the soil. As a general rule, the tines should not penetrate deeper than the seeds are planted. Since the tines bounce and seed depth varies a little, this means the tines should be adjusted to run at least 0.5 inch above the seeding depth (even shallower in rough or crusted conditions).

In small grains, tine weeders are most commonly used pre-emergence or at emergence, and again when the crop has two to four leaves. Many experienced operators suggest waiting until oats have two to three leaves and until barley, wheat or spelt have four leaves before a post-emergence tine weeding. Several extensive studies have demonstrated, however, that earlier weeding can be conducted without stand loss. Whether early post-emergence weeding is safe probably depends on soil type, soil moisture and how carefully the weeder is adjusted and operated. Waiting for the crop to become well established is the safest course of action, but this also allows the weeds to become larger and more resistant to damage and soil covering. Generally, a tine bent at 45–60 degrees works best with small grains. An 85-degree tine tends to uproot the grain. A straight tine works well pre-emergence but tends to bury more plants during early post-emergence weeding than a bent tine. Some growers also use tine weeders to comb sprawling and vining weeds like common chickweed and catchweed bedstraw out of small grains shortly before the stem of the grain elongates. Several studies in small grains have shown a close relation between the percentage of weed control and the percentage of grain plants covered with soil. Consequently, you may want to increase planting density to compensate for plant loss during tine weeding. Close spacing of plants in the row also helps deflect the tines out of the row, thereby causing more damage to weeds and less damage to the crop. Because of the tradeoff between crop damage and weed control, most studies in small grains and legumes planted at narrow row spacings have found no consistent increase in yield with tine weeding relative to weedy control plots, except occasionally when the weeds are dense. This does not mean, however, that tine weeding small grains is useless. Generally, weed seed production in grains is

directly related to the density of mature weeds. Tine weeding may not improve yield of this year's grain crop, but it will reduce seed input and help prevent a dangerous buildup of weeds that can affect yields in later years.

To prevent damage to your tine weeder, always lower it to the ground when the tractor is moving forward. Also, avoid turning with the implement in the ground, as this can bend the tines.

Tines wear faster on the sides than on the front or back, leading to sharpening over time. A sharpened tine can cut the crop. Also, it will not push enough soil to bury weeds. Moreover, the flattening will prevent the tine from wiggling from side to side, which is important for uprooting small weeds. When tines become sharpened by wear, they can be clipped off. Note, however that this changes the length of the tine and therefore the action of the weeder. Remember that a shorter tine is stiffer and more aggressive. Be cautious after clipping tines until you understand how your rejuvenated weeder will perform.

Rod Weeders

These consist of a rotating square or triangular rod that is turned by a chain or gear drive connected to the implement's gauge wheels. When the rod encounters a weed, the turning motion of the rod pushes the plant upward out of the soil (Figure 4.15). The implement is commonly used for fallow weed control in dryland farming systems where land is regularly fallowed to restore soil moisture. It is ideal for post-harvest mechanical weed control because it preserves and even increases surface organic matter. However, in humid conditions, many weeds will often reroot.

OTHER PHYSICAL WEED CONTROL DEVICES

Mowers

Mowers can be used for weed management in several ways. They are clearly useful for preventing seed production and suppressing perennial weeds around field margins and in grassed alleys, particularly when these are planted between permanent vegetable beds. A study in Missouri showed that a mower using multiple PTO driven string trimmers effectively controlled annual weeds between the rows in a no-till system. This implement may be useful as growers and researchers continue to develop increasingly reduced tillage

Figure 4.15. Rod weeder. The arrows indicate the direction of travel and rod rotation. Illustration by Vic Kulihin.

organic cropping systems. We also know a group of growers that put together an elevated, multiple blade rotary mower for topping weeds that emerge through soybeans. Such a mower would be useful for reducing seed production by in-row escapes even in situations where yield is not threatened. Mowers designed for removing corn tassels during hybrid corn seed production are ideal for cutting off weeds that emerge above crop canopies. However, they will generally be too expensive for most growers to purchase specifically for weeding. Although most weed species will recover to some extent after being topped, the reduced shade cast on the crop can improve yield, and removing the tops of the weeds will often substantially reduce weed seed production.

Flame Weeders

Flame weeders briefly expose weeds to a propane or butane flame at 1,500–1,800°F. This damages cell membranes and leads to rapid dehydration. A bank of burners can flame a wide area to kill weeds before crop planting or before crop emergence, or to defoliate plants prior to harvest. Flame weeding is currently the most common way to create a true stale seedbed in an organic cropping system. A canopy over the burner that contains the heat increases efficiency when flaming a whole bed. If you irrigate a few days before planting, the first flush of weeds will emerge in time to flame before the crop is up.

Burners directed toward the row can control in-row weeds in crops that have a protected terminal bud, like corn and leeks, and in cotton and sunflowers, which have tough stems (Figure 4.16). Although the range of crops that can be flame weeded after emergence is limited, the value of flame weeding for those crops is great. Flame weeding can effectively control weeds that escape in-row cultivation and can make the difference between a weedy and a weed free crop. Since most of the crops that tolerate flaming require several months to mature, they tend to be crops in which weeds go to seed. Flame weeding thus has consequences for weed management in subsequent crops. Flaming has little effect on weed seeds that have already fallen to the ground, however, since these are protected by surrounding soil particles.

Shields can also be used to protect flame sensitive crops from the direct heat of the flame and to allow highly effective weed control close to the crop row. Shields similar to those used for close cultivation of small crops can be fitted on tractor mounted flame

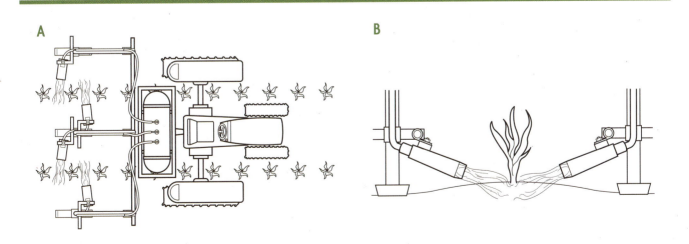

A   B

Figure 4.16. In-row flame weeder. (A) Top view; (B) rear view. Note that the burners are staggered to avoid forcing flame up onto the crop canopy. Illustration by Vic Kulihin.

weeders. With a shield in place, the flame should be directed rearward instead of toward the row as when weeding flame tolerant crops. Fixing a wheel and a shield to the flame wand of handheld and push type flame weeders allows precision flaming of small acreages.

The length of time weeds are exposed to the flame is usually adjusted by changing the speed of movement through the field. A slow speed kills better but also uses more gas. The amount of gas required for 95% control varies substantially with weed species and weed size. For example, control of white mustard (similar to wild mustard) in the two- to four-leaf stage required 1.5–2 times more propane per acre than seedlings in the zero- to two-leaf stage (Ascard 1994). Many species such as common lambsquarters, common chickweed and common groundsel are well controlled by gas doses of less than 45 pounds per acre when young (Ascard 1995b). In contrast, grasses, which will generally resprout from ground level, and broadleaf species in which the bud is protected by tightly clustered leaf bases (for example, corn chamomile and common purslane) may regrow after flaming. Controlling such species may require multiple treatments or large gas doses (Rahkonen and Vanhala 1993, Ascard 1995b).

Flame weeding is most effective when the plant surface is completely dry. Otherwise, vaporization of surface moisture absorbs some of the heat and protects the weeds. Soil moisture is not critical, but the soil surface should be smooth. A lumpy surface will shelter some weeds and, during in-row flaming, reflect heat up onto sensitive parts of the crop plants.

Starting a fire is a risk with flame weeding. Particular care should be taken to avoid igniting dry grass or brush when turning near the ends of a field. Bits of crop residue may also catch fire, even in a relatively clean tilled field. These usually quickly burn out without damaging crops but under windy conditions may blow away to ignite flammable materials outside the target area. Particular care also should be taken in small scale intensive vegetable systems where dry crop residue or a bed mulched with straw may be near a flamed area. Other methods of killing weeds with heat or cold that have reduced fire danger have been studied. These include freezing the weeds with a jet of liquid nitrogen or with carbon dioxide snow, and cooking the weeds with infrared radiation. The latter simply involves running a very hot metal plate near the weeds. These approaches all use substantially more energy than direct flame weeding and have not been commercialized.

Steam and Hot Water Weeders

Like flame weeders, steam and hot water weeders kill by disrupting plant surfaces, leading to subsequent desiccation. Their advantage over flame weeders is that the flame is enclosed in a boiler, which minimizes risk of starting a fire. Their energy use, however, is several fold higher per acre than typical flame weeders. Hot water weeders are more energy efficient than steam weeders because they do not bring water to a boil, but steam weeders require a smaller volume of water. Although hot water and steam weeders are finding some acceptance for urban weed management, adoption rates in agriculture have been low due to their high fuel usage and the potential soil compaction from hauling large volumes of water over tilled fields.

Electrical Discharge Weeders

The type of electric discharge weeder currently marketed in the United States is used primarily to kill weeds that escaped cultivation earlier in the season in relatively low growing row crops like beets and beans. They operate by bringing a high voltage electrode into contact with weeds above the crop canopy. Electrical resistance of the weeds causes sap to vaporize, which disrupts tissues. The proportion of weeds that are controlled decreases with increasing weed density because many pathways for electricity to reach ground results in a lower energy dose per plant.

The equipment consists of a horizontal charged bar carried by a front three-point hitch or loader arms. This electrode is powered by a PTO driven generator and transformer carried in a cart behind the tractor. Large tractors of 125 hp or more are required. Energy use increases with weed density, which makes charged bar electrical discharge weeders impractical as a primary weed management tool. However, they are cost effective for low density populations that escape other management measures. Operating speeds vary with weed density and the type of weeds, from 3 to 6 mph. Generally, slower speeds are required for grass relative to broadleaf weeds because the electricity has to travel down many leaves rather than a few primary stems. Ground speed will be slow and fuel use excessive if the weeds are wet because the moisture on the plant surfaces provides alternative routes for the current to travel to ground.

One study showed that an electric weeder, an herbicide wiper and a mower all controlled seed production of sugar beet bolters equally well, but that the electric weeder killed more bolters than the mower and could be used in a wider range of weather conditions than the wiper without damaging the crop. Electric discharge weeders are important in an integrated weed control program because they are one of the few implements that can remove weeds from the crop row after the crop grows large. Removing the largest weeds potentially improves yield in the current crop, but it also reduces weed seed production. This can ease weed management in subsequent crops. Because the electrical discharge kills the weeds and disrupts development of immature seeds, charged bar electrical discharge weeders have advantages over other late season weed management tools like raised mowers and weed pullers. Electrical discharge weeders are, however, substantially more expensive to purchase and operate.

A radically different type of electrical discharge weeder is sold in Europe. In this type of weeder, a row of electrodes is pulled through the soil or brush over all vegetation in an area. Models are available for inter-row weeding and to kill weeds prior to no-till planting of crops. It has potential for continuous no-till organic cropping systems. However, few data are available on horsepower requirements, fuel usage, ground speed or effects of weed density, residue and soil conditions on performance.

Weed Pullers

Mechanical weed pullers are another tool for removing tall weeds that have overtopped the crop. The only machine of this sort currently on the market consists of hydraulically driven pairs of wheels that rotate upward over the crop row (Figure 4.17). The wheels consist of a rubber tire either backed or fronted with a metal wheel covered with rubber mesh. When a weed contacts the wheels, it is drawn into them and pulled out of the ground. The machine can be front or rear mounted but works best in a front mounted position. It can be attached to a front-end loader, however, so a front three-point hitch is not required. Moist, loose soil and dry foliage improve action. The device is most effective against tall, fibrous stemmed species like waterhemp,

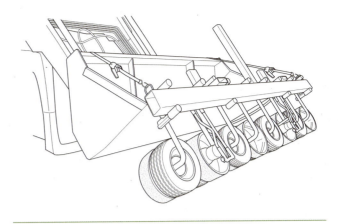

Figure 4.17. Weed puller. Illustration by Vic Kulihin.

shattercane, volunteer corn and sugar beet bolters. The tool is potentially effective for reducing weed seed production if it is used before seeds have begun to form. However, in many cases, by the time the weeds have gotten tall enough for the puller to selectively pull the weeds without damaging the crop, seeds will have begun to form and will continue to develop on the dying weeds. Crop root damage is likely to be slight in taprooted crops like soybeans and carrots, but the potential for root damage of fibrous-rooted crops like tomatoes needs to be evaluated.

Abrasion Weeders

Abrasion weeders are still in the development stage but offer potential for weed management without soil disturbance. These weeders work like miniature sand blasters, with an air stream directing grit at small weeds, particularly in the crop row. The grit of choice appears to be ground corn cobs. Trials with prototypes have shown good control of broadleaf weeds with minimal damage to corn. One potential application for this type of weeder might be supplemental weed control in crops no-till planted into rolled cover crops.

CULTIVATOR GUIDANCE SYSTEMS

Cultivation is an exacting task that wearies the tractor operator and can kill crop plants if you are not careful. These problems are multiplied when you use tools that work in or very near the crop row. Fortunately, great progress has been made recently in automation of implement and tractor guidance.

The simplest approach is purely mechanical. Wheels mounted on the cultivator guide the implement by rolling along the sides of raised beds or ridges, or else travel in furrows that are laid down by the planter. These systems are sufficiently accurate for cultivating at high speeds with in-row tools. They are best adapted to rear mounted machines since the implement must have some lateral sway relative to the tractor. These systems are inexpensive relative to the electronic guidance systems discussed below. Furrow guidance requires implements that are six rows or wider, however, since two wheels are needed for stability, and the tractor tires must not obliterate the furrows. Also, if you are working with a planter-made furrow rather than with ridges or raised beds, you need to preserve the furrow through early season full-field cultivations. We have done this by removing rotary hoe wheels or raising tine weeder tines over the inter-rows where the furrows are located (Mohler et al. 1997). This does not reduce weed control since you will later clean up the inter-rows with a row-crop cultivator. The furrow also has to be recreated by the row crop cultivator for subsequent passes. You can do that by mounting a furrower instead of the central sweep in the appropriate inter-rows. Placing the guide wheels onto the cultivator is usually easy, but mounting the furrower on the planter may require substantial shop work. In loams and clay soils, the guidance furrows can be stabilized by running a wheel behind the furrower. An appropriately shaped packing wheel can be constructed by welding two disks together at the rims. In sandy soils, instability of the furrow can limit the usefulness of furrow guidance.

Three general types of electronic guidance systems are currently in use. They are distinguished by the type of sensors that determine the position of the crop rows: 1) wands that sense the crop rows by physically touching them, 2) global positioning systems (GPS) that create an electronic map of the crop rows and 3) machine vision systems that find the crop rows optically. Once the position of the rows is sensed, a computer sends a signal to a steering mechanism that adjusts the position of the cultivator. Steering devices correct the cultivator's position either by 1) shifting it laterally relative to the tractor's three-point hitch, 2) turning it slightly using disk wheels or 3) turning the tractor steering wheel. The last approach results in a longer delay between error and correction on rear mounted machines but allows the driver to watch for jamming and other problems. It is also the only approach that

is well adapted to belly or front mounted machines, although in principle the first approach could work if the implement is attached to the tractor by a laterally sliding carriage.

The oldest electronic guidance systems use a pair of wands that touch a pair of rows. If the cultivator strays to one side, then one of the wands loses contact with its row and a signal is sent to correct the cultivator position. Generally, the wands cannot sense crops like corn and sorghum that are flexible when young until the crop is 5–6 inches tall; beans can be detected at 3–4 inches. Since guidance is most critical when the crop is small, the crop size limitation on wand-based systems restrict their usefulness. In particular, by the time the crop is large enough to sense, many weeds will be too large for control with in-row tools. Some wand systems can guide electronically off of a planter-made furrow when the crop is too small to be sensed. However, if an adequate furrow is available, one might as well guide the cultivator directly from this without the wands. High weed density can also limit the performance of wand guidance systems by causing the wands to lose track of the crop rows.

Real-time kinematic (RTK) GPS can accurately determine the position of a sensor to within less than 0.5 inch and can be used to guide tractors and cultivators. These systems greatly improve GPS accuracy by analyzing the carrier waves from the satellites in reference to a ground station. For cultivator guidance, a map of the crop rows is made during planting. Sensors on the tractor or cultivator then allow a computer to direct the tractor and cultivator through the precise electronic map of the field. If side hills are present on the farm, a dual guidance system, in which the tractor is guided between the crop rows and the cultivator is independently adjusted to the proper position relative to the crop rows, is useful. In theory, RTK GPS can determine position with great accuracy. However, suboptimal satellite positioning, unfavorable weather conditions for signal transmission and the inevitable error in translating a computer position into positioning a large piece of machinery reduce the accuracy of positioning an actual implement to within 1–1.5 inches of the crop row. This level of accuracy is sufficient for high-speed inter-row cultivation with minimal fatigue for the tractor operator but is not sufficiently accurate for use with most in-row cultivating tools without risk of crop damage.

Most machine vision guidance systems use black and white video cameras operating in the visual spectrum coupled with artificial intelligence image-recognition software to distinguish the regular pattern of plants in the crop row from weeds. These systems work best when the crop is taller than the weeds and the weeds have low to moderate density. Performance is markedly better with two camera systems relative to those with a single camera, because a single camera can be fooled by shadows. Good machine vision systems provide enough guidance accuracy to allow use of in-row weeding tools like rubber star wheels at speeds up to 7 mph. For such high-speed weeding, sweeps will usually be the most practical tool for cleaning out the inter-row weeds. The most sophisticated tractor-mounted machine vision systems distinguish individual crop plants from weed plants and cultivate the spaces between crop plants using reciprocating knives, horizontal cultivating disks or rotating tines that move in and out of the crop row.

Machine vision systems that use hyper-spectral analysis to distinguish between hundreds of colors are under development. Coupled with computation-intensive visual recognition software, these systems can distinguish between closely related species like tomatoes and black nightshade. These systems could be very useful in situations where weed density is high or where the weeds have over-topped the crop. Another system under development that could be useful in high weed density situations involves the measurement of x-ray absorption by the crop-weed canopy. Absorption of this short wavelength electromagnetic radiation is usually greatest in the crop row where biomass is greatest, and thus the sensors can find the crop row even when it is over-topped by weeds. Whether the sophistication of hyper-spectral and short wavelength absorption systems has value for routine cultivation remains to be determined.

Rapid strides in artificial intelligence are leading to the development of autonomous, self-driving robot weeders. Multiple models of such machines should be commercially available by the mid-2020s or sooner. Some of these machines simply allow inter-row cultivation without the supervision of an operator. More sophisticated machines can locate individual crop plants and remove in-row weeds from between them. Some current prototypes use a solar panel to allow continuous operation during daylight hours without the need for recharging or refueling. The only robot weeders commercially available as of the writing of this book are small devices that wander about randomly, deflecting off crop plants taller than about 1 inch while cutting weed seedlings with a tiny string trimmer. They are only suitable for small areas, such as a home garden.

MATCHING THE IMPLEMENT TO THE TASK

Effective mechanical weed control typically requires several machines. These need to be appropriate for the type of crop, timing of crop development, tillage practices and type of weed problem. That is, you need to integrate various mechanical weed management practices into a program appropriate for the crop and your farm. Moreover, you need to integrate the mechanical methods with the cultural management methods discussed in Chapter 3.

A mechanical weed management program commonly used by organic corn and soybean growers in the midwestern United States consists of two to three rotary hoeings followed by two cultivations with sweeps or shovels. It is well adapted to both ridge tilled and flat tilled fields. In this system the rotary hoe reduces weed density and delays establishment of the weeds relative to the crop. At the first inter-row cultivation the crop is usually protected from burial by shields, though some soil may be allowed to roll under the shields to bury small weeds. At the second inter-row cultivation, the grower uses the cultivator to throw more soil around the plant bases to bury more weeds. The machines used are simple, robust and pulled at high speeds, allowing rapid cultivation of large fields. The weed control may be less complete than that achieved by more sophisticated devices, but some weeds can be tolerated in competitive field crops.

Well managed mechanical weed control programs in high value vegetable crops vary enormously in detail, but many share common elements. Often the grower uses a short-tilled fallow or stale seedbed to reduce initial weed density. Tine weeding of large seeded crops or flaming of small seeded crops can then be used to further reduce weed density before crop emergence, but surprisingly few growers use these options. After the crop is up, the emphasis is often on frequent cultivation close to the crop row using a basket weeder or vegetable knives. After the crop gets large, most growers cultivate the inter-rows with duck-foot shovels or sweeps. The high value of the crop often makes hand hoeing of weeds in the crop row economically viable. Consequently, the fields may be very clean, and this facilitates weed management and minimizes the cost of hand weeding in subsequent years.

You need to find the right mix of implements to meet the particular situations presented by the soils, climate, crops and weed species on your farm. Given the idiosyncrasies of many cultivation tools, the equipment should also match your personality and your proclivity for adjusting and experimenting with machinery. You will probably need multiple implements to meet the diversity of weeding tasks that you encounter; additional machines may be useful for saving crops in unusual circumstances.

HOEING WEEDS

Hand hoeing is often cost effective for in-row weeding of high value crops, especially if they are poor competitors (figures 4.18, 4.19 and 4.20). Hoeing is easiest when the weeds are still small. For example, Maine farmer and author Eliot Coleman finds that multiple passes hoeing small weeds is less time consuming than hacking out large weeds in a single pass. In addition, hoeing when the weeds are small allows you to hoe

shallowly, which prevents bringing more seeds to the soil surface where they can establish.

Traditional chopping type hoes (figures 4.18 E, F, H, and 4.19 D) tend to dig too deep, damage crop roots and bring up more weed seeds. They also tend to leave an uneven soil surface that exposes more weed seeds to germination cues (see "Seed Germination: Why Tillage Prompts Germination"). Stirrup hoes (Figure 4.18 K) (also called scuffle or shuffle hoes), onion hoes (Figure 4.18 G), sweep hoes (Figure 4.18 N, O) and diamond hoes (Figure 4.18 M) are designed to weed shallowly and move less soil. They can cover more ground with less effort than traditional hoe designs if weeds are small, but they have difficulty digging out large weeds if hoeing is delayed. Sweep hoes and diamond hoes (Figure 4.18 N, O, M) have long handles and a blade angle that allows reaching across a typical vegetable bed so that the whole bed can be weeded from one side without walking on it. They can easily cut off a crop plant, however, if not aimed carefully. Stirrup hoes (Figure 4.18 K), in contrast, can easily be used close to the crop without damaging it since the sides of the stirrup help prevent the crop from being nicked by the blade. The wobble in the blade of a stirrup hoe helps keep the blade horizontal as the

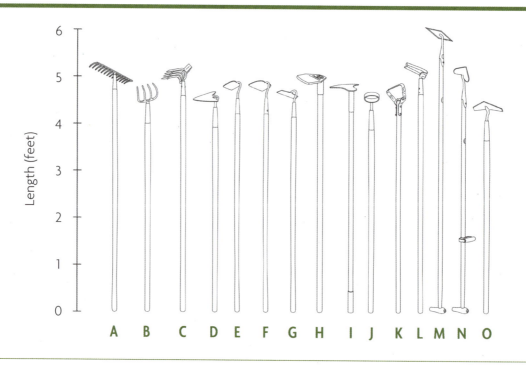

Figure 4.18. Various types of long handled weeding tools. From left to right: (A) garden rake, (B) weeding rake, (C) weeding rake-shuffle hoe combination, (D) furrowing hoe, (E) narrow garden hoe, (F) wide garden hoe, (G) onion hoe, (H) heavy chopping hoe, (I) fine-point hoe, (J) circle hoe, (K) stirrup hoe, (L) another type of weeding rake, (M) diamond hoe, (N) narrow sweep hoe, (O) wide sweep hoe.

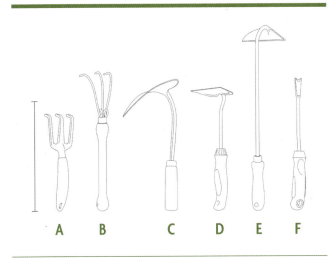

Figure 4.19. Some short handled weeding tools. From left to right: (A) weeding claw, (B) spring tine weeding claw, (C) swan weeder, (D) short handled hoe, (E) draw knife hoe, (F) asparagus knife. Scale is 1 foot.

hoe slides toward and away from the user. The circle hoe (Figure 4.18 J) is similarly effective for precision weeding around delicate crop plants, but its area of coverage is small. Weeding forks, particularly those with spring steel tines (Figure 4.18 B, C) can be used like a precision tine weeder in taprooted crops when they are young. An ordinary garden rake (Figure 4.18 A) can sometimes be used in the same way. Garden rakes are also useful for loosening soil to kill small weeds when planting a small area is delayed, for example, when sowing parts of a bed with different herbs at various times. Handheld, battery powered, variable speed mini-rototillers are now available for power assisted hoeing.

Using short handled tools for more than a few hours is potentially damaging to the ergonomic health of workers and should be avoided. Situations sometimes arise, however, when use of these tools is unavoidable without substantial crop loss. For example, we have used swan neck weeders (Figure 4.19 C) and short handled hoes (Figure 4.19 D) to remove weeds from cabbage after the leaves had become too large and brittle to allow use of stirrup hoes (our preferred tool for this crop). If the farm manager participates in weeding with short handled tools, he or she is likely to take precautions to avoid the need for short handled tools in the future!

Wheeled hoes can be used for inter-row weeding in place of a tractor mounted cultivator on small acreage farms (Figure 4.21). They are several times faster than conventional hoeing. Since hand guidance allows

operation very close to the crop row, they can remove weeds that tractor drawn sweeps and knives cannot get without an expensive guidance system. A wide range of tools can be mounted on wheel hoes, from sweeps, to narrow tines, to stirrup hoes. The double wheel type wheel hoe allows very close, simultaneous cultivation on both sides of the crop. A disadvantage of wheel hoes is that the operator generally walks close to the crop row, creating compaction. Not only can this inhibit crop growth, but it also helps bring on the next flush of weeds and makes subsequent hoeing more difficult. Battery powered wheel hoes are now available that reduce the labor involved in pushing the hoe. If soil in the bed is loose, the battery powered wheel hoe can be guided from the side of the bed, thereby eliminating the compaction problem.

Crops that you expect to hoe should be planted with hoeing in mind. Space the crop 1–3 inches farther apart in the row than the hoe is wide. If you plant several rows to the bed, stagger the middle row relative to the outer rows so that you can hoe in a cross hatched

Figure 4.20. Proper hand positions for using an onion hoe, collinear hoe or stirrup hoe. Illustration by Vic Kulihin.

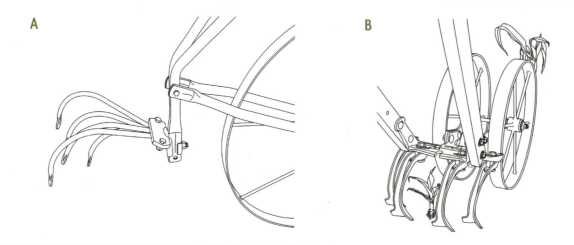

Figure 4.21. (A) Typical wheel hoe with cultivating tines. (B) Two-wheel wheel hoe for cultivation close to the row on both sides of the row simultaneously. Other types of tools including sweeps and stirrup blades can be mounted on either type of implement.

pattern without having to try to straddle the bed. Having hoes of several widths and types helps you match the hoe's action to the crop and situation. Many of the basic principles of mechanical weeding apply to hand hoeing as well.

CULTIVATION AND TILLAGE IN THE DARK

As explained in Chapter 2, light stimulates seed germination in many weed species. During cultivation or tillage, seeds may be exposed to a brief flash of light and then buried again. Consequently, tillage and cultivation at night, or with implements that are covered with light-excluding canopies, can result in a modestly lower weed density. Nevertheless, some seeds, even of generally light sensitive species, do not require light for germination. Others will end up near enough to the surface to satisfy their light requirement regardless of how or when you do the operation. Thus, dark cultivation only reduces but does not eliminate weed emergence. Probably for the same reasons, the degree of weed reduction by dark cultivation has varied greatly across experiments. Reductions in weed density of 20–50% relative to tillage in light are typical, but in some experiments, no reduction was observed.

The optimum strategy for using dark cultivation may be to perform primary tillage in the light, wait for emergence, and then prepare the final seedbed and plant in the dark. So far, no studies have reported on the effectiveness of dark cultivation after planting. A possible strategy for post-planting cultivation would be to perform tine weeding or rotary hoeing in the dark to minimize weeds in the crop row. Early

inter-row cultivation could then be done in the light to help clean out the seed bank, since usually 100% of young weeds in the inter-row can be killed by subsequent operations.

Some species, like the pigweeds (Gallagher and Cardina 1998), require only a very tiny amount of light to stimulate germination. Potentially, moonlight or light reflected from tractor headlights could be sufficient to stimulate germination, and a canopy over the implement is unlikely to provide a sufficient level of darkness. Given that the method is ineffective against species that are not light sensitive or those that are hypersensitive, and that the procedure usually provides a relatively low reduction in weed density, the extra bother of cultivating at night will rarely be worthwhile.

SOIL TILTH AND CULTIVATION

Good soil tilth is critical for weed management. Good tilth helps the cultivator break weed roots free from the soil. It also reduces the chances of knocking over crop plants with clods. Moreover, shallowly working tools like tine weeders is relatively ineffective in cloddy soil because 1) seedlings emerge from greater depths in cloddy soil, and the tools cannot reach them without damaging the crop, 2) when clods are moved, seedlings emerge that otherwise could not reach the soil surface, and 3) seedlings in clods may be rolled around by the cultivator but continue growing if rain or irrigation eventually allows the clods to merge back into the soil. All of these factors argue for practices that improve soil structure, including cover crops, manuring, rotation with sod crops and controlling wheel traffic. They also argue for delaying tillage until soil moisture conditions are appropriate, even if this entails a delay in planting.

Even in soil with good structure, clods will form if the seedbed preparation is inadequate to eliminate them. For many large seeded crops, a coarse seedbed is not detrimental to establishment and may be beneficial in reducing erosion. However, for the reasons mentioned above, a coarse seedbed is rarely advantageous during cultivation. If the seedbed is too fine, however, the probability of wind erosion may increase, and a crust may form when the soil dries after rain.

Four elements are key to obtaining good soils structure:

Avoid Working the Soil When Wet

Tillage or cultivation when the soil is wet smears the soil below the tool, creating a sealed layer that roots and water have difficulty penetrating. The soil chunks that are thrown up by the implement later bake into clods that interfere with cultivation. Moreover, pressure of the implement on wet soil squeezes out pore space, creating a massive structure that will be difficult to work into a seedbed in the future.

Working the soil when it is wet is also a poor practice from a weed management perspective. Soil will not flow properly off of sweeps and tines, and it will gum up spyders and rolling gangs. Wet soil will tend to stick to weed roots, preventing desiccation. Moreover, weeds are more likely to reroot if the soil is wet.

Control Wheel and Foot Traffic

In vegetable systems, you can control wheel traffic by creating permanent beds with paths between the beds. The beds never have to be driven on. You can sow the paths with grass and then mow periodically to prevent weed growth. The sod will compete with weeds between the beds, help support the weight of the tractor and reduce muddy working conditions for the field crew. You will, however, need to regularly edge sod alleys to prevent grass from encroaching on the beds. Raised beds have advantages, including early warming in the spring and less stooping by workers during picking. Weeding the sides of the beds will require equipment that can disturb soil on the sides of the beds while simultaneously pushing it up out of the alley. Rolling cultivators and disk hillers are commonly used for this task. Avoid walking on the beds. Foot traffic can create an amazing amount of compaction,

especially when the beds have been recently tilled.

In grain crops, the traditional way of controlling wheel traffic was to use a ridge till system. The tractor tires always travel in the furrows between the ridges, so that the area where the crops grow never gets compacted by wheel traffic. Maintaining the ridges through a rotation into small grain crops requires planting arrangements that skip planting rows in the furrow. Innovative growers are increasingly using high accuracy GPS systems to maintain the same wheel tracks (see profiles of Scott Park and Carl Pepper). This approach is applicable to complex crop rotations and a wide range of crops. At present, however, the systems are expensive, and this restricts use to larger farms.

Regardless of what system you use, avoid walking and driving on the soil when it is wet. Soil strength declines with increasing moisture content, and compaction from traffic thus increases with increasing soil moisture. Combine harvesting of crops like corn and soybeans is a frequent cause of compaction because the soil often dries slowly in cool autumn weather and combines are heavy machines. If combine harvesting frequently causes ruts in a field, avoid planting late harvested crops there. Shifting these fields to perennial forages or early harvested row- and small-grain crops will decrease compaction and improve mechanical weed management.

Add Organic Matter to Soil

When compost, cover crops and mulch materials from the previous year are incorporated into the soil, the decomposing organic matter and the beneficial fungi that grow on it bind soil particles into small, stable crumbs. The spaces between crumbs allow air and water to easily penetrate the soil and crop roots to grow rapidly. Good crumb structure also facilitates cultivation and formation of dust mulches. The decomposing organic matter provides a source of nutrition for crop growth. Since much of the benefit of organic matter additions to the soil derive from the activity of rapid-acting microbial populations, regular annual additions of organic matter are better than occasional heavy pulses. Cover crops are a particularly valuable source of soil organic matter because they improve the soil while growing as well as after incorporation. The fibrous roots of grasses help bind soil into crumbs, and root secretions from legumes also bind soil into aggregates. Finally, taprooted cover crops like red clover and sweet clover penetrate and loosen the subsoil. Killing the cover crop and leaving it on the surface as a mulch for weed suppression will be more beneficial for surface soil structure than incorporating it, since tillage tends to disrupt aggregation. But either way, the soil benefits from a cover crop.

Keep Soil Covered

Keeping the soil continuously covered with a crop, cover crop or mulch prevents raindrops from breaking up soil crumbs. It also prevents the soil from baking hard in the sun. Moreover, organic materials from cover crops and mulches provide food and cover for earthworms. Earthworms avoid hot, dry soil and are much more active near the surface when the soil is covered by an organic mulch or a dense crop canopy. Earthworms create soil pores with their burrowing and cement soil particles into crumbs with their slime. They also consume many weed seeds.

Keeping the soil covered has additional benefits for weed management. Fewer weed seeds will be prompted to germinate under a cool, dark cover of crop plants or residue, and the weeds that do emerge from the soil will grow poorly in the low light conditions.

ENERGY USE IN PHYSICAL AND CHEMICAL WEED MANAGEMENT

Physical weed management is often assumed to require more energy than chemical management. This is sometimes the case, but depending on the implements and chemicals compared, the two approaches are often surprisingly similar in energy intensity, and sometimes the physical approach compares favorably. Primary tillage is unquestionably energy intensive (Table 4.4), and if perennial weeds are not an issue, then reducing tillage will result in substantial energy savings regardless of the overall weed management strategy. If perennial weeds are a problem, however, then in a chemical no-till system a glyphosate containing product like Roundup PowerMax® will probably be used, whereas in a non-chemical system the field will probably be moldboard plowed and then disked or finished with a harrow before planting. Plowing plus disking uses approximately 2.02 gallons per acre of diesel fuel (Table 4.4). However, control of perennial weeds with Roundup® typically requires 1–3 quarts per acre (Monsanto 2010), and manufacturing the herbicide is an energy intensive process (Table 4.5). Even

assuming a relatively low application rate for perennial weed control of 1 quart per acre, manufacturing the active ingredient (glyphosate) requires the energy equivalent of 1.69 gallons per acre of diesel. With an additional 0.27 gallons per acre of diesel required to spray the herbicide (Table 4.4), chemical control of the perennials in a no-till system requires the equivalent of 1.96 gallons per acre of diesel, which is scarcely different from the mechanical management. If a higher rate of Roundup® or an additional herbicide is needed, then the mechanical management is likely to be the less energy intensive approach. Conversely, if multiple passes with a tillage implement are required, as in a bare fallow, then the mechanical approach may require more energy.

Chemical and mechanical weed management in the crop can also have similar energy requirements. For example, a common cultivation regimen for corn and soybeans is to rotary hoe twice and then cultivate with a row-crop cultivator twice (Mohler et al. 1997). From Table 4.4, this requires approximately 1.38 gallons per acre of diesel fuel. If triazine resistant weeds are not present in the field, a reasonable herbicide program for field corn might be 1 pint per acre of an atrazine product plus 1.33 pints per acre of Dual II Magnum® (Table 4.5). This requires the energy equivalent of 1.22 gallons per acre of diesel for manufacturing the active ingredients in the herbicides plus 0.27 gallons per acre of diesel for spraying, for a total energy usage of 1.47 gallons per acre, slightly

Table 4.4. Energy Requirements for Tillage and Weed Management Machinery, Given in Equivalents of Gallons per Acre of Diesel Fuel Used[1,2]

Tool	Fuel Use
	Gallons per Acre of Diesel
Moldboard plow	1.32
Chisel plow	0.99
Disk	0.7
Field cultivator	0.43
Inter-row cultivator	0.38
Rotary hoe	0.31
Sprayer	0.27
Flame weeder	7.93

[1]All figures are from Clements et al. (1995) except the flame weeder, which is from Ascard (1995a).
[2]Fuel required for use of a harrow or tine weeder is probably similar to that for an inter-row cultivator or rotary hoe.

Table 4.5. Energy Embodied in the Active Ingredients of Various Common Herbicides, Given in Equivalents of Gallons per Acre of Diesel Fuel Used per Application[1]

Product[2]	Active Ingredient	Group	Typical Rate per Acre[3]	Energy per Application
Solve 2,4-D LVE	2,4-D	Phenoxy acid	1 pt	0.13
Banvel	Dicamba	Benzoic acid	4 fl oz	0.11
Fusilade DX	Fluazifop-butyl	Aryloxyphenoxy propionate	8 fl oz	0.43
Dual II Magnum®	Metolachlor	Chloroacetimide	1.33 pt	0.3
Glean	Chlorsulfuron	Sulfonyl urea	0.66 oz	0.01
Aatrex	Atrazine	Triazine	1 pt	0.31
Treflan	Trifluralin	Dinitroaniline	1.5 pt	0.37
Gramoxone	Paraquat	Bipyridylium	3.5 pt	1.33
Roundup PowerMAX®	Glyphosate	Organophosphorus	22 fl oz	1.16
Lorox	Linuron	Substituted urea	1 pt	0.48

[1]Energy embodied in herbicides from Green (1987), converted to diesel fuel equivalents assuming one liter of diesel = 36.4 megajoules of energy (NCCE 2009). The embodied energy in a chemical is the cumulative energy in the form of heat and electricity required to transform and purify simple precursor molecules into the final product through many steps.
[2]The same active ingredient may be present in several commercial products. The herbicides listed are used as examples and do not imply endorsement of any particular product.
[3]Application rates vary with crop, timing and target weeds. The application rates listed are typical of those used for major crops in which the herbicides are commonly used.

more than the standard cultivation practice. The usual management for conventional soybeans is 22 ounces per acre of Roundup®, which requires the equivalent of 1.16 gallons per acre of diesel for manufacturing the glyphosate plus 0.27 gallons per acre to spray the herbicide, for a total energy usage equivalent of 1.43 gallons per acre of diesel. Again, this is slightly greater than the standard cultivation program.

These calculations leave out some energy costs such as the energy required to make additives like wetting agents and to package, transport and market the herbicides. They also leave out the energy costs of manufacturing the cultivators and transporting them from the factory to the farm. All of these energy costs are relatively small on a per acre basis (Green 1987, Clements et al. 1995), however, and would not substantially affect the comparisons above.

Obviously, both herbicides and physical weed management programs vary greatly with crop and region. Tank mixing multiple herbicide products and repeat applications are common for some crops, and some crops may be cultivated many times. Flame weeding has very high energy costs compared with both mechanical and chemical approaches (Table 4.4). In contrast with these energy intensive approaches, the sulfonyl urea herbicides are applied at very low rates and thus have negligible energy costs (e.g., Glean in Table 4.5). But these herbicides are commonly used in conjunction with others to achieve a broader spectrum of control and are more prone than most classes of

herbicide to select for resistance. Consequently, they cannot be considered a panacea for the problem of energy use in weed management. Although particular chemical and physical weed management practices use more energy whereas others use less, neither approach appears generally superior to the other with regard to energy use.

As more weeds become resistant to particular herbicides, however, additional herbicides will need to be added to herbicide programs to obtain full spectrum control. This seems likely to improve the energy use advantage of mechanical weed management programs. This will probably result in a greater integration of herbicide-based systems with the cultural and mechanical systems described in this book.

SUMMARY

- Tillage and post-planting cultivation affect growing weeds by cutting them up, burying them and uprooting them so that they desiccate. They affect seeds in the soil by changing soil properties and the position of seeds in the soil profile.
- The effect of tillage or cultivation on a weed population depends on the interaction between the nature of the soil disturbance and the ecological characteristics of the weed.
- The timing of a tillage or cultivation event relative to season, weather, and the growth stage of the weeds and the crop largely determines the effectiveness of the procedure.

- Mechanical weed management is most effective when it follows a well-considered sequence of events using a variety of implements, with each one appropriate to the size and species of the weeds present and the crop being grown.

- A primary way to control perennial weeds is the repeated removal of the shoots to exhaust the storage roots or rhizomes. The process is most effective if the storage organs are first cut into small pieces and subsequent shoot removal occurs at the point carbohydrate reserves in the storage organs are at a minimum. Other tactics that may be effective against certain perennial species include drying out the storage organs, exposing the storage organs to freezing and physically removing the storage organs.

- Delayed tillage and planting of spring planted crops often reduces subsequent weed density because the early flushes of weeds will be eliminated by the tillage and many species of spring germinating weeds enter secondary dormancy during hot weather. Delayed tillage before planting winter grains may similarly reduce the density of winter annuals in the crop. Delayed tillage and planting before summer planted crops is rarely effective for reducing subsequent weed density.

- Tillage redistributes seeds in the soil profile, which can change the proportion of both seeds that survive and seedlings that emerge. Moldboard plowing buries most seeds on the soil surface too deeply for subsequent emergence. If seed survival is high, moldboard plowing the next year will bring many seeds back to the upper part of the soil profile. Chisel plowing, disking and rotary tillage keep a large proportion of surface seeds in the upper part of the soil profile. A consistent no-till system with good weed control can exhaust the near surface seed bank, leading to low weed pressure. If the field is regularly tilled, then moldboard plowing will tend to have lower weed emergence than other tillage systems, particularly if it is used after seasons with high weed seed production.

- Ridge tillage reduces tillage intensity while achieving adequate weed control. Ridges built the previous year are scraped by an attachment on the planter and rebuilt during inter-row cultivation. Ridge scraping moves seeds shed the previous year into the inter-row, where emerging seedlings are easily controlled by cultivation.

- Tilled fallows can flush seeds out of the seed bank. Integrating tilled fallow with good cover crop management can maintain soil tilth.

- In the stale seedbed technique, the soil is tilled and firmed as for a tilled fallow, but then the weeds are killed with a propane flame or herbicide without further soil disturbance before planting. The absence of soil disturbance reduces the emergence of weeds, giving the crop a strong head start and minimizing expensive hand weeding in high value crops.

- Cultivation for weed management is guided by several simple principles:

  1. The planter and any inter-row cultivator should work the same number of rows.

  2. Cultivation timing becomes increasingly critical the closer to the row that the cultivator operates.

  3. Cultivation is most effective if a size difference between the crop and the weeds is created and maintained.

  4. Over a wide range of weed densities, cultivation kills the same proportion of the weeds present, but at a very high weed density, the proportion of weeds killed declines.

  5. When you intend to use an implement that randomly kills a small proportion of the crop, like a rotary hoe or tine weeder, you should plan ahead by increasing your planting density to compensate.

  6. Good soil drainage and careful timing with regard to changing weather and soil conditions can improve the effectiveness of cultivation.

  7. Effective cultivation requires good soil tilth and careful seedbed preparation.

  8. Using cultivators to create a 1-inch-thick layer of loose crumbs on the soil surface (a "dust mulch") keeps the zone from which most weed seedlings arise too dry for successful weed emergence.

  9. Use cultivators to weed early, shallow and often.

  10. To weed effectively without damaging the crop, cultivators must be adjusted to the conditions at hand. Generally, the importance of adjustment is greater the closer the implement works to the crop row.

- A wide range of implements is available for killing weeds within a crop. Inter-row cultivators can be equipped with sweeps, knives, horizontal disks and spider gangs. PTO powered tines rotating on either horizontal or vertical axes can create intense soil disturbance to kill weeds. Machines that work especially close to the crop row include cultivating disks, spyders, basket weeders and weeders that use plastic brushes. Tine weeders and rotary hoes kill weeds regardless of their position relative to crop rows but are aimed primarily at removing weeds from within the row. Additional devices that kill small weeds in the crop row include torsion weeders, spring hoes, spinners, rubber finger weeders, rubber star wheels and air jet weeders. All of these implements kill weeds through soil disturbance that uproots, breaks or buries weeds.

- Additional weed control implements that do not rely on soil disturbance include mowers, weed pullers, electric discharge weeders, abrasion weeders, and flame and other thermal weeders.

- Each of the many machines available for managing weeds within crops has its particular strengths and weaknesses. Acquiring a complementary set of implements appropriate for the types of weeds, crops, soils and weather conditions on the farm is critical for good weed management.

- Guidance systems ranging from the simple and mechanical to sophisticated computer-based systems are available to facilitate operating cultivation equipment. Good guidance allows higher cultivating speeds with less stress on the tractor operator. Highly precise guidance facilitates the use of in-row cultivating tools.

- Hand hoeing is frequently an economically viable option for high value crops. The long-handled garden hoe is the most widely used type of hoe, but it is neither the fastest nor the most effective tool. Short handled weeding tools provide greater precision, but due to health problems associated with prolonged stooping, they should only be used when other management options are ineffective.

- Good soil tilth helps cultivators shake soil loose from crop roots and decreases crop damage caused by clods thrown into the crop row. Good tilth can be achieved by regular incorporation of crop and cover crop residue, rotation with soil building perennial forage crops, keeping the soil covered with vegetation as much of the time as possible, and controlling wheel and human traffic.

- Although tillage during seedbed preparation is highly energy intensive, energy use during common mechanical post-planting weed management programs is similar to or even lower than the energy use required for manufacturing and applying common herbicide programs. As additional herbicide applications become necessary to control herbicide-resistant weeds, the energy use advantage of mechanical weed management is likely to increase.

REFERENCES

Anbari, S., A. Lundkvist and T. Verwijst. 2011. Sprouting and shoot development of *Sonchus arvensis* in relation to initial root size. *Weed Research* 51: 142–150.

Ascard, J. 1994. Dose-response models for flame weeding in relation to plant size and density. *Weed Research* 34: 377–385.

Ascard, J. 1995a. Thermal weed control by flaming: biological and technical aspects. *Institutionen für lantbruksteknik, (Swedish University of Agricultural Sciences), Alnarp Report* 200: 1–61.

Ascard, J. 1995b. Effects of flame weeding on weed species at different developmental stages. *Weed Research* 35: 397–411.

Buhler, D.D., K.A. Kohler and R.L. Thompson. 2001. Weed seed bank dynamics during a five-year crop rotation. *Weed Technology* 15: 170–176.

Cussans, G.W., S. Raudonius, P. Brain, and S. Cumberworth. 1996. Effects of depth of seed burial and soil aggregate size on seedling emergence of *Alopecurus myosuroides*, *Galium aparine*, *Stellaria media* and wheat. *Weed Research* 36: 133–141.

Clements, D.R., S.E. Weise, R. Brown, D.P. Stonehouse, D.J. Hume and C.J. Swanton. 1995. Energy analysis of tillage and herbicide inputs in alternative weed management systems. *Agriculture, Ecosystems and Environment* 52: 119–128.

Gallagher, R.S. and J. Cardina. 1998. Phytochrome mediated Amaranthus germination II: development of very low fluence sensitivity. *Weed Science* 46: 53–58.

Green, M.B. 1987. Energy in pesticide manufacture, distribution and use. In: *Energy in Plant Nutrition and Pest Control*, eds. B.A. Staut and M.S. Mudahar. pp. 165–177. Elsevier: Amsterdam.

Grubinger, V. and M.J. Else. No date. *Vegetable Farmers and their Weed-Control Machines*. DVD 75 min.

Mirsky, S.B., E.R. Gallandt, D.A. Mortensen, W.S. Curran and D.L. Shumway. 2010. Reducing the germinable weed seed bank with soil disturbance and cover crops. *Weed Research* 50: 341–352.

Mohler, C.L., J.C. Frisch and C.E. McCulloch. 2006. Vertical movement of weed seed surrogates by tillage implements and natural processes. *Soil and Tillage Research* 86: 110–122.

Mohler, C.L., M. Liebman and K. Staver. 2001. Mechanical management of weeds. In Liebman, M., C.L. Mohler and C.P. Staver. 2001. *Ecological Management of Agricultural Weeds*. Cambridge University Press: New York.

Mohler, C.L., J.C. Frisch and J. Mt. Pleasant. 1997. Evaluation of mechanical weed management programs for corn (*Zea mays*). *Weed Technology* 11: 123–131.

Mohler, C.L. 1993. A model of the effects of tillage on emergence of weed seedlings. *Ecological Applications* 3: 53–73.

Mt. Pleasant, J. and R. Burt. 1994. Time of cultivation in corn: effects on weed levels and grain yields. *Proceedings of the Northeastern Weed Science Society* 48: 66.

Monsanto. 2010. Roundup PowerMAX® Label. Monsanto Company: St. Louis, Missouri.

NCCE. 2008. *Conversion Factors for Bioenergy*. North Carolina Cooperative Extension.

OSU. 2005. *Weed 'Em and Reap. Part I. Tools for Non-Chemical Weed Management in Vegetable Cropping Systems*. Oregon State University, Department of Horticulture: Corvallis, OR. DVD 36 min.

Rahkonen, J., and P. Vanhala. 1993. Response of a mixed stand to flame and use of temperature measurements predicting weed control efficiency. *International Conference I.F.O.A.M. Non-Chemical Weed Control Proceedings, Dijon*: 177–181.

CHAPTER 5

Profiles of Farms with Innovative Weed Management Practices

The farms included in this chapter were chosen for their interesting, innovative and integrated approach to weed management. These farms are also innovative in a variety of other ways. Most notably, all achieve a substantial reduction in tillage relative to otherwise similar organic farms.

The farms include both large and small farms, and they represent both field crop and vegetable production systems. The farms span a wide geographical range.

THE MARTENS
Penn Yan, N.Y.
Grain, forage and processing vegetables

Klaas Martens. Photo by Jack Waxman.

Klaas, Mary-Howell and Peter Martens grow a wide range of grain, forage and, in some years, processing vegetable crops on 1,700 acres of certified organic land in the Finger Lakes region of central New York. The Martens are widely acknowledged as leaders in the organic farming community, and their advice is widely sought by growers in New York and beyond. Among the more notable aspects of their farm is the high level of weed control. Their son, Peter, an expert in weed

management in his own right, is playing an increasing role in the management of the farm. In addition to farming, the Martens operate Lakeview Organic Grain, a feed mill and organic seed company that is at the heart of a growing organic grain and dairy industry in central New York. Another son, Daniel, is active in the management of the mill and seed business.

Consistently good weed management is difficult to achieve on a large organic farm in the humid Northeast, where rain in early summer often interferes with timely cultivation. Although cultivation is a key element in the Martens' weed management, other elements of their farming system complement the cultivation and help keep their fields clean. These include a diverse crop mix, sound crop rotation, cover cropping, timing of field operations and careful nutrient management.

The Martens grow a wide variety of crops including food grade soybeans, dry beans, corn, spelt, wheat, triticale, barley, oats and, depending on market conditions, cabbage, snap beans and sweet corn. Recently, they began growing perennial and annual forages for neighboring dairy farms. They regularly experiment with new crops like hull-less oats and edamame soybeans. Their clean fields and careful management allow them to sell much of the harvest as certified seed to other growers through their seed company. The high diversity of crops they grow spreads out the workload, making timely cultivation more feasible. In addition, the wide range of crops provides a variety of opportunities for disrupting weed life cycles. For example, the cabbage is sufficiently high value that they can justify hiring a crew to hoe it, and this provides a level of weed control that is rare on a farm growing only grain crops.

For many years, their basic crop rotation was soybeans, then a small grain overseeded with red clover, followed by a heavy feeding row crop like corn or cabbage. They sometimes departed from this sequence due to weather, market demands and other considerations, but it provided a framework for planning their assignment of crops to fields each year. This rotation helped control weeds. The alternation of winter grains with spring sown row crops helps control both spring and fall germinating weeds. By the time spring germinating weeds begin to establish, the winter grains are growing vigorously and competitively suppress these weeds. The Martens usually overseed the winter grains with red clover while the ground is still frozen. The clover thus establishes early and competes with weeds as the grain begins to dry down and becomes less competitive. After the grain is harvested, they wait until the weeds begin to flower and then mow the weeds and clover at 4–6 inches. By waiting until the weeds start to mature, they reduce the weeds' ability to sprout from below the mowing height while ensuring a high vigor in the clover. Consequently, few weeds produce seeds during a year with winter grain.

By the time the clover is plowed under for planting corn or another heavy feeding crop, it has had a full year to fix nitrogen. As a result, the Martens only need a little organic fertilizer in the seed box to get high corn yields. The soybeans usually do not receive fertilizer, but based on results of a Cornell University organic cropping systems experiment (Caldwell 2016), they have been applying manure before planting winter grains.

The disruption of life cycles created by high crop diversity and crop rotation, coupled with minimal use of soluble fertilizers, has prevented any one weed species from becoming a severe problem on the farm. In 2002 and 2003 several scientists, including the authors, studied most aspects of the Martens' operation, including the weeds in their fields. A list of the three most abundant weeds in each of five key crops (soybeans, corn, spelt, cabbage and snap beans) included 15 species, an unusually diverse community for a farm. Common lambsquarters, quackgrass, ragweed and giant foxtail were among the most commonly encountered species, yet none of these weeds were very productive. Weed dry weight in the field crops rarely

exceeded 300 pounds per acre and was often much less, though a few fields had moderately high weed dry weight in the 500–1,200 pounds per acre range. Weeds in the vegetables never exceeded 100 pounds per acre and were usually less than 10 pounds per acre.

In recent years, the Martens developed innovative cooperative agreements with three small dairy farms in the neighborhood. In this arrangement, the Martens supply all of the forage, grain concentrate and bedding needed by the dairy farms in exchange for a percentage of the milk check. The arrangement gives the Martens an incentive to provide the highest quality feed for the cows and ensures high productivity of the organic dairy herds. The addition of forage crops to the farm has further increased crop diversity and provided new crop rotation opportunities. For example, black turtle beans can be grown after triticale harvested for forage, and brown mid-rib (BMR) sorghum-sudangrass can be grown after a winter barley crop. Double cropping forages with grains not only increases the profitability of the farm, but it also provides new opportunities for interrupting weed life cycles. The Martens avoid growing corn for silage. Klaas believes that the annual forages they grow produce higher quality forage and that, by double cropping them with grains, the profitability is greater. With the increased crop diversity, they now largely avoid growing soybeans after corn, which he says sets the soil up for erosion.

As an additional consequence of the Martens' close relationship with the neighboring dairy farms, they now raise 70 replacement heifers. Raising heifers allows them to keep three fields in a creek floodplain in permanent pasture. Rather than losing soil during flood events, they now capture soil with permanent grass sod. Any new weed species arriving in the floodwater is much less likely to establish in a grazed and trampled pasture than in an arable field. They also graze the heifers on cover crops. This allows them to incorporate the cover crops without mowing first.

Since the Martens take back the manure from the dairy barns, they are able to close the nutrient cycles on the farm in a way that was impossible when they concentrated on cash crops. The grain going to their mill and seed company slowly draws down the excess P and K from an earlier era. This reduces the vigor of many weed species relative to farms that regularly apply high rates of off-farm manure as a nitrogen source.

The great diversity of their crop mix allows them

opportunities to plant a rye cover crop before soybeans. They then roll down the rye with a roller-crimper on the front of the tractor and drill the soybeans into rolled down rye in a single pass. This practice was not feasible when the soybeans followed corn since the corn was harvested too late to plant the rye. The absence of soil disturbance and the temperature-moderating effect of the rye avoids triggering weed seed germination. Their Great Plains drill has a 5-inch spacing between the openers, which leads to rapid canopy closure to suppress any weeds that break through the rye mulch. In the occasional year with a spring drought, the rye can dry out the soil enough to prevent good soybean emergence. In dry springs, they cut the rye for silage to avoid this problem.

Although the Martens use a multi-tactic approach to weed management, their innovative mechanical methods distinguish them from other growers in the Northeast. Early season in-row weed management is critical for successful weed control in field crops, and Klaas and Peter have become experts in the art of tine weeding. They believe that different situations require different machines. For soybeans and beans, which have a deep taproot, they often use a Lely, because the close spacing of the 85-degree tines allows aggressive uprooting of weeds, but the flexibility of the thin Lely tines allows them to bend around the relatively brittle beans. After dissatisfaction with the performance of the Lely and a straight tine Kovar in corn, Klaas asked the Kovar Co. to make him a weeder with a 45-degree tine. The shallower bend avoids uprooting of young corn but is more aggressive than the straight tine. The 45-degree Kovar weeder is now marketed nationally. When soil is relatively wet, the Martens shift to the straight tine Kovar. Klaas points out that uprooted weeds do not dry out when the soil is damp, so the objective then is to bury small seedlings. The straight tine more effectively throws a wake of soil than a bent tine. They also bought a set of Einboch gangs that they have adapted to mount on a Kovar frame. Multiple machines and careful set up adjustments optimize tine weeding in a variety of crops and soil conditions.

The Martens have a good farm shop with substantial capacity for fabrication. And they use it. Among their more important creations are several 6-row belly-mounted cultivators they have fabricated out of older machines. These were the mainstay of their row-crop cultivation for many years. The belly mounted machines carry two gang bars per row, with each gang pulling just a single half sweep next to the row. This configuration allows each sweep to have its own gauge wheel, so if a gauge wheel goes over a rock, only one sweep comes up out of the ground. More importantly, however, the belly mounted cultivators allow the Martens to cultivate close to young crops early in the season. Such exact cultivation would be impossible with a rear-mounted cultivator alone. To rip out weeds in the inter-row area, however, they do pull a three-point hitch mounted cultivator with heavy C-shanks. Klaas prefers C-shanks to S-shanks because he says weeds can slip between the shovels on the more flexible S-shanks.

Although a lot of the Martens' soybeans are now no-till planted into rolled down rye, they still cultivate many row crops. After poor satisfaction with RTK GPS guidance for cultivation, they purchased a machine vision system for a 6-row rear mounted cultivator. The machine vision system has two cameras that focus on different rows. Klaas says that this arrangement prevents a single camera from becoming confused by shadows, weedy patches and thin places in a row. The high accuracy of the machine vision system allows Klaas and Peter to run Kress rubber star wheels in the crop row. Klaas says that this often allows them to skip the final tine weeding. Although they still use the belly mounted cultivators, Klaas says that the machine vision system allows them to cultivate four times faster and requires less skill to operate.

Although they do not currently use RTK GPS when cultivating, they find it very useful for planting accurately. Much of their planting is done with a 240 horsepower, GPS guided Fendt tractor. They have fitted an in-line disk and power harrow with three-point hitches, which carry either their Great Plains drill or Case IH planter units, depending on the crop being planted. To balance the weight, they feed the drill or planter from Amazone air-seeder boxes on the front of the tractor. The tractor is sufficiently heavy and powerful to pick up both the tillage implement and drill or planter. This system allows them to till and plant in a single pass. Although the tractor is heavy, they reduce potential compaction using dual rear wheels and lowering the tire pressure to 13 pounds per square inch. More critically, their single pass tillage and planting system avoids the compaction associated with driving over tilled ground for secondary tillage and planting.

GPS guidance during planting allows precise placement of successive passes, which facilitates subsequent good cultivation of the guess rows. GPS guidance also allows precise planting along strip edges. Before they had the GPS guidance, the strip edges tended to become weedy because the edge row would get off into the adjacent strip. That problem has now been solved. But Klaas warns that the tradeoff for the accuracy of GPS and machine vision is the considerable effort of programming computers and maintaining the effectiveness of sensors.

In the rare event that an unacceptable number of weeds survive cultivation and break through the canopy of a soybean or bean crop, the Martens use a homemade "weed-topper" to cut off the flowering stalks before they can set seed. This tool consists of a hydraulically adjustable front-mounted toolbar equipped with rotary mower blades. Each blade is driven by a separate hydraulic motor, and alternation of shorter and longer shafts allows overlap of adjacent blades. The weed topper reduces competition with the crop, debris in the harvested beans and the number of weeds that have to be managed in the succeeding crops.

A critical component in the Martens' weed management program is vigorous crops. They maintain and adjust their planters to get solid stands without skips. This denies weeds gaps where they could proliferate. They also usually delay corn planting until the second half of May. This allows more time for clover to grow. In addition, the warm soil of late spring reduces seed rot of their untreated organic seed and promotes quick emergence and fast early growth. This allows a good head start on the weeds and maximizes the effectiveness of their mechanical management.

The Martens' feed mill and seed business, Lakeview Organic Grain, is a key part of their farm operation. It provides a market for the wide range of crops they produce, thereby contributing to the high diversity of their cropping system. As an independent business, Lakeview now dwarfs the farm in sales and its grain purchases provide a stable market for organic grain growers across New York State. Lakeview ships the feed it produces in bulk throughout most of New York and much of Pennsylvania, and they ship palleted bag feed throughout the Northeast. The seed wing of the business provides growers with an extensive range of certified organic and untreated conventional grain, forage and cover crop seed. They also sell animal health care products. The Martens feel they have a mission to foster organic agriculture in their region. As a consequence, they try hard to maintain relatively stable prices both for their customers and for the farmers they buy from.

Through a variety of good farm management practices, coupled with careful attention to the details of cultivation, the Martens achieve a level of weed control that makes their advice sought after by farmers across the continent. Their enthusiasm for organic farming and generosity of spirit leads them to answer dozens of questions a week from farmers and agricultural professionals. The Lakeview newsletter includes not just the usual company price list but also provides many pages of useful information about feeding practices, animal health, the agronomy of various crops and the results of the many experiments they conduct on their farm. They invite people to call and ask questions. Working while talking on a cell phone has become a way of life. They were also the guiding force behind the creation of New York Certified Organic (NYCO), a large group of growers that meets three times each winter to hear guest speakers and discuss a wide range of farming issues. When a grower is facing a particularly difficult problem, one of the Martens frequently has a helpful suggestion.

PAUL MUGGE
Sutherland, Iowa
Soybeans, corn and small grains

Photo by Gene Lucht, *Iowa Farmer Today.*

Paul Mugge grows 300 acres of certified organic soybeans, corn and small grains in northwest Iowa. After decades following low-spray practices and relying largely on organic fertility sources, he decided to take the farm organic in the late 1990s. His first acreage was certified in 2000, and the whole farm has been certified since 2002. When he decided to get out of pork production in 2007, he expanded his acreage from the 300-acre home farm and began organically custom cropping an additional 500 acres owned by a local egg producer. Mugge discontinued operating that farm a few years later to concentrate on his home farm.

Mugge rotates tillage practices as well as crops. He believes that changing between spring and fall planted crops and between row- and solid-seeded crops helps him suppress weeds. His basic crop rotation is corn-soybeans-small grain-corn. The corn and soybeans are planted on 30-inch centers, which allows effective inter-row cultivation, and the fall planted cereals suppress spring germinating weeds. The small grain slot in the rotation is often fall sown triticale or wheat. He also grows spring wheat, oats and barley. He overseeds 12–14 pounds per acre of medium red clover onto his winter grains with a spinner as early as possible in the spring. He tries to broadcast the clover while the ground is still frozen, but as often as not the weather takes his fields directly from snow to mud. He says that even sowing later when the ground is drying consistently produces a good stand. The spring grains are also seeded with clover. For the spring crops, he puts the clover in the grass seed box of his John Deere drill and plants the grain and clover in a single pass.

Although his crop rotation is not unusual among organic grain growers in the northern half of the United States, the tillage methods Mugge uses within this rotation are highly innovative. He had trouble killing the red clover varieties available in Iowa in the spring before corn planting without a moldboard plow. So, he began planting a southern clover variety, "Cherokee," which he later replaced with "Southern Belle," a variety that has better seed availability. These southern clover varieties weaken during the cold Iowa winters. With the clover winter damaged, he found he could incorporate it with just a disk and field cultivator.

The backbone of Mugge's tillage rotation, however, is ridge-tillage planting of soybeans. He begins the year before by building ridges in the corn with the hilling disks on his Buffalo single sweep cultivator. This not only buries weeds in the corn row but also provides a 6-inch ridge for the following year. He added spacers to the front and rear wheels on his combine so that the wheels travel in the valleys between ridges. The 26-inch-wide tires on the combine nip the edges of the ridges a bit, but for the most part the ridges are

well preserved. Mugge believes that ridge-tillage has reduced grass weeds on his farm.

The fall grain is no-till drilled immediately following the soybean harvest. The no-till drill is adjusted so that the height of the leading coulters matches the height of the ridges left behind from cultivating the soybeans. In this way the seeding depth of the grain is uniform across the width of the drill, and the ridges and residue cover are maintained. The grain is overseeded in the spring with the red clover. Corn follows the small grain. The clover is incorporated with disks and a field cultivator, which prepares a seedbed for corn. This tillage rotation has allowed Mugge to move completely away from high-draft chisel and moldboard plowing. He finds the ridge tillage soybean system particularly beneficial. "Ridge till saves a lot of fuel and a lot of time, and it's good for soil conservation," Mugge says. "On the contours it's like a mini-terrace every 30 inches."

All Mugge's implements are 30 feet wide, with the planters and row crop cultivators set up for 12 rows on 30-inch spacing. In addition to his Buffalo planter and cultivator, he has a White planter that he uses for corn and an International cultivator set up with three shanks per row. He has a Yetter rotary hoe with staggered, self-cleaning wheel arrangement that allows it to be used in high residue conditions. He also has an Einbock tine weeder that he uses primarily in corn.

Mugge usually uses his Einbock tine weeder two to three times in corn, both before and after emergence. If the soil crusts, he uses the rotary hoe instead of the tine weeder. His first inter-row cultivation is with the International, which throws some soil off the front sweeps into the row to bury small weed seedlings. At the second and last cultivation, he switches to the Buffalo to really hill up around the corn to bury even some larger weeds and create the ridge for next year's soybeans. Mugge has found that staggering the disk hillers on the Buffalo by 2 feet really improves weed control in the row. The forward disk knocks the seedlings over, and then the rear disk covers them up.

In the soybeans, Mugge rotary hoes once and, if possible, twice before emergence at three- to four-day intervals. Then after the soybeans are up, he rotary hoes again. In contrast with the corn, his first two cultivations in the soybeans are with the Buffalo cultivator because it is better able to handle the heavy corn residue that is in the inter-row areas. To make the corn residue easier to handle, he has installed Calmer stalk rolls on the combine corn head that chop up the stalks. At the final cultivation, the residue has broken down, and he uses the International because it can cultivate closer to the crop row. Mugge uses RTK GPS guidance to steer the tractor between the crop rows. He also has cultivating mirrors on both sides of his tractor that allow him to see the position of the cultivator well enough that he can cultivate within one inch of the crop row. If the cultivator drifts out of optimal position, he manually overrides the GPS guidance to re-center the sweeps on the row. Mugge warns, however, that if the wind is from behind, dust raised by the cultivator can obscure the mirrors.

Mugge has a flame weeder that he uses on both corn and soybeans if some extra weed control is needed. He flames the soybeans at the cotyledon stage. The soybeans get a little singed but recover. He flames the corn when it is 18–30 inches tall. By this stage, the leaves are above the flames and the stalks resist the heat. This treatment kills many of the weeds that escaped cultivation and reduces both competition with the corn and production of weed seeds that could infest subsequent crops. He says that flame weeding has a lot of potential, but learning to do it well is difficult.

Mugge does not tine weed or rotary hoe his small grain crops because that would disturb the clover. Annual weeds rarely get a chance to go to seed in his grain fields, however, because weeds that bounce back after combining get mowed again in August. The rapid regrowth of the clover suppresses them after that.

Mugge teaches physics and advanced mathematics courses part-time at the local high school. "This keeps my mind active," he says. Teaching also lets him get to know a lot of high school students, some of whom he hires in the summer to pull weeds in his soybeans. "Back before herbicides, most farmers used to walk their beans, but people have gotten away from that now." Mugge's soybeans are all high protein, food grade beans. Keeping the fields free of weeds reduces staining of the beans during harvest and allows him to sell a premium product to highly discriminating customers. The hand weeding also leads to very low weed seed production during the soybean phase of the crop rotation. This leads to a small weed seed bank on the soil surface, which makes a successful no-till planted grain possible.

Mugge's training as an engineer and his interest in science led him to regularly experiment with new farming practices. He has been participating in on-farm trials with Practical Farmers of Iowa for more than 25 years. These experiments are frequently well-designed trials that are suitable for the statistical analyses he teaches his students. In 2010 he collaborated with Iowa State University plant breeders on a trial of an aphid resistant soybean variety. "It wasn't a bad year for aphids, so it wasn't a good test," Mugge says, "but the new variety hardly had any aphids, whereas the strain without the aphid resistant genes had around 100 per plant." All his soybeans now have the two-gene resistance trait, and much of his crop is sold as seed to other growers.

The increase in extreme weather events in recent years has been a challenge for farmers across the country, and Iowan farmers have suffered more than most. Mugge has been more fortunate than some with flood prone land. But he notes that the 5 inches of rain that he got in early October 2019 made harvesting high-quality soybeans for seed difficult. No farmer can make the weather cooperate with his plans, but Mugge's innovative methods and continuing willingness to tinker with his systems increase the resiliency of his operation.

**ERIC AND
ANNE NORDELL**
Beech Grove Farm
Trout Run, Penn.
*Fresh market vegetables
and herbs*

Photo courtesy the Nordells.

Eric and Anne Nordell summarize their weed management philosophy as "weed the soil, not the crop." When they bought their 90-acre farm in the rolling hills of northern Pennsylvania in 1982, they made several decisions: They would farm with horses; they would farm organically; and they would avoid hiring laborers. They quickly realized that avoiding hired labor meant that they needed to get the weeds under control at the outset and keep them under control permanently. Otherwise, they would either work themselves to exhaustion or have to hire in help.

Since the ground where they planned to grow vegetables had been in hay for many years, the field was thick with quackgrass and other perennials. So, they planted high value herbs in a half-acre field and fallowed the six-acre area where they planned to grow vegetables. Plowing followed by repeated harrowing exhausted and dried out the quackgrass rhizomes while flushing out and killing a lot of annual weed seeds. To compensate for a potential decline in soil quality due to the fallowing, and to smother out remaining quackgrass, they planted a thick cover crop of rye early that fall.

The success of the fallowing/cover crop treatment led to development of a weed suppressive crop rotation scheme in which cash crops alternate with years of cover crops and fallow. "Instead of relying on the cultivator or the hoe to save the crop from the weeds," they say, "we use cultural practices, including cover cropping, bare fallow periods, rotation and shallow tillage to reduce the overall weed pressure in the soil" (Nordell and Nordell 2007). This approach to weed management has reduced weed pressure on the farm to extraordinarily low levels. One of the authors crawled down 40 feet of un-hoed lettuce about ready for harvest in 2003 and only found a single common chickweed and one volunteer rye plant, each about 2 inches tall. The weeds are so well controlled that the Nordells now commonly plant onions without tillage, a practice that would be unthinkable on most vegetable farms.

The key to their weed management program is alternating cash crops with bare fallows and weed suppressive cover crops in a regular but flexible four-year crop rotation (Figure 5.1). They have divided their vegetable field into 12 half-acre strips. Thus, in any given year, they plant each of the four sets of crop categories in the rotation in three field sections (Figure 5.2). Each year, the crop categories move to the right in Figure 5.2 by one section. The rotation is deliberate and systematic, but it is also highly flexible. They can adjust the species and amount of various crops within any of the three early sections to respond to weather conditions and market opportunities. Similarly, they can adjust the late crops as necessary.

Following a bare fallow in the summer of year one, they plant a rye/hairy vetch cover crop. The absence of a cash crop in year 1 allows them to plant the vetch in

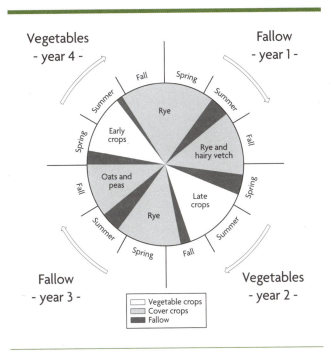

Figure 5.1. The crop rotation clock at Beech Grove Farm.

mid-August for optimum growth before winter. Good growth in the fall means that they can mow and plow under the rye/hairy vetch cover crop early in year two and still build soil structure. For early summer crops, they incorporate the cover crop in April and use some compost for supplementary nitrogen. For later planted crops they incorporate the cover crops in May after the vetch has fixed substantial nitrogen. Although the surface soil is inverted with a moldboard plow, when incorporating the rye and vetch in early spring, they disturb only about the top 3 inches, an operation they call "skim plowing." They use disks to incorporate the heavier cover crop biomass in late spring. They then fallow the field for about six weeks before planting their late planted crops. This fallow period provides an opportunity to flush out cool-season-germinating weeds like common chickweed. A wide range of crops can go into the late crop slot, from squash or a summer planting of lettuce to fall harvested broccoli. After the late crops are harvested, they till shallowly and plant rye. If they plan to harvest the crop late in the fall, they interplant a row of rye between the crop rows after the last cultivation. This does not compete with the crop but tillers and eventually sprawls out to provide good soil cover over the winter.

Year three of the rotation is devoted to fallow and cover crops. The Nordells mow the rye two to three times in the spring before it begins to head. This

Rotation direction →

Field	1	2	3	4	5	6	7	8	9	10	11	12
Initial cover crop	Rye›	Rye & hairy vetch›	Rye›	Dead oats & peas› fallow›	Rye›	Rye & hairy vetch›	Rye›	Dead oats & peas› fallow›	Rye›	Rye & hairy vetch›	Rye›	Dead oats & peas› fallow›
Year type	Fallow›	(Fallow)› late cash crops	Fallow›	Early cash crops› fallow›	Fallow›	(Fallow)› late cash crops	Fallow›	Early cash crops› fallow›	Fallow›	(Fallow)› late cash crops	Fallow›	Early cash crops› fallow›
Final cover crop	Rye & hairy vetch	Rye	Oats & field peas	Rye	Rye & hairy vetch	Rye	Oats & field peas	Rye	Rye & hairy vetch	Rye	Oats & field peas	Rye
Tillage depth	Deep	Shallow	Deep	Shallow	Deep	Shallow	Deep	Shallow	Deep	Shallow	Deep	Shallow
Example cash crops		Potatoes		Lettuce peas spinach		Squash celery kale		Onions		Fall cole crops lettuce spinach		Herbs flowers

Figure 5.2. Field layout showing cash crops, cover crops, fallow periods and tillage depth for incorporation of the cover crops in the spring for the 12 half-acre field sections at Beech Grove Farm. Crops move to the right one section each year. The cash crops in any given position vary flexibly, and the specific crops indicated are only typical examples. Fallow in parentheses indicates that the fallow period may or may not take place depending on the timing of the planting of the subsequent cash crop.

promotes rapid regrowth and prolongs the growth of the rye well into June. In late June or early July, they moldboard plow in the rye. They then shallowly till with a spring tooth harrow if the residue is light or with a field cultivator if the residue is heavy, at about 10- to 14-day intervals until the time comes to plant the fall cover crop. In the first year the Nordells farmed the field, they maintained the fallow for three months. As the weeds declined, they have reduced the summer fallow period to as little as three weeks. When they were still fighting quackgrass, they used a spring tooth harrow to work the rhizomes to the surface and shake the soil free from the roots. In later years they have often pulled a cultipacker behind the spring tooth harrow or field cultivator. This combination produces a firm seedbed that promotes germination of weed seeds so they can be eliminated with the next cultivation. After the fallow period, the farm broadcasts and works in oats and field peas in early August. If done poorly, broadcast seeding can produce a patchy stand. In the hands of the Nordells, however, broadcast seeding produces a more uniformly spaced stand than drilling could, and consequently the cover crop canopy quickly closes to smother late emerging weeds. Increasingly

they now create ridges with disk hillers after broadcasting the oats and peas, in preparation for ridge planted crops the next year. The oats and peas produce a dense, lush growth before being killed by frost in late fall.

They grow early planted crops in year 4 of the rotation. These include peas, spinach and early lettuce but especially onions, which are one of the farm's specialties. If the field section was ridged the year before, they scrape off the tops of the ridges to create narrow bands of bare soil for planting. The raised ridge warms quickly and promotes rapid early growth for the early planted crops. They disk lightly field sections that are planted "on the flat" to break up the oat and pea residue, and they then undercut any weeds with broad, widely spaced sweeps on a field cultivator. This shallow tillage avoids bringing up weed seeds into the surface zone that was cleaned by the fallowing the previous year. They follow the early planted crops with rye or a mixture of rye and spelt. The winter grain cover crop is mowed two or three times the following spring (the beginning of year one of the rotation). They eventually incorporate the cover crop by plowing and then begin the fallow period.

The Nordells adjust both the length and timing of the fallow as necessary to arrest potential weed problems before they develop. If weeds are increasing, they will lengthen the fallow period. In contrast, if weeds are virtually absent from a strip, the fallow will be short and cover crops will be left on the field for more of the season. They shift the fallow period from summer to spring if cool season annuals appear to be increasing.

Although the Nordells cultivate both early and late planted vegetables, they emphasize that this is largely to conserve soil moisture. The cultivation creates a shallow "dust mulch" that interrupts capillary flow of water to the soil surface. Also, they usually stop cultivating two to six weeks after planting to interseed a single row of rye or hairy vetch between the rows. Often, no cultivation for weed control would be needed to achieve high yields.

Eliminating any weeds that do emerge is important, however, to prevent seed production. If necessary, the Nordells hand rogue out any weeds that threaten to produce seeds. The weeds are so well controlled by other means that the time required for hand weeding is minimal, but the Nordells are very consistent in preventing weeds from going to seed. Their system depends on this consistency. The fallow periods eliminate a large percentage of the surface seed bank. If many seeds were allowed to enter the soil during vegetable years, their weed problems following the fallow year, though reduced, would still require much cultivation and hoeing in later crops.

Note that the Nordells do not use a nitrogen fixing cover crop like hairy vetch before the long summer fallow periods (Figure 5.1). No crops are present during the fallow to take up the nitrogen. In addition to creating potential environmental problems, excess nitrogen would tend to promote weeds. Instead, they strategically position the nitrogen fixing cover crops prior to the vegetable crops that can use the nitrogen. By supplying most of their crops' nitrogen requirements with legume cover crops, the Nordells avoid the need for large amounts of compost. Low compost rates have prevented the build up of excessive levels of phosphorus and potassium that would promote weeds like purslane and hairy galinsoga. Research at Cornell has shown that the Nordells' farming system maintains a balance between inputs and exports of N, P and K.

The Nordells' whole-farm approach to weed control has led to an elegantly integrated management system. The economics of the farm operation is supported by careful attention to the biological life of the farm. For example, the four-year rotation sequence (figures 5.1 and 5.2) makes long lags before replanting a particular crop family relatively easy. This prevents build up of soilborne diseases. The potentially destructive impact of the bare fallow periods is balanced by large and repeated inputs of soil building cover crops, and by reduced tillage during the vegetable years of the cycle. Despite the fallow periods, the Nordells' soil scores very well on various measures of soil health (Gugino et al. 2007). Their approach to farming provides a whole new meaning to the phrase "feed and weed."

Photo courtesy Scott Park.

SCOTT PARK

Park Farming Organics
Meridian, Calif.
Tomatoes and other vegetables, vegetable seed, rice and other grain crops

Scott Park grows 900 acres of vegetables and 800 acres of grain in the Central Valley of California near Sacramento. He has been farming since 1973 and farming organically since 1986. In that time, he has developed an integrated but flexible strategy for dealing with weeds. "Weeds are my major battle," he says. "They are 10 times more of a problem than nutrients, water management and other production issues."

With an operation his size, Park needs a lot of help. He has his son, Brian, and 15 full-time employees. "Everybody knows how to do all the different tasks. We don't specialize," he says. That allows everyone to keep busy all year around. Park's wife Ulla handles the bookkeeping, which is fully computerized. "We can track each of our 27 fields as if it were a separate enterprise," he says. He says that the careful recordkeeping really helps with organic certification; it also helps them evaluate profitability of farm practices over multiple years.

Park approaches weed management from a long-term perspective. "A lot of crop rotation decisions are made for weed management rather than for what the net profit will be this year." Developing crop rotations that effectively suppress weeds is facilitated by the diversity of crops produced on the farm. Processing tomatoes are his biggest money maker and typically cover about 20% of his acreage. In addition, he produces substantial acreage of lettuce, peas, dry beans and edible sunflower seed. He also typically grows more than 200 acres of vegetable crops for seed, including various cucurbits, brassicas, lettuce, herbs and others. The vegetable crops are balanced in the rotation by rice, wheat and field corn. In fields where he can grow rice, a typical rotation would be rice-dry beans or peas-winter wheat-tomatoes-rice. Flooding the rice is good at suppressing field bindweed and johnsongrass, which Park rates at the top of his "misery index" of weeds. Other major weeds on the farm include pigweeds, common lambsquarters and watergrass (barnyardgrass). Since the rice depletes nitrogen and produces a lot of residue that ties up the remaining N, Park usually follows rice with a nitrogen fixing legume crop. Wheat comes next to prepare the field for tomatoes. The wheat is harvested in June, which leaves plenty of time for a tilled fallow to flush weeds out of the soil. He will typically irrigate after wheat harvest to bring the weeds up and then tills shallowly through the summer to kill successive flushes. If the field bindweed is still bad at wheat harvest, he may even flood the field for a few weeks before beginning the fallow. In the fall, he plants a legume cover crop of purple vetch, Magnus peas or bell beans in a mixture with a small amount of wheat or other grain. The cover crop helps suppress weeds through the rainy winter months and supplies some N for the following tomato crop, though he also spreads chicken litter before planting the heavy feeding tomatoes. Some fields have too much slope for rice or are adjacent to neighboring walnut orchards that cannot tolerate flooding. On these fields, a typical rotation might be wheat/cover crop–tomatoes/cover crop–vegetable–winter wheat, or corn–cover crop–tomatoes. In this sequence, winter wheat works well after a summer harvested vegetable, whereas corn works well after a fall harvested one. Although the crop sequences are typical of the sort Park uses, he emphasizes that his choice of crop depends on the weed pressure in a field. He plans ahead one to two years so that weeds will be manageable in each succeeding crop.

In addition to flooding for control of perennial weeds, the rotation sequences Park uses suppress weeds in several important ways. First, the alternation between spring, summer and fall planted crops and cover crops interrupts weed life cycles so that the same weed species rarely can prosper for two years in a row. Second, drilled rice and wheat make dense canopies, which tend to smother out weeds, whereas other species are planted in wider rows and can be cultivated. Thus, the basic weed management strategy varies from one crop in the sequence to the next, and this also helps prevent the buildup of particular species. Finally, Park tries to plan his crop sequences so that the ground stays covered except when he is using a tilled fallow to deplete the seed bank. When growing tomatoes, he aims to have the plants completely cover the beds to provide competition against weeds and protect the soil. With crops or cover crops continuously present on the fields, either desirable plants are there to compete with the weeds or a high value vegetable crop is present to provide an incentive for attacking the weeds with cultivators and hoes.

Building soil quality is a critical part of Scott's farming practice. The farm's silt loam and clay loam soils could become massive and difficult to work if not carefully managed. The entire farm is laid out in permanent beds on 60-inch centers. In much of the country, growing grain crops in permanent beds would seem peculiar, but most of Park's fields are furrow irrigated, and the furrows between the beds provide a route for water to get to the crops, whether they are vegetables or grains. The furrows provide permanent drive tracks for the tractors, so that the soil in the beds is never compacted. This, coupled with high residue inputs from the grain and cover crops, gives his soil excellent aggregation. The soil has good waterholding capacity, and this really helps with weed control. "In the spring, we transplant the crop, and its roots are moist," Park says. "But the soil surface dries out so that the weeds don't germinate. We may not need to irrigate for 40 days. If the soil structure is bad, then you have to irrigate right away, and this brings on the weeds." In rice production, Park pre-irrigates, then scratches the soil and drills the rice seed into moisture 2.5 inches deep. The light tillage eliminates the weeds germinated from the irrigation, plus dries out the top 2 inches to curtail further weed germination. The rice can grow 6 inches tall before permanent flood is established. Waiting to start permanent flood also provides an opportunity to scratch out some weeds that come after planting.

To manage tillage and cultivation on multiple beds at a time, Park and his staff have purchased and built a wide range of specialized equipment. First, the beds are laid out exactly with a GPS system. This means that every unit of a multi-bed machine is squarely centered on the bed. "The GPS steers the tractor," Park says. "The tractor operator is there to turn the tractor around at the end of the field and in case something goes wrong." Key implements include ground driven bed mulchers, furrow rippers and Lilliston rolling cultivators. Much of the tillage is shallow, and Park commonly uses row crop cultivators as tillage tools by swapping around knives and sweeps. For example, before wheat, he incorporates chicken litter with a Lilliston rolling cultivator equipped with 46-inch sweeps in addition to the spider gangs. Then he just scratches the soil surface with a Lilliston before planting. After harvest, he disks in the wheat residue. Deeper tillage with a chisel plow usually occurs on a given field only once every couple of years, usually before a high value vegetable crop.

Park Farming Organics has 20 different cultivators, many of which they built themselves. These are set up on sleds, which provide precise depth control on fields that have been leveled for furrow irrigation. They also make their own cultivating knives, which allows them to cheaply replace worn out knives. Since they sharpen their cultivating knives to cut weeds more effectively, knives need to be replaced regularly. Making their own knives also allows them to customize cultivators for any row spacing. This allows them to cultivate effectively while using the space on the beds optimally. "We have cultivators set up to cultivate one row, two rows, up to 10 rows per bed depending on the crop," Park says. Many of the knives are made with a pipe welded to the back side. This ensures that the weeds are not just cut off below ground but are lifted so that they dry out and do not reroot. Commonly a cultivator will be set up with a standard blade in front followed by a blade with a pipe behind. Soil is sometimes deliberately thrown into the crop row to bury small weeds if the crop can tolerate it. But an advantage of the special blades is that they provide good disturbance of the weeds with little of the lateral movement of the soil that one gets with a steeper pitched sweep. This is critical when cultivating close to small vegetable plants. And they do cultivate close! "With the GPS steering the tractor, we can cultivate to within 1.5 inches of the row," Park says.

They also have flame weeders set up for both in-row flaming of crops and flaming of beds to create stale seedbeds. "We don't use the flame weeders a lot," Park notes, "but they are useful for some crops and situations." Over the years he has tried other weed control methods but found them unsuitable for his farm. "All our crops are harvested mechanically, and plastic mulch interferes with the mechanical harvesters," he says. "It is also too much work to dispose of after harvest." In conjunction with University of California researchers, he also experimented with natural product herbicides, including high concentration vinegar, but found these too expensive for use on a commercial scale.

In addition to precision cultivation, Park hires crews to hoe the vegetable crops. The tomatoes get hoed one to three times depending on the weed pressure. He times hoeing relative to irrigation so that the surface soil is dry. That way weeds do not germinate after the soil is moved by the hoes. The crews use conventional, long-handled, 4-inch-wide garden hoes for the work. That kind of hoe moves more soil around than he would like. "But that is the kind of hoe they prefer," he says. "I guess it's what they are used to."

All of the crops are harvested mechanically, and this kills most of the weeds present at harvest. Fields are cleaned up and planted with a cover crop or another crop quickly after harvest to prevent weed regrowth and seed set. Park also takes measures to prevent weeds from setting seeds in the crop. He once burned a wheat field rather than let an infestation of canary grass go to seed. That was before his management methods had been refined, and in any case, field burning is no longer allowed. He has equipment set up so that he can drive over the tomato beds and mow off the flowering stalks of weeds that poke out above the crop. He will also do an extra hoeing to prevent weeds from setting seed if necessary.

Park believes in building strong relationships. With such a diversity of crops he needs to work with several different buyers, but he has built up a history of trust with each of them. "I have been working with most of these companies for decades," he says. "We don't plant unless I have some parameters on what we can get for the crop, but often the contract doesn't get signed until after delivery. I know they will buy what they agreed to, and they know I will come through with the goods. In 46 years, I've only had to pay a lawyer for four hours of work."

Whether it is marketing or weed management, avoiding problems is a big part of Park's strategy for successful farming. "I am out in the fields every day," he says, "and I keep my eyes open." When he saw fiddleneck coming into one field, he had it hoed up and removed the plants from the field. He and his crew also clean machinery before moving to a different field to avoid spreading weeds from one part of the farm to another. Weeds may be Park's major battle and perhaps one that can never be entirely won. But with this sort of care, Park Farming Organics can hold their own in the struggle and ensure that the farm continues to prosper.

CARL PEPPER
O'Donnell, Texas
Cotton

Photo by Kayla Pepper.

Carl Pepper grows organic cotton on the High Plains of west Texas, a region that has sometimes been called "the world's largest cotton patch." By paying attention to the biology of his system and developing equipment specially adapted to his operation, he grows one of the world's most chemical intensive crops without any chemicals at all. When he began farming on his own in 1992, he began experimenting with an organic approach on 160 acres. He currently grows organic cotton on 3,200 acres of his own farm and another 300 acres on his sister-in-law's farm.

Compared with most organic farms, Pepper's crop rotation is remarkably simple. All of his land is planted to cotton each summer and is planted with a cover crop in the fall. The usual cover crop seeding consists

of 10–12 pounds per acre of a mix consisting of 70% rye, 20% tillage radish and 10% hairy vetch. Despite the minimal diversity of the rotation, he has seen no buildup of diseases, insects or weeds over the 25 plus years he has been farming this land. He attributes the low disease and insect pressure to the arid climate of west Texas. He has also noticed that if he leaves the aphids alone early in the season, they attract beneficial insects that attack more serious pests later. Often bollworm populations in his fields crash by natural processes about the same time his neighbors are spraying for this pest. Since cotton matures earlier farther south in Texas, he gets an early warning in years when bollworms are particularly bad, and, if necessary, he releases Trichogramma wasps to help suppress the pest.

Since Pepper does not irrigate, much of his attention is focused on water conservation and ensuring that soil moisture is sufficient to support his cotton. This includes preventing the growth of weeds. And on the windy plains of Texas, preventing soil erosion is similarly critical. To simultaneously conserve water, keep the soil in place and manage weeds, he has developed an integrated tillage/cultivation system that relies on machinery that he and his full-time crew of four employees designed and built themselves. The beauty of his system can be best understood by following the annual crop cycle.

The cover crop is sown into the unpicked cotton before the last cultivation in late September or early October. The low seeding rate of 10–12 pounds per acre is just sufficient in a wet winter to provide good cover, but in a dry winter, it is sparse enough to prevent the cover crop from using moisture the cotton will need later. He begins cultivating the cover crop to suppress it and kill weeds beginning in early March. For this purpose, he uses low-pitch 24-inch sweeps on 20-inch centers, with alternate sweeps centered on the future crop row and on the center of the inter-row (Figure 5.3). A high accuracy GPS system allows him to keep the tractor on the same tire tracks throughout this and all subsequent operations. The sweeps travel only 2 inches deep and undercut the cover crop while leaving it on the soil surface to protect the soil. Another one or two such operations eventually kill the cover crop while keeping weeds suppressed until planting time.

Pepper's soils range from sandy loams to clay loams. Most have a moderate number of rocks, but in a few places the soil is shallow over bedrock. The finer textured soils have produced good cotton yields for decades with very little added fertility. On the sandier soils, Pepper used to apply 1–2 tons per acre of composted cattle manure prior to planting, but for the last several years he has found that the compost is unnecessary. The lack of fertility inputs probably helps limit weed growth. Since the harvested crop is mostly carbohydrate in the form of cotton lint, most nutrients remain in the field. Possibly due to the consistent use of cover crops and a conservation tillage system, he has seen an increase in soil organic matter of 0.3.–0.8% over the years. Since a 0.1% increase in organic matter represents one ton per acre of additional organic matter in the soil, this increase represents an important additional source of nutrients.

Pepper begins planting cotton between May 15 and June 10. He uses a plant eight, skip one row planting pattern. The skipped row provides access for ATVs during hand weeding. Pepper uses a high seeding rate because he expects his aggressive weeding later in the season will kill some plants. Thus, he plants at three seeds per foot of row, hoping for 2.25–2.5 plants per foot at harvest.

His post-planting weeding begins with a cultivator composed of rotary hoe gangs alternating with sand fighters on the same toolbar. The sand fighters are metal triangles with extensions on the sides. These are welded on an axle in a staggered fashion so that, looked at from the side, they look like a six-pointed star. The sand fighters flip divots of soil to create surface roughness that reduces wind erosion. His rotary hoe units over the crop line consist of an angle iron frame holding two offset lines of hoe arms. The front line has three arms and 1.5-inch spacers on the ends. The second line has four arms. Thus, the hoe wheels on each section are 3 inches apart, but the sections are staggered so that the overall spacing of wheels for the whole implement is 1.5 inches. He runs this implement at 10–16 mph, which covers his extensive acreage in two days. "I'd estimate we kill 90% of the annual weeds with each pass of the hoe," Pepper says. He continues using the rotary hoe after each rain event until the cotton reaches the two- to four-leaf stage. At that point he switches to a specially modified row crop cultivator. This has low pitch 32-inch sweeps that keep the cover crop and weed residue on the soil surface to prevent wind erosion. Over the crop rows he mounts a gang of rotary hoe wheels spaced 2 inches apart to continue the in-row weeding. Perhaps the most unique feature of the cultivator, however, is the spring-steel hay baler pick-up fingers he has bolted to the wings of the sweeps. These reach out toward the crop row at a 30-degree angle to cut off or uproot weeds close to the crop row. The forward motion of the cultivator causes the springs to deflect away from the cotton, but this creates pressure toward the row, causing the soil to boil and disturb small weed seedlings. Pepper begins with the spring tips spaced about 1–1.5 inches apart and runs the cultivator at 5 mph. A GPS controlled

Figure 5.3. Sweeps with wire weeders.

mounting on the three-point hitch keeps the cultivator well centered on the crop rows.

Pepper cultivates after each rain to keep the soil surface loose. This prevents capillary draw of moisture from deeper in the soil. During long periods without rain, he cultivates every 30 days to keep perennial weeds suppressed. After the cotton reaches the six-leaf stage, the rotary hoe gangs are removed from the cultivators. But as the bark on the roots toughens, he is able to move the spring fingers on the sweeps ever closer to the row, until, by the end of the season, he has the fingers crossed so that the row is thoroughly cultivated. With the help of his GPS guidance system, he is able to cultivate at 8 mph by the end of the season. Although he makes many passes over the field with one sort of cultivator or another, he avoids compaction by consistently following the same wheel tracks, and the low draft of his implements results in a typical fuel use of only 0.25 gallons per acre of diesel. To successfully manage soil moisture, the machinery needs to move fast, and Pepper and his crew can cultivate the entire 3,500 acres in six to seven days.

Pepper's major annual weeds are kochia, devil's claw and Palmer amaranth, which in his region is called careless weed. Silverleaf nightshade, lakeweed (a spurge) and field bindweed are perennials that are problems in parts of some fields. To control the perennials, he breaks from his standard minimum tillage and plows to disrupt the root systems. He only plows, however, in the part of a field where the perennials are a problem. To ensure that the weeds do not spread, he and his crew scrape down the plow and cultivators after they are used in areas infested with perennials. Although Pepper's tillage/cultivation system does a good job of eliminating most weeds, he still finds some hand hoeing necessary. For fields where escapes are more plentiful, he hires in an eight-person hoeing crew. On most of the land, where the weeds are sparser, his regular full-time staff spends about two hours per acre hoeing. To clean up any escapes, they also roam through the skip rows on 4-wheel ATVs to hand rouge out escapes. He estimates that hand weeding cost an average of about $30 per acre, which he considers a good investment. The hand weeding helps conserve moisture for the cotton, and it also reduces seed production by the annuals and the build-up of root reserves in the perennials. "My fields are generally cleaner than my neighbor's GMO cotton," Pepper notes.

Pepper broadcasts the cover crop seed into the standing cotton near the end of September, give or take about two weeks. He then cultivates for a final time to eliminate any remaining weeds and to incorporate the cover crop seed. He times the seeding and cultivation so that moisture is adequate for establishment of the cover crop. By harvest in November, the field is green. Since Pepper does not use defoliants, he waits until a freeze kills the leaves before harvesting the cotton. The first freeze comes about November 6, on average, and he usually begins harvesting around Thanksgiving.

Pepper's integrated system for conserving moisture, protecting the soil and managing weeds produces 300–400 pounds per acre of cotton lint, a harvest that regularly matches the county average. One of the keys to Pepper's success is good preparation. "We spend a lot of time in the shop making sure that everything is ready to go when the time is right." This preparation is critical since most operations are timed to rainfall events. Pepper also recognizes that his crew of four full-time employees is one of the keys to his success. "They have bought into the system," he says. "We work really hard when something needs to be done, but when there is a slack period, they get to spend time with their families. Family comes first, but the farm feeds the family."

REFERENCES

Caldwell, B., C.L. Mohler, Q.M. Ketterings and A. DiTommaso. 2014. Yields and profitability during and after transition in organic grain cropping systems. *Agronomy Journal* 106: 871–880.

Gugino, B.K., O.J. Idowu, R.R. Schindelbeck, H.M. van Es, D.W. Wolfe, J.E. Thies and G.S. Abawi. 2007. *Cornell soil health assessment training manual edition 1.2.* Cornell University, Geneva, New York.

Nordell, A. and E. Nordell. 2007. *Weed the Soil, not the Crop: A Whole Farm Approach to Weed Management.* Available from the authors.

PART II

Major Agricultural Weeds of the United States and Canada

HOW THE SPECIES CHAPTERS WERE DEVELOPED

The species treated here include the most common and difficult to control agricultural weeds in the United States and southern Canada. The species were chosen largely through a systematic analysis of Bridges and Bauman (1992), which lists the 10 most abundant weeds in each of the major crops grown in each of the 50 states. Space limitations and lack of information prevents us from covering all of the less common, though not necessarily less troublesome, species encountered by farmers. The common names used in this book are those of the Weed Science Society of America's Composite List of Weeds (http://wssa.net/wssa/weed/composite-list-of-weeds). However, we also provide a list of other common names by which the species may be known in various parts of North America. Scientific names follow the USDA Plants Database (https://plants.usda.gov). Taxonomic controversies, such as whether waterhemp is one species or two, and the proper family in which to place common milkweed, have been bypassed by following the USDA Plants Database.

Information on non-herbicidal methods for managing most individual weed species is sparse. Published sources were consulted when these could be located, but many of the management recommendations for particular species were developed directly from the ecological behavior of the species. Thus, many recommendations for individual species have not been field tested or have been tested only in a preliminary manner. They should be considered informed suggestions as to how to cope with particular problems. In any case, you should always try a new management method on a small area before applying it to the whole farm. We, the authors, will appreciate your contacting us directly with accounts of successes and failures using the procedures we suggest.

The ecological information compiled for each species came from a variety of sources. Where literature reviews of a species were available, we relied heavily on these but consulted the primary sources when the information supplied in the review was ambiguous or contradictory. For many of the species, no general review was available, and the entire account was constructed from primary sources. For many species, information on one or two aspects of the species' ecology could not be located. In particular, information on plant response to disturbance and soil conditions (drought, fertility and physical conditions) was often sparse or anecdotal. We give what information could be found, but in some cases, the information is not the aspect of the species that one might most want to know. In chapters that cover two or three related species, sometimes information on a topic is only available for some but not all of the species, and then the lack of information is particularly noticeable.

References have been reduced to a few further readings. However, the taxonomy chapters with full documentation are available at https://weedecology.

css.cornell.edu/. In addition to these formal sources, however, we relied substantially on our 100+ years of personal observations on these weed species as well as on observations of many farmers and colleagues.

Some categories of information have inherent problems of interpretation. The tolerance to drought and shade has been studied for few weed species. Usually, only the general impression of those who have worked with the species is available. When quantitative data could be located, summarizing these in terms that are useful for a grower has sometimes been difficult.

Most seed survival studies have placed seeds deep in the soil and left them undisturbed. Such studies provide an upper limit on how long the seeds last under optimal conditions for survival. Typically, seeds die off at a constant rate, and the depletion of the seed bank is faster near the surface and faster when the soil is disturbed by tillage (see Chapter 2). Where data on depletion of seeds in disturbed soil were available, we have converted them to percentage loss per year. This provides a more direct way to think about how management of the seed bank will affect weed pressure. For some species, however, survival data vary substantially between experiments.

Usually, seed production (described for each species in the "Reproduction" section) has been measured on plants that emerged at the optimal time of year and grew with little or no competition. Under extreme competitive stress from crops and/or other weeds, most weed species produce very few seeds per plant. Most agricultural situations are intermediate between these extremes. Early flushes of weeds are often controlled, and the crop provides some competition. Thus, most weeds in an agricultural field often produce only 0.1–5% of the published maximum seed output. We have tried to specify when data are derived from plants growing with minimum competition and when data are derived from plants growing in typical cropping environments.

We give weights and measures in American units. The one exception is that we consistently give seed weights only in milligrams (abbreviated "mg"). The smallest commonly used American unit of weight is the ounce, but the seeds of most weed species are in the range of 0.00001 to 0.0001 ounces. Such numbers are difficult to read and even more difficult to comprehend. As explained in Chapter 2, however, seed weight is a very important property of a weed species,

and the difference between 0.00001 ounces (0.28 mg) and 0.0001 ounces (2.8 mg) has substantial implications for management. Most varieties of lettuce seeds weigh close to 1 mg, so if the mg is an unfamiliar unit, just think of seed weights as multiples or fractions of a lettuce seed.

HOW TO FIND ECOLOGICAL INFORMATION AND DEVELOP A MANAGEMENT PLAN FOR SPECIES NOT COVERED IN THIS BOOK

Many species that have limited geographical range or are problems only in a few crops or in particular circumstances could not be covered in this book. Nevertheless, some of those species pose substantial problems to particular growers. The basic concept of this book is that understanding the ecological characteristics of a weed provides insight into how to manage the species. Consequently, when faced with managing a weed not covered here, begin by gathering critical information about its ecology.

Begin by identifying the weed species. Each taxonomic level (family, genus and species) potentially provides additional information about the weed. For example, species in the mallow and morningglory families usually have hard seeds that do not germinate until the seed coat softens. The genus potentially provides additional information—for example, the genus *Amaranthus* (pigweeds, amaranths and waterhemps) are heat loving C_4 plants. But the species within a genus often differ in ways that affect management, and having a positive identification to species will be helpful. Many resources are available for helping identify weeds (Box 3.2). Your local Cooperative Extension office or crop consultant can also help.

Ecological information about individual weed species is not widely available. That is one reason we have compiled so much of it in this book! A search of internet sites may locate a Cooperative Extension fact sheet or scientific papers on the species that contain useful information. Information on some weeds can also be found in the series "Intriguing World of Weeds," originally published in the journal *Weed Technology* but available also at the WSSA website: www.wssa.net/weed/intriguing-world-of-weeds. Some weeds found in the northern United States and Canada have been thoroughly described in the series "Biology of Canadian Weeds," published in the *Canadian Journal of Plant Science*. These have been collected into five volumes by the Agricultural Institute of Canada (Mulligan 1979, 1984, Cavers 1995, 2000, 2005). They are likely available in the library at your state land grant university or at a major agricultural college but may be difficult to locate otherwise. Two large compilations of information on particular weeds also contain some ecological information (Holm et al. 1977, 1997). Again, these are likely to be available only in the library of a land grant university or major agricultural college.

Fortunately, you can obtain much useful information by observation. Record your observations in a notebook or in a computer document so that all the information is organized in one place and does not get lost. Careful observation over time will require some effort and persistence. Note, however, that the weed problem likely did not develop in a single year, and resolving the problem will likely require several years. So, investing a little time over the course of a growing season to systematically observe the weed will likely save effort in the long run. The quality and quantity of the information you obtain and your ability to think about it are likely to improve if you share the work and results with a friend. Children can also be recruited to collect the information as a hands-on educational activity. Information you are likely to find useful and methods for obtaining it follow.

- **What sort of weed is it?** Is it a summer annual, winter annual, stationary perennial or creeping perennial? Does it grow upright, sprawl across the ground or twine up crops? Do the plants begin life as a low growing rosette that later develops a flower stalk, or does it begin vertical growth immediately? How tall does it grow in your crops? These basic attributes affect a wide range of management considerations.

- **How big are the seeds?** Collect some seeds and compare them with the seeds of species discussed in this book. Seed size affects depth of emergence, ability to emerge through mulch, ability of seedlings to grow despite shade from a crop, and likely modes of dispersal (for example, plants that produce many small seeds often move with soil on tires, machinery and livestock).

- **What time of year does it emerge most abundantly in your fields?** What other times of year does it emerge in lesser numbers? This will tell you how rotating between crops with different planting times may affect the weed. It will also indicate the best times of year for using a tilled fallow to weed the soil.

- **From what depth do seedlings emerge?** You can discover this by carefully excavating seedlings shortly after they emerge. On most grassy weeds, the seed remains attached to the seedling for several weeks after germination, and you can measure the distance from the seed to the base of the shoot. On broadleaf weeds, measure the distance from the primary root to the green part of the shoot. Do not wait too long to make these observations, since most weeds will produce secondary roots near the base of the shoot. Measure 50–100 seedlings and count how many fall into various depth categories: 0–0.5 inch, 0.5–1 inch, etc. Knowing the depth of emergence will tell you what percentage of the seedlings you can hope to uproot with a tine weeder or rotary hoe. If a substantial proportion of seedlings emerges from below this depth, you will need to aim for physical damage and burial of the seedlings.

- **For creeping perennials, where are the storage roots or rhizomes located?** Carefully dig down through a patch of the weed and observe where the thickened horizontal roots or rhizomes are located. Are they within the plow layer? Do many lie below the plow layer? This information will tell you whether you can effectively break up the storage roots or rhizomes, or whether you will have to focus on repeatedly killing shoots to exhaust the plant's storage reserves. As you are digging up the plant, see how long the new growth on the storage roots or rhizomes is and use this to guess how fast the plant spreads without the aid of tillage. Test the strength of the storage roots or rhizomes. Can they be worked to the surface to dry during hot weather, or do they fragment too easily for this?

- **When does the weed produce seeds?** Does it begin producing seeds the same time of year regardless of when it emerges (then flowering is tied to day length), or does it begin producing seeds at a specific interval after the plants emerge? Knowing when seeds are produced will help you plan rotations that suppress the weed and will help in thinking about how to prevent seed production. For some species, the time of seed maturity is not obvious. Generally, however, if seeds fall out when you shake the plant over a white cloth or into a white bucket, then the seeds are mature. Knowing when mature seeds are shed from the plant can suggest whether they can be collected and removed from the field during combining.

- **Does the species die as the seeds are maturing, or does it continue to flower and set seed for many weeks after the first seeds are dropped?** Mark a few plants that are just beginning to flower so you can find them again later. Then visit them every two weeks and note whether they are flowering or shedding seeds. This information will help you plan ways to limit seed production.

- **Are seeds dormant when they are shed from the parent plant?** Spread some freshly collected seeds on a stack of two or three moist paper towels on a plate. Cover with another moist towel. Put them in a clear plastic bag and set them in a light, warm, location where the temperature will fluctuate between day and night. A north facing windowsill is often a good choice. Observe whether few or most germinate within a week or two. If few germinate, then the seeds are mostly dormant when shed. This information will help you plan crop rotations and fallows to control the weed.

- **How does the species respond to fertility?** Take some relatively poor, exhausted soil of the same soil type as yours (perhaps from a neighbor's farm). Knowing how the chemical analysis of this soil compares with yours will be helpful, but it is not essential. Add various amounts of your usual fertility sources and place the resulting mixes into medium sized flowerpots. To ensure that your fertilization rates are within a reasonable range, compute the area of the pot in acres (a small number!). Then calculate how much compost, manure or other amendment would correspond to a typical application. Try mixes that have 0 (no addition), 1x and 5x your typical application rate. You will want about three pots with each mix. Bury the pots to within an inch or two of the rim in a place where they will not get disturbed but where you are sure to observe them periodically. The edge of a garden may work well. Sow a few seeds of your weed into each pot, or, if it is a creeping perennial, plant two sections of the rhizome or storage root in each pot. After plants emerge, thin down to one or a few plants, depending on the potential size of the species. Also pull out competing weeds. Grow the plants until they

flower. Do the plants in highly fertilized soil flower earlier? How does plant size compare in the three treatments? This information will help you discover whether excess or imbalanced fertilization is aggravating your weed problem. Note that the high (5x rate) may correspond most closely with your fields since organic amendments often build up organic matter and stored nutrients.

- **What management operations likely allowed your weed to thrive?** Think closely about those situations where your weed is most problematic and those where it is least. What crops are most favorable for it? What cropping practices are associated with that crop that may favor your weed? Think about the degree of soil disturbance, the fertility source, the growth form and competitiveness of crops, the length of the growing season before harvest, and the timing of when all of these occur. An understanding of what cropping practices contribute most to the growth of your weed will provide a good clue as to what practices may control it.

You will likely have several inspirations about how to manage the weed by the time you have collected the information above. To add to these and help integrate them into a comprehensive control strategy, read the accounts of similar species provided in this book. Reading about species in the same genus or family that have similar growth habits will be most informative. Note how the ecology of your target weed is similar and different from each of the weeds you are reading about. Where the characters are similar, see how those aspects of the weed's ecology are translated into management recommendations. Lastly, compile your management tactics into an overall plan of attack on the weed. While composing your plan, recognize that the weed problem is a result of your overall farm management, and that its solution will likely require some departure from business as usual. The solution may be relatively easy, but more likely, you will have to balance the costs of tolerating the weed against costs of changing machinery, the range of crops grown, sources of fertility etc. Although your problem weed is unlikely to disappear entirely, we are confident that you can create an overall management strategy that will allow you to keep the weed in check and improve the profitability of your farm operation.

REFERENCES

Bridges, D.C. and P.A. Baumann. 1992. Weeds causing losses in the United States. In *Crop Losses Due to Weeds in the United* States. D. C. Bridges (ed.). pp. 75–147.

Cavers, P.B. ed. 1995. *The Biology of Canadian Weeds.* Contributions 62–83. Agricultural Institute of Canada: Ottawa.

Cavers, P.B. ed. 2000. *The Biology of Canadian Weeds.* Contributions 84–102. Agricultural Institute of Canada: Ottawa.

Cavers, P.B. ed. 2005. *The Biology of Canadian Weeds.* Contributions 103–129. Agricultural Institute of Canada: Ottawa.

Holm, L.G., D.L. Plucknett, J.V. Pancho and J.P. Herberger. 1977. *World's Worst Weeds: Distribution and Biology.* East-West Center, University Press of Hawaii: Honolulu.

Holm, L., J. Doll, E. Holm, J. Pancho and J. Herberger. 1997. *World Weeds: Natural Histories and Distribution.* John Wiley & Sons: New York.

Mulligan, G.A. ed. 1979. *The Biology of Canadian Weeds.* Contributions 1–32. Publication 1693. Agriculture Canada: Ottawa.

Mulligan, G.A. ed. 1984. *The Biology of Canadian Weeds.* Contributions 33–61. Publication 1765. Agriculture Canada: Ottawa.

Weed Characteristics Summary Tables

The following tables summarize the characteristics of the weed species described in Part Two. Species are organized by summer annual weeds, winter annual weeds and perennial weeds.

Summer annual weeds: Species that primarily emerge in spring and set seed in summer. Some species that are more frost tolerant may emerge in the fall and overwinter, particularly in warm regions.

Winter annual weeds: Species that primarily emerge in fall and set seed in spring. Most of these species also can behave as spring emerging annuals.

TABLE HEADING NOTES—ANNUAL WEEDS

General: The designation "–" signifies that data is not available or the category is not applicable.

Weed: Weed common name as listed in the Weed Science Society of America Composite List of Weeds, presented in alphabetical order.

Growth habit: A two-word description. The first word indicates relative height (tall, medium, short, prostrate) and the second word indicates degree of branching (erect, branching, vining).

Seed weight: Range of reported values in units of "mg per seed."

Seed dormancy at shedding: "Yes" if most seeds are dormant when shed; "Variable" if dormancy is highly variable; "No" if most seeds are not dormant.

Factors breaking dormancy: The principle factors that are reported to break dormancy and facilitate germination. The order of listing does not imply order of importance. Abbreviations are:

> scd = seed coat deterioration
> cms = a period subjected to cold, moist soil conditions
> wst = warm soil temperatures
> li = light
> at = alternating day-night temperatures
> ni = nitrates

Optimum temperature range for germination: Temperature (Fahrenheit) range that provides for optimum germination of non-dormant seeds. Germination at lower percentages can occur outside of this range. The dash refers to temperature range, and the slash refers to alternating day/night temperature amplitudes.

Seed mortality in untilled soil: Range of mortality estimates (percentage of seed mortality in one year) for buried seeds in untilled soil. Values were chosen where possible for seeds placed at depths below the emergence depth for the species and left undisturbed until assessment. Mortality primarily represents seed deterioration in soil.

Seed mortality in tilled soil: Range of mortality estimates (percentage of seed mortality in one year) for seeds in tilled soil. Values were chosen for seeds placed within the tillage depth and subjected to at least annual tillage events. Seed losses are the result of dormancy-breaking cues induced by tillage, germination and deterioration of un-germinated seeds.

Typical emergence season: Time of year when most emergence occurs in the typical regions of occurrence for each weed. Some emergence may occur outside of this range.

Optimum emergence depth: Soil depths (in inches below the soil surface) from which most seedlings emerge. Lower rates of emergence usually will occur at depths just above or just below this range.

Photosynthesis type: Codes "C$_3$" or "C$_4$" refer to the metabolic pathway for fixing carbon dioxide during photosynthesis. Generally, C$_3$ plants function better in cooler seasons or environments and C$_4$ plants function better in warmer seasons or environments.

Frost tolerance: Relative tolerance of plants to freezing temperatures (high, moderate, low).

Drought tolerance: Relative tolerance of plants to drought (high, moderate, low).

Mycorrhiza: Presence of mycorrhizal fungi. "Yes" if present; "no" if documented not to be present, "unclear" if there are reports of both presence and absence; "variable" if the weed can function either with or without, depending on the soil environment.

Response to nutrients: Relative plant growth response to the nutrient content of soil, primarily N, P, K (high, moderate, low).

Emergence to flowering: Length of time (weeks) after emergence for plants to begin flowering given typical emergence in the region of occurrence. For species emerging in fall, "emergence to flowering" means time from resumption of growth in spring to first flowering.

Flowering to viable seed: Length of time (weeks) after flowering for seeds to become viable.

Pollination: "Self" refers to species that exclusively self-pollinate; "cross" refers to species that exclusively cross-pollinate; "self, can cross" refer to species that primarily self-pollinate, but also cross-pollinate at a low rate; and "both" refers to species that both self-pollinate and cross-pollinate at relatively similar rates.

Typical and high seed production potential: The first value is seed production (seeds per plant) under typical conditions with crop and weed competition. The second value, high seed production, refers to conditions of low density without crop competition. Numbers are rounded off to a magnitude that is representative of often highly variable reported values.

TABLE HEADING NOTES—PERENNIAL WEEDS

(Headings unique to perennial tables are described here. Headings in common with annual weed tables are described under annual weeds.)

Perennial overwinter organ: Principal plant organ that survives winter and from which growth resumes in subsequent years.

Emergence period from perennial organs: The time of year when most emergence occurs from perennial overwintering organs in the typical regions of occurrence for each weed. Some emergence may occur outside of this range.

Optimum emergence depth from perennial organs: Soil depths (in inches below the soil surface) from which most shoots emerge from perennial organs. Lower rates of emergence usually will occur at depths above or below this range.

Time/stage of lowest reserves: Time of year and/or weed growth stage at which carbohydrate reserves are lowest. This usually corresponds to the time when the weed is most susceptible to weed management operations.

Frost tolerance: Relative tolerance of aboveground shoots to freezing temperatures (high, moderate, low).

Drought tolerance: Relative tolerance of aboveground plants to drought (high, moderate, low).

Importance of seeds to weediness: The relative importance of seeds to dispersal, genetic diversity and survival of the species as a weed in agricultural environments (high, moderate, low).

Emergence to flowering: Length of time (weeks) after emergence from perennial organs to the beginning of flowering in the typical regions of occurrence. Note that this refers to established perennial plants, recognizing that some species may not flower in their initial year of establishment.

Weed	Growth habit	Seed weight (mg)	Seed dormancy at shedding	Factors breaking dormancy	Optimum temperature range for germination (F)	Seed mortality in untilled soil (% per year)	Seed mortality in tilled soil (% per year)
Broadleaf weeds							
Common cocklebur	tall, erect	50–75	Variable	scd, at	86/68–91/77	50	–
Common groundsel	short, branched	0.16–0.25	No	li	50–68	45	100
Common lambsquarters	tall, erect	0.5–0.72	Yes	cms, li, at, ni	64–77	8–51	31–52
Common purslane	prostrate	0.08–0.15	No	li, at, ni	86–95	60 (hot climate)	17–29 (cool climates) 76–87 (hot climate)
Common sunflower	tall, branched	4.3–8.7	Yes	cms	68–77	26–47	–
Field pennycress	medium, branched	0.8–1.5	Yes	cms, wst, li, at, ni	59/43–95/68	10	50
Galinsoga species	short, branched	0.17–0.27	No	li	54–97	50–99	–
Hemp sesbania	tall, erect	6–15	Yes	scd	58–104	27	65
Horseweed	medium, erect	0.03–0.07	No	li	68/50–86/68	76	–
Jimsonweed	tall, branched	6–12	Yes	scd, cms, li, at	68–95	6–50	–
Kochia	tall, branched	0.2–0.85	No	–	68–77	97–100	–
Morningglory species	vining	19–35	Yes	scd, wst	59–95	36–70	58
Nightshade species	short, branched	0.4–1.3	Variable	cms, li, at, ni	77–86	15	28–45
Palmer amaranth	tall, erect	0.44–0.49	Yes	li, at, ni	86–99	39–80	–
Pigweed species	medium, branched	0.25–0.54	Variable	li, at, ni	86–104	39–88	36–41
Prickly lettuce	medium, erect	0.45–0.62	No	li, at	54–75	40–85	–
Prickly sida	short, branched	2.3	Yes	scd	86–104	60	88
Ragweed, common	medium, branched	1.2–7.7	Yes	cms, li, at	77/68–95/86	7–12	–
Ragweed, giant	tall, branched	17–45	Yes	cms, at, ni	50–75	66–95	–
Russian-thistle	medium, branched	1.1–1.7	Yes	at	59/32–77/41	99	–

Typical emergence season	Optimum emergence depth (inches)	Photosynthesis type	Frost tolerance	Drought tolerance	Mycorrhiza	Response to nutrients	Emergence to flowering (weeks)	Flowering to viable seed (weeks)	Pollination	Typical & high seed production potential (seeds per plant)
mid-spring to early summer	0.4–4	C_3	low	high	yes	high	10–16	3–4	self, can cross	1,800 & 10,000
early spring to early summer	0–0.8	C_3	high	low	yes	low	4–5	1–2	self, can cross	1,500 & 38,000
early spring to summer	0.1–0.2	C_3	low	moderate	no	high	5–12	2–3	both	30,000 & 300,000
late spring to summer	0–0.1	C_4	low	high	no	high	2–8	1–2	self	10,000 & 100,000
spring	0–4	C_3	high	moderate	yes	variable	9–17	8	cross	– & 5,000
spring and fall	0–0.8	C_3	high	low	no	moderate	5–7	1–2	self, can cross	500 & 14,000
late spring to summer	0–0.1	C_3	low	low	yes	high	4–8	1–2	both	10,000 & 100,000
mid-spring to summer	0.4–1.2	C_3	low	moderate	yes	low	6–7	6	both	2,000 & 20,000
fall and spring	0–0.1	C_3	high	high	yes	low	8–12	2–3	self, can cross	50,000 & 300,000
mid-spring to early summer	0.4–2	C_3	low	low	–	high	5–9	4	self, can cross	1,500 & 30,000
spring	0–0.4	C_4	moderate	high	no	high	8–16	–	self, can cross	20,000 & 100,000
summer	1–2	C_3	low	low	yes	high	4–8	4	both	300 & 15,000
mid-spring to summer	0–1.6	C_3	low	low	yes	high	5–10	4–8	self, can cross	10,000 & 500,000
late spring to summer	0–0.5	C_4	low	high	no	moderate	3–8	2–3	cross	40,000 & 400,000
late spring to summer	0.2–0.8	C_4	low	moderate	no	high	3–8	3–8	self	50,000 & 500,000
fall and spring	0–0.1	C_3	high	high	yes	moderate	8–12	3	self, can cross	5,000 & 50,000
mid-spring to summer	0.2	C_3	low	moderate	yes	high	8–12	2–3	self, can cross	2,000 & 8,000
early to late spring	0–1	C_3	low	high	yes	high	10–20	–	both	3,000 & 60,000
early to late spring	0.5–2	C_3	moderate	low	probably	moderate	8–18	3	cross	200 & 2,000
early spring	0.4–1	C_4	low	very high	no	low	10	–	both	15,000 & 150,000

Weed	Growth habit	Seed weight (mg)	Seed dormancy at shedding	Factors breaking dormancy	Optimum temperature range for germination (F)	Seed mortality in untilled soil (%/year)	Seed mortality in tilled soil (%/year)
Broadleaf weeds (cont.)							
Sicklepod	medium, branched	23–28	Yes	scd	68–97	28–40	46
Smartweed, ladysthumb Smartweed, Pennsylvania	short, branched	1.4–4 3.6–6.8	Yes	cms, at	86/59–95/68	25–50	24–43
Sowthistle, annual species	medium, branched	0.27–0.42	No	li	41–95	53–55	48–65
Velvetleaf	tall, erect	6–12	Variable	scd	75–86	3–17	32–53
Waterhemp	tall, erect	0.19–0.27	Yes	cms, li, at	68–91	30–78	40
Wild buckwheat	twining	4.7–7	Yes	scd, cms, at	68–77	20–52	32–50
Wild mustard	medium, branched	1–2.3	Yes	cms, li, at, ni	50–68	22–45	20–52
Wild radish	medium, branched	2–12	Yes	cms, at	39–68	32–33	29
Grass weeds							
Barnyardgrass	tall	1.7–2.1	Yes	cms, li	77–100	37–42	–
Fall panicum	tall	0.2–0.9	Yes	cms, li, at	68/50–95/68	39–56	–
Foxtail, giant Foxtail, green Foxtail, yellow	prostrate to tall	1.5–1.6 0.6–1.5 1.9–4.2	Yes	cms, at	68–86	72–93	–
Goosegrass	prostrate	0.4–0.5	Yes	cms, li, at, ni	86/68–104/86	18–44	–
Large crabgrass	prostrate	0.46–0.59	Yes	cms, at	68–86	45–54	–
Sandbur, field Sandbur, longspine Sandbur, southern	short, tufted	1.7–3.4 6.8 2.4–7.9	Variable	scd, cms, at, ni	77/50–95/77	19–67	46
Shattercane	tall, erect	19–22	Variable	scd, at	77–95	57–99	–
Wild oat	medium	14–24	Yes	wst, ni	59–82	70–90	–
Wild-proso millet	tall	3.8–7.2	Variable	cms	68–86	10–60 (black seed)	–
Witchgrass	short	0.15–0.65	Yes	cms, li, at, ni	86/59–95/68	–	–

Typical emergence season	Optimum emergence depth (inches)	Photosynthesis type	Frost tolerance	Drought tolerance	Mycorrhiza	Response to nutrients	Emergence to flowering (weeks)	Flowering to viable seed (weeks)	Pollination	Typical & high seed production potential (seeds per plant)
late spring to summer	0–3	C_3	low	moderate	yes	moderate	6–12	–	both	5,000 & 16,000
early to mid-spring	0–2	C_3	low	moderate low	unclear	moderate	6–9	4	both	100 & 2,000 20,000 & 100,000
spring, summer, fall	0–0.4	C_3	high	low	yes	low	6–9	1	self	– & 20,000
mid-spring	0.5–1	C_3	low	high	yes	high	11–12	2	self	1,000 & 10,000
late spring to summer	0–1	C_4	low	low	no	high	3–7	3–4	cross	200,000 & 1,000,000
early spring	0.4–1.6	C_3	low	moderate	no	high	6–12	3	self	– & 20,000
spring	0–0.8	C_3	high	low	no	high	3–6	5–6	cross	3,000 & –
spring and fall/winter	0.4–1.2	C_3	high	low	no	high	3–7	3–4	cross	500 & 10,000
mid-spring to early summer	0.5–2	C_4	low	low	yes	high	5–8	3	self, can cross	10,000 & 100,000
mid-spring to mid-summer	0–1	C_4	low	low	probably	moderate	7–13	3–4	cross	10,000 & 100,000
mid-spring to early summer	0.5–2	C_4	low	moderate	yes	high	5–13	2–3	self, can cross	1,000 & 10,000
spring to summer	0–0.8	C_4	low	moderate	yes	high	5	5	self, can cross	5,000 & 50,000
spring	0–0.8	C_4	low	high	yes	high	8–10	–	self, can cross	1,000 & 145,000
spring	0.4–4	C_4	moderate	high	yes	–	7–13	0	self	1,000 & 100,000
late spring to summer	1–2	C_4	low	moderate	yes	moderate	8–11	1–2	self, can cross	3,000 & –
spring and early fall	0.8–2.8	C_3	low	low	yes	high	7–8	3–4	self, can cross	100 & 500
late spring	1–2	C_4	low	high	probably	moderate	2–4	4–5	self, can cross	500 & 90,000
late spring to early summer	0–0.5	C_4	low	high	yes	high	8–13	3–4	self, can cross	– & 11,000

Weed	Growth habit	Seed weight (mg)	Seed dormancy at shedding	Factors breaking dormancy	Optimum temperature range for germination (F)	Seed mortality in untilled soil (% per year)	Seed mortality in tilled soil (% per year)
Broadleaf weeds							
Catchweed bedstraw	sprawling	4–17	No	cms, ni	33–72	41	51–65
Common chickweed	short, matting	0.36–0.51	Yes	wst, li, at, ni	54–68	17–30	33–72
Chamomile species	medium, branched	0.4–1.2	Yes	scd, li, at, ni	68–86	11–37	42–51
Flixweed	medium, branched	0.12	Yes	wst, li, at, ni	59/43–68/50	25	23–33
Henbit Purple deadnettle	short, sprawling	0.5–0.6 0.65–0.92	Yes	wst, li, at	41–68	20	39–60
Shepherd's-purse	short, erect	0.09–0.14	Yes	cms, li, at, ni	59/43–86/59	11–24	35–52
Grass weeds							
Annual bluegrass	short, erect to prostrate	0.19–0.48	Variable	cms, wst, li, at, ni	41–68	17–26	26–50
Downy brome	medium	2.5–3.7	Variable	ni	59–68	99	–
Italian ryegrass	medium	1.3–2.6	Variable	li	50/41–77/41	58–64	–

Typical emergence season	Optimum emergence depth (inches)	Photosynthesis type	Frost tolerance	Drought tolerance	Mycorrhiza	Response to nutrients	Emergence to flowering (weeks)	Flowering to viable seed (weeks)	Pollination	Typical & high seed production potential (seeds per plant)
fall and spring	0.8–2.4	C_3	high	low	unclear	high	6–10	4	self, can cross	350 & 1,500
fall and spring	0–0.4	C_3	high	low	no	high	6	2	self	1,500 & 15,000
fall and spring	0–1	C_3	high	moderate	unclear	moderate	8–12	–	cross	2,000 & 20,000
fall, some early spring	–	C_3	high	low	no	low	4–6	6	self	2,000 & 76,000
late summer to fall, spring	0–1	C_3	high	low	yes	moderate	2–5	2–4	both	1,000 & 50,000
fall and spring	0–0.5	C_3	high	moderate	no	moderate	4–16	2–3	self, can cross	3,000 & 50,000
late summer to fall, spring	0–1	C_3	high	low	yes	high	1–8	1–2	self, can cross	2,000 & 20,000
fall and spring	0–2	C_3	high	high	variable	high	4	4	self	50 & 500
fall and spring	0.25–0.5	C_3	moderate	low	yes	high	4–8	3	cross, can self	– & 300

Weed	Growth habit	Perennial overwinter organ	Emergence period from perennial organs	Optimum emergence depth (inches) from perennial organs	Time/stage of lowest reserves	Photosynthesis type	Frost tolerance
Broadleaf weeds							
Bindweed, field Bindweed, hedge	twining	thickened roots	mid-spring to summer	0–6	12–28 inch stems, 4–6 leaves	C_3	moderate
Canada thistle	medium, erect	thickened roots	mid-spring to fall	4	12-inch shoots, flower bud set	C_3	low
Common milkweed	medium, erect	thickened roots	late spring to early summer	1–12	mid-summer	C_3	low
Dandelion	rosette	taproot	early spring	0–4	spring, flowering	C_3	high
Dock species	rosette	taproot	early spring	0–3	–	C_3	high
Horsenettle	short, branched	roots	mid-spring to summer	0–12	early summer, flowering	C_3	low
Plantain, blackseed Plantain broadleaf Plantain, buckhorn	rosette	basal stem	mid-spring	–	–	C_3	high
Sowthistle, perennial	medium, erect	thickened roots	mid-spring	1–8	5–7 leaves	C_3	low
Woodsorrel, yellow	short, branched	rhizomes	mid-spring to summer	–	–	C_3	moderate
Grass, sedge and allium weeds							
Bermudagrass	prostrate grass	rhizomes	late spring	2–4	mid-summer	C_4	low
Johnsongrass	tall grass	rhizomes	mid-spring	0–4	6–12 inches tall, 4–8 leaves	C_4	low
Nutsedge, purple	short sedge	tubers	summer	1–6	–	C_4	low
Nutsedge, yellow	short sedge	tubers	late spring to early summer	4–8	late summer, flowering	C_4	low
Quackgrass	short grass	rhizomes	early spring	1–6	3 leaves	C_3	high
Wild garlic	short allium	bulbs	fall and early spring	1–4	2 leaves	C_3	high

Drought tolerance	Mycorrhiza	Fertility response	Importance of seeds to weediness	Seed weight (mg)	Dormancy of shed seeds	Factors breaking dormancy	Optimum temperature range (F) for seed germination	Seedling emergence period	Emergence to flowering (weeks)
high	yes	low	moderate	8–20 28–34	yes	scd	95/68 77/59	late spring to early summer	6–9
high	yes	moderate	moderate	1–1.7	no	cms, at, ni	77–86	mid-spring	8–10
high	yes	moderate	moderate	3.5–7.4	yes	cms, at, ni	68/50–95/68	spring	6–8
high	yes	moderate	high	0.34–0.54	no	li, at, ni	59/41–77/59	early to mid-summer	4–5
high	no	high	high	0.7–3	variable	cms, li, at	68/50–95/68	spring and fall	4–8
high	yes	moderate	moderate	1.1–1.9	variable	at, ni	68–86	late spring	5–8
high	yes	low	high	0.35–0.7 0.06–0.34 0.8–2.9	variable	cms, li, at, ni	59–86	late spring to summer	6–10
low	unclear	low	moderate	0.38–0.69	no	li, at	77–86	late spring to mid-summer	10–14
moderate	yes	moderate	high	0.13–0.15	no	li	60–80	mid-spring to summer	4–6
moderate	yes	high	low	0.23–0.36	yes	li, at, ni	–	–	10–15
moderate	yes	high	high	2.6–6.2	yes	cms, li, at, ni	95/59	late spring through summer	7
high	yes	moderate	low	0.22–0.3	yes	scd, wst	–	–	3–7
high	yes	low	low	0.13–0.31	yes	li	95	spring	8–12
moderate	no	high	low	2	no	at	77/59	spring	8–12
high	yes	low	low	0.5–1.5	yes	cms	–	–	mid-spring

Grass Weeds and their Relatives

Annual bluegrass

Poa annua L.

Annual bluegrass seedling
Scott Morris, Cornell University

Annual bluegrass panicle
Scott Morris, Cornell University

Annual bluegrass stand
Randall Prostak, University of Massachusetts

IDENTIFICATION

Other common names: low speargrass, six-weeks grass, annual meadowgrass, annual spear grass, dwarf spear grass, dwarf meadow grass, causeway grass, speargrass, poanna

Family: Grass family, Poaceae

Habit: Short winter or summer annual bunchgrass, sometimes surviving through a second growing season. Two major variants have been identified, *P. annua* var. *annua*, an annual with an erect habit, and *P. annua* var. *reptans*, a short-lived perennial with a prostrate growth habit. The perennial variants are more common in turfgrass than are the annual variants. Perennial variants survive by growing roots from stem nodes in contrast with perennial *Poa* turfgrasses which spread by rhizomes.

Description: The linear **seedling** leaf (0.2–0.35 inch long by 0.04–0.1 inch wide) emerges perpendicular to the ground. Young true leaves unfold from the bud into a V with a tip shaped like a boat prow; they lack auricles and have 0.1–0.2 inch long, translucent, slightly pointed ligules. Collar regions are green and hairless. Light green blades are 0.75–2 inches long by less than 0.04–0.1 inch wide, crinkled or wavy on the surface, and

smooth edged. Blades may be hairless or have scattered soft hairs. Hairless, flattened sheaths open down the stem. **Mature plants** grow into 1- to 12-inch tall, highly tillering, upright or prostrate clumps. Stems are bright green to yellow green. Sheaths open nearly to the base and easily pull away from the stem. Sheaths are transparent and pale, hairless, V-shaped and without auricles; mature leaves have a ligule similar to that of seedling leaves. Blades are light green, linear, 0.5–5.5 inches long by 0.1 inch wide, nearly hairless, with the distinctive boat prow-shaped tip. The root system is shallow, fibrous and matting. Spontaneous rooting occurs at tiller bases. Numerous 1- to 3-inch tall, pyramid shaped **inflorescences** develop. White-green immature seedheads are open or airy, with several small flower clusters (spikelets) loosely grouped at the ends of branches within the pyramid. Individual spikelets are 0.2 inch long and contain two to six flowers. As with other grasses, the apparent **seed** includes a thin, tightly adhering layer of fruit tissue. Yellow-gray seeds are 0.1 inch long, elliptical and with one blunt end.

Similar species: Three perennial bluegrass species resemble annual bluegrass. Canada bluegrass (*Poa compressa* L.) is 1–2 feet tall, has flat, wiry stems and narrow, blue-green blades. Kentucky bluegrass (*Poa* *pratensis* L.) is 1–3.25 feet tall, with dark green leaves 2–30 inches long. It is a coarser looking plant whose inflorescences consistently have a whorl of two to five small branches at the base. Roughstalk bluegrass (*Poa trivialis* L.) is 1–3 feet tall, with yellow-green leaves 1–8 inches long. Leaves are rough and folded in the bud, with a large tapering membranous ligule. Italian ryegrass [*Lolium perenne* L. ssp. *multiflorum* (Lam.) Husnot] resembles annual bluegrass but can be distinguished by its long, narrow auricles.

MANAGEMENT

Annual bluegrass can be a severe problem in turfgrass. In agriculture it is primarily a problem in vegetables and small grain crops. Close mowing shifts the population in grassy areas to the creeping perennial form that is more easily controlled by tillage when it invades tilled fields. If annual bluegrass is abundant at the time of spring tillage, use a moldboard plow to invert the soil, which will bury the plants and kill them. Subsequent shallow secondary tillage should rely on disks or harrows to avoid bringing clumps back to the soil surface. Rotary tillage is not recommended as it will leave many clumps only partially buried. These will have a good head start on your crop and may cause problems if the crop is not highly competitive.

Severe infestations are usually the result of leaving soil undisturbed from late summer/early fall through the spring. In this situation, a few plants can produce many seeds before late spring tillage, and the large clumps are likely to survive tillage. To prevent problems with annual bluegrass the following year, till seedlings under late in the fall after most have germinated and then plant a winter cereal cover crop at high density to compete with late emerging individuals. When cultivating to destroy young annual bluegrass, burial is more effective than uprooting since this weed is very good at rerooting. Cultivating annual bluegrass after it forms clumps is not effective unless performed in very hot and dry conditions.

In pastures, annual bluegrass usually indicates over grazing or excessive trampling. It tends to invade where other grasses have died or been damaged rather than displacing them by competition.

ECOLOGY

Origin and distribution: Annual bluegrass originated in southern Europe and has been introduced into more than 80 countries in North Africa, North Asia, Australia, North and South America, and even Antarctica. It is a weed of temperate and subarctic climates and occurs in the tropics only at high elevations.

Seed weight: Mean seed weights from field grown plants range from 0.19–0.48 mg, though seed weights as high as 0.59 mg have been observed from greenhouse grown plants.

Dormancy and germination: Annual bluegrass produces both dormant and non-dormant seeds. The proportion of dormant seeds varies between populations and is affected by conditions during seed set. For example, a dormant Louisiana population that required warm temperatures to break dormancy functioned as a winter annual, whereas non-dormant Wisconsin populations functioned as summer annuals. Annual forms generally have a higher percentage of dormant seeds while perennial types have minimal dormancy. Dormancy declines over a period of months. Thus, non-dormant seeds produced in the spring germinate in the summer, whereas the dormant seeds are ready for germination by fall. Dormant seeds collected in Louisiana required moist heat to break dormancy whereas seeds from Australia required chilling. A few seeds can germinate at temperatures as low as 36–41°F, but high germination percentages require temperatures of 41–68°F. In ecotypes from Alabama, germination was best at 66/50°F (day/night) and declined at higher temperatures. Temperatures over 86°F decrease germination of non-dormant seeds and can reduce viability of dormant seeds. Many seeds that get incorporated into the soil before germination (for example, if they are cultivated into dry soil) become dormant and require specific environmental cues to stimulate germination. For such seeds, germination is stimulated by light, fluctuation in day/night temperatures and the presence of nitrate. Germination is reduced in dry conditions more so than for other winter annual grass species. In the coastal area of central California, long day length in the spring was hypothesized to contribute to inducing dormancy in spring and early summer, whereas shortening day length was hypothesized to contribute to decreasing dormancy in fall and early winter.

Seed longevity: Few seeds of annual bluegrass survive in the soil longer than five years. In soil stirred four times per year, seed density declined by 26% per year, and in undisturbed soil, it declined by 22% per year. In another study, seed mortality was 29–50% per year in frequently disturbed soil and 17–26% per year in undisturbed soil.

Season of emergence: Most emergence occurs in the late summer and fall, but some emergence occurs during spring and summer.

Emergence depth: This does not appear to have been studied, but given the small seed size, most seeds probably emerge from the top 1 inch of soil.

Photosynthetic pathway: C_3

Sensitivity to frost: Annual bluegrass tolerates frost well. Exposed, fully frozen plants will die in about two weeks, but the species commonly overwinters under snow or in the shelter of cover crops or crop residue.

Drought tolerance: Annual bluegrass is intolerant of drought and high soil temperatures. Plant leaf length, tillering and mass were greater when grown at 72°F than when grown at 90°F. Plants, however, can tolerate short exposure to high air temperatures, with plants surviving a 3-hour exposure to 117°F.

Mycorrhiza: Annual bluegrass is mycorrhizal.

Response to fertility: Annual bluegrass size responds directly to the amount of N, P or K applied. Nitrogen favors leaf production, whereas P and K promote seed production. However, herbage growth also increases with increasing levels of P and K. It grows best on soils with pH from 5.5–6.5 and is only rarely found on soils with pH less than 5.3.

Soil physical requirements: Annual bluegrass can grow in different types of soils, from clays to sands. It tolerates extreme compaction and waterlogged conditions. Fine textured soils that maintain higher water content in winter tend to have higher annual bluegrass populations than sandier, well drained soils.

Response to shade: The species grows well in moderate shade, and it often thrives under trees and in pastures and hay meadows dominated by taller grasses. It can sustain vegetative and reproductive growth at 80–92%

shade, explaining its common occurrence in fields of highly competitive cereal crops. It is associated with shaded cracks in sidewalks.

Sensitivity to disturbance: Annual bluegrass is difficult to uproot with hoes and cultivators because part of the spreading root system often remains in the ground. The dense roots hold soil and allow uprooted plants to survive and quickly reroot. Once the plants have made side shoots (tillers), this species is among the hardest weeds to kill by uprooting. It is very tolerant of trampling, and it will flower and set seed when kept mowed to a height of 0.25 inch.

Time from emergence to reproduction: Overwintering plants begin growth in the very early spring and flower almost immediately. Peak seed production occurs in May to early June but also can occur throughout the year. Cold vernalization accelerated heading, which occurred at the 11-leaf stage with vernalization and 15-leaf stage without vernalization. Spring germinating plants of the annual form set seeds in 44–55 days, whereas the perennial form required 81–93 days. Shade may delay time to first flowering by up to 20 days. Cut flowering panicles can still produce viable seeds, suggesting that seeds mature quickly.

Pollination: Annual bluegrass primarily self-pollinates but occasionally cross-pollinates by wind dispersed pollen.

Reproduction: Annual bluegrass plants typically produce 1,050–2,250 seeds per plant. In a British study, annual bluegrass produced about 80 seeds per stalk when the plants were grown at low density. Pasture populations dominated by the perennial form averaged 30 seed stalks per plant in the first year and 450 the second. Populations from a mixture of other sites dominated by the annual form averaged 100 seed stalks per plant the first year and 180 in the second. Under ideal conditions, seed production could exceed 20,000 seeds per plant. In the perennial form of the species, the prostrate stems root at the nodes. The stems then decompose over the winter so that this form of the plant has true vegetative reproduction.

Dispersal: Annual bluegrass disperses short distances by rain wash and wind. People, however, contribute most to the dispersal between fields and farms. The seeds stick readily to wet shoes, tires and mowing machinery, and since seeds often reach high density in soil, they probably also frequently disperse in soil clinging to tillage and cultivation machinery. Mowing can be an important factor dispersing this species. Seeds also pass through cattle and are spread with the manure. Birds may occasionally disperse the seeds.

Common natural enemies: Bluegrass billbug larvae (*Calendra parvulus*) eat the rootlets. Several species of *Carambus* moth, including the bluegrass webworm (*C. teterellus*) attack annual bluegrass; many fungal diseases also attack it. Large earthworms eat annual bluegrass seeds and digest about 70% of what they consume (Hutchinson and Seymour 1982).

Palatability: People do not eat any part of annual bluegrass. Livestock, however, find it highly palatable.

Note: Annual bluegrass pollen causes hay fever symptoms in many people.

FURTHER READING

Hutchinson, C.S. and G.B. Seymour. 1982. Ecological flora of the British Isles. *Poa annua* L. *Journal of Ecology* 70: 887–901.

Mitich, L.W. 1998. Annual bluegrass (*Poa annua* L.). *Weed Technology* 12: 414–416.

Vargas, J.M. Jr. and A.J. Turgeon. 2004. *Poa annua*. Physiology, Culture, and Control of Annual Bluegrass. John Wiley & Sons, Inc. 184 pp.

Warwick, S.J. 1979. The biology of Canadian weeds. 37. *Poa annua* L. *Canadian Journal of Plant Science* 59: 1053–1066.

Barnyardgrass

Echinochloa crus-galli (L.) P. Beauv.

Barnyardgrass seedling
Joseph Neal, North Carolina State University

Barnyardgrass plant
Randall Prostak, University of Massachusetts

Barnyardgrass inflorescence
Joseph DiTomaso, University of California, Davis

IDENTIFICATION

Other common names: water grass, panic grass, cockspur grass, cocksfoot panicum, barn grass, summergrass, billion dollar grass, Japanese millet

Family: Grass family, Poacea

Habit: tall, upright, summer annual grass

Genus variability: The genus *Echinochloa* is composed of many distinct species and ecotypes that are difficult to classify. Two distinct clusters of weedy *Echinochloa* can be distinguished by molecular and structural markers, namely barnyardgrass (*E. crus-galli*) and late watergrass [*E. oryzicola* (Vasinger) Vasinger]. Both are found in North America; the former is a widespread weed found in many upland habitats whereas the latter, along with early watergrass [*E. oryzoides* (Ard.) Fritschis], are rice mimics that are found in permanently flooded rice fields in California. The watergrass species have larger seeds, less seed dormancy, shorter longevity in the soil and greater ability to emerge in deeply flooded fields. Junglerice [*E. colona* (L.) Link] is a predominant *Echinochloa* species in rice and other crops in Arkansas but requires several vegetative, reproductive and seed traits to distinguish from

barnyardgrass. The discussion in this chapter pertains to *E. crus-galli* unless otherwise noted.

Description: The first leaf of the **seedling** is longer (2.5–5.5 inches) and more upright than most weedy annual grasses but shorter and more parallel to the soil surface than later leaves. All other leaves are vertical and rolled in the bud; thus, they uncurl as they come out of the stem to form a flat or slightly V-shaped surface. No ligules or auricles are present. Leaf blades are hairless and rough on the edges. Seedlings often feel "flat" when held. The stem may be reddish near the base. **Mature plants** have many tillers and grow in clumps. Plants can reach 5 feet in height. Near the base, stems may have a few hairs and may be red-purple in color. Sheaths are split open near the top so that the underlying sheath of the next leaf is visible. The collar region is white, wide, hairless and with no ligule or auricle. Leaf blades are 4–8 inches long by 0.13–0.75 inch wide and have a prominent midvein. Barnyardgrass has a fibrous root system and can grow new roots where tillers touch the ground. The **inflorescence** is a green to purple, lumpy panicle. Short stalks are widely spaced along the stems, although this feature can vary. **Seeds** are teardrop shaped, brown-gray and flat on one side, measuring about 0.13 inch in size. As with other grasses, the apparent seed includes a thin, tightly adhering layer of fruit tissue. Seeds have ridges along their length, and the attached chaff has awns of 0.13–0.5 inch.

Similar species: Aside from the similar *Echinochloa* species discussed above, similar species occur in other genera. Fall panicum (*Panicum dichotomiflorum* Michx.) can be confused with barnyardgrass, but fall panicum has a ligule and coarser foliage than barnyardgrass. Yellow foxtail [*Setaria pumila* (Poir.) Roem. & Schult.] has a similar growth habit and red stem bases, but it has a hairy ligule and a somewhat soft inflorescence that resembles a bottlebrush or a fox's tail.

MANAGEMENT

Since the seeds are only moderately persistent, rotation with perennial forage crops can help reduce density of barnyardgrass. Because barnyardgrass germinates in warm soil, winter grains or grain crops planted in early spring will be highly competitive by the time the weed emerges. Also, shade by the crop will cool the soil and reduce further germination. For summer planted crops, a lag of two weeks or more between tillage and seedbed preparation will destroy many seedlings, provided the soil is warm and moist.

For crops planted in late spring and summer, cultivate frequently and close to the row to control this fast growing weed. Although timely tine weeding or rotary hoeing can greatly reduce the population, some individuals in the row will emerge from deeper than these implements can safely reach. Barnyardgrass may re-emerge following shallow burial but not if completely buried by 1.6 inches or more of soil. Therefore, hill up soil in the crop row to the fullest extent that the crop will tolerate.

Organic mulch placed early will keep the soil cool and suppress germination, but some barnyardgrass may emerge anyway. Mulch placed after the soil warms must be exceptionally dense or thick to be effective. Dense planting of summer cover crops or cash crops that will tolerate high density helps control this shade-intolerant weed.

Removing escapes before plants set seeds is useful for long-term control in intensively managed vegetable systems. Promptly cleaning up fields after harvest of small grains and vegetables is similarly helpful. In a pair of experiments in Arkansas, 41% of seeds were still on the plant at soybean harvest, indicating that a substantial proportion of seeds could be captured or destroyed during combine harvesting.

Since the seeds do not tolerate high soil temperatures when moist, solarization with clear plastic for 40 days kills barnyardgrass seeds in the top 1.2 inch of soil and reduces seed density deeper in the soil.

ECOLOGY

Origin and distribution: Barnyardgrass is native to South Asia and Europe. It ranges around the world from 50° N to 40° S latitude, except for tropical Africa. It occurs across southern Canada and in all states except Alaska.

Seed weight: Mean seed weight of 10 populations varied from 1.7–2.1 mg, with most near the lower end of the range.

Dormancy and germination: Seeds are dormant when shed from the parent plant. Three to eight months of either dry or moist after-ripening will break dormancy. Barnyardgrass seeds undergo an innate annual dormancy cycle that is not induced by soil conditions, with germinability reaching a peak in May–July and falling to near zero germinability in September–November. If the seeds are moist, alternate freezing and thawing for four days breaks dormancy of over 65% of seeds. Two weeks at 40°F or soaking for four days will each break dormancy of over 30% of seeds. Seeds of barnyardgrass will germinate from 60–100°F, with the optimum temperature for germination varying among populations in the range 77–100°F. Exposure to small amounts of light promotes germination of some seeds, especially after burial over the winter, but several days of exposure may be required for full effect. Temperature of 115°F for as little as half an hour removes the light requirement of most seeds. Nitrate only induces a small increase in percentage germination. Unlike most agricultural weeds, barnyardgrass will germinate in completely anaerobic conditions and when fully immersed in water.

Seed longevity: Barnyardgrass seeds can retain high viability for at least three years when buried in the soil, and a few seeds can persist for up to 13 years. One study, however, found less than 6% of seeds survived for six months. Another found approximately 37–42%

mortality after one year but with extreme variation in subsequent seed survival among replicates. Thus, local conditions can have a large effect on seed survival. Most seeds near the soil surface that do not emerge will lose viability within one year.

Season of emergence: Barnyardgrass begins emerging in mid-spring, reaches peak emergence in late spring or early summer, and a few plants continue to emerge until late summer or early fall. In California, barnyardgrass emergence is more closely associated with irrigation than with season.

Emergence depth: Seedlings emerge best from the top 1 inch, but substantial emergence occurs from 1–2 inches. Seedlings rarely emerge from deeper than 3 inches. One report indicated poor emergence from seeds on the soil surface and best emergence began from the 0.5-inch depth. In flooded conditions, emergence is from the top 0.8 inch.

Photosynthetic pathway: C_4

Sensitivity to frost: The species is frost sensitive.

Drought tolerance: Growth of young barnyardgrass is reduced more by dry soil than many other weed species and is about as drought tolerant as corn. Roots of well-developed plants reach to more than 40-inch soil depths, which allows them to tap deep soil moisture during dry periods.

Mycorrhiza: Barnyardgrass is mycorrhizal.

Response to fertility: Barnyardgrass is highly responsive to nitrogen fertility, with plant size increasing rapidly with N application rates up to 143 pounds per acre. Its growth response to increasing P is relatively flat compared to other weeds and crops, but it continues to increase in size up to application rates of 122 pounds per acre of P_2O_5. It can take up and concentrate phosphorus at the expense of crops. In Arkansas field margins, it occurred most frequently in soils of pH 6.4–6.8.

Soil physical requirements: Barnyardgrass does best on soils that hold water well but can tolerate soil textures as coarse as loamy sands. The species is highly tolerant of waterlogged and flooded soils, and is one of the worst weeds of rice paddies. Soils with relatively large waterholding capacity and high fertility provide an ideal substrate for this species. It is moderately tolerant of salinity and suffers less than rice in saline conditions.

Response to shade: Barnyardgrass tolerates 50% shade well. Plant weight, but not height, is greatly reduced by shade of 73% or more.

Sensitivity to disturbance: Some seedlings will re-emerge following complete burial, provided the soil remains moist. The fibrous root system of medium to large plants makes them difficult to fully uproot with hoes or cultivators. Barnyardgrass regrows readily after clipping.

Time from emergence to maturation: Plants take longer to mature and produce more seeds when they emerge in spring when day-length is increasing but mature sooner and produce fewer seeds when they emerge in summer when day-length is decreasing. Also, plants grow most rapidly at temperatures of about 95°F in the day and 77°F at night but mature more slowly than when temperatures are warmer or cooler. Plants typically flower five to eight weeks after emergence, with late emerging plants flowering most quickly; but very late emerging plants (September) are also slow to flower. A few seeds may become viable at the milk-soft dough stage as early as five days after flowering. Most seeds are viable by three weeks after flowering.

Pollination: Barnyardgrass is primarily self pollinated, but some wind mediated cross-pollination also occurs. Late in the season, flowers may self-pollinate while still wrapped in the sheath.

Reproduction: A large plant may produce over 100,000 seeds, whereas a highly stressed individual may produce less than 10. More typically, seed production is several thousand to several tens of thousands of seeds per plant, but seed production of over 200,000 per plant has been observed in a California sugar beet field. The intensity of competition from the crop greatly affects seed production. For example, plants emerging with corn produced 11,200 seeds per plant, whereas those emerging at the four-leaf stage produced only

200 seeds per plant. Other studies have shown similar large reductions in seed production for plants emerging after the crop.

Dispersal: Barnyardgrass has no inherent dispersal mechanism, and seeds become scattered around the mother plant. As a weed, the primary dispersal mechanism is on farming equipment or in crop seeds. The seeds are a common contaminant of crop seed, manure and cotton gin waste, and can be dispersed when these materials are used to fertilize fields. Barnyardgrass seeds can float for several days and disperse in irrigation water, and probably also in flood water. Seeds are dispersed by ducks and other birds, and by the awns sticking to animal fur.

Common natural enemies: A weevil, *Hyperodes humilus*, has been reported to attack the growing point of barnyardgrass in Massachusetts and kill young plants, but generally, this weed has no important natural enemies.

Palatability: The seeds are edible, and the plant is generally a good forage for livestock. Japanese millet (*E. frumentacea* Link) is a cultivated variety of barnyardgrass that is grown for grain in tropical Asia and Africa, and is used as a fast growing forage and cover crop in the United States. Under highly fertile conditions, nitrate may accumulate to levels that are toxic to livestock.

FURTHER READING

Holm, L.G., D.L. Plucknett, J.V. Pancho and J.P. Herberger. 1977. *The World's Worst Weeds: Distribution and Biology*. The University Press of Hawaii: Honolulu.

Maun, M.A. and S.C.H. Barrett. 1986. The biology of Canadian weeds. 77. *Echinochloa crus-galli* (L.) Beauv. *Canadian Journal of Plant Science* 66: 739–759.

Norris, R.F. 1992. Case history for weed competition/population ecology: barnyardgrass (*Echinochloa crus-galli*) in sugar beets (*Beta vulgaris*). *Weed Technology* 6: 220–227.

Vengris, J., M.E.R. Hill and D.L. Field. 1966. Clipping and regrowth of barnyardgrass. *Crop Science* 6: 342–344.

Bermudagrass

Cynodon dactylon (L.) Pers.

Bermudagrass growth habit
Jack Clark, University of California

Bermudagrass shoots growing from stolons
University of California Division of Agriculture and Natural Resources

IDENTIFICATION

Other common names: scutch grass, dogs-tooth grass, wire grass, couchgrass

Family: Grass family, Poaceae

Habit: Long-lived, prostrate, fine-leaved perennial grass that spreads by runners and rhizomes

Description: Seedling leaf sheaths and collars are smooth. Leaves are rolled in the bud and are smooth on both surfaces. The ligule is membranous, fringed and very short. Auricles are absent. The **mature plant** has erect stems up to 1.6 foot tall that arise from runners and rhizomes. Leaves are linear, 2–6.3 inches long by 0.08–0.2 inch wide, sometimes lightly hairy on the upper surface, and hairy on the lower surface. The collar region has an inconspicuous membranous ligule with a fringe of 0.01-inch-long hairs. Sheaths are sometimes hairy on lower leaves and have long, tufted hairs on their margins in the collar region. The plant has fibrous roots developing at nodes on the runners and rhizomes. **Inflorescences** are located at the end of stems in an arrangement of 3–9 spikes that are 1.2–4 inches long. Spikelets are 0.08–0.1 inch long and awnless; they consist of one flower that produces one light brown, egg-shaped **seed**.

Similar species: Creeping bentgrass (*Agrostis stolonifera* L.) also spreads through runners and rhizomes. However, it can be distinguished by its long, membranous ligules, smooth leaves and sheaths, and dense, open panicles.

MANAGEMENT

Bermudagrass is considered the world's worst grass weed, and it infests 40 crops in over 80 countries. The tenacity with which it survives inhospitable conditions has led to its development as a forage and turfgrass as well as to its success as a weed. Management generally requires considerable effort and persistence since intermittent tillage only serves to spread the extensive network of rhizomes and runners. Consequently, preventing its invasion into non-infested fields is critical, for example by cleaning machinery before entering a field and digging out any clumps when they first appear.

Near the northern limits of its range where the soil freezes in winter, plow the field in the fall and work the rhizomes to the soil surface, where they will freeze. Plants will emerge from rhizomes that escape, but this can reduce the infestation. In regions with predictable drought, moldboard plow during drought, leaving a rough surface so that air penetrates the soil. One plowing during a protracted drought period will substantially reduce bermudagrass. For a more rapid and thorough kill, repeat tillage at four- to seven-day intervals until the soil is completely dry to the full depth of the plow layer, and then continue this for at least an additional two weeks. This will kill rhizomes to the depth of tillage. When rains return or the field is irrigated, follow up with cultivation and hand hoeing in the row to eliminate sprouts from deep rhizomes.

Competitive cover crops are effective for suppressing bermudagrass. Plant a dense stand of rye, winter oats or winter barley in the fall. Harvest this for grain or forage and plow under the stubble. Plant a highly competitive summer cover crop like cowpeas or velvetbeans. The one year of dense shade and early summer soil disturbance will greatly suppress the bermudagrass. Heavy competition from taller plants is particularly useful against bermudagrass due to its short stature. Also, the shade will lower leaf and soil temperatures, which will slow growth of this heat loving species. High density planting of the cover crops is critical to ensure rapid canopy closure and dense early shade.

ECOLOGY

Origin and distribution: Bermudagrass is native to tropical Africa or possibly southern Asia but now occurs in most places between 45° N and 45° S latitude. It occurs throughout the United States except for some of the most northern parts of the Midwest and New England, but it is a serious weed primarily in the southern half of the country.

Seed weight: 0.23–0.36 mg.

Dormancy and germination: Seeds with hulls have low germination of 6–9%. Bermudagrass has negligible germination in the dark at any constant temperature. Fluctuating temperatures promote germination, and day/night temperature differences of more than 30°F are more effective than more moderate fluctuations. Light and nitrate further increase germination in a fluctuating temperature environment but are ineffective if the temperature is constant. Seeds will germinate in anaerobic conditions if other factors are favorable.

Seed longevity: No information on survival of bermudagrass seeds could be located. If the species is like most perennial grasses, the seeds disappear from the soil within a few years.

Season of emergence: In a California study, shoots began emerging from transplanted plugs when soil temperature at 2 inches reached 63°F. The rate of shoot emergence was slow in March and April but increased rapidly after mid-May.

Emergence depth: Seedlings emerge best from about 0.25 inch but emerge reasonably well from the top 1 inch of soil. None emerge from deeper than 2 inches. The rhizomes can occur as deep as 40 inches, and shoots can emerge from below the plow layer. However, shoots are less likely to emerge from rhizome fragments below 6 inches than from rhizome fragments at 2–4 inches.

Photosynthetic pathway: C_4

Sensitivity to frost: Bermudagrass grows poorly in cold weather and quickly ceases growth after frost. Exposed rhizomes are killed by freezing. In contrast, the species thrives in climates where the temperature commonly reaches 100°F.

Drought tolerance: Bermudagrass becomes dormant during extended droughts and then sprouts from rhizomes when moisture is again available. However, its productivity and rate of clonal expansion respond strongly to water availability. Rhizomes can recover from short periods of low soil moisture but will not resprout after seven days of desiccation in completely dry soil. The species can also survive prolonged flooding.

Mycorrhiza: Bermudagrass is mycorrhizal.

Response to fertility: Bermudagrass is capable of surviving low fertility conditions. However, the growth of plants established from seeds responds strongly to N, P or K applied alone, and growth is even more rapid if the fertility is balanced. Established plants respond strongly to N fertilizers but respond to P and K only if N is not limiting.

Soil physical requirements: Bermudagrass can grow on any soil type from sand to heavy clay, but it grows best on medium to heavy textured soils that are well drained but remain moist.

Response to shade: Bermudagrass does not tolerate shade. In a series of experiments, the size of bermudagrass clones increased exponentially in response to the amount of sunlight received.

Sensitivity to disturbance: Bermudagrass can withstand frequent, intensive defoliation, and selections of the species are used for golf course greens, which are mowed daily to less than 0.375 inch. In temperate regions like the United States, carbohydrate reserves reach a peak in winter and decline to a low in mid-summer. Consequently, the species is most sensitive to disturbance of the rhizomes during mid-summer.

Time from emergence to reproduction: Small sod plugs, similar to what might be dispersed by tillage machinery, produced flowers in 10–15 weeks when planted in March to May in southern California. Those planted in June through September 1 flowered in as little as four weeks and produced seeds within eight weeks. A planting on October 1 did not flower.

Pollination: Bermudagrass is wind pollinated.

Reproduction: Most reproduction is by fragmentation of rhizomes and runners during tillage. Small four-node rhizome fragments buried at about 2.5 inches in the spring produced 3.3–6.6 feet of runners and 12–28 inches of rhizome in four weeks. These had approximately 20–40 runner nodes and 10–23 rhizome nodes. Heavier four-node fragments of both types produced a greater length of runners and rhizomes. Small sod plugs planted 8 inches apart in a row in southern California produced up to 85 rhizome nodes per square foot with the greatest density produced by plantings in March and July. Plugs planted in March produced 6,800 seeds per square foot, with seed production declining with later plantings through the spring and summer.

Dispersal: The seeds can survive at least 50 days of submergence in water. They disperse in irrigation water and probably also by overland flow and along streams. Seeds pass through cattle and are spread with the manure. The small seeds of bermudagrass can lodge in the fur of animals and thereby can be dispersed over large areas. Seeds do not shatter easily and are moved about with hay. Most dispersal, however, occurs by movement of vegetative parts. Tillage implements can drag rhizome and runner fragments long distances within fields. Rhizomes and runners get carried in mud on cattle and on farm machinery. Pieces of sod can float down streams. Runner fragments may be included in hay or lawn clippings used as mulch. The species has been dispersed between ports in ship ballasts and in packing materials.

Common natural enemies: Domestic cultivars of bermudagrass are attacked by a mirid bug (*Trigonotylus doddi*), fall armyworm (*Spodoptera frugiperda*), leafhoppers (mainly *Carneocephala flaviceps*, *Exitianus exitiosus*, *Graminella nigrifrons*, *G. sonora*, *Draeculacephala balli* and *Cuerna costalis*) and planthoppers (mainly *Delphacodes propinqua* and *Sogatella kolophon*). The species probably has similar susceptibility when growing as a weed.

Palatability: Bermudagrass is a high quality forage when young but quality deteriorates substantially as it matures. Mature bermudagrass can cause poisoning of cattle, horses and goats, possibly due to a fungal infection in the foliage, but cases are rare.

FURTHER READING

Holm, L.G., D.L. Plucknett, J.V. Pancho and J.P. Herberger. 1977. *The World's Worst Weeds: Distribution and Biology.* The University Press of Hawaii: Honolulu.

Keeley, P.E. and R.J. Thullen. 1989. Influence of planting date on growth of bermudagrass (*Cynodon dactylon*). *Weed Science* 37: 531–537.

Mitich, L. 1989. Bermudagrass. *Weed Technology* 3: 433–435.

Phillips, M.C. 1993. Use of tillage to control *Cynodon dactylon* under small-scale farming conditions. *Crop Protection* 12: 267–272.

Downy brome

Bromus tectorum L.

Downy brome young plants
Randall Prostak, University of Massachusetts

Downy brome plant
Joseph DiTomaso, University of California, Davis

IDENTIFICATION

Other common names: downy brome grass, slender chess, early chess, downy chess, cheatgrass, drooping brome, wall brome, cheatgrass brome, slender brome, drooping brome grass, thatch grass

Family: Grass family, Poaceae

Habit: Winter annual grass, sometimes a summer annual

Description: A very long (3 inches) and narrow (0.125 inch) seed leaf emerges perpendicular to the soil. Young, true leaves of the **seedling** are rolled in the bud, unrolling as they emerge from the stem in an upward twisting or spiral pattern. Blades have soft, fuzzy hairs at the tip and on their undersides. Sheaths are white with reddish or maroon bases and short, fuzzy hairs. **Mature plants**, reaching 4–24 inches tall, are coated throughout with short, downy hairs. Ligules are translucent, 0.1 inch long and slightly jagged or frayed. Auricles are absent. Blades are 1.25–8.25 inches long by 0.1–0.3 inch wide, flat and sharply pointed at the tips. Sheaths are open at the top and pink-veined. Roots are fibrous; unlike many other grasses, roots do not form at the nodes of tillers. **Inflorescences** are 1.5–8 inches long, branched, nodding, oat-like panicles. Panicles have numerous soft, white hairs and are frequently purple-hued and shiny, especially when the seed is ripe. Individual spikelets are large (0.5–2 inches long) and contain three to 10 flowers. The fuzz-coated, papery outer chaff of each spikelet has one ridge; the outer chaff of each flower within the spikelet is wooly and has a 0.4–0.7 inch-long awn. Awns become sharp and stiff with seed maturity. **Seeds** are 0.2–0.3 inch long if the outer chaff is removed or 0.4 inch if it is still tightly attached. Seeds are linear or lanceolate, yellow to red-brown and grooved.

Similar species: Field brome (*Bromus arvensis* L.), which is also called Japanese brome, is best distinguished from downy brome by the outer chaff of the spikelets, which has three ridges rather than one. Unlike downy brome, cheat (chess) (*Bromus secalinus* L.) is largely hairless and has awns no longer than 0.2 inch. Common velvetgrass (*Holcus lanatus* L.) has sheaths that overlap one another and hairs on the back of the ligule.

MANAGEMENT

Reduced tillage practices that retain surface residue create a favorable environment for downy brome in fall-seeded crops. Crop rotation with spring-planted crops helps control downy brome in winter wheat cropping systems. Because downy brome is a winter annual, seedbed preparation for a spring-planted crop will destroy the previous year's seedlings, and relatively few new seedlings will germinate in the spring crop. Similarly, delaying planting of fall-sown crops allows more time for downy brome to germinate, so that more seedlings are destroyed during seedbed preparation and density is reduced in the subsequent crop. Dryland crop rotations that include summer fallow should be cultivated after wheat harvest or should begin early enough in the spring of the fallow year to prevent seed set.

Downy brome seeds are relatively short lived in the soil and need near-surface conditions for emergence. As a consequence, plowing the current year's seed crop under before planting is an effective management tool. Few seeds are dormant when shed from the parent plant, and many will germinate too deeply to emerge following fall plowing. Seeds that are dormant will die before later tillage events return them to the surface.

The dense mat of fine roots produced by downy brome can reduce soil moisture levels below the tolerance point of perennial forage species like bluebunch wheatgrass [*Pseudoroegneria spicata* (Pursh) Á. Löve ssp. *spicata*]. Consequently, reducing downy brome populations before seeding new pastures is advisable. In existing pastures, avoid grazing practices that produce gaps larger than 24 inches, since uniform stands prevent establishment of downy brome. Fire and overgrazing are highly associated with invasion of this species.

ECOLOGY

Origin and distribution: Downy brome is a native of the Mediterranean region of Europe, North Africa and the Middle East. It has been introduced widely in other parts of Europe and Asia, and in Australia, New Zealand and South Africa. In North America, it occurs from Alaska and southern Canada throughout the United States and into Mexico, except in the extreme southeastern corner of the United States. As an agricultural weed, it is primarily a problem in the western half of the continent where rainfall occurs primarily in fall and winter followed by dry summers.

Seed weight: Mean population seed weight is 2.5–3.7 mg, including the chaff.

Dormancy and germination: A substantial but variable proportion of downy brome seeds are able to germinate when shed from the parent plant. Dormancy of the remaining seeds is broken by one to three months of after-ripening. The rate of after-ripening increases with increasing soil moisture. In the field, germination is typically delayed by dry soil conditions until the onset of fall rains, and by then all seeds are able to germinate. Optimal germination temperature increases with seed age, generally ranging from 59–68°F during the normal fall germination period. However, germination can occur at temperatures as low as 32°F. Temperature fluctuation does not promote germination. Some seeds will germinate in as little as two days, but complete germination of apparently non-dormant seeds may require five weeks or longer. Unlike most weeds, light has either no influence on or can actually inhibit germination of downy brome. Nitrate, however, stimulates germination.

Seed longevity: Maximum seed longevity in undisturbed soil has been variously reported at two to five years. The great majority of seeds germinate the first fall after being shed. Only a few of the remainder survive until the following fall, and survival of any remaining seeds is negligible. Seeds buried 1 inch in the soil have a very low (less than 1%) survival rate. Fire can reduce the seed bank to less than 3% of unburned areas, but subsequent seeding from surviving seed can reestablish the seed bank after only one season.

Season of emergence: Most seedlings emerge in the fall following significant rains. If conditions remain mild, emergence may continue into the winter. Some seedlings emerge in the spring, but seed production by spring emerging plants is highly variable. Flexibility in reproductive success over a wide range of emergence periods adapts this species to environments with highly variable conditions.

Emergence depth: Most seeds produced seedlings from depths of less than 2 inches, but a few seedlings emerged from more than 3 inches.

Photosynthetic pathway: C_3

Sensitivity to frost: Downy brome is a highly frost-tolerant winter annual that survives temperatures of 10°F with little damage. Although aboveground growth ceases during cold weather, root growth continues in soil as cold as 37°F. Overwinter survival rates are generally high, with over 50% of plants surviving from seedling to maturity, but death from frost-heaving can occur when temperatures alternate above and below freezing.

Drought tolerance: Downy brome does not develop a deep root system in the fall and thus is sensitive to fall drought. By flowering time, however, roots penetrate from 20–50 inches or more. Under highly variable soil moisture conditions, downy brome usually has an advantage over competing species, thus accounting for its high invasive potential.

Mycorrhiza: Downy brome can be mycorrhizal but does not require a fungal associate. This characteristic allows establishment in disturbed sites with minimal mycorrhizal fungi.

Response to fertility: Downy brome tolerates extremely infertile soil conditions. Nevertheless, the species is highly responsive to N and P. Growth continues to increase with additional P, even at high P rates. The growth rate in soil already invaded by downy brome is higher than in uninvaded soil, a phenomena partially explained by increased availability of N, P, Mn and S in downy brome infested soil.

Soil physical requirements: Downy brome thrives on a wide range of soil types, but not on saline soils or very dry soils. It will grow on badly eroded land.

Response to shade: Downy brome is moderately shade tolerant. Moderate shade can promote growth.

Sensitivity to disturbance: Seed production by downy brome can be greatly reduced by mowing the plants between flowering and the dough stage of seed development. Mowing cannot completely eliminate the weed at any stage in plant development, however, because some plants can resprout if mowed before the dough stage, and seeds will mature on the clippings if the plant is cut at the dough stage or later. Mowing or grazing downy brome at any stage in the life cycle reduces ultimate plant size and seed production.

Time from emergence to reproduction: Plants typically emerge in the fall and flower in mid- to late spring,

and seeds mature about four weeks later. Flowering is earlier in populations from hot, arid regions compared to those from cool, moist forested regions.

Pollination: Downy brome flowers self-pollinate.

Reproduction: Plants typically produce all their seeds in a single burst of reproduction before dying. Occasionally plants will rejuvenate and produce a second burst of reproduction following rain. Plants growing in favorable conditions can produce more than 500 seeds each, but most plants in more typical situations produce 10–80 seeds each.

Dispersal: Downy brome spread through the western United States primarily in bedding and manure of transported cattle, but also in straw used for packing dry goods and in contaminated crop seed. The awns stick in clothing and in the skin and fur of animals. Seeds are also dispersed by rodents. Seeds disperse in irrigation water.

Common natural enemies: Head smut (*Ustilago bullata*) is very common in some areas and can greatly reduce seed production. The fungus *Pyrenophora semeniperda* kills a large proportion of dormant seeds, particularly in dry sites and years, and infects a greater proportion of seeds as they age.

Palatability: Downy brome is not eaten by people. Immature plants are excellent forage for livestock and make up a substantial part of spring forage for cattle in much of the west. The stiff awns of mature downy brome irritate the mouths and digestive tracts of livestock.

FURTHER READING

Mack, R.N., and D.A. Pyke. 1983. The demography of *Bromus tectorum*: Variation in time and space. *Journal of Ecology* 71: 69–93.

Upadhyaya, M.K., R. Turkington and D. McIlyride. 1986. The biology of Canadian weeds. 75. *Bromus tectorum* L. *Canadian Journal of Plant Science* 66: 689–709.

Young, F.L., A.G. Ogg, Jr. and J.R. Alldredge. 2014. Postharvest tillage reduces downy brome (*Bromus tectorum* L.) infestations in winter wheat. *Weed Technology* 28: 418–425.

Fall panicum

Panicum dichotomiflorum Michx.

Fall panicum seedling
Randall Prostak, University of Massachusetts

Fall panicum plant and inflorescence
Antonio DiTommaso, Cornell University

Fall panicum plant
Randall Prostak, University of Massachusetts

IDENTIFICATION

Other common names: spreading witch grass, sprouting crab grass, smooth witchgrass, western witchgrass

Family: Grass family, Poaceae

Habit: Large, erect, summer annual grass

Description: Seedlings often have a purple hue. Seed leaves and first true leaves are parallel to the ground, lanceolate to linear in shape and densely hairy on their undersides. True leaves are rolled in the bud and hairless after the fourth or fifth leaf stage. Hairy ligules are 0.06 inch long; auricles are absent. Blades are rough edged. Sheaths are compressed and hairy. **Mature plants** are largely hairless, reaching up to 5 feet tall. Stems are waxy with irregularly spaced, swollen nodes that give the stem a zigzagged appearance. Red-purple sheaths are compressed, smooth, hairless and open at the top. Collars are wide; ligules are densely hairy and 0.06 inch long. Blades are dull green with red edges, glossy undersides and lanceolate to linear, 4–20 inches long by 0.25–0.75 inch wide. Leaves have a pale green to white, prominent midvein. The root system is fibrous with spontaneous roots developing at lower stem-sheath joints. The **inflorescence** is a widely spaced, 4–16 inch long panicle with long, subdividing branches. Panicles are purplish at maturity and are located at stem ends and in leaf joints. Narrow, stiff stems support stalked, narrow spikelets at the branch tips. Spikelets are dull yellow to purple-hued, 0.10–0.12 inch long by 0.08 inch wide, and oval to thumb shaped. Spikelets each produce one dull, yellow-brown, 0.06 inch long **seed**.

Similar species: Yellow foxtail (*Setaria pumila* (Poir.) Roem. & Schult.) is similar at the seedling stage, but it has hairless sheaths and long wispy hairs near the blade base. Witchgrass (*Panicum capillare* L.) seedlings have hairs on both blade surfaces. Wild-proso millet (*Panicum miliaceum* L.) is distinguished from fall panicum by its half membranous, half hairy ligule.

MANAGEMENT

Crop rotation is a key element in controlling fall panicum. Since it does not germinate until the soil warms, either winter grains or early planted spring grains are highly competitive with this species. Also, leaf cover from well-established grain crops will keep soil temperatures cool and relatively constant, which will decrease germination. Moreover, both fall and spring sown grains are harvested before fall panicum sets seed. Post-harvest cleanup or a competitive interplanted cover crop like red clover can prevent seed set, particularly if the cover crop is mowed when fall panicum begins to flower. Set the mower at about 4–5 inches; the goal is to sever the young flowering stalks but leave the cover crop enough tissue to regrow quickly and smother the weed. Since the seeds do not persist well in the soil, rotation with hay is helpful in suppressing this species. In vegetables, any short season crop that is harvested before fall panicum begins to set seed in late summer will disrupt the life cycle, provided the field is cleaned promptly after harvest.

For row crops, tine weed pre- and post-emergence, and throw soil into the row to cover seedlings if the crop will tolerate such a procedure. If possible, cultivate shallowly close to the row. Severing the root from the shoot with minimal soil disturbance may be more effective than excavating or burying this weed. Unless cultivation is effective, long season row crops are likely to favor fall panicum.

Avoid soil compaction, as this favors fall panicum. Similarly, if the field has drainage problems, drain tile or other measures to improve drainage will increase the competitiveness of crops relative to the weed.

ECOLOGY

Origin and distribution: Fall panicum is native to North America. It occurs throughout the United States, Mexico and the West Indies, and it has been introduced throughout the more humid parts of southern Canada. Its status as a native in much of the western United States seems doubtful.

Seed weight: 0.2–0.9 mg.

Dormancy and germination: Seeds are dormant when shed due to the tight, hard chaff. Four to five months of after-ripening or two weeks of cold, moist conditions breaks this dormancy. Even when this first level of dormancy is relieved, the seeds still require light to germinate and either high temperatures or day/night temperature fluctuations in the range of 68/50°F to 95/68°F to germinate. Seeds are only capable of dark germination when dug up in mid-summer months and are exposed to a day/night temperature fluctuation of 95/68°F. Fall panicum seeds require high soil moisture for germination. For example, germination of this weed is more sensitive to moderately dry soil than is

corn or common lambsquarters.

Seed longevity: Two percent of seeds germinated after 10 years of burial at 9 inches under sod in Nebraska. Seeds buried at 8 inches in silt loam and gravelly loam in Ontario declined by 39% and 56% per year over a 4.5-year period. Seeds buried more shallowly disappeared faster, partially due to seedling emergence.

Timing of emergence: Fall panicum emerges primarily in mid- to late-spring in the South and mid-spring to mid-summer in the North, but some plants emerge throughout the growing season.

Emergence depth: Fall panicum emerges best from the soil surface to 1 inch, but a few seedlings can emerge from as deep as 3 inches.

Photosynthetic pathway: C_4

Sensitivity to frost: This species is sensitive to frost. Seeds fall from the plant after the first frost.

Drought tolerance: This species tends to be favored by moist conditions. Emergence and seedling survival were enhanced by supplemental irrigation that alleviated droughty conditions.

Mycorrhiza: There are no reports of the mycorrhizal status of fall panicum, but other *Panicum* species are mycorrhizal.

Response to fertility: Increasing rates of 10-10-10 fertilizer promoted growth of fall panicum between corn rows where competition was reduced. Conditions were favorable for corn growth in this experiment, and competition from the crop prevented a noticeable fertility response by the weed within rows. Potassium concentration was lower in plants growing in the row than between the rows.

Soil physical requirements: Fall panicum occurs on soils with a wide range of textures, but it does best on fine textured and organic soils. Fall panicum is an indicator of anaerobic, compacted soils. It tolerates waterlogged and salty soils.

Response to shade: Shade from crops can substantially

decrease biomass, height and seed production and can increase maturation rate.

Sensitivity to disturbance: Cutting seedlings just below the base of the shoot kills them, so shallow tillage that severs seedlings should be effective at controlling this species.

Time from emergence to flowering: Plants emerging in early May in Wisconsin flowered in about three months. Time to flowering and seed set decreases with later emerging plants. For example, in Massachusetts, plants emerging in late May required 116 days to set seed, whereas plants emerging in late July required only 74 days.

Pollination: Fall panicum is wind pollinated.

Reproduction: Typical plants growing in the field in Massachusetts produced 500,000 seeds per plant. Production of 10,000–100,000 seeds per plant may be more usual for larger plants. Seed production under competition from a mixed community of weeds will range from 100–300 seeds per plant.

Dispersal: As plants commonly grow near water and the seeds float, the seeds probably disperse in waterways. Fall panicum seeds are spread with cattle manure. Given the high density of seeds sometimes found in the soil, seeds probably also spread with soil clinging to tires, shoes and machinery.

Common natural enemies: The specialist smut fungus (*Ustilago destruens*) can greatly reduce seed production of fall panicum.

Palatability: People do not eat fall panicum. At the vegetative stage, fall panicum can provide ruminants with nutrient levels similar to cultivated grasses. The species can cause nitrate poisoning and photosensitivity in livestock.

FURTHER READING

Brecke, B.J. and W.B. Duke. 1980. Dormancy, germination, and emergence characteristics of fall panicum (*Panicum dichotomiflorum*) seed. *Weed Science* 28: 683–685.

Foxtails

Giant foxtail, *Setaria faberi* Herrm.
Green foxtail, *Setaria viridis* (L.) P. Beauv.
Yellow foxtail, *Setaria pumila* (Poir.) Roem. & Schult. = *S. glauca* (L.) P. Beauv.
= *S. lutescens* (Weigel) F.T. Hubbard

Giant foxtail stand
Antonio DiTommaso, Cornell University

Giant foxtail leaf surface
Scott Morris, Cornell University

Giant foxtail inflorescence
Scott Morris, Cornell University

Green foxtail seedling
Scott Morris, Cornell University

Green foxtail inflorescence
Scott Morris, Cornell University

Yellow foxtail seedling
Jack Clark, University of California

Yellow foxtail hairs at the base of a leaf blade
Scott Morris, Cornell University

Yellow foxtail plant with inflorescence
Randall Prostak, University of Massachusetts

IDENTIFICATION
Other common names:

Giant foxtail: Faber's foxtail, Chinese millet

Green foxtail: bottle grass, pigeon grass, wild millet, green brittle grass, pussy grass, green bristlegrass, green bottle grass

Yellow foxtail: pigeon grass, summer grass, golden foxtail, wild millet, glaucous bristly foxtail, yellow bristle grass

Family: Grass family, Poaceae

Habit: Erect or sprawling summer annual grasses

Description: Seedlings are rolled in the bud, uncurling roughly parallel to the ground. Ligules of all three species are very short and hairy. Auricles are absent.

Giant foxtail: The seed leaf is oval, 0.5–1 inch long by 0.1–0.2 inch wide. Young true leaves have short hairs on the sheath margin and upper blade surface.

Green foxtail: The seed leaf is 0.5–0.8 inch long by 0.1–0.16 inch wide. Young true leaf blades are hairless, rough and narrow, and light green with red near the collar; sheaths are densely hairy, slightly compressed and open at the top.

Yellow foxtail: The seed leaf is elongated, up to 2 inches long by 0.13 inch wide. Leaf blade edges are smooth to slightly rough, and the rest of the blade is hairless and flat. The collar region is green and smooth with a scattering of long, silky hairs. The sheaths are hairless, ridged and compressed. The lower seedling stem may be red.

Mature plants have hairy, 0.04–0.13 inch-long ligules. The root system of these tillering plants is fibrous.

Giant foxtail: Stems can reach up to 6.5 feet but are usually 2.5–4 feet tall. Stems are upright and hairless. Sheaths are light green and open, with a hairy edge. Ligules are densely hairy. Blades are arching, 12–20 inches long by 0.0.3–0.8 inches wide and light green with a white midrib; blades have short rough hairs covering their upper surface.

Green foxtail: Thin, round stems are 0.5–3 feet tall and radiate vertically from the center of the plant. Sheaths are concentrated at the stem base, slightly flat, smooth to rough, self-wrapping and hairy edged. The collar may be reddish. Light green blades are flat, hairless and 10–12 inches long by 0.25–0.38 inch wide.

Yellow foxtail: Stems can reach 2–4 feet in height. Sheaths are flattened, hairless, split open at the top, red at the base and collar, and have a ridged midvein. Ridged blades are 10–12 inches long by 0.25–0.5 inch wide, with 0.13–0.4 inch long, silky hairs on the upper surface near the collar. Plants are sometimes red at the base.

Inflorescences, located at stem tops, are bristly, spike-like panicles. Spikelets are set in densely packed, short-branched whorls. Spikelets are one-flowered. What we refer to as seeds in this chapter include a thin, dry tightly adhering layer of fruit tissue.

Giant foxtail: Panicles are nodding, 1.5–8 inches long by up to 1.25 inches wide, green-purple, and are the largest of the three species. Each 0.13 inch-long spikelet has three to six yellow green, 0.2–0.4 inch-long awns. Seeds are 0.1 inch long by 0.06 inch wide and wrinkled.

Green foxtail: Panicles are erect, 1.25–3.25 inches long by 0.5–1 inch wide, and greener and broader than yellow foxtail. Each 0.06–0.08 inch-long spikelet has 1–3 green-purple, slightly barbed, 0.25–0.5 inch-long awns. There are several pale green to purple spikelets per branch. Seeds are 0.09 inch long by 0.05 inch wide and slightly cross-wrinkled, with one flat side and one round side.

Yellow foxtail: Panicles are erect, 0.75–6 inches long by 0.5 inch wide, yellowish and hairy along the central stem. Each 0.08–0.1 inch-long spikelet has five or more yellowish-brown, 0.16–0.35 inch-long awns. Each branch has one spikelet. Seeds are 0.13 inch long by 0.08 inch wide, horizontally ridged, dull gray-yellow, pointy tipped and ellipse shaped. Seeds are the largest of the three species.

Similar species: Bristly foxtail [*Setaria verticillata* (L.) P. Beauv.] is a weak-stemmed, lodging plant with blue-green leaves. Its spikelets are widely spaced and cling to objects. Fall panicum (*Panicum dichotomiflorum* Michx.) and the crabgrasses (*Digitaria* spp.) both have dense hairs on the sheaths and blades at the seedling stage. Mature fall panicum has swollen, angled stem joints, no hairs and a white midvein. Crabgrass tillers heavily and has a translucent, jagged ligule.

MANAGEMENT

The key to managing the foxtails lies in the timing and depth of tillage. Since the seeds (particularly of yellow foxtail) are relatively large and highly palatable, leaving them on the soil surface over winter will encourage seed predation by birds, rodents and insects, and will thereby decrease the population. Shallow cultivation in late spring to incorporate the seeds followed by a short fallow will cause a flush of emergence. The seedlings can then be killed by tillage for seedbed preparation. Moldboard plowing prior to seedbed preparation will place seeds from the previous year too deep for emergence. Since few foxtail seeds persist for more than two years, most of the seeds buried by plowing will die before they return to emergence depth. Planting short cycle vegetable crops in mid-spring will similarly decrease the population. Summer fallows are relatively ineffective against foxtail because the seeds enter secondary dormancy in hot weather. Avoid moldboard plowing in late summer and fall after seed drop when most seeds will be dormant.

Foxtails form a thin thread of tissue that feeds energy from the seed into the base of the shoot. This can be easily broken by a rotary hoe or tine weeder just before or shortly after emergence. Once crown roots form, however, plants are much more difficult to kill. Because green and yellow foxtail elongate rapidly, even young plants are difficult to bury during cultivation, so inter-row cultivation should throw soil into the row as soon as the crop can tolerate it. Foxtails reroot readily if the soil is moist.

Giant foxtail is well-adapted to no-tillage cropping systems because of its capacity to germinate at the soil surface, particularly in the presence of crop residue that can provide a moist microclimate for establishment. This species is particularly adapted to germination and establishment in organic roll-kill systems, so maintaining a low giant foxtail seed bank is important for the success of these systems. Organic mulch is relatively ineffective for suppressing yellow foxtail because of its ability to elongate greatly when shaded and because of substantial food reserves in the seed. Giant foxtail is more susceptible to mulching but can thrive where gaps in the mulch or sparse soil coverage occur. Even if emergence is not suppressed by mulch, growth of foxtail species can be delayed and reduced by mulches, thereby giving crops a competitive advantage.

Foxtails are susceptible to crop competition and especially shade. Winter grains suppress all three species well due to an early head start in the spring. Oats and spring barley will establish at lower temperatures than any of the foxtails, so early planting gives these crops a head start and a competitive advantage.

When competing with yellow foxtail, corn yield losses tend to decrease as nitrogen fertility improves, even though the foxtail becomes more vigorous. This apparently occurs because the N stimulates deep penetration of corn roots, which allows corn to access water unavailable to the foxtail during late-season dry periods. Giant foxtail, however, continues to increase in size at compost rates above those that maximize corn yield.

Because the foxtails usually do not set seed until after harvest of both winter and spring sown grain crops, cleanup of fields after harvest can prevent seed set and interrupt the species' life cycle. Cleanup after early harvested vegetables is similarly useful. Consequently, crops that allow mid-summer harvest should be planned into the rotation if foxtails are a problem.

ECOLOGY

Origin and distribution: Giant foxtail originated in eastern Asia and now occupies most of the eastern two-thirds of North America. It is particularly a problem in the Corn Belt. Green foxtail is native to Europe and is now widespread in the temperate regions of the world. It is most troublesome in the northern Midwest. Yellow foxtail is native to Eurasia and occurs throughout Asia, Europe, North America and the wetter parts of Australia. It also occurs in southern Africa, the Caribbean and the Andean countries of South America.

Seed weight: Giant foxtail: 1.6 mg; green foxtail:

0.6–1.5 mg; yellow foxtail: 1.9–4.2 mg.

Dormancy and germination: Newly dispersed seeds are usually completely dormant, though some varieties of green foxtail apparently produce non-dormant seeds. Seeds after-ripen and lose dormancy most quickly when exposed to cool, moist conditions. Consequently, many are ready to germinate when weather warms in the spring. Significant germination begins at temperatures in the range of 50–59°F. The range of temperatures suitable for germination widens as the seeds after-ripen. Thus, fully after-ripened seeds may germinate well at 68–95°F, but such temperatures will often maintain dormancy in seeds that have not fully after-ripened. High temperatures (higher than 86°F) frequently induce secondary dormancy. Thus, foxtail seeds commonly cycle out of dormancy in winter and into dormancy in the summer. Day-night alternation in temperature promotes germination, but light usually does not.

Seed longevity: When deeply buried in undisturbed soil, a few seeds will last for 10–39 years. In agricultural situations, however, the seeds die rapidly. In Saskatchewan, the great majority of green foxtail seeds died or emerged during the first one to two years, with a similar result for giant foxtail in a multi-state experiment in the midwestern United States. In a cropped field with simulated tillage, very few giant foxtail seeds survived two years, and none survived for three. Annual seed mortality during the first year in these experiments ranged from 83–93%. In New York, annual mortality rates of giant foxtail seeds buried 6 inches ranged from 72–86%. No green or yellow foxtail seeds buried at 1.7 inches under turf grass survived longer than three years.

Season of emergence: Most emergence occurs mid-spring to early summer, but a few plants continue to emerge throughout the growing season.

Emergence depth: Considering the difference in seed size between the three species, their ability to emerge from various depths in the soil is remarkably similar. Emergence tends to be best with shallow burial (0.5 inch), but emergence is still high down to 2 inches. Emergence from deeper in the soil is poor and is negligible from depths greater than 4 inches.

Photosynthetic pathway: All three species have the C_4 pathway.

Sensitivity to frost: Seedlings are frost sensitive: When exposed to mild frosts they die or show substantial damage from which they may recover. Most adult plants are already dying by the first frost in autumn.

Drought tolerance: Yellow foxtail recovers rapidly from drought periods of over one month, even when the plants are young.

Mycorrhiza: Green foxtail has been observed with mycorrhizae. Giant and yellow foxtail are considered weak hosts to mycorrhizal fungi.

Response to fertility: Giant foxtail tends to perform more poorly than other foxtails on nitrogen deficient soil. In contrast, its growth continues to increase up to compost rates that supply 320 pounds per acre of N and 250 pounds per acre of P_2O_5. Giant foxtail tends to out-compete yellow foxtail on highly fertile soils. Shoot growth of green foxtail increases greatly up to N application rates of 71 pounds per acre and increases only slightly more with higher rates, whereas root growth increases steadily up to N application rates of 437 pounds per acre. In contrast, shoot growth of green foxtail responds greatly to P, but response of roots to P is slight. Yellow foxtail tolerates infertile soil but responds to higher fertility with profuse growth and high seed production. It grows on soils with a pH from 6.1–8.

Soil physical requirements: All the foxtails will grow on a wide range of soil textures, but green and giant do best on sandy to loamy soils. Yellow foxtail grows on soil types ranging from clay to river gravel and can tolerate poor drainage. In a compaction study, giant foxtail produced a dense, vigorous stand of plants on heavily compacted soil. In the same study, yellow foxtail emerged poorly on the compacted soil, but plants that did establish grew well. In another study, green foxtail emerged better from compacted soil in one of two years.

Response to shade: Foxtail growth and seed production declines with increased shading by neighboring plants, and giant foxtail is completely suppressed by

98% shade, the amount produced by a high corn population. Moderate shading, however, causes plants to grow taller, though with less tillering. This helps them penetrate into and through the crop leaf canopy.

Sensitivity to disturbance: Yellow foxtail cut at 2 inches form many new tillers and short seed stalks, but close grazing of grain stubble by sheep can prevent seed production. Uprooting or burial readily kills young foxtail seedlings. Larger plants are also susceptible to hoeing or cultivation, especially shallow cultivation that severs the roots from the shoot. New shoots cannot sprout from the roots alone. However, giant and yellow foxtail shoots, and even single tillers, will readily root into moist soil.

Time from emergence to reproduction: Plants emerging in the spring usually begin flowering in July as day length decreases. Since flowering is triggered by shortening day length, the time from emergence to flowering varies from a few weeks for late emerging plants to three months for early emerging plants. The first seeds mature about two weeks after flowering, but seeds on a head continue to mature and disperse over a two- to three-week period.

Pollination: Foxtails normally self-pollinate but occasionally cross-pollinate by wind.

Reproduction: Flowering usually begins mid-summer for spring germinating plants, and plants often continue to flower and set seed until the weather becomes cold. Very stressed plants may produce only a single seed. In corn and soybean crops in Minnesota, single large inflorescences of giant, green and yellow foxtail produced about 1,500, 1,500 and 250 seeds each, but averages for the three species were 246, 242 and 52 each. In corn and soybeans, giant foxtail production ranged from 100 to 2,500 seeds per plant depending on experimental treatments, while production of 4,000 and 16,000 seeds per plant were observed in soybeans in favorable years.

Dispersal: The foxtails spread in contaminated seed grain. All three species occur in cattle manure. The ability to survive unharmed in the ruminant gut implies that the seeds move with the animals when they are sold. Birds also disperse foxtails. All three species

disperse in irrigation water. Seeds also probably disperse in soil adhering to tires and machinery.

Common natural enemies: Field crickets were apparently responsible for high consumption rates of giant foxtail seeds (58% per day) in wheat underseeded with red clover in Iowa. Rodents caused significant overwinter losses of giant foxtail seeds on the soil surface, ranging from 31% to 91%.

Palatability: People occasionally gather and eat seeds of green and yellow foxtail. Both species have been domesticated as grain crops (green foxtail as foxtail millet; yellow foxtail as korali), though foxtail millet (*Setaria italica*) is no longer fully interfertile with green foxtail. Young foxtail plants have good forage value, but plants become unpalatable to grazers as they mature due to increased fiber and decreased protein content. Yellow and green foxtail were found to be palatable, but giant foxtail unpalatable to sheep. The awns of mature seed heads damage the mouths of cattle and horses.

Note: Yellow foxtail apparently produces allelopathic toxins: Water extracts of the species inhibit germination of alfalfa, cabbage, radishes and soybeans, but not several other crops.

FURTHER READING

Buhler, D.D. and T.C. Mester. 1991. Effect of tillage systems on the emergence depth of giant (*Setaria faberi*) and green foxtail (*Setaria viridis*). *Weed Science* 39: 200–203.

Davis, A.S. and M. Liebman. 2003. Cropping system effects on giant foxtail (*Setaria faberi*) demography: I. Green manure and tillage timing. *Weed Science* 51: 919–929.

Dekker, J. 2003. The foxtail (*Setaria*) species-group. *Weed Science* 51: 641–656.

Douglas, B.J., A.G. Thoas, I.N. Morrison and M.G. Maw. 1985. The biology of Canadian weeds. 70. *Setaria viridis* (L.) Beauv. *Canadian Journal of Plant Science* 65: 669–690.

Steel, M.G., P.B. Cavers and S.M. Lee. 1983. The biology of Canadian weeds. 59. *Setaria glauca* (L.) Beauv. and *S. verticillata* (L.) Beauv. *Canadian Journal of Plant Science* 63: 711–725.

Goosegrass

Eleusine indica (L.) Gaertn.

Goosegrass seedling
Joseph Neal, North Carolina State University

Goosegrass young plant
Antonio DiTommaso, Cornell University

Goosegrass collar
Scott Morris, Cornell University

Goosegrass panicle
Scott Morris, Cornell University

IDENTIFICATION

Other common names: wiregrass, yard grass, crowfoot grass, crows foot grass, silver crabgrass, bullgrass, white crabgrass, Indian goosegrass

Family: Grass family, Poaceae

Habit: semi-prostrate summer annual grass

Description: The **seedling** is folded in the bud and its first leaf opens parallel to the ground. The collar region is broad, white and lightly hairy, lacks auricles and has a small (less than 0.04 inch long), membranous, uneven, centrally notched ligule. Blades are 0.8–1.8 inches long by 0.1–0.2 inch wide, light green, smooth and distinctly veined. Leaf sheaths are smooth and light green to white at the base. **Mature plants** form a rosette, with sometimes erect but typically prostrate stems. Stems have branches and nodes; they can reach 28 inches in height. Blades are 2–8 inches long by 0.1–0.3 inch wide, occasionally hairy on blade surfaces and rough-edged. Sheaths are smooth, light green to white at the base and hairy near the collar region. Ligules and collar regions are similar to those of seedlings. The root system is fibrous. The multiple **inflorescences** are more upright than the leaves. Each stalk has 1–13 spikes radiating from the end. Spikes are 1.6–6 inches long by 0.1–0.3 inch wide. Spikelets are arranged on the spikes in two rows; they each have 3–6 fertile flowers. Flowers are 0.1–0.3 inch long; each contains one dark brown, 0.04–0.07 inch-long **seed** enclosed in white to light-tan chaff. Like all grasses, the apparent seed is covered by a thin, tight layer of fruit tissue.

Similar species: Crabgrass (*Digitaria*) species have the same mat-forming rosette habit but can be distinguished from goosegrass by their leaves being rolled in the bud. Orchardgrass (*Dactylis glomerata* L.) is folded in the bud but can be distinguished by its long (0.2–0.3 inch) ligule and panicle-like inflorescence.

MANAGEMENT

Since goosegrass seeds die out of the soil relatively quickly, rotation with perennial sod crops should reduce populations substantially. Winter grains are well established by the time goosegrass germinates and will compete well against the weed. Since seed survival in the soil is limited and seedlings only establish from seeds close to the soil surface, moldboard plow the spring following heavy seed production events. Relatively few seeds will survive to find their way back to the soil surface during subsequent tillage events.

Goosegrass has a long period of emergence throughout the Southeast, and late emerging plants of this fast growing weed can compete with the crop and build the seed bank. However, its germination is strongly promoted by tillage. Thus, before planting relatively slow growing summer crops on goosegrass infested land, use a fallow period with shallow tillage at intervals to deplete the surface seed bank before planting. Tine weed row crops with increasing aggressiveness as the crop establishes, and do the last inter-row cultivation as late as possible. Cultivate shallowly close to the crop rows with sharp, flat pitched half sweeps or vegetable knives to sever shoots from the roots.

In vegetable crops, hoe out plants missed by cultivation before they flower, preferably when the weather will assure rapid drying so that the plants do not reroot. Rogue out large escapes to prevent seed production. Straw mulch at 2.7 tons per acre has been shown to reduce goosegrass emergence by 90% relative to bare soil, and heavier rates would be even more effective.

In all cropping systems, avoid soil compaction and over-fertilization, as both are likely to promote goosegrass.

ECOLOGY

Origin and distribution: Goosegrass probably originated in south or east Asia.Its present distribution includes the Middle East; south, southeast and east Asia; sub-Saharan Africa; New Guinea; Australia; New Zealand; and North, Central and South America. It occurs in most of the United States except for parts of the Pacific Northwest and northern Rocky Mountains. It also occurs in Hawaii and southern Quebec and Ontario. It is most common in the southeastern and central states.

Seed weight: Mean population seed weight ranges from 0.4–0.5 mg.

Dormancy and germination: Freshly matured seeds of goosegrass are dormant. Fluctuating temperatures, light, nitrate, chilling, scarification and aging all promote germination. Germination is negligible at 68°F or lower. Few fresh seeds will germinate at constant temperatures of 68–104°F, but most will germinate if daily temperatures fluctuate within this range by at least 18°F. Light increases germination at constant temperatures and at fluctuating temperatures. Nitrate also stimulates germination. About half of scarified seeds will germinate at a constant 86°F or 104°F but none at 122°F. Chilling wet seeds for 40 days or letting the seeds age for several months increases germination, and light is no longer needed to promote germination of seeds that have aged. Seeds germinate well at pH from 5 to 10 and in saline conditions, but the species cannot germinate under even moderate moisture stress.

Seed longevity: Studies of goosegrass seed survival in the soil differ, with some finding few seeds surviving beyond three years, whereas others have found moderate death rates with some seeds remaining as long as nine or 10 years. Annual mortality rate computed from a study in Mississippi was 44%. However, an estimated 82% of seeds survived one year buried under turf. The species thus forms short to moderately long-lived seed banks, with the variation probably due to differences in soil, climate and characteristics of the local population.

Season of emergence: In the mid-South, goosegrass emerges from April through August. Seedlings emerged primarily in late spring to early summer in a Mediterranean climate similar to that of California.

Emergence depth: Goosegrass emerges best at the soil surface or from the top 0.8 inch, only a few seeds can emerge from 2 inches, and none can emerge from deeper than 2.8 inches.

Photosynthetic pathway: C_4

Sensitivity to frost: Goosegrass cannot tolerate more than brief exposure to subfreezing temperatures.

Drought tolerance: It can tolerate moderate periods of hot, dry conditions if well rooted, but it is not truly drought tolerant.

Mycorrhiza: Goosegrass is mycorrhizal.

Response to fertility: Goosegrass showed a positive growth response to N application rates up to 206 pounds per acre, whereas in the same experiment, rice showed no growth response above 138 pounds per acre. Nevertheless, goosegrass was less competitive against upland rice as N application increased. N, P and K were applied at rates of 45, 45 and 36 pounds per acre alone or in combination to a nutrient poor soil. N alone more than doubled the growth of a goosegrass dominated weed community. P alone substantially increased weed cover but had little effect on weed dry weight unless N was also applied. K only increased weed dry weight if N and P were also applied. Goosegrass is most commonly observed on soils with relatively high P, K, Ca and Mg. Thus, goosegrass appears to be highly responsive to N and to be generally favored by high fertility conditions. High pH also benefits goosegrass, and it can tolerate pH of 8.6.

Soil physical requirements: Goosegrass occurs on many types of soil. Although soil compaction decreases root growth moderately, it has little effect on shoot growth, and the species is highly competitive in compacted soil. Although the species is more common on well drained soils, it tolerates wet soils and grows well under poorly aerated conditions. It is moderately salt tolerant and excretes excess salt through special glands.

Response to shade: Shade substantially reduces growth. Plants grow more upright when shaded.

Sensitivity to disturbance: Goosegrass tolerates traffic well and responds to it by developing shorter, wider leaves. The species is highly tolerant of soil contaminated with petroleum and heavy metals.

Time from emergence to flowering: Under favorable conditions, plants will flower in 30 days and shed seeds in 70 days. When planted in an experiment, plants began flowering at six weeks after planting, and the number of inflorescences per plant increased progressively over the next six weeks.

Pollination: Plants readily self-pollinate, and genetic analysis indicates that selfing is common but that cross-pollination also occurs.

Reproduction: A single plant can produce up to 140,000 seeds in favorable conditions, but average seed production in low competition conditions is more typically 40,000–50,000 seeds per plant. Plants growing in an oil palm plantation produced an average of 4,800 seeds per plant.

Dispersal: Seeds blow in the wind. They also disperse with mud on the feet of animals, shoes, motor vehicles and probably on machinery as well. Goosegrass seeds occur in manure and are dispersed when it is spread.

Common natural enemies: The leaf-spotting fungus *Bipolaris setariae* sometimes causes substantial damage.

Palatability: Goosegrass has been occasionally grown for grain in parts of Africa and Asia, and it is also sometimes grown for hay. It is one of the two ancestors of the tetraploid crop finger millet (*Eleusine coracana* subsp. *coracana*).

FURTHER READING

Ampong-Nyarko, K. and S.K. De Datta. 1993. Effect of nitrogen application on growth, nitrogen use efficiency and rice-weed interaction. *Weed Research* 33: 269–276.

Arrieta, C., P. Busey and S.H. Daroub. 2009. Goosegrass and bermudagrass competition under compaction. *Agronomy Journal* 101: 11–16.

Holm, L.G., D.L. Plucknett, J.V. Pancho and J.P. Herberger. 1977. *The World's Worst Weeds: Distribution and Biology*. The University Press of Hawaii: Honolulu.

Italian ryegrass

Lolium perenne L. ssp. *multiflorum* (Lam.) Husnot = *Lolium multiflorum* Lam.

Italian ryegrass seedling
Jack Clark, University of California

Italian ryegrass plant in flower ·
Joseph DiTomaso, University of California, Davis

Italian ryegrass spike
Jack Clark, University of California

IDENTIFICATION

Other common names: annual ryegrass, Australian ryegrass, Australian rye

Family: Grass family, Poaceae

Habit: Summer annual, winter annual or biennial bunch grass, or in mild climates like the Pacific Northwest, a short-lived perennial

Description: The seed leaf emerges vertically. The first true leaves of the **seedling** are rolled in the bud, dark green, have a 0.1 inch translucent ligule and small inconspicuous auricles, and leaf blades are 4–5 inches long by 0.1 inch wide with obvious, raised veins and glossy undersides. Blades and sheaths both lack hairs. Sheaths are red at the base. **Mature plants** of upright, sturdy Italian ryegrass reach 20–40 inches in height and have a coarse texture. Stems are pale yellow to red at the base; sheaths are round and vary in texture. Older leaf blades measure 4–8 inches long by 0.2–0.4 inch wide and are rough with prominent, raised veins, smooth edges and glossy undersides. The broad collar region has long, slender, claw-like auricles and a flat, translucent ligule that may exhibit light tearing or fringing. Blades are rough with prominent, raised

veins, smooth edges and glossy undersides. **Inflorescences** are flat, with 4–15 inch-long spikes located at the end of stems. Spikelets 0.4–0.7 inch long, containing 10–20 flowers that are attached to the spike alternately so that the length of the spikelet runs parallel to the stem. Spikelets have one thin, papery, 0.3–0.5 inch-long outer chaff attached at the base of the spikelet. Each individual flower has two, 0.4 inch chaffs and one long awn. As with all grasses, **seeds** are covered by a thin, tight layer of fruit tissue. They are oval and 0.3 inch long, including the chaff but not the long awn. The flattened seed base is encircled by a raised ridge.

Similar species: Perennial ryegrass (*Lolium perenne* L.) has spikelets that consist of only 6–10 flowers and lacks awns. While the leaves of Italian ryegrass are rolled in the bud, those of perennial ryegrass are folded. Quackgrass (*Elymus repens* (L.) Gould), a rhizomatous perennial, also has auricles and a flat, terminal spike inflorescence. However, it is distinguishable by its sharp, pointed rhizome tips, longer seedling blades (4–8 inches), very short ligule and two outer chaffs at the base of each spikelet in its inflorescence.

MANAGEMENT

Italian ryegrass is a serious weed of winter wheat

throughout the Southeast, southern Midwest and far West. It grows relatively poorly in hot weather, and it can be controlled effectively in summer row crops with a combination of tine weeding, inter-row cultivation and hilling up of the crop at lay by. Consequently, rotation of winter grains with summer row crops tends to interrupt the life cycle of Italian ryegrass and reduces populations.

Since the seeds die off rapidly and must be close to the soil surface for successful emergence, plowing under the seeds can also aide in long-term control. Tilling under ryegrass seeds on the soil surface greatly reduced density of the weed in a subsequent winter grain crop relative to use of a minimum tillage regimen. However, the seeds produced by a bad infestation may be so numerous that rotating into a hay crop for several years may be required to substantially reduce the seed bank, provided the mowing regimen prevents seed production.

To reduce infestations within grain crops, tine weed before emergence and again when the crop has three leaves. Tine weeding can be done more aggressively if the crop is planted at a consistent depth and if the density of plants in the row is sufficient to deflect the weeder's tines out of the row. A weeder with stiff, straight or 45-degree-bent tines will work best to break and bury young ryegrass at the post emergence weeding. A relatively high crop seeding rate also has been shown to increase wheat yield and decrease dockage, especially when wheat was planted in a narrow row spacing (3 inches versus 8 inches).

At harvest, use a chaff collection system on the combine or set the combine to collect as many of the ryegrass seeds as possible, and then clean them out of the grain afterward. Although some seeds will shatter to the ground during combining, Italian ryegrass tends to mature later than winter wheat, so most seeds will still be on the plants. Collecting seed with the combine can tip the balance toward long-term control of the weed. The cleanings can be fed to cattle but only after thorough grinding, since a large proportion of any undamaged seeds will pass through the animals' guts. Incorporate ryegrass stubble immediately after harvest to arrest further seed production. If the field is cut higher than 1.5 inch, graze the stubble to prevent regrowth of the ryegrass. When all other measures fail, some growers harvest small grains infested with Italian ryegrass as a high quality forage to prevent ryegrass seed production.

Many of the tillage, cultivation and crop rotation strategies discussed above apply to vegetable crops as well as to grains. Flush seeds out of the soil with a short fallow before planting cool season vegetable crops. Plastic mulches effectively suppress this weed. Italian ryegrass has been used as a cover crop, but this should be avoided in areas where it is a weed problem. It is adapted to coexist with other winter annual cover crops, but it will continue to grow and set seeds after these other cover crops flower and are mechanically terminated.

ECOLOGY

Origin and distribution: Italian ryegrass is native to southern Europe, southwest Asia and northwest Africa but has been widely introduced throughout the temperate zones and upland tropics. It now occurs in North America, Australia, New Zealand and South America. In North America, it occurs throughout the United States and Canada, and in Mexico.

Seed weight: 1.3–2.6 mg, with a mean of 2 mg.

Dormancy and germination: Most seeds of Italian ryegrass lack primary dormancy, although some biotypes produce up to 55% dormant seeds. These seeds become germinable after six months of after-ripening. Usually, seeds germinate soon after entering the soil if moisture is sufficient. Seed viability can be as high as 90–95%, although germination rates in the field are often lower than this. Seeds will germinate at temperatures as low as 36°F, with the greatest germination occurring at day/night temperatures from 50/41°F to 77/41°F. When after-ripened, seeds can germinate at daytime temperatures as high as 86–104°F. Light promotes germination of previously buried seeds, but substantial germination can still occur in the dark. Cold (40–50°F), saturated soil conditions tend to preserve seeds in a viable but non-germinating state and induce secondary dormancy. Germination and establishment proceed more rapidly than that of other grasses under cool conditions.

Seed longevity: In undisturbed soil in Oregon, seeds of Italian ryegrass died off at a rate of about 58% per year, and depletion of the seed bank occurred more rapidly in well drained than in poorly drained soil. An experiment in Wales also resulted in an average loss rate of 58% per year. In another experiment in Oregon, an average of only 36% of seeds survived burial for 180 days from October to April, indicating a more rapid rate of seed bank decline.

Season of emergence: Spring seedlings can emerge anytime from March to May, depending upon local conditions. In California and Oregon, emergence usually occurs with the onset of the fall rainy season. Fall emerging plants will function as winter annuals in climates where they survive winter.

Emergence depth: Italian ryegrass emerges best from depths of 0.25–0.5 inch.

Photosynthetic pathway: C_3

Sensitivity to frost: Italian ryegrass is sensitive to severe cold but grows as a winter annual in plant hardiness zone 6 and south.

Drought tolerance: Italian ryegrass grows poorly during drought. Growth is also reduced when temperatures exceed 77°F, even when moisture is available. On non-irrigated sites, the fibrous root system can grow to over 3 feet deep, and well-established plants have the potential to tap deep soil water reserves to persist through drought periods. Unlike many other grass species, roots will not regrow after drying out.

Mycorrhiza: Annual ryegrass is a mycorrhizal species.

Response to fertility: Nitrogen is the most critical nutrient for growth of Italian ryegrass, and the species responds favorably to very high N application rates. Annual ryegrass grows well at pH 5–7.8 but does best from pH 5.5–7.

Soil physical requirements: Italian ryegrass is well adapted to a wide range of soil types, from fine to coarse textured. It tolerates saline soil conditions. It tolerates wet soil conditions and can survive flooding for 15–20 days, provided temperatures are below 81°F.

Response to shade: Although some sources list Italian ryegrass as being shade intolerant, the species also has been described as moderately shade tolerant. We have found that it can establish and grow within a cover crop canopy as well as under a closed corn canopy.

Sensitivity to disturbance: Trampling by livestock can kill the plant when the foliage is frosted. Italian ryegrass regrows quickly after mowing if the weather is cool and if the plant is cut above 3 inches. Cutting stems below the growing point, for example at 1 inch, causes the plant to weaken or die. Plants should be mowed or killed before early bloom to prevent seed set.

Time from emergence to reproduction: Fall emerging plants flower in late spring and early summer in Colorado and Montana, but flower in early spring in many other areas. More than 11 hours of daylight are required to initiate flowering. The embryo becomes viable 10 days after pollination, and seeds mature 20 days later.

Pollination: Cross fertilization by windborne pollen is normal, but self-pollinated plants often produce a few seeds.

Reproduction: Drilled, fertilized stands of Italian ryegrass produced 113–121 seeds per inflorescence and roughly 300 seeds per plant. Lower seed production could be expected from plants growing with a competitive crop. Plants do not spread vegetatively.

Dispersal: The chaff has short bristles that probably aid in dispersal on animal fur. About one-third of seeds pass through cattle unharmed and could be dispersed with manure; however, holding the manure in a tank for three months kills the seeds. The species is a contaminant of grass seed unless the seed is produced under strict standards.

Common natural enemies: Mites sometimes damage Italian ryegrass, particularly during dry weather. Crown rust (*Puccinia coronata*) and gray leaf spot (*Pyricularia* sp.) are common diseases of the species. Stem rust (*Puccinia graminis* ssp. *graminicola*) is common in Oregon but rare in the Southeast. Ryegrass blast (*Magnaporthe grisea*) can cause substantial damage in the Southeast.

Palatability: Humans do not eat any part of Italian ryegrass. The species makes nutritious forage for livestock, and several million acres are planted with the species, primarily for winter pasture in the Southeast and Northwest. Italian ryegrass can cause annual ryegrass toxicosis, but the disease is rare. The toxic compounds are only produced when the species is infected by the nematode *Anguina agrostis* and only when this nematode is infected by a bacteria (*Clavibacter toxicus*), which is itself infected with a virus.

FURTHER READING

Beddows, A.R. 1973. Biological flora of the British Isles. *Lolium multiflorum* Lam. (*L. perenne* L., ssp. *multiflorum* (Lam) Husnot, *L. italicum* A. Braun). *Journal of Ecology* 61: 587–600.

Justice, G.G., T.F. Peeper, J.B. Solie and F.M. Epplin. 1994. Net returns from Italian ryegrass (*Lolium multiflorum*) control in winter wheat (*Triticum aestivum*). *Weed Technology* 8: 317–323.

Trusler, C.S., T.F. Preeper and A.E. Stone. 2007. Italian ryegrass (*Lolium multiflorum*) management options in winter wheat in Oklahoma. *Weed Technology* 21: 151–158.

Johnsongrass

Sorghum halepense (L.) Pers.

Johnsongrass rhizomes
Jack Clark, University of California

Johnsongrass plant in flower
Joseph DiTomaso, University of California, Davis

Johnsongrass panicle
Scott Morris, Cornell University

IDENTIFICATION

Other common names: Egyptian millet, means grass, Egyptian grass, false guinea grass, millet grass, Morocco millet, Aleppo grass, grass sorghum, Cuba grass, St. Mary's grass, evergreen millet, maiden cane, Arabian millet, false guinea, Syrian grass

Family: Grass family, Poaceae

Habit: Perennial grass spreading by thick rhizomes

Description: The **seedling** leaf is roughly horizontal and 0.6–1 inch long by 0.16–0.24 inch wide. The collar region of true leaves is narrow, pale and absent of auricles. The membranous ligule is translucent with fine teeth. Seedling true leaf blades are 1.6–7 inches long by 0.1–0.2 inch wide, hairless and rolled in the bud. The midrib is strong and white near the base of the blade. Sheaths are green, hairless, smooth, open and sometimes red at the base. Seedlings closely resemble domestic corn or sorghum. **Mature plants** produce tillers that grow 2–8 feet tall. Stems are pale yellow-green and 0.8 inch in diameter at maturity. Sheaths are flattened, hairy, open, ribbed and slightly toothed; sheaths are maroon to red-brown at the base and pale green near the collar region. The collar region is broad, smooth and pale, with a translucent 0.2 inch-long membranous ligule on newer leaves and a fringed, hairy ligule on older leaves. Some scattered hairs are present above the collar on the upper surface of rough edged blades. Blades are lanceolate (8–24 inches long by 0.2–2 inches wide, but typically less than 1 inch) wide, with a strong, wide, white midrib. Roots are fibrous and grow from an aggressive rhizome system that can exceed 6 feet; rhizomes are white with purple or red splotches and long, scaly, brown sheaths at nodes. The **inflorescence** is a purplish, 6–20 inch-long panicle. Several branches, reaching up to 10 inches long, are widely spaced and whorled about the nodes. Branches are further divided into spikelets attached in pairs along the length of the subdivision. Each spikelet pair has a longer, sterile flower with no awn and a shorter, broader, fertile flower with a twisted and kinked awn. Fertile flowers produce one 0.1 inch-long, oval shaped, dark chestnut brown **seed**. As with other grasses, a thin, dry layer of fruit tissue adheres tightly to the seed coat.

Similar species: Young plants may be confused with fall panicum (*Panicum dichotomiflorum* Michx.). However, johnsongrass does not have a hairy ligule or dense hairs on leaf undersides, sheaths or collars.

Shattercane (*Sorghum bicolor* (L.) Moench ssp. *verti-cilliflorum* (Steud.) de Wet ex Wiersema & J. Dahlb.) is closely related and seedlings differ from johnsongrass only by their half membranous, half hairy ligule and the presence of hairs on both blade surfaces near the collar. Mature shattercane lacks rhizomes, has purple spotted stems and can reach 13 feet high.

MANAGEMENT

Johnsongrass is considered one of the world's 10 worst weeds because it is widely distributed and difficult to control. No-till systems tend to favor the establishment and growth of johnsongrass populations. Conversely, repeated tillage to break up and exhaust rhizomes dramatically limits the growth of this weed and is the most common method of non-chemical control. The rhizome buds will sprout any time the soil is warm, so tillage is effective during late spring, summer and early fall. Tillage during cool weather in early spring or late fall is useless since it will not prompt the buds to sprout. Before you begin an eradication attempt, determine how deep the rhizomes go in your soil by digging a few test pits. Deep, rotary tillage with slow forward motion will chop the bulk of the rhizomes into very small pieces that are easier to exhaust. For treating large areas, heavy disks are probably the best implement. If you have many deep rhizomes, follow this with deep chisel plowing. Two passes at a 90-degree angle to each other will work best. Do the rotary tillage or disking first, since otherwise the chisel plowing will loosen the rhizomes and prevent them from getting cut into small pieces. Carbohydrate storage in the rhizomes reaches a minimum 10–30 days after the shoots emerge and when plants are 6–12 inches high with 4–8 leaves, so successive tillage events at this time will eliminate shoots before the rhizomes recharge. A field cultivator is ideal for suppressing regrowth after the initial attack on the rhizome system.

Johnsongrass populations can also be badly damaged by overheating or freezing the rhizomes. Naturally, the former works better in subtropical portions and the latter works better in temperate portions of this weed's range. For either procedure, use tillage that breaks the rhizome into coarse fragments and work the pieces to the soil surface with a spring tooth harrow. To freeze kill the rhizomes, the frost must penetrate into the soil. If you are treating a small area, push snow off to ensure deep frost penetration. When the soil freezes to a crust, chisel plow to break up the surface and allow deeper penetration of the frost and thorough freezing of the surface chunks. Freeze killing johnsongrass depends entirely on your coldest snow free winter weather conditions. The procedure may fail completely one year but work well the next.

Heat killing works best on dark colored soils. Do this during the hottest, sunniest part of the summer. Direct sunshine is more critical than high air temperatures for heating the soil. After the rhizomes have cooked for a few days (see "Drought tolerance"), stir the soil with a field cultivator or spring tooth harrow to bring new rhizomes to the surface. Remember that summer soil temperatures drop rapidly with depth. Small areas can be treated more effectively by covering the soil with black or clear plastic. Remove the plastic periodically so you can stir the soil to expose more rhizomes to killing temperatures.

Persistence is the key to johnsongrass management. Once you have set the population back to manageable levels with the procedures above, continue to hammer it whenever a break in your cropping allows a clean tilled fallow period or heat killing.

ECOLOGY

Origin and distribution: Johnsongrass is native from southern Europe through India. It was introduced into the southeastern United States in the early 1800s as a forage crop and subsequently spread through most of the United States. It has been introduced into most of the temperate and tropical areas of the world but is best adapted to the humid summer rainfall areas of the subtropics. A variety that overwinters only as seeds was reported to be spreading northward into southern Canada.

Seed weight: Mean seed weights of various populations range from 2.6–6.2 mg.

Dormancy and germination: Johnsongrass seeds are dormant when shed from the plant due to persistent chaff and tannins in the seed coat that prevent uptake of water. Water stress of the mother plant during seed development reduces seed dormancy through enhanced permeability to oxygen diffusion. After four to five months or a period of cold treatment, for example 20 days at 50°F, seeds tend to lose dormancy. Germination response to temperature, light and nitrate varies between populations. Extreme day/night temperature fluctuations, such as 95/59°F, stimulated the highest germination, although after-ripening reduced the magnitude of the response to temperature fluctuation. In the absence of dormancy, the seed germination rate was highest at 97°F, whereas rhizome bud break was fastest at 90°F. In one study, percent germination was highest after chilling, followed by fluctuating temperatures of 104/75°F, light and a weak nitrate solution. Johnsongrass germination is turned on by red light (such as direct sunlight) and off by far red light (such as light passing through a plant leaf canopy). A leaf canopy also reduces germination in soil by reducing the difference in soil temperature between day and night. Dormancy in johnsongrass rhizomes is disputed, but some evidence indicates winter dormancy and hormonal suppression of lateral bud sprouting by the terminal bud. The temperature optimum for rhizome sprouting was 82°F.

Seed longevity: Johnsongrass seeds can survive for up to five years. In one study, johnsongrass seeds were able to survive for six years in undisturbed soil when buried at a depth of 9 inches but less than two years when buried more shallowly. The majority of overwinter seed losses when seeds remain on the soil surface was attributed to seed predation. Viability of buried seeds decreased an estimated 12–17% per year in Mississippi but 52% in one year in Arkansas.

Season of emergence: Johnsongrass sprouts begin emerging in the spring when soil temperatures reach about 59°F. Seedling emergence is greatest in late spring but continues throughout the growing season. In Texas, johnsongrass seedlings began emerging at 61°F. The minimum temperature for seedling emergence of johnsongrass in Argentina was 47°F, which is similar to the minimum temperature of 49°F required for emergence of shoots from rhizomes in Italy.

Emergence depth: Seedlings emerge well from the top 2 inches of soil, and a few seedlings can emerge from as deep as 4–6 inches. Reduction of day/night temperature differences with depth was identified as an important factor in limiting germination of seeds deeper in the soil profile. Shoots from rhizome fragments emerge best from the surface 0–4 inches but can emerge from 6 inches.

Photosynthetic pathway: C_4

Sensitivity to frost: The shoots are frost sensitive. Rhizomes die if they freeze. Freezing typically occurs if rhizomes are exposed to temperatures of 21–27°F for eight hours in the lab or to daily minimum soil temperatures of less than 16°F in the field.

Drought tolerance: Long rhizomes resist dehydration more than short rhizomes, and long rhizomes can withstand long periods of drought. Johnsongrass responds to drought by allocating more resources to growth of fine roots that create a higher root surface area for uptake of water. Exposure of rhizome pieces to high temperature on the soil surface, however, kills them within a few days even in moist soil. They die in one to three days at 122–140°F and in about seven days at 86–95°F.

Mycorrhiza: Johnsongrass can form mycorrhizal associations.

Response to fertility: Johnsongrass is highly responsive to nitrogen. For example, dry weight nearly doubled when 48 pounds per acre of N was applied. The species tolerates pH from 5 to 7.5.

Soil physical requirements: Johnsongrass tolerates a wide range of soil conditions but does best on porous lowland soils and least well on poorly drained clay soils. Compaction restricts growth. The species tolerates one to four weeks of flooding. Rhizomes grow closer to the surface in clay soils than in sandy loam.

Response to shade: Johnsongrass can grow rapidly in shade, and the shoots are tall enough to overtop most crops, thereby avoiding shade. When shaded, the plants put more energy into leaf growth and less energy into storage in the rhizomes.

Sensitivity to disturbance: Most johnsongrass rhizomes lie in the plow layer where they can be disturbed by tillage, though one study found more than 10% of rhizomes below 12 inches. Short rhizome fragments (1–4 inches) were less vigorous and required 20–30 days longer to initiate new rhizome growth than fragments 6–8 inches. See also "Drought tolerance" and "Sensitivity to frost."

Time from emergence to reproduction: Johnsongrass begins flowering about seven weeks after emergence and continues flowering until frost. Seed shed begins three month after planting and can continue for an additional three to four months. Johnsongrass flower development was most rapid at 90°F but was nil at 54°F or 104°F. The minimum temperature for rhizome formation was between 59°F and 68°F. New rhizome initiation begins approximately 30–60 days after planting, depending on the size of initial rhizome fragments.

Pollination: Johnsongrass normally self-pollinates but can also cross-pollinate by wind.

Reproduction: Johnsongrass reproduces by both seeds and rhizome sprouts. Two-year-old plants grown in a garden experiment in Mississippi produced 28,000 seeds per plant. Seed production by individual plants decreased from 2,350 to 87 seeds as plant densities increased from moderate to high levels. Plants in a dense stand of johnsongrass produced 66 seeds and 40 rhizome buds per plant. New rhizomes grow from the overwintering rhizomes in early summer and form new shoots. Rhizomes formed after flowering will overwinter. Single plants of johnsongrass produced 18 pounds and 200–300 feet of new rhizomes per year. In an Israeli study, patches from single individuals covered 183 square feet with 18 shoots per square foot.

Dispersal: Johnsongrass seeds are dispersed by water, on machinery, in contaminated grain and hay, and by wind for short distances. They pass through cattle digestive tracts and are spread with manure. They are also dispersed by birds. Within a field, johnsongrass disperses from the primary source by movement of rhizomes in the direction of tillage.

Common natural enemies: In one study, *Sphacelotheca cruenta* (loose kernel smut) infected 55% of johnsongrass panicles but had little impact on the competitiveness of this weed.

Palatability: Johnsongrass is highly palatable to livestock and is grown as a forage species in the southeastern United States. Young growth and plants stressed by frost or drought may develop toxic levels of cyanide producing compounds.

Note: Johnsongrass produces allelopathic compounds that suppress the growth of many crops and other weeds.

FURTHER READING

Lolas, P.C. and H.D. Coble. 1980. Johnsongrass (*Sorghum halapense*) growth characteristics as related to rhizome length. *Weed Research* 20: 205–210.

McWhorter, C.G. 1972. Factors affecting Johnsongrass rhizome production and germination. *Weed Science* 20: 41–45.

Monaghan, N. 1979. The biology of Johnson grass (*Sorghum halepense*). *Weed Research* 19: 261–267.

Warwick, S.I., and L.D. Black. 1983. The biology of Canadian weeds. 61. *Sorghum halepense* (L.) PERS. *Canadian Journal of Plant Science* 63: 997–1014.

Large crabgrass

Digitaria sanguinalis (L.) Scop.

Large crabgrass seedling
Scott Morris, Cornell University

Large crabgrass young plant
Randall Prostak, University of Massachusetts

Large crabgrass rooting at stem nodes
Randall Prostak, University of Massachusetts

Large crabgrass panicle
Scott Morris, Cornell University

IDENTIFICATION

Other common names: purple crab grass, finger grass, Polish millet, crowfoot grass, pigeon grass, hairy crabgrass, hairy finger grass, northern crabgrass

Family: Grass family, Poaceae

Habit: Sprawling summer annual grass with stems rooting at the nodes

Description: Seedlings have an upright habit and begin tillering when they have only four to five true leaves. Seed leaves are parallel to the ground, oval to elongated in shape (only three to four times longer than wide) and rounded at the tip. True leaves are rolled in the bud and uncurl as they emerge. Ligules are membranous, translucent and jagged. Auricles are absent. Collars are broad with hairy edges. Blades are linear, 2–4.75 inches long by 0.1–0.2 inch wide and tapered to a point. Stiff, erect hairs are present on both blade surfaces and on the sheath. **Mature plants** have dozens of tillers; stems are 12–48 inches with erect central stems and sprawling outer stems that root at the nodes. Mature stems are flattened and red at the base. Sheaths are compressed, hairy and closed. Leaves are similar to young seedling leaves, but blades are up to 8 inches long by 0.6 inch wide. Fibrous roots extend up to 6.5 feet deep. **Inflorescences** are terminal, composed of up to 13 (usually three to five) flat, fingerlike spikes arranged in a spiral near the stem tip. Two rows of 0.1 inch spikelets are arranged alternately along each spike. **Seeds** are glossy, yellow-brown, elliptical or lanceolate.

Similar species: Smooth crabgrass (*Digitaria ischaemum* (Schreb.) Schreb. ex Muhl.) is closely related and has a non-jagged, translucent ligule and few hairs on leaf blades or sheaths. Goosegrass (*Eleusine indica* (L.) Gaertn.) is also mat-forming, but goosegrass leaves are folded in the bud. Southern crabgrass (*Digitaria ciliaris* (Retz.) Koeler) has hairs on the sheaths but none on the blades. Yellow foxtail (*Setaria pumila* (Poir.) Roem. & Schult.) and witchgrass (*Panicum capillare* L.) seedlings are also rolled in the bud and lack auricles, but they have a hairy ligule.

MANAGEMENT

The tiny, fragile seedlings of large crabgrass are easily killed by tine weeding or rotary hoeing just before or shortly after they emerge. Once the seedlings have developed roots at the base of the shoot in addition to the primary root from the seed, they are harder to kill.

Early planted corn may grow too large for in-row weed management when most crabgrass emerge. In crops that tolerate hilling-up, try to do so before the first true leaf elongates since an incompletely buried plant can recover. When cultivating to kill larger plants, set up cultivators to cut flat and shallow so that the shoots quickly dry on the soil surface. Burying the plants is less effective because complete burial of larger plants is nearly impossible, and large crabgrass reroots exceptionally well.

Since most large crabgrass seeds do not persist long in the soil, rotation into a sod crop for a few years greatly decreases the seed bank, provided seed production is prevented during establishment of the sod. Cutting sod as high as possible will facilitate competitive suppression of crabgrass by these crops, including turfgrass. The prostrate growth habit and tolerance of hot, dry summer conditions make this species well-suited to gaps in regularly mowed hay crops, so good hay stands are essential during the hay phase of rotations. Vegetable rotations that include early, short-season crops tend to reduce the population by allowing many seedlings to emerge shortly before harvest. Subsequent tillage for the next crop or cover crop then destroys the young crabgrass before they can set seeds. A tilled fallow period during the first four to six weeks of warm weather can effectively decrease severe infestations since most seeds will germinate if provided with warm, moist conditions.

Straw mulch applied in the spring keeps the soil too cool for germination and is highly effective. The same mulch material applied after the soil warms will be relatively ineffective because large crabgrass seedlings can worm up through at least 3 inches of straw. Green manure cover crops in the mustard family incorporated in the spring provided 48–79% control of large crabgrass at four weeks after transplanting bell peppers in South Carolina, but no control was detected later in the season.

ECOLOGY

Origin and distribution: Large crabgrass probably originated in Asia or Africa and was introduced into North America during European settlement. It occurs in both temperate and tropical climates, and is currently distributed from 50° N latitude to 40° S latitude around the world.

Seed weight: 0.46–0.59 mg

Dormancy and germination: Seeds are dormant when shed from the parent plant, but they gradually lose dormancy over a period of several months. Cold treatment accelerates the loss of dormancy, as does puncturing or damaging the seed coat. Daily alternating temperatures between 68°F and 86°F maximize germination. Light is not essential, but it can enhance germination. Nitrate modestly increased laboratory germination and field emergence of fresh (dormant) seeds, but otherwise had no effect on germination or emergence of after-ripened seeds. Seeds undergo a periodic annual cycle in soil whereby dormancy is low in spring following cold after-ripening, and secondary dormancy is induced in mid-summer when temperatures exceed 81–86°F.

Seed longevity: No seeds survived longer than three years when buried at 1.7 inch under turfgrass, but approximately 55% survived for one year. The annual mortality rate of large crabgrass seeds buried in an agricultural field was computed to be 54%. Seed survival is probably poorer in tilled soil.

Season of emergence: In Canada, large crabgrass emerged primarily in June and July. In Ohio and the Mid-Atlantic area, most seedlings emerged in May and June, with some emergence continuing through the growing season. In warm climates such as Florida, soil tillage in February to April led to maximum establishment. In all cases, these emergence periods generally follow the last frost for the respective area. Emergence occurs when rainfall exceeds 0.2 inches of rain and when soil temperature exceeds 50–58°F.

Emergence depth: Large crabgrass emerges best within the top 0.8 inch, with some emergence from as deep as 3 inches.

Photosynthetic pathway: C_4

Sensitivity to frost: Large crabgrass is frost sensitive.

Drought tolerance: Large crabgrass grows best at high temperatures and thrives in conditions that cause heat or moisture stress in many other plant species. It develops a disproportionally larger root system very quickly, which enhances its ability to compete with neighboring species. The roots penetrate to 6 feet, which allows the plants to tap residual soil moisture that is unavailable to many crops.

Mycorrhiza: Large crabgrass has mycorrhizae.

Response to fertility: Large crabgrass tolerates infertile soils, but it is highly responsive to fertility, both when growing alone and in competition with a crop. It is a strong concentrator of most major nutrients but particularly K, which may reach more than 6% of foliage dry weight. Nevertheless, strong competition from corn prevented a response to nutrient rate in one experiment. It tolerates low pH soil, with maximum emergence and growth at pH 4.8–5.8. Magnesium carbonate was more effective than calcium carbonate at reducing emergence and growth of large crabgrass when raising soil pH to neutral levels.

Soil physical requirements: Large crabgrass tolerates a wide range of soil conditions.

Response to shade: Large crabgrass is partially shade tolerant. In field conditions it has been found to tolerate up to 63% shade without reduction in productivity or seed set. Greater degrees of shade, however, greatly reduce growth.

Sensitivity to disturbance: Seedlings are tiny and easily damaged. Completely uprooting larger plants is difficult due to the extensive, fibrous root system. Medium to large plants reroot readily in moist soil. Like all grasses, the growing points are near the soil surface, so simply slicing off the plant at ground level will result in vigorous resprouting. Covering the lower portion of stems with soil during cultivation or hoeing promotes rooting of the stems at the nodes.

Time from emergence to reproduction: Large crabgrass generally flowers from July to September. Plants that emerged in May first flowered in July, about 10 weeks after emergence, whereas plants that emerged in June flowered eight weeks after emergence.

Pollination: Large crabgrass is primarily self pollinated, but it sometimes cross pollinates by wind.

Reproduction: Reproduction is principally from seeds, but large crabgrass is one of the few annual weeds that also can reproduce vegetatively. The outer stems of the plant lie on the ground and tend to form roots where the nodes (especially the first node) come into contact with the soil. Consequently, single plants up to 10 feet in diameter have been observed. The species flowers in response to short day length. Once a plant begins to flower, it will continue to flower and set seed until frost. Well-spaced plants emerging in May in Connecticut produced an average of 145,000 seeds per plant, but seed production at high plant densities range from 100–6,800 seeds per plant.

Dispersal: Large crabgrass seeds have no apparent adaptations for dispersal. However, because they often reach high densities in soil, they are easily spread from one site to another in soil clinging to tires and machinery. Seeds also disperse in irrigation water. The species is common on waste areas and in lawns, and it frequently has to travel only short distances to invade a field.

Common natural enemies: Larvae of the flea beetle, *Chaetocnema denticulate*, bore into the top of the seed head while it is in the boot stage and feed on the stems and immature seeds. In Virginia, 26% of seed heads were infested, and the insect completed three generations per year. Various species of *Oscinella* fruit fly larvae mine the stems. Loose smut (*Ustilago syntherismae*) can significantly reduce the proportion of reproductive plants. *Drechslera gigantea* leaf spot can cause significant injury and has been explored as a bioherbicide.

Palatability: The seeds are edible and can be ground for flour or cooked as porridge. The species is excellent cattle fodder and is sometimes harvested for hay.

FURTHER READING

Holm, L.G., D.L. Plucknett, J.V. Pancho and J.P. Herberger. 1977. *The World's Worst Weeds: Distribution and Biology.* University Press of Hawaii: Honolulu.

Peters, R.A. and S. Dunn. 1971. Life history studies as related to weed control in the Northeast. 6 – Large and small crabgrass. Storrs Agricultural Experiment Station, The University of Connecticut: Storrs, CT.

Purple nutsedge

Cyperus rotundus L.

Purple nutsedge shoots emerging from tubers
Joseph DiTomaso, University of California, Davis

Sprouting purple nutsedge tubers, with new shoots and roots
Joseph DiTomaso, University of California, Davis

Purple nutsedge inflorescence
Joseph DiTomaso, University of California, Davis

IDENTIFICATION

Other common names: nut grass, coco grass, nut sedge, coco sedge, coco, purple nutgrass

Family: Sedge family, Cyperaceae

Habit: Erect, grass-like perennial herb with triangular stems and three-ranked leaves that forms extensive networks by rhizomes and tubers.

Description: Seedlings and young **vegetative shoots** are upright, light green with a white midvein and somewhat stiff. The short stem is triangular with a solid center. **Mature plants** have dark green leaves and triangular stems, typically about 6 inches, but can reach up to 1–2 feet tall. Leaves do not have collar regions; they transition smoothly from blade to a sheathing base that completely overlaps the one below it, broadening the basal stem region. Waxy leaves are hairless, 0.13–0.25 inch wide, bluntly pointed at the tip, sturdy and thick at the base, and they have prominent paralleled veins. Leaves emerge in groups of three, alternating around the triangular base. The triangular flower stalk, growing from the plant base, is typically as tall as or taller than the leaves. The **inflorescence** consists of several spikes that originate from

a common point in a spoke-like manner. The spikelets are closely clustered purple with a red or brown cast. Several long, thin, pointed, leaf-like bract structures are present immediately below the inflorescence; they are roughly horizontal and of equal to or greater length than the inflorescence. Dark **seeds** develop in three-sided, dark brown to black, 0.08–0.13 inch long, dry, elliptical fruit. Seeds are frequently non-viable; propagation occurs primarily via underground rhizomes, concentrated in the first foot of soil, that give rise to **tubers**. Tubers are white, turning dark brown or black as they age; they are covered by hard, rough, dark-red scales and ridges, and they give rise to new vegetative shoots, roots and rhizomes.

Similar species: Yellow nutsedge (*Cyperus esculentus* L.) is a slightly taller plant with a yellowish cast throughout. Its leaf tips are long and tapered into narrow points; its inflorescences are yellow and more condensed and bottlebrush-like. Yellow nutsedge rhizomes produce tubers only at their ends, and tubers lose their scales as they mature.

MANAGEMENT

Purple nutsedge is considered a highly competitive and persistent weed in a wide range of crops throughout the world, and extreme measures are generally required to manage it. The most effective method for controlling purple nutsedge is to desiccate the tubers by thoroughly tilling the soil to the depth of the deepest tubers at the beginning of a period of hot, dry weather. For this approach to be effective, the tillage must break the tubers free of deep roots. Tubers in the upper few inches of hot, dry soil die in about eight days. Although a single tillage operation at the beginning of a hot, dry period of several weeks may be sufficient to kill all the tubers, multiple deep tillage operations will dry the soil and tubers faster, and help ensure successful control.

Attempting to exhaust an intact tuber-rhizome system is essentially futile due to dormant tubers and huge underground reserves. Consequently, when attempting to exhaust tubers, begin by thoroughly tilling to break apart tuber chains so that tubers are induced to sprout. Subsequent cultivation to kill shoots should occur at less than three-week intervals since longer intervals will allow formation of new tubers. Even with prior deep tillage, exhausting tubers may require two growing seasons. Some tubers may remain dormant below the depth of tillage; the few tubers that sprout subsequent to the eradication effort should be dug out by hand. Although fallowing a field for nutsedge eradication requires taking the land out of production during the summer, winter grain planted in October and harvested in June will not interfere with the eradication campaign since purple nutsedge grows poorly in cool weather and the grain is highly competitive. A comparison of several integrated systems showed that tilling at three-week intervals from March to August was cheaper and was as effective for purple nutsedge control as was covering the ground with green or clear polyethylene film for the whole period or by preceding tillage with a turnip cover crop. All of these systems reduced tuber density, provided the subsequent fall pepper crop was either straw mulched or hand weeded. Inter-row cultivation reduces nutsedge competition with the crop but cannot by itself provide complete control. For example, two early season cultivations in cotton provided between 65% and 80% control, but the remaining infestation still significantly reduced yields.

The sharp tips of purple nutsedge rhizomes readily pierce black polyethylene agricultural film and paper mulch, and black film actually increases infestations by warming the soil and thereby encouraging sprouting

and rapid growth. Heavier opaque materials can prevent shoot emergence, but some tubers will persist under the cover, and dense shoot clumps will form along the edges. Purple nutsedge emergence was severely limited through heavy paper mulch. Clear polyethylene mulches can be used to kill purple nutsedge by heating the soil to lethal temperatures (solarization). Some rhizome tips will, however, penetrate thin clear plastic films, particularly if the film is in contact with the soil. However, thicker film (e.g., 4 mil) with a 0.2–0.4 inch gap over the soil causes the leaves to emerge and become trapped under the mulch where they are scorched. In the desert regions of the Southwest, black plastic achieves lethal temperatures to a depth of 6 inches, but in the Southeast, infrared-type plastic that promotes capture of infrared radiation is most effective at achieving lethal temperatures. Although lethal temperatures only penetrate to about 4 inches in the Southeast, the increase in temperature and the increase in the difference between day and night temperatures at greater depths promote tuber sprouting, and the shoots that appear under the plastic are then scorched. Thus, translucent mulches provide some control even if temperatures lethal to the tubers are not achieved. Unlike most perennial weeds, purple nutsedge is significantly suppressed by dead mulches of organic material.

Swine are fond of the tubers, and 24–30 pigs can remove the tubers from one acre in a day. Chickens at a stocking rate of one bird per 125 square feet can eradicate a severe infestation in two years if they can be made to graze uniformly. At a stocking rate of 32 birds per acre, geese can keep a severely infested cotton field free of purple nutsedge competition, and a second year of goose grazing at a reduced stocking rate can eradicate the weed.

Although purple nutsedge reduces production of all crops, some more competitive crops like green beans and transplanted cabbage are less affected than poor competitors like garlic and okra. Since purple nutsedge is sensitive to shade, crops that rapidly develop a leaf canopy are more suppressive of purple nutsedge. Sweet potatoes inhibit purple nutsedge by allelopathy, but its production is substantially reduced by the weed anyway. Dense grass cover inhibits the establishment of purple nutsedge. Even when tubers are planted, many fail to establish shoots in dense sod. However, the few weak plants that do establish send out horizontal rhizomes that explore the soil for openings where new, more vigorous shoots establish and form tubers. Consequently, disturbed areas in the sward are likely to become heavily infested if any purple nutsedge plants are nearby. Cover crops can slow the rate of tuber formation relative to land without a cover crop, but the population may increase anyway.

ECOLOGY

Origin and distribution: Purple nutsedge is probably native to India, but it has been introduced to tropical and subtropical regions throughout the world. It is a serious pest in the Southeast from Virginia to central Texas, and in parts of Arizona and California; it occurs sporadically in the north central and northeastern United States.

Seed and tuber weight: Seed lots contain many light, non-viable seeds. Heavier seed fractions, which contain at least some viable seeds, have seeds that weigh from 0.22–0.3 mg.

Tuber weights can vary considerably, ranging from 160–440 mg in Alabama, 400–1,400 mg in Japan and 200–2,000 mg in Israel. Tubers forming near the soil surface tend to be substantially smaller than those deeper in the soil, though very deep tubers may be relatively small.

Dormancy and germination: Purple nutsedge seed production is rare in the United States. Seeds are dormant when shed and slowly lose dormancy over several years of after-ripening. Damage to the seed coat (scarification), microbial action in the soil or exposure to high temperature (e.g., 140°F) increases germination.

Most tubers have one to eight buds, but a few large tubers have up to 13. The bud farthest from the mother plant on a tuber or tuber fragment sprouts first and tends to suppress sprouting of the other buds. If that shoot is destroyed, the next lower bud will sprout. Similarly, the most recently formed tuber in a chain of tubers inhibits sprouting of tubers farther back on the chain, though these will sometimes sprout anyway. Breaking up chains experimentally or by tillage causes immediate sprouting of most tubers. The optimal temperature for sprouting is 77–95°F, and little sprouting occurs below 50–59°F. Short term chilling of tubers at 40°F for four days can accelerate tuber sprouting and subsequent vegetative and reproductive development.

Extended chilling for two months promotes subsequent sprouting at suboptimal temperatures of 59–68°F, which promotes early emergence in temperate climates. Although many tubers will sprout at 68°F, shoots may fail to elongate or may elongate slowly. Fluctuating temperatures increase percentage sprouting, speed of sprouting and speed of shoot elongation, especially when temperatures are suboptimal. Temperatures above 109°F usually inhibit sprouting, and temperatures over 122°F kill the tuber within two days to less than an hour, depending on the temperature. Sprouting is inhibited at 10–20% soil moisture, but tubers sprout readily at 30–40% soil moisture. At greater than 50% soil moisture, sprouting is suppressed but the tubers survive. Inhibition of tuber sprouting in waterlogged soil is probably due to low oxygen concentration. Short day lengths prompt production of dormant tubers, whereas long day lengths favor formation of tubers that sprout immediately. Light speeds sprouting and increases the number of sprouts per tuber, but 100% of tubers will sprout in complete darkness.

Tuber longevity: In an experiment in Costa Rica, tubers died off at a slowly increasing rate, and less than half were alive after 18 months. Depth of burial did not affect tuber survival. Tubers can survive 200 days of immersion.

Season of emergence: Shoots emerge continuously whenever the mean temperature is greater than about 59°F. In Georgia, emergence peaks in July.

Emergence depth: Seedlings emerge best from about 0.5 inch, with fewer emerging from near soil surface and none emerging from deeper than 1 inch.

Studies differ regarding how well shoots can emerge from tubers placed at various depths, but they generally agree that tubers throughout the plow layer can produce emerged shoots. Some experiments found that shoots emerged best from tubers in the top 1–6 inches of soil, and also can emerge from 6 to 12 inches. Another pair of experiments found 50–70% emergence from tubers at 2 feet and 0–6% from 3 feet.

Photosynthetic pathway: C_4

Sensitivity to frost: Tubers exposed to a temperature of 32°F survived for seven to 10 days; those exposed to

23°F or lower died within two hours. Another experiment, however, showed that tubers could survive eight hours of exposure to 25°F.

Drought tolerance: Tubers die when their moisture content drops below 15%. Tubers left at the surface of dry soil exposed to full sun were killed in four days. Under simulated field conditions, tubers at 2 inches and 4 inches in dry soil exposed to sunlight and protected from rain were killed after 12 and 16 days, respectively. In another experiment, tubers buried at 6 inches in dry soil for four weeks did not survive. Roots penetrate to 54 inches and are numerous and well branched deep in the soil; these help avoid water stress in an intact rhizome-tuber system and, consequently, undisturbed plants can survive six months of drought.

Mycorrhiza: Mycorrhizal fungi infect purple nutsedge roots but decrease productivity. Presence of onion, a mycorrhizal crop, increases infection.

Response to fertility: Increasing N fertility increases the competitive ability of purple nutsedge in a wide range of crops. In radish, a poor competitor, nutsedge tuber production peaked at 98 pounds per acre of N, while shoot production peaked at 196 pounds per acre of N; radish yield declined linearly with increasing N application rates. For several other crops, application of low rates of N increased the weed's growth and tuber production but still increased crop yield, whereas higher N application rates frequently decreased yield due to excessive competition from the weed. In experimental conditions, purple nutsedge grew best at pH 3.5–7, but it also thrives on soils with pH 8.7 or higher.

Soil physical requirements: Purple nutsedge thrives in soil ranging in texture from sand to heavy clay. Shoot growth is inhibited by a dense, poorly aerated layer in the soil; although roots penetrate such layers, rhizomes do not, and tubers will not form there.

Response to shade: Shoot and tuber production decreases linearly with shade. At 60–80% shade, tubers are still produced, but they are small. Under high shade, plants allocate more energy to roots and rhizomes and less energy to tuber production. Screens producing 72–73% shade reduced tuber production by 10–70%. Severe shading of 99% by a crop or cover crop causes the top-growth to die back.

Sensitivity to disturbance: The proportion of tubers below the 8-inch plow layer varies greatly depending on location; in most temperate climate sites, at least some tubers lie below the normal depth of tillage, but few or none form deeper than 12 inches. Clipping shoots appears to promote sprouting of dormant buds on the tubers. Although mowing three times per week at 0.5 inch greatly reduced spread and tuber production, plants still produced 13–19 feet of rhizomes and 59–103 tubers per plant in about four months, and other experiments have similarly demonstrated the inability of clipping to control purple nutsedge.

Time from emergence to reproduction: Planting tubers may simulate what happens when tubers become independent of one another during tillage. Following planting in spring and summer, flowers and formation of dormant tubers usually occur within three to four weeks but may take as long as seven weeks. Flowering usually occurs with tuber formation or follows it by about one week. Formation of new shoots with tuber-like basal bulbs can occur within three weeks. Following planting in late fall and winter in a Mediterranean climate similar to California, the first tubers formed in about two months, and flowering was delayed until the following spring.

Pollination: Flowers are cross pollinated, mainly by wind, but most pollen is shriveled and non-viable.

Reproduction: In the United States, purple nutsedge reproduces almost exclusively by tubers. Seed producing populations have been found in moist river valley fields in Alabama, but only if they were tilled in the spring. Generally, however, seed production is rare and, when seeds are produced, the proportion that is viable is low. A genetic study showed that all of the populations evaluated from the United States and Caribbean were likely a single clone; a lack of cross pollination in most locations may explain the rarity of seed production.

Tuber production is rapid and continuous during spring and summer but slows during fall and winter. A single tuber can produce 99 tubers in 90 days. A tuber planted in late March produced a clone roughly 15 feet in diameter in one year, with shoots forming a nearly continuous sod in the occupied area. In a similar study, a single tuber produced a patch 238 square feet in area in 60 weeks. Stands of purple nutsedge grown from well-spaced tubers produced 1,000–3,000 tubers per square yard within 1.5–2 years.

Dispersal: Tubers are spread by erosion events and by moving water in streams. They are dispersed in nursery containers. They can be collected along with the crop during the harvest of potatoes and sweet potatoes and then dispersed to new locations when the crop is sold. Clones can spread more than 6 feet per year by rhizome expansion.

Common natural enemies: Moths in the genus *Bactra* bore into the basal bulb and damage the plant. Population increase of the moth is slower than that of the weed, so the moths do not control purple nutsedge unless populations are augmented by early season releases of larvae or adults. The nematode *Tylenchus similiis* attacks tubers, but the damage is insufficient to control the weed.

Palatability: Purple nutsedge has some medicinal uses but is not eaten as food by humans. The foliage quickly becomes fibrous and is poor forage, but it can serve as low quality forage in the absence of more desirable species. Swine are fond of the tubers.

FURTHER READING

Bangarwa, S.K., J.K. Norsworthy, P. Iha and M. Malik. 2008. Purple nutsedge (*Cyperus rotundus*) management in an organic production system. *Weed Science* 56: 606–613.

Chase, C.A., T.R. Sinclair and S.J. Locascio. 1999. Effects of soil temperature and tuber depth on *Cyperus* spp. control. *Weed Science* 47: 467–472.

Hershenhorn, J., B. Zion, E. Smirnov, A. Weissblum, N. Shamir, E. Dor, G Achdari, H. Ziadna and A. Shilo. 2015. *Cyperus rotundus* control using a mechanical digger and solar radiation. *Weed Research* 55: 42–50.

Holm, L.G., D.L. Plucknett, J.V. Pancho and J.P. Herberger. 1977. *The World's Worst Weeds: Distribution and Biology*. The University Press of Hawaii: Honolulu.

Smith, E.V. and G.L. Fick. 1937. Nut grass eradication studies: I. Relation of the life history of nut grass, *Cyperus rotundus* L., to possible methods of control. *Journal of the American Society of Agronomy* 29: 1007–1013.

Smith, E.V. and E.L. Mayton. 1938. Nut grass eradication studies: II. The eradication of nut grass, *Cyperus rotundus* L., by certain tillage treatments. *Journal of the American Society of Agronomy* 30: 18–21.

Quackgrass

Elymus repens (L.) Gould = *Elytrigia repens* (L.) Desv. ex Nevski = *Agropyron repens* (L.) P. Beauv.

Quackgrass, excavated rhizome and tillers
Antonio DiTommaso, Cornell University

Quackgrass auricles
Antonio DiTommaso, Cornell University

Quackgrass spikes
Randall Prostak, University of Massachusetts

IDENTIFICATION

Other common names: quitch grass, couch grass, wheat grass, shelly grass, knot grass, scotch grass, quick grass, twitch grass, witchgrass, devil's grass, bluejoint, pondgrass, Colorado bluegrass, false wheat, dog grass, seargrass, quickens, wickens, stroil

Family: Grass family, Poaceae

Habit: Perennial grass, spreading by shallow rhizomes

Description: Seedlings are upright, and the long and narrow seed leaf grows vertically. Leaves of the seedling are rolled in the bud, smooth below and hairy to smooth above, and 3.5–8 inches long by 0.08–0.1 inch wide. The ligule is less than 0.02 inch long and membranous, and auricles are small and often inconspicuous. The sheath is smooth to hairy, and the collar is prominently veined and whitish to pale green. **Mature** quackgrass is upright, unbranched and can reach 4 feet in height. Leaves are rolled in the bud, and a pair of narrow auricles clasps the stem at the base of each leaf blade. The ligule is translucent, membranous and less than 0.04 inch long. The sheath is open at the top and hairless to hairy. Often the sheath of the bottom-most leaf is hairy whereas the sheaths of other leaves are nearly hairless. Leaf blades are 6–16 inches long and 0.12–0.4 inch wide. The root system is composed of spreading rhizomes with pointed tips and fibrous roots at the nodes. The rhizomes can spread up to 24 inches horizontally and extend to 8 inches below the soil surface. The **inflorescence** is a 2–8 inch-long, flattened spike with two alternating rows of numerous 0.4–0.6 inch-long spikelets. The spikelets may have 0.07–0.4 inch-long awns at the tips. The inflorescence turns from greenish blue to straw colored as it matures. Each spikelet produces four to six tan, narrow, oblong, 0.04–0.2 inch-long **seeds**. As with all grasses, the apparent seed includes a tight, thin outer layer of fruit tissue.

Similar species: Tall fescue [*Schedonorus arundinaceus* (Schreb.) Dumort.], ryegrass (*Lolium*) species and smooth brome (*Bromus inermis* Leyss.) may all be mistaken for quackgrass. Tall fescue is a clump forming grass that lacks rhizomes and has blunted auricles that do not wrap around the stem like those of quackgrass. Ryegrasses may have similar auricles to quackgrass but do not have rhizomes. Smooth brome develops rhizomes, but it lacks auricles and has closed sheaths.

MANAGEMENT

Quackgrass tends to increase when the crop rotation includes frequent spring cereal grains and legume forage crops. The former provide little competitive pressure during quackgrass' prime spring growth period, and the latter provide nitrogen and a long period for undisturbed growth of the rhizomes. Since control of severe infestations of quackgrass without the use of herbicides largely requires repeated or intensive tillage, preventing buildup of populations is critical for the maintenance of soil health.

The fastest way to rapidly control a severe quackgrass infestation is to thoroughly dry out the rhizomes. The traditional English method is to plow the soil when it is wet so that it becomes cloddy. Then stir the clods repeatedly during dry weather so that the whole plow layer becomes thoroughly dry. A less effective method that does not require such a long fallow begins by tilling the soil with a chisel plow or field cultivator at the beginning of a dry period. The aim of the tillage is to loosen the soil around the rhizomes while leaving the rhizome pieces as long as possible. Then use a spring tooth harrow to work the rhizomes to the soil surface to dry. Use the harrow every few days to speed drying of the soil and bring more rhizomes to the surface so that most dry thoroughly before the next rain. If the infestation is particularly dense and the field is small, you can remove a substantial percentage of the rhizomes by bringing them to the surface as explained above and then raking them onto the field margins. A window of opportunity for this approach is often available in late summer after harvest of a cereal grain or short season vegetable crop. Beginning tillage immediately after harvest and preventing fall quackgrass growth is more important than repeated cultivations later in fall.

The principal approach for managing moderate quackgrass populations involves regular depletion of storage reserves in the rhizomes. Rhizome reserves are at a minimum when the shoot has developed three leaves, so try to target tillage and cultivation at that growth stage, but disturbance is useful whenever the shoot is showing. The first tillage of the year should be aimed at cutting the rhizomes into the smallest pieces possible. Thus, beginning with rotary tillage or disking is more effective than moldboard or chisel plowing. Following the initial tillage with moldboard plowing, however, is often useful for placing the small rhizome pieces deep in the soil. They will then have to use up substantial reserves to reach the surface. Buds on quackgrass rhizomes will sprout at any time during the growing season, so many opportunities to deplete reserves occur before and after a summer crop. For example, before soybeans or a summer vegetable crop, till the soil when quackgrass begins to grow and then delay a few weeks before preparing a seedbed. Unlike a re-worked seedbed for depleting a surface seed bank, however, tillage for seedbed preparation should work the top 6 inches to disturb as many rhizome fragments as possible. If necessary, this procedure can be repeated in the fall after harvest, followed by winter grain as a cash or cover crop. A dense, vigorous stand of winter grain like wheat will competitively suppress quackgrass. When row crops are in the ground, continue to cultivate as long as you can get between the rows safely. The more times the quackgrass shoots are cut off at about the three-leaf stage, the more you will weaken the rhizomes.

Obviously, the intensive tillage discussed above will tend to deplete soil organic matter and pose a potential erosion risk. Consequently, they should be used only when quackgrass has become a severe problem or appears likely to become so. Moreover, when using these practices, you should compensate for the reduced soil organic matter by incorporating cover crops into the crop rotation. Also, proper equipment can reduce the impact of quackgrass control practices on soil health. For example, a field cultivator equipped with broad, shallow sweeps can be run shallowly after crop harvest to cut off quackgrass shoots while maintaining crop residue on the soil surface. When cultivating row crops, shallow low-pitch blades will sever the quackgrass shoots with less oxidation of soil organic matter and disturbance of crop roots than more conventional sweeps and shovels.

Additional approaches for controlling quackgrass that do not include tillage and are not destructive of soil quality include mowing and interseeded cover crops. Repeated post-harvest mowing reduced rhizome production and increased grain yield in the subsequent year. Under-sown cover crops also have the potential to control quackgrass if they can attain growth of at least 900 pounds per acre. Interseeding red fescue with a winter wheat cash crop at planting will suppress quackgrass, but red clover can actually benefit rhizome production.

Quackgrass density usually increases in hay and

pasture. Although quackgrass is nutritious and causes negligible economic loss in the hay crop, you will likely have increased quackgrass when rotating back to annual crops. Consequently, try to get the quackgrass well under control before sowing the forage.

ECOLOGY

Origin and distribution: Quackgrass is native to Europe, but it has spread throughout the temperate regions of the world, including North and South America, Asia and Australia. It is found throughout the United States (including parts of Alaska) and in all Canadian provinces but is primarily a problem weed only in the northern states. It is uncommon in dry regions.

Seed weight: 2 mg.

Dormancy and germination: Quackgrass seeds germinate best when exposed to day/night temperature fluctuations of 77/59°F. Few seeds will germinate at any constant temperature from 41–86°F. Newly shed quackgrass seeds are not dormant and require no after-ripening. Rhizome buds can become dormant in late spring and early summer. This condition can be broken by nitrate or when a rhizome segment is severed from the parent plant.

Seed/bud longevity: A few seeds may remain viable in undisturbed soil for 10 years or more, but in normal field conditions few will survive longer than three years. In Alaska, undisturbed seeds declined by 59% per year. Rhizome bud viability declines rapidly in the first year, but it takes two years for all to die.

Season of emergence: Seedlings emerge in spring into summer, but these are uncommon. In undisturbed sod, shoots emerge in early spring, whereas shoots arising from the tips of new rhizomes emerge in late summer and fall. If the rhizomes are broken by tillage or hoeing, however, new aboveground shoots will arise from the terminal bud of rhizome fragments within a few days regardless of the season. However, in Sweden, shoot development from exhumed rhizomes declined in fall in response to shortening day length, but increased again in late fall to spring after exposure to freezing temperatures.

Emergence depth: Peak seedling emergence occurs from seeds that are within 0.5 inch of the soil surface, but an occasional seedling may come up from as deep as 4 inches. Shoots can arise from rhizomes at 10 inches or deeper, though few rhizomes occur naturally at such depths. In general, from smaller rhizome segments, shoots emerged in highest proportions from 1–2 inches, whereas from larger segments, shoots emerged from depths to 6 inches.

Photosynthetic pathway: C_3

Sensitivity to frost: Quackgrass is highly frost tolerant, and green leaves commonly survive the winter in the northern United States. Rhizome bud populations can increase at temperatures above 32°F, so control late into fall is advised. Quackgrass rhizomes are killed by freezing at temperatures below 20°F. Since soil and snow are both good insulators, the rhizomes generally need to be on the soil surface during cold, snow free weather to freeze.

Drought tolerance: The species is moderately drought tolerant. Rhizomes can regenerate after up to 60% desiccation. Therefore, to kill rhizomes, they need to be left on the soil surface during hot, dry weather to thoroughly dry.

Mycorrhiza: Mycorrhiza have been reported as present on quackgrass, but mycorrhiza is absent from some populations and has been classified as a weak host.

Response to fertility: Quackgrass is highly responsive to fertilization with N but responds little to P or K. It is a luxury consumer of nutrients. It can persist in low fertility soils but is not vigorous under such conditions.

Soil physical requirements: Quackgrass does well on a variety of soil textures from coarse sands to heavy clays and on drainage classes from well to poorly drained. It is also tolerant of salt.

Response to shade: Quackgrass is slightly shade tolerant. It will persist (without vigor) in meadows of taller vegetation like goldenrod but is absent from forests. A reduction in level of light produces an increase in the percentage of rhizome buds developing as shoots. Only when light is reduced by 97% does rhizome formation cease.

Sensitivity to disturbance: The rhizomes are mostly distributed in the top 4 inches of an untilled sod where they tend to be cut up during tillage or spading. This increases the number of shoots but decreases the vigor of each shoot.

Time from emergence to reproduction: Shoots begin growing vigorously in early spring and flower from June through August. Long 16-hour days promote formation of elongated stems and inflorescences, but as day length shortens to 12 hours or less, plants shorten and stop producing inflorescences. New rhizomes develop when plants have three to four leaves if they develop from established rhizomes but at six to eight leaves when they develop from seedlings.

Pollination: The species is wind pollinated and self-sterile.

Reproduction: Quackgrass reproduces both vegetatively and by seeds. It typically produces 25–40 seeds per stem. However, seedlings are rarely observed, probably because many seeds are sterile. The species requires cross-pollination, and many populations apparently consist of just one to a few clones, making effective pollination difficult. Seed production serves mainly to disperse the species between sites rather than to maintain the population at a particular site. Most reproduction in a given field or garden is vegetative, with shoots arising from the tips of each rhizome branch. Rhizomes can spread up to 10 feet per year from the parent plant. A single rhizome node planted in late fall in Pennsylvania produced 14 rhizomes with a total length of 458 feet by the following fall. If tillage or deep hoeing breaks rhizomes, most segments will produce an aboveground shoot. These shoots become independent of the rhizome fragment at the five-leaf stage and begin producing new rhizomes at the three- to four-leaf stage. In heavily infested fields, rhizomes can weigh as much as 7–9 tons per acre. Production of new rhizomes is favored by a moderate temperature of 70°F with long day length typical of May or June conditions. In contrast, production of rhizomes that grow upward to establish new shoots is favored by either a low temperature of 50°F, such as occurs in April or November, or by a high temperature of 90°F, as occurs during mid-summer.

Dispersal: Seeds are dispersed in hay used for mulch, in manure and as a contaminant of grain seed (e.g., used for cover crops). The rhizomes are commonly dispersed on tillage machinery. Rhizomes are also sometimes dispersed in the soil surrounding the roots of nursery stock.

Common natural enemies: *Agropyron* mosaic virus causes yellow streaking of younger leaves but does little to reduce the vigor of the plant. Cattle, sheep and horses find quackgrass desirable forage. Pigs will grub out the rhizomes and eat them.

Palatability: Quackgrass is a good forage species with comparable nutritive value, fiber, digestibility and palatability to other perennial forage grasses.

Note: Quackgrass has demonstrated allelopathic effects, which may partly explain the competitiveness of this species.

FURTHER READING

Brandsæter, L.O., M. Goul Thomsen, K. Wærnhus and H. Fykse. 2012. Effects of repeated clover undersowing in spring cereals and stubble treatments in autumn on *Elymus repens*, *Sonchus arvensis* and *Cirsium arvense*. *Crop Protection* 32: 104–110.

Majek, B.A., C. Erickson and W.B. Duke. 1984. Tillage effects and environmental influences on quackgrass (*Agropyron repens*) rhizome growth. *Weed Science* 32: 376–381.

Palmer, J.H. and G.R. Sagar. 1963. *Agropyron repens* (L.) Beauv. (*Triticum repens* L.; *Elytrigia repens* (L.) Nevski). *Journal of Ecology* 51: 783–794.

Ringselle, B., G. Bergkvist, H. Aronsson and L. Andersson. 2015. Under-sown cover crops and post-harvest mowing as measures to control *Elymus repens*. *Weed Research* 55, 309–319.

Werner, P.A. and R. Rioux. 1977. The biology of Canadian weeds. 24. *Agropyron repens* (L.) Beauv. *Canadian Journal of Plant Science* 57: 905–919.

Sandburs

Field sandbur, *Cenchrus spinifex* Cav. = *C. pauciflorus* Benth. = *C. incertus* M.A. Curtis)
Longspine sandbur, *Cenchrus longispinus* (Hack.) Fernald
Southern sandbur, *Cenchrus echinatus* L.

Field sandbur growth habit
Joseph DiTomaso, University of California, Davis

Field sandbur inflorescence
Joseph DiTomaso, University of California, Davis

Longspine sandbur growth habit
Joseph DiTomaso, University of California, Davis

Longspine sandbur inflorescence
Joseph DiTomaso, University of California, Davis

Southern sandbur inflorescence
Joseph DiTomaso, University of California, Davis

IDENTIFICATION
Other common names:

Field sandbur: coastal sandbur, mat sandbur

Longspine sandbur: bur grass, sandbur grass, bear grass, hedgehog grass, field sandbur

Southern sandbur: hedgehog grass

Family: Grass family, Poaceae

Habit: Tufted, usually annual grasses with stems rooting at the lower nodes but curving upward to become erect. Field sandbur sometimes persists through winter in the southern United States.

Description: Seedlings are rolled in the bud and have membranous, hairy ligules less than 0.06-inch long. The bur from which the seedling emerged usually remains attached at the base of the plant.

Field sandbur: Hairs are present only on the blade base and sheath margin. Seedlings are often purple.

Longspine sandbur: The seed leaf is purple toward the base. Early leaves are rough, narrow and often have long hairs at the base. The collar is narrow, pale and distinct. Sheaths are flat, smooth, often red and have fine hairs on the edges.

Southern sandbur: Seedlings are upright and hairless except on the sheath margin and at blade base. Blades are green, flat and abrasive. Blades and sheaths may redden with age.

Mature plants tiller readily and root at stem nodes, forming clumps of prostrate to upright stems. **Sheathes** are compressed and open. Collars are angular. Ligules are very short and membranous, with a fringe of hairs as long as or longer than the membrane. Blades are flat, rough and sometimes folded. Roots are fibrous and shallow.

Field sandbur: Stems are 8–32 inches long. Sheaths are hairless or sometimes hairy on the margin. Ligules are 0.02–0.06 inch long. Blades are 1.5–12 inches long by 0.1–0.25 inch wide, hairless or with a few long, straight hairs at the base.

Longspine sandbur: Stems are 4–30 inches long and light green but sometimes red at base. Sheaths are ridged, hairless to sparsely hairy, with translucent edges. Ligules are 0.02–0.07 inch long. Blades are 2.5–7.5 inches long by 0.1–0.3 inch wide, light green and hairless, but rough on upper surface.

Southern sandbur: Stems are 9–36 inches long. Sheaths are dark green to red, ridged and generally hairless. Ligules are 0.02–0.08 inch long. Blades are 2–12 inches long by 0.12–0.37 inch wide and occasionally hairless but usually with very short, rough hairs on the upper surfaces and no hairs on the lower surfaces.

Inflorescences are narrow, unbranched terminal spikes of spiny burs. Spikes of southern and longspine sandbur may be partially contained in the last leaf. Burs are green when young, turning yellow or brown as they mature. Spines of unripe burs are often purple. Each bur contains one to four spikelets, and each spikelet has two flowers. Burs detach readily when mature and contain one or more brown, oval to egg-shaped seeds.

Field sandbur: Inflorescence is 1.25–5.5 inches long by 0.4–0.8 inch wide, with six to 12 tightly clustered burs. Burs are oval to spherical, 0.2–0.4 inch long by 0.1–0.2 inch wide, hairless to moderately hairy, with eight to 45 spines. Spines are 0.08–.0.22 inch long. Each bur contains two spikelets. Seeds are 0.1 inch long by 0.04–0.08 inch wide.

Longspine sandbur: Inflorescence is 1.25–4 inches long by 0.4–0.8 inch wide, with four to 20 burs. Burs are spherical and 0.2–0.4 inch long. Spines are 0.13–0.25 inch long, numerous (45–75) and arranged irregularly. Each bur contains one to four spikelets. Seeds are 0.08–0.16 inch long by 0.06–0.1 inch wide.

Southern sandbur: Inflorescence is 1–4 inches long by 0.4–0.7 inch wide, with five to 50 burs. Burs are 0.2–0.4 inch long, with a ring of flexible spines at the base. The upper spines are 0.08–0.2 inch long. Each bur contains two to four spikelets. Seeds are 0.06–0.13 inch long by 0.06–0.09 inch wide.

Similar species: Distinctive collars paired with rough foliage and inflorescences composed of burs set these species apart from most other grasses. Young seedlings, however, may be confused with foxtail species (*Setaria* spp.) or barnyardgrass [*Echinochloa crus-galli* (L.) P. Beauv.]. Seedlings of foxtails have new leaves that uncurl as they emerge, while sandbur leaves unfold. Barnyardgrass seedlings do not have a ligule, while sandbur seedling ligules are a short fringe of hairs. Sandbur seedlings can often be distinguished from other young grasses by the presence of a bur at the base of the plant.

MANAGEMENT

The timing of sandbur emergence depends on the species and region, but in many places most seedlings emerge relatively early in the spring. If this is the case, then delaying tillage for planting row crops will help suppress the weed. Similarly, rotating with summer planted crops will help manage sandburs. Shallowly incorporating the burs in the fall will increase seed mortality and promote earlier emergence from the remaining seeds. If small mammals that feed on seeds are abundant, however, leaving the burs on the soil surface through the winter may be a better approach for encouraging seed mortality. In this case, shallowly incorporate the burs early in the spring to promote germination, and then kill the young plants later in the season when preparing the seedbed for crop planting. Incorporation of a high glucosinolate rapeseed green manure crop reduced longspine sandbur emergence and greatly suppressed growth in a greenhouse experiment, but this approach may be more effective on broadleaf species in a mixed weed community under field conditions.

Sandbur species compete poorly with well-established crops. Consequently, a strong stand of a winter grain will suppress sandbur. In Australia, interseeding alfalfa into winter wheat further reduced longspine sandbur seed reproduction if the alfalfa seeding rate was higher than 5.1 pounds per acre. These species will tend to emerge with spring grains, decrease crop yield and produce many burs, so spring grains should be avoided if sandbur is a problem.

Because these species can emerge from deep in the soil, rotary hoeing may be relatively ineffective at reducing density. A tine weeder with stiff tines that break or bury the sandbur seedlings may be better. Cultivate between crop rows before the plants begin to root at the nodes. In irrigated regions, delaying irrigation immediately before and after a cultivation will reduce the amount of rerooting of the dislodged seedlings.

A dense, well managed pasture resists invasion by sandbur species, but overgrazing allows establishment and seed produoction. Hand weed local infestations, for example in disturbed ground around feeding and watering stations, before the sandbur spreads. If sandbur is more widespread, use rotational grazing to intensively but not excessively utilize pasture in the spring; if the forage stand is good, the perennial grasses should recover more rapidly than the young sandbur plants and competitively suppress them. Then mow the pasture when the sandbur flowering stalks are still in the boot to reduce seed production. Additional mowing may be needed when secondary inflorescences begin to form. If sandbur is a severe problem in the pasture, consider replanting using a winter grain nurse crop. If necessary, take the grain as forage before the sandbur flowering stalks are out of the boot. Prevent contamination of hay with unpalatable burs by mowing the hay while the inflorescence is still in the boot.

ECOLOGY

Origin and distribution: Field sandbur is native to southern North America and occurs throughout the southern and southwestern states. It has been introduced into southern South America, the Middle East, southern Asia, Australia, Europe and South Africa. Longspine sandbur is native to southern North America but has spread into most of the United States and parts of southern Canada. It has been introduced into Australia. Southern sandbur originated in tropical America and occurs from the southern United States to Argentina and Chile. It has been introduced into tropical Africa, Madagascar, India, Israel, Hungary, southeast Asia, Australia and islands of the Pacific Ocean. In the United States it occurs in the Atlantic and Gulf Coast states from Maryland to Texas and across the Southwest from Oklahoma to California.

Seed weight: Field sandbur, 1.7–3.4 mg, mean 2.8 mg;

longspine sandbur, 6.8 mg; southern sandbur, 2.4–7.9 mg, mean 5.4 mg. The burs of longspine sandbur contain one to three seeds. The central, uppermost (primary) seed is larger than the lower (secondary) seeds.

Dormancy and germination: Seeds are normally retained in the burs until they germinate, and the surrounding bur and chaff partially inhibit germination. Three months of 32–39°F temperatures were sufficient to permit germination under favorable conditions. Secondary seeds of longspine sandbur tend to be more dormant than primary seeds. For example, in one experiment with whole burs, about twice as many primary seeds germinated in 14 days (71%) than did secondary seeds. Seeds germinate best with fluctuating daily temperatures where the high temperature is 77–95°F and the low temperature is 50–77°F. Temperatures over 104°F inhibit germination and induce secondary dormancy, and prolonged exposure of moist seeds to such temperatures kills a high proportion of seeds. Unlike many weed species, light tends to reduce germination. Germination of southern sandbur is stimulated by nitrate and by minor damage to the seed coat (scarification).

Seed longevity: Few seeds of longspine sandbur persist in the soil for more than a few years and in contrast with most weed species, the seed bank is depleted more rapidly when the burs are buried. In an experiment in Australia, 75% of longspine sandbur seeds buried at 1 inch or 4 inches produced seedlings within three years. The remaining seeds died off at an average rate of 67% per year. Fewer seedlings emerged from burs left on the soil surface (62%), and the mortality rate for the remaining seeds was 52%. Thus, burial of burs reduces the seed bank both by facilitating seedling emergence and by promoting seed mortality. In a second experiment, only 2% of primary seeds buried at 1.6 inch in whole burs remained after 10 months in undisturbed grazing land, and no primary seeds remained after 10 months in tilled soil. In contrast, 81% of secondary seeds were still present in undisturbed land and 54% in tilled soil.

Season of emergence: Field sandbur emerges in early spring. The timing of emergence in longspine sandbur varies regionally. In eastern Washington, over 98% of seedlings emerged between mid-April and June 1, with

a few seedlings continuing to emerge until October. In Colorado, most seedlings emerged between May 25 and June 15, with a few more continuing to emerge until August.

Emergence depth: Fewer seedlings emerged from longspine burs left on the soil surface (62%) than from those planted at 1 inch or 4 inches (78 and 72%), while minimal emergence occurs from 12 inches. A contradictory study showed that 79% of pre-germinated longspine seeds placed at 0.4 inch emerged, and emergence percentage declined smoothly to about 2% at 4.3 inches. Despite the apparent contradiction, both studies indicate that many seedlings of longspine sandbur can emerge from seeds buried several inches deep. Southern sandbur can emerge from as deep as 3.5 inches in clay soil and 4 inches in loamy soil. Seedlings emerged from burs buried at 4 inches about one month later than from burs on the soil surface.

Photosynthetic pathway: Southern and field sandbur are C_4 species, and the whole genus appears to possess this pathway.

Sensitivity to frost: These species generally die during the winter but may survive a mild winter and produce additional burs the following spring.

Drought tolerance: All three species prosper in excessively drained habitats in moderately low rainfall regions. However, they also emerge primarily in the spring when soil moisture is relatively high. Authors disagree on whether field sandbur becomes semi-dormant during drought conditions and then greens up and resumes seed production after rain.

Mycorrhiza: Field sandbur, southern sandbur and other species in the genus are mycorrhizal.

Response to fertility: No information located.

Soil physical requirements: All three sandbur species prefer sandy soils but will grow on fine-textured soils as well.

Response to shade: These species are shade intolerant. For example, southern sandbur died out of a new coffee plantation in Nicaragua as the canopy closed and was only found in open, sunny habitats in a vegetation survey in Brazil.

Sensitivity to disturbance: All three species root at the nodes, and large plants form tussocks that are difficult to bury with plowing and hard to control with cultivation. Repeated mowing or heavy grazing before the plants flower reduces but does not completely prevent seed production. Plants are most sensitive to mowing when in the boot stage of development.

Time from emergence to reproduction: Heads of longspine sandbur emerge seven to 13 weeks after seedling emergence, with the time to flowering decreasing as the season progresses. Warmer June temperatures appear to speed flowering of spring emerging plants, and warmer September temperatures appear to speed flowering of late emerging plants. In longspine sandbur, 20% of the seeds are viable by heading and over 40% by flowering.

Pollination: Field sandbur can self-pollinate. Many flowers of longspine sandbur apparently self-pollinate prior to opening.

Reproduction: Burs contain one to three seeds. Well-watered longspine sandbur plants grown without competition from seeds sown in May, June, July and August in eastern Washington produced an estimated 133,000, 49,000, 5,000 and 40 seeds respectively. Longspine plants emerging in late May in an irrigated Colorado corn field produced 1,120 burs, and the number of burs produced declined rapidly with later emergence. Four-week-old seedlings of longspine sandbur planted into a drying pasture of winter annual forages in southeastern Australia produced only 1–133 seeds per plant, and production decreased with the density of skeleton weed.

Dispersal: The burs of all three species disperse by clinging to fur, clothing and tires. The burs float and disperse along streams and irrigation canals. These species also disperse in contaminated hay.

Common natural enemies: Many species of mice eat substantial quantities of sandbur seeds.

Palatability: Sandbur plants are palatable to livestock when young, but once the inflorescence emerges, the burs irritate the mouths and throats of grazers that ingest them.

FURTHER READING

Anderson, R.L. 1997. Longspine sandbur (*Cenchrus longispinus*) ecology and interference in irrigated corn (*Zea mays*). *Weed Technology* 11: 667–671.

Holm, L.G., D.L. Plucknett, J.V. Pancho and J.P. Herberger. 1977. *The World's Worst Weeds: Distribution and Biology*. The University Press of Hawaii: Honolulu.

Parsons, W.T. and E.G. Cuthbertson. 2001. *Noxious Weeds of Australia, 2nd ed.* CSIRO Publishing: Collingwood, Victoria, Australia.

Twentyman, J.D. 1974. Control of vegetative and reproductive growth in sandbur (Cenchrus longispinus). *Australian Journal of Experimental Agriculture and Animal Husbandry* 14: 764–770.

Shattercane

Sorghum bicolor (L.) Moench ssp. *verticilliflorum* (Steud.) de Wet ex Wiersema & J. Dahlb.

Shattercane immature plant
Antonio DiTommaso, Cornell University

Shattercane unopened panicles
Joseph DiTomaso, University of California, Davis

Shattercane open panicle
Scott Morris, Cornell University

IDENTIFICATION

Other common names: sorghum, black amber cane, wildcane, milo, chicken corn

Family: Grass family, Poaceae

Habit: Tall, upright annual grass

Taxonomic note: The taxonomic status of shattercane is controversial. Some recent authors have used *Sorghum bicolor* (L.) Moench nothosubsp. *drummondii* (Steud.) de Wet ex. Davidse for shattercane, but USDA Plants uses *Sorghum bicolor* (L.) Moench ssp. *drummondii* (Nees ex Steud.) de Wet & Harlan to refer to sudangrass. All competent authorities agree that shattercane is differentiated from domestic sorghum only at the subspecies level.

Description: Seedlings are similar in appearance to corn. Leaves are rolled in the bud, and the ligule is a translucent fringed membrane. The collar is whitish, and auricles are absent. Early leaves are 0.08–0.2 inch wide by 1.6–7 inches long and usually hairy at the base and otherwise hairless. Sheaths may or may not have hairs. **Mature plants** tiller to form clumps and can reach 8 feet in height. The stems are erect,

unbranched, hairless and may have purple spots. Sheaths are smooth and open, with occasional scattered hairs near the collar. Auricles are absent. The ligule is a fringed, translucent membrane up to 0.2 inch long. Blades are 1–2 feet long by 1–4 inches wide, hairless to sparsely hairy, and have a prominent, pale green to white midvein. The roots are fibrous and branching, and brace roots develop at the base of the stem. The **inflorescence** is a compact to open, 3–20 inch-long panicle with whorled and upright branches that turn from green to reddish brown as it matures. Spikelets are paired, and each spikelet has one fertile and one infertile flower. The fertile flower is stalkless and 0.16–0.24 inch long with a 0.25–0.5 inch-long awn, whereas the infertile flower is stalked and lacks an awn. As with all grasses, **seeds** are covered with a thin, tight, hard layer of fruit tissue. Seeds are red to black, 0.1–0.17 inch wide, oval and flattened. The seeds fall readily when ripe, and panicle branches may become fragile and break when mature.

Similar species: In domestic sorghum [*Sorghum bicolor* (L.) Moench ssp. *bicolor*], the short stem (rachilla) that attaches the seed to the main stem of the inflorescence breaks to free the seed during combining. In most biotypes of shattercane, a special layer of cells forms at the tip of the rachilla, which causes the seed to fall off spontaneously when ripe. In a few biotypes the rachilla itself becomes fragile at maturity and breaks spontaneously. Johnsongrass [*Sorghum halepense* (L.) Pers.] is similar to shattercane but is a perennial with large, scaly rhizomes and narrower (0.5–2 inches wide) leaves than shattercane. Seedlings of shattercane and johnsongrass are nearly indistinguishable, but if removed from the soil the attached seed coat may be used to differentiate them. Johnsongrass seeds are much smaller than shattercane seeds, just 0.06–0.08 inch wide. Corn (*Zea mays* L.) seedlings may be distinguished from shattercane by the larger attached seed of corn.

MANAGEMENT

Prevent movement of shattercane into uninfested fields by cleaning combines and tillage equipment. If shattercane is not already present in a field, avoid growing sorghum of any sort on that field. Essentially all sorghum seed contains at least a few seeds of shattering off-types that will introduce the weed to that field. If you have a compelling reason to grow sorghum on an uninfested field, plan to hand rogue out off-type individuals after the inflorescence has emerged but before they set seed. If plants have flowered, remove the cut plants from the field.

Where the weed is already present, management of the shattercane seed bank is critical. A season-long tilled fallow can reduce shattercane seed by 91%, but this may not be sufficient to make this approach worthwhile. In regions with mild winters, leave seeds on the soil surface until spring to maximize seed predation. In regions where the soil freezes, shallowly incorporate the seed to ensure they imbibe and thereby become subject to damage by freezing. Avoid deep burial in the fall as that will decrease the number of freeze-thaw cycles the seeds experience and may prevent freezing altogether. In the spring, bury the seeds deeply by moldboard plowing to reduce emergence. Although a few will return to the surface with next year's tillage, the mortality rate of shattercane seeds is sufficiently high that most will not. Although shattercane continues to emerge through the summer, most seedlings appear in late spring and early summer, so delaying tillage and planting can reduce shattercane density in the crop.

Since many shattercane seedlings will emerge from below the working depth of a rotary hoe or tine weeder, the objective of in-row weeding must be to break or bury the seedlings rather than to uproot them. A tine weeder is better for these purposes. Plant soybeans at high density to allow for more aggressive weeding and consequently greater soybean stand loss. If conditions are favorable for continued shattercane emergence, a late tine weeding at the first to second trifoliate stage of the soybeans can be useful for removing new seedlings close to the row. Since shattercane can emerge late into the summer, delay the lay-by cultivation as long as is practical. If shattercane is a severe problem on the farm, consider obtaining a weed puller, electric discharge weeder or a mower to top the shattercane after it emerges above the crop. Since shattercane generally overtops grain sorghum, at least the first two options can be effective for managing the weed in sorghum. A mower that tops the weed will be most effective after the inflorescence has emerged but before the flowers have opened.

A dense, vigorous crop of winter grain or early planted spring grain will be well established by the time shattercane emerges and will suppress the weed. Since small grains are harvested before shattercane goes to seed, clean-up fields with tillage or mowing after harvest to prevent seed production and, in conjunction with good seed bank management, reduce the severity of an infestation. Harvesting corn or sorghum for forage before shattercane sets seed can also interrupt the life cycle of the weed. Although some shattercane seeds will likely survive through rotation of a field into alfalfa, many will not, and in any case, repeated mowing will prevent further seed production by the weed during the alfalfa portion of the rotation.

ECOLOGY

Origin and distribution: The first shattercane varieties originated in northern sub-Saharan Africa along with domesticated sorghum. It probably emerged multiple times and has further evolved through multiple hybridizations between it, domesticated sorghum and wild *Sorghum* species around the world. It has been introduced into semi-arid areas of Asia, Europe, Australia and North America. It occurs throughout the United States and in southern Ontario and Quebec but is most problematic in the Midwest.

Seed weight: Mean population seed weights of

shattercane range from 19–22 mg.

Dormancy and germination: A moderate proportion of the seeds is capable of germination immediately after falling from the parent plant. Late emerging plants produce a higher proportion of dormant seeds, and the proportion of dormant seeds probably also varies among populations. Removal of outer chaffs or slight damage to the seed coat (scarification) decreases dormancy. The optimal temperature for germination is 77–95°F, and germination is poor below 59°F. Temperature fluctuations promote germination and consequently, shade inhibits emergence by reducing temperature fluctuations. After burial, light promotes germination of seeds in some populations but not in others. Nitrogen fertilization does not promote germination.

Seed longevity: Shattercane is genetically diverse, and seed survival in soil varies with the biotype present. Long-term survival in soil depends on the tight outer chaffs that enclose the seed of most populations. Seeds buried at 8.7 inches under sod remained viable for as long as 13 years. Over 30% remained viable for eight years, but viability declined rapidly after that. In a subsequent experiment, viability of seeds buried 8 inches declined to 37–43% after the first year, and 0–3% were viable after five years. In another experiment, 1% of seeds from plants with open inflorescences survived for three years at 3–6 inches, but survival increased with burial at greater depths. Seeds from compact inflorescences did not survive more than one year unless deeply buried. In experiments in Nebraska, only 11–17% of seeds buried 1–5 inches in November survived until March, and most of the remaining seeds died within the next year. Covering seeded areas with clear plastic increased winter soil temperatures and seed survival, so overwinter seed survival is probably better in milder climates. Freezing greatly decreases viability. Progeny of outcrossed populations derived from grain sorghum and weedy shattercane had smaller, less tightly adhering outer chaffs than shattercane and were unable to survive one Nebraska winter on the soil surface.

Season of emergence: Shattercane begins to emerge in late spring and can continue to emerge throughout the summer.

Emergence depth: Shattercane emerges best from the top 1–2 inches, but 10–50% of seeds placed at 4 inches can successfully produce seedlings. Emergence has been observed from depths of more than 6 inches.

Photosynthetic pathway: C_4

Sensitivity to frost: Shattercane is killed by frost and is injured by temperatures below 55°F.

Drought tolerance: Shattercane is moderately drought tolerant. If seed set is reduced by drought during flowering, side inflorescences can form to partially compensate for the loss.

Mycorrhiza: The species is mycorrhizal.

Response to fertility: Recommended rates of balanced N-P-K increased productivity of forage sorghum by 45% relative to no fertilizer. In competition with corn, shattercane took up significant amounts of nitrogen and subsequently reduced corn yields.

Soil physical requirements: Shattercane grows best on irrigated bottomlands. It does well on soil textures ranging from very fine sandy loam to silty clay loam and heavy clay.

Response to shade: The species is intolerant of shade, but due to rapid early growth supported by large seed reserves and its potentially tall height, it can often avoid shade by overtopping competing crops.

Sensitivity to disturbance: In Nebraska, plants cut before August will regrow and produce viable seeds before frost.

Time from emergence to reproduction: Shattercane in Wisconsin flowered 11 weeks after emergence. In Nebraska, shattercane produced seeds in eight to nine weeks. Seeds became viable 10 days after anthers emerged, and shattering occurred 9–12 days later. Shattercane plants that emerged as late as July in Nebraska were still able to produce viable seeds before a killing frost.

Pollination: Plants are primarily self pollinated, but cross pollination can also occur. The pollen can blow for a mile or more, which makes production of genetically pure sorghum seed difficult.

Reproduction: A shattercane inflorescence produces 500–1,500 seeds, so a typical plant with three heads produces about 3,000 seeds. In Nebraska, shattercane grown in rows similarly to grain sorghum produced 1.5–1.9 panicles and 2,900–3,800 seeds per plant.

Dispersal: Tillage machinery and harvest equipment can move seeds between fields. Even certified sorghum seed with a 99.95% genetic purity can introduce as many as 20 shattercane plants per acre if the seed is sown at a typical 40,000 per acre. Thus, even "clean" sorghum seed is a potential source of shattercane introduction. Shattercane seeds pass through cattle in a viable condition and may thus be present in manure or spread with the movement of livestock. Shattercane seeds float and can be carried long distances in runoff and irrigation water.

Common natural enemies: Many birds and small mammal species feed on shattercane seeds.

Palatability: The seeds are commonly eaten during famines in Africa. The plant is generally palatable for ruminants but can be poisonous to horses. Nitrate can accumulate to poisonous levels during dry summers. Avoid grazing livestock on young stands as the seedlings contain a compound that releases cyanide when consumed. Shattercane is not recommended for feeding poultry because its high tannin content slows growth.

FURTHER READING

Defelice, M.S. 2006. Shattercane, *Sorghum bicolor* (L.) Moench ssp. *drummondii* (Nees ex Steud.) de Wet ex Davidse – black sheep of the family. *Weed Technology* 20: 1076–1083.

Horak, M.J. and L.J. Moshier. 1994. Shattercane (*Sorghum bicolor*) biology and management. *Reviews of Weed Science* 6: 133–149.

Wild garlic

Allium vineale L.

Wild garlic plant
William Ferlatte

Wild garlic plants
Randall Prostak, University of Massachusetts

Wild garlic hollow stem
Randall Prostak, University of Massachusetts

Wild garlic bulblets on ripe inflorescences (center) in a wheat field
Joseph DiTomaso, University of California, Davis

IDENTIFICATION

Other common names: field garlic, wild onion, crow garlic, scallions, ramp

Family: Lily family, Liliaceae

Habit: Perennial grass-like herb arising from a bulb

Description: Seedling leaves are slender, hollow, upright, hairless, round and grass-like in appearance. Seedlings smell of garlic or onion when crushed. **Mature plant** leaves are similar to seedling leaves.

Leaves are 6–24 inches tall by 0.1–0.4 inch wide. Belowground bulbs are egg shaped, have papery coverings and can also develop in segments from the main bulb. A fibrous root system emerges from bulb bases. Stalks are unbranched, round, smooth, waxy, leafless and solid cored; the top of the stem gives rise to either aerial **bulblets** or **flowers** that emerge from a sheathed globular structure. Aerial bulblets are small and teardrop shaped, with a thin, green leaf emerging from the top. Maroon, pink or white-green flowers may develop on the globe above the bulblets on 0.25–1 inch-long stalks. Occasionally, flowers produce three-chambered, egg-shaped capsules that contain up to six flat, wrinkled, black 0.1 inch-long seeds.

Summary of reproductive structures and terminology used:

Wild garlic reproduces primarily by production of four types of bulbs. Plants not sufficiently mature to produce a reproductive stalk form a **terminal bulb** in early summer that resumes growth in the fall. Plants producing reproductive stalks usually form a single large **soft offset bulb** below ground in the axil of an inner leaf. These generally sprout in the fall replacing the former terminal bulb. Larger plants forming reproductive stalks typically produce one to four **hard offset bulbs** in the axils of outer leaves that can remain dormant for two or more seasons and produce new plants. Aerial **bulblets** are small bulbs formed at the top of reproductive stalks. Most plants emerge from bulbs and bulblets; seed production is rare in much of the United States. When flowers occur, they form in the same heads as bulblets and can produce **seeds**.

Similar species:

Domestic garlic (*Allium sativum* L.) plants generally have considerably larger stalks, bulbs and flowers than wild garlic. The flowering stalk of domestic garlic typically curves in a loop, whereas that of wild garlic is straight. Wild onion (*Allium canadense* L.) leaves are flat and solid when cut. Wild onion bulbs do not become hard when dormant. The white, star-shaped flowers and prominent, white midvein of star of Bethlehem (*Ornithogalum umbellatum* L.) distinguish it from wild garlic. Star of Bethlehem also lacks the strong characteristic odor of wild garlic.

MANAGEMENT

Wild garlic is not a competitive weed and rarely causes noticeable yield loss, but has a large impact by tainting farm produce, thereby reducing its value. When consumed by pastured livestock, it can produce a disagreeable taste in milk and meat. In addition, the bulblets are harvested with cereal grains and are difficult to clean out of the grain. Flour ground from contaminated grain has an unpleasant taste and odor. Moreover, even a low percentage of bulblets in the grain (0.01%) can cause caking on the mill, which necessitates frequent cleaning. Consequently, mills reject contaminated grain shipments or levy heavy docking fees. Wild garlic is less of a problem in vegetable crops, but separating the long thin leaves from salad and braising greens can substantially slow harvest and processing of these crops.

Wild garlic is best managed by late fall to early spring tillage. The optimal time for tillage is when two foliage leaves are well formed (do not count the short sprout leaf, which quickly withers). Tillage at this time destroys growing plants after food has been transferred to leaves but before new offsets and bulblets have formed. Since development is not synchronized across all plants, several years of appropriately timed tillage will be needed to bring an infestation under control. Repeated harrowing in fall and spring without primary tillage also greatly reduces the population but generally is no better than fall plowing followed by spring harrowing and a spring planted crop. Tillage in late spring will allow formation of offsets, but they will not have had time to form a hardened covering scale and, therefore, will not be dormant. Mowing or tillage when the stalks have started to lengthen but are not full height will prevent or reduce production of bulblets. Plants will, however, still produce dormant offset bulbs at the base of the plant. Summer fallow is useless for controlling wild garlic. Plowing in the fall before plants have two long foliage leaves will kill a few plants by deep burial, but many will resprout and develop normally.

Rotation of badly infested land to spring planted row crops is beneficial as it allows spring tillage and repeated cultivation to damage plants during the formation of offset bulbs and bulblet bearing stalks. If the land is rotated to a summer planted crop, precede the crop with a spring fallow or with early plowing followed by a rapidly growing cover crop. Spring sown cereal grains can also decrease wild garlic populations if tillage can be timed appropriately.

Like its domesticated relatives, wild garlic is a poor competitor that will be suppressed by several years of competition from a dense stand of perennial grasses or legumes. Hence, in lightly infested pastures, use good management practices to ensure a vigorous growth of grasses and legumes between grazing episodes. This will also dilute the percentage of wild garlic in the forage and thereby reduce the risk of tainting milk and meat. To reduce the potential for tainting of animal products, avoid early grazing of infested fields since wild garlic will present fodder to the animals before grasses and legumes begin growing. Badly infested pastures should be tilled in late fall or early spring and, if possible, rotated to crops in which wild garlic is more easily controlled for a few years. If the land is easily eroded, plant to a perennial hay crop and fertilize it well after establishment to ensure lush, vigorous growth.

Avoid planting winter grains on infested land. If the infestation is sparse, an occasional winter grain crop may be grown successfully if the stand is good and the crop is vigorous. Consequently, use a high seeding rate, and fertilize the crop well. These measures should greatly reduce the number of bulblet bearing stalks and hence contamination in the harvested grain. If the stand turns out poor, consider harvesting the crop for hay: the garlic flavor does not persist in dried hay as it does in fresh forage or silage.

ECOLOGY

Origin and distribution: Wild garlic is native to Europe and western Asia, where it is widespread. It occurs also in North Africa and has been introduced into North America, Australia, New Zealand and Chile. In North America, the species is present throughout most of the eastern and central states, Ontario and the moister parts of the far west, including British Columbia and Alaska.

Seed and bulblet weight: Bulblets weigh 3–70 mg and occasionally more, with the median around 17 mg. Larger bulblets have a higher rate of sprouting than smaller bulblets. Seeds weigh 0.5–1.5 mg with a mean of 1.07 mg.

Dormancy and germination: All types of bulbs are dormant when produced, but terminal bulbs, soft offsets and most bulblets lose dormancy and begin sprouting by the summer or fall of the year they are produced. Bulblets lose dormancy after several months at 72°F but not at 33°F or 40°F, a pattern that favors plant establishment in early fall and a growth advantage when temperatures warm in early spring. Sprouting of bulblets was higher at 68°F than at 86°F. The hard, waxy scale on hard offsets tightly surrounds the bulb and, despite completion of after-ripening, typically delays germination for a year or longer. Eventually, these coverings split or decay, and the hard offset sprouts. Sprouting of non-dormant hard offsets was best at 50°F. Seeds require two to three months of cold, moist conditions to germinate.

Seed and bulb longevity: Terminal bulbs and soft offset bulbs usually die if they do not sprout the autumn after production. In one study, seeds and bulblets buried in the soil did not survive for a full year. In other studies, however, 4–5% of bulblets survived more than one year, but not longer. Hard offset bulbs can remain dormant in the soil for five years or more, though all but a few sprout within the first two to four years after production.

Season of emergence: Bulbs sprout from August until the ground freezes. Some shoots do not reach the soil surface until spring, however, so shoots emerge in both fall and early spring. Emergence in spring can also occur from bulbs at greater depth in the soil. Seeds would be expected to germinate in spring based on their need for a period of cold conditions (see the "Dormancy" section), but this has never been observed in the field.

Emergence depth: The ability of a bulb to produce a shoot depends on both depth and the size of the bulb. Placement on the soil surface reduced establishment and growth of plants compared to placement 0.5–0.75 inches deep. Bulblets emerge best from the top 2 inches and generally cannot emerge from deeper than 4–6 inches. Offset bulbs emerge best from the top 4 inches with minor emergence from 8 inches. The largest bulbs can emerge from the bottom of the plow layer (8–10 inches) or deeper (16 inches). The likelihood of a shoot producing a reproductive stalk, and the number of belowground daughter bulbs that a plant produces, declines with the depth of the parent bulb, particularly when the parent bulb is below 4–6 inches. Soft offset bulbs buried below this depth in fall or early spring produced few stems or inflorescence the following spring. In long established grass swards, the largest bulbs and the most vigorous plants have their bases at about 2.75 inches. This is apparently the optimum depth for terminal bulbs and hard offsets, and plants that first establish near the soil surface eventually pull themselves down to this depth by root contraction.

Photosynthetic pathway: C_3

Sensitivity to frost: Wild garlic is very resistant to even hard frosts. The leaves sometimes brown and die back to the soil surface but continue to lengthen again in early spring along with additional new leaves.

Drought tolerance: Wild garlic is highly resistant to drought. Plant growth slows during dry weather, but the bulbs will survive long drought periods, even when uprooted and lying on the soil surface following tillage.

Mycorrhiza: Mycorrhiza have been observed. Mycorrhizal association with wild garlic is more important when nutrients are limiting.

Response to fertility: Balanced fertilizer application appropriate for winter grain production moderately increases the likelihood of flowering, the number of foliage leaves per plant, the number of offset bulbs and the average weight of the offset bulbs. Plants arising from bulblets, however, had slower growth and produced fewer hard offsets when balanced fertilizer was applied, probably due to osmotic effects. The strength of garlic's flavor was reduced by sulfur deficiency in the nutrient medium.

Soil physical requirements: Wild garlic grows successfully on a wide range of soil textures from heavy clay to coarse sand. Bulbs can survive an entire growing season in waterlogged soil.

Response to shade: Wild garlic tolerates moderate shade, but like its domestic relatives it is easily suppressed by competition from more robust plants. In particular, competition from a vigorous crop greatly decreases formation of reproductive stalks, though in most cases the population will maintain itself by production of belowground bulbs.

Sensitivity to disturbance: Wild garlic plants can be completely killed by burial at the two-leaf stage to early three-leaf stage (the little sprout leaf does not count in this regard). Shallow burial before this stage will result in resprouting and normal growth. Deep burial at greater than 4 inches, however, will kill most younger plants. Burial after the second- to early third-leaf stage may appear to have killed the plants, but they will form bulbs underground from energy stored in the leaves. Cutting the flowering stalk when it first forms prevents production of bulblets but increases the size of offset bulbs by as much as three-fold. Bulblets will continue to mature on a cut off or buried reproductive stalk once the stalk has reached its full length, even if the head has not begun to swell. Plants survive multiple defoliations by mowing or grazing. Plants cut in late spring or summer wait to resprout until the usual time in the fall. Dormant bulbs are essentially immune to management practices other than digging and removal.

Time from emergence to reproduction: Shoots emerge throughout the fall, and reproductive stalks typically appear in mid-spring. Plants require a cold treatment to induce stalk formation. Stalks produce mature bulblets by late spring or early summer. Plants establishing from bulblets generally require two to four seasons to mature.

Pollination: The occasional flowers that are produced are cross pollinated by bumblebees and flies.

Reproduction: Wild garlic reproduces primarily by production of the various bulb structures described above, but occasional plants produce flowers and seeds as well. Most reproduction occurs by formation of bulblets on the reproductive stalk, where flowers would be expected on domesticated garlic. Four populations in Sweden produced 75–141 bulblets per plant. Production of seeds varies greatly with season, locality and among plants, ranging from three to 70 seeds per inflorescence. Seed production is rare in the northeastern United States. Some genotypes of wild garlic tend to allocate more resources toward belowground asexual offsets and, occasionally, sexually produced seeds, whereas other genotypes allocate more to aerially produced bulblets and less to offsets or seeds.

Dispersal: Bulblets and seeds disperse similarly short distances, with the majority found within 14 inches of the inflorescence. Bulblets move in contaminated seed grain and especially in rye and winter wheat used as cover crop seed since this is rarely certified for seed quality. They are sometimes found in straw. They also travel in mud on tires, farm implements, shoes and the feet of livestock. Bulblets will float for many hours and likely disperse with overland flow off of fields, along waterways and by blowing across ponds and lakes.

Hard offsets disperse little distance from the parent plant but can remain dormant several years, allowing them to disperse in time. In contrast, bulblets do not survive more than a year but can disperse long distances by the mechanisms discussed above, although most fall within a meter of the parent plant. Bulblet performance is maximum when dispersed 10 inches and declines with increasing dispersal distance from the parent plant, suggesting that local adaptation occurs at the scale of natural dispersal, thereby favoring asexual reproduction. The evolutionary contribution of seeds to survival of the species has yet to be clarified.

Common natural enemies: Several fungi cause extensive damage to belowground bulbs. These include onion white rot (*Sclerotium cepivorum*) and various species of *Penecillium* and *Fusarium*. Slugs and rabbits feed on the foliage.

Palatability: The bulbs are occasionally used to flavor cooked dishes, but most people consider the flavor much inferior to that of domestic garlic. The bulbs do, however, contain substantial concentrations of the same beneficial antioxidants found in domestic garlic.

Livestock readily eat wild garlic leaves and shoots, but consumption of fresh material can taint milk and meat. The flavor persists through the silage-making process but is not present in dried hay. Consumption of large quantities of wild garlic can cause poisoning in both people and livestock.

FURTHER READING

DeFelice, M.S. 2003. Wild garlic, *Allium vineale* L. – little to crow about. *Weed Technology* 17: 890–895.

Håkansson, S. 1963. *Allium vineale* L. as a weed, with special reference to the conditions in south-eastern Sweden. *Växtodling (Plant Husbandry)* 19: 1–208.

Lazenby, A. 1961. Studies on *Allium vineale* L.: I. The effects of soils, fertilizers and competition on establishment and growth of plants from aerial bulbils. *Journal of Ecology* 49: 519–541.

Lazenby, A. 1962b. Studies on *Allium vineale* L. IV. Effect of cultivations. *Journal of Ecology* 50: 411–428.

Peters, E. J. and S. A. Lowance. 1981. Effects of date and depth of burial on wild garlic (*Allium vineale*) plants. *Weed Science* 29: 110–113.

Wild oat

Avena fatua L.

Wild oat young plant
Joseph DiTomaso, University of California, Davis

Wild oat stand
Jack Clark, University of California

Wild oat inflorescence
Joseph DiTomaso, University of California, Davis

Wild oat spikelet
Jack Clark, University of California

Wild oat ligule
Jack Clark, University of California

IDENTIFICATION

Other common names: wheat oats, oat grass, flax grass, poor oats, drake, haver corn, hever, black oats, oats

Family: Grass family, Poaceae

Habit: Summer annual grass in most of its range; winter annual in California

Description: The **seed leaf** is oriented vertically and round tipped. True leaves are rolled in the bud, usually emerging with a counter-clockwise twist that is retained to maturity. The ligule is membranous, relatively big and roundly pointed; auricles are absent. Hairs are sometimes present on the dark green blades; when present, small hairs are evenly distributed and large hairs are scattered. The upright, sturdy **mature plants** have three to five tillers and can reach 1–3.5 feet in height. Flattened sheaths have transparent edges and scattered hairs; they are split open near the collar region. The mature ligule is 0.1–0.25 inch long and pointed, with an uneven, slightly toothed edge. Blades are linear, flat, 2.75–15.75 inches long by 0.25–0.75

inch wide, pointy tipped and rough, with sparse, long hairs present on leaf edges at the base. An extensive fibrous root system is present. The **inflorescence** is an open, pyramidal, 4–15 inch-long by 8 inch-wide or less panicle. The widely spaced panicles each have nodding spikelets of two to three flowers per branch node. Each spikelet is covered by papery, 1 inch-long outer chaffs, and each flower is encased by a 0.75 inch-long, **hairy inner chaff**. Inner chaff hairs are white to brown, bristly and scattered. Each inner chaff has a distinctive, 1.25–1.5 inch-long, dark brown to black awn that is bent at a 90-degree angle and twists when in contact with water. Seeds fall off the panicle at maturity with the inner chaff and its long awn still on the seed, leaving a round scar where they were attached to the stem; this scar is sometimes surrounded by short, bristly hairs. **Seeds** are 0.4 inch long, range in color from yellow-brown to rust to nearly black. The apparent seed includes a thin, tight covering of fruit tissue.

Similar species: Cultivated oats (*Avena sativa* L.) are most easily differentiated from wild oats when the seedhead is present. Cultivated oats have a narrower, more compact panicle than wild oats, and the non-shattering seed has an awn that is short enough to sometimes appear absent. Vegetatively, the two are nearly impossible to tell apart, especially when the plants are young.

MANAGEMENT

Although wild oats can be a problem weed in many crops, typically it is a weed of cereal grain-dominated cropping systems. It is particularly problematic in systems in which spring cereal crops are rotated with other drilled crops like canola and flax. Rotation of drilled crops with row crops that can be cultivated between rows like dry beans, potatoes and sugar beets will help decrease wild oat infestations. This will be especially true if precision cultivation tools are used to work close to the crop row.

The tillage regimen greatly affects wild oat management. Wild oat seeds survive poorly on the soil surface. Consequently, unless other weed species are about to go to seed, avoid summer tillage after grain harvest. Even a few weeks of exposure on the surface will decrease the number of viable seeds present. Very shallow harrowing (less than 2 inches) in the fall will stimulate up to half of the current crop of seeds to germinate and subsequently die over the winter in cold climates. In the spring, tillage options depend on the amount of seed produced in the previous crop; if seed production was high, then plowing in the spring will likely bury more seeds than it brings to the surface, whereas if seed production was low, then shallow spring tillage may be preferable. In any case, delaying tillage and planting of either a fall or spring sown crop will reduce the density of wild oat by allowing more seeds to germinate before seedbed preparation. Note that rotary hoeing or tine weeding after crop emergence is not very effective because a large proportion of wild oat seedlings emerge from below the operating depth of these implements. Nevertheless, some control can be achieved using a tine weeder with stiff tines to bury young wild oat seedlings. Not planting every third row and then cultivating the skipped rows has achieved substantial control in European studies.

Since options for mechanical management of wild oat in grain cropping systems are limited, crop competition is critical for management of this weed. In general, winter grains are more competitive against this spring germinating species than spring grains. Among spring drilled crops, the ranking of competitiveness against wild oat is barley-canola-wheat-flax. Usually, long strawed varieties and species (for example, spelt or rye) are more competitive than semi-dwarf varieties. Narrow row spacing and increased seeding density will likely improve yield if wild oat is a problem and can be guaranteed to decrease seed production of the weed.

Wet soils favor wild oat, so tile drainage and practices that avoid compaction have long-term benefits for managing this weed.

Inspect seed grain carefully before you buy it! Wild oat is a common contaminant of most grain crop seed, and even a low percentage of contamination can result in planting thousands of wild oat seeds per acre. Although less than 1% of wild oat mill screenings fed to cattle pass unharmed into the manure, this can still be an enormous number of seeds. Consequently, grinding wild oat grain before feeding and/or composting manure before spreading is advisable.

ECOLOGY

Origin and distribution: Wild oat probably originated in central Asia, and it presently occurs throughout Europe and Asia except mainland southeast Asia. It has been introduced into Australia, New Zealand, East

Africa, North America and most of South America. In North America, wild oat occurs in Alaska and across southern Canada and the United States except for most of the Southeast and the southern Great Plains. It is particularly problematic in the northern prairie region.

Seed weight: Mean population seed weights range 14–24 mg.

Dormancy and germination: Usually, wild oat seeds are dormant when shed from the parent plant, but they lose dormancy over time when exposed to warm, dry conditions. The duration of after-ripening is dependent on complex interactions between wild oat genotype and temperatures during both seed development and after-ripening. Generally, warm, dry conditions hasten after-ripening, and cool, moist conditions delay it. Immersion in water, or alternate wetting and drying, induces secondary dormancy in previously germinable seeds. Once seeds become non-dormant, the optimum temperature for germination is 59–82°F. Nitrate promotes germination of dormant seeds, and applications of N fertilizer have been shown to increase emergence of wild oat in the field. Light or a low oxygen atmosphere inhibits germination.

Seed longevity: The annual mortality rate of wild oat seeds is high, typically in the range of 70–90% per year, but a small fraction of seeds remains viable even after five to nine years. However, in Colorado, no wild oat seeds survived burial for two years, leading to speculation that seeds of Colorado biotypes had less intense dormancy and that, consequently, wild oat was a less serious weed problem south of the 43rd parallel. Wild oat seeds survive poorly on the soil surface, but survival increased when seeds were buried deeper than 4 inches below the soil surface.

Season of emergence: Most emergence occurs in the spring and early fall, but some seedlings emerge throughout the growing season. In Minnesota and North Dakota, wild oat emergence was concentrated primarily in a 28–42 day period after initial emergence began in spring. In Canada, emergence occurred earlier in the spring under conservation tillage than under conventional tillage.

Emergence depth: Optimum recruitment depth is 0.8–2.8 inches, but emergence can occur from seeds as deep as 6–9 inches. Few seedlings emerge from seeds on the soil surface.

Photosynthetic pathway: C_3

Sensitivity to frost: Seedlings are frost sensitive but grow well at cool temperatures. Fall emerging plants in the northern prairies are killed by hard frost before they mature, but they can survive the winter in California.

Drought tolerance: Wild oat is relatively sensitive to drought. Even within a field, it tends to be more common in wet depressions than on dry knolls.

Mycorrhiza: Wild oat is mycorrhizal.

Response to fertility: Growth of wild oat is highly responsive to N; in particular, it is more responsive than wheat and similar in responsiveness to canola. Its aboveground growth is moderately responsive to P, being more responsive than wheat but less responsive than canola. Wild oat tolerates soil pH down to 4.5.

Soil physical requirements: Wild oat tolerates a wide range of soils but is most common on clay and loam soils.

Response to shade: Wild oat does not grow in shady environments. In response to shade such as from crop competition, plants grow to the same height, but the number of tillers and seed production are greatly reduced.

Sensitivity to disturbance: Wild oats do not reproduce well from vegetative parts and can be killed easily by cultivation. However, damaged seedlings can regrow in the top 1 inch of soil, presumably if the seed and roots remain intact, allowing access to large seed reserves and soil moisture.

Time from emergence to reproduction: Heading of wild oat plants begins seven to eight weeks after emergence and is controlled more by temperature than by day length. Wild oat plants mature seeds three to four weeks after heading, with seeds at the tip of the main axis of the panicle ripening and shedding first. The capacity of wild oat to mimic the development of the grain crops that it infests, and to shatter mature seeds before grain harvest, ensure its success in these crops.

Pollination: Wild oat mostly self-pollinates, but 1–2% of seeds are produced through cross pollination by wind.

Reproduction: Wild oat produces up to 500 seeds per plant, but 100–150 seeds per plant is more typical. Plants emerging early produced 120 seeds per plant whereas plants emerging three weeks later produced 61 seeds per plant in the Minnesota Red River Valley. Wild oat plants emerging before emergence of a barley crop produced 68–81 seeds per plant, whereas plants that emerged after barley produced 2–58 seeds per plant, depending on the barley growth stage at which the wild oats emerged. Plants subjected to intense competition from a vigorous barley crop produced 20–30 seeds per plant. Seeds that mature in the crop understory are 10–30% less viable than seeds that mature above the crop leaf canopy.

Dispersal: Seeds fall on the ground without any natural means of dispersal. For thousands of years, the main mode of spread has been in contaminated seed grain. The species also commonly moves about on combines.

Only a low percentage of wild oat seeds passes through cattle in a viable condition.

Common natural enemies: The natural enemies of wild oat are similar to those of grain crops.

Palatability: The seeds are edible. Wild oat mill screenings are commonly sold as a feed grain and have about 90% of the nutritional value of domestic oat. Wild oat is palatable forage, and extensive tracts of annual grasslands dominated by wild oat are harvested for hay in California.

FURTHER READING

Bullied, W.J., A.M. Marginet and R.C. Van Acker. 2003. Conventional- and conservation-tillage systems influence emergence periodicity of annual weed species in canola. *Weed Science* 51: 886–897.

Harker, K.N., J.T. O'Donovan, R.B. Irvine, T.K. Turkington and G.C. Clayton. 2009. Integrating cropping systems with cultural techniques augments wild oat (*Avena fatua*) management in barley. *Weed Science* 57: 326–337.

Sharma, M.P. and W.H. Vanden Born. The biology of Canadian weeds. 27. *Avena fatua* L. *Canadian Journal of Plant Science* 58: 141–157.

Wild-proso millet

Panicum miliaceum L.

Wild-proso millet immature inflorescence
Joseph DiTomaso, University of California, Davis

Wild-proso millet inflorescence
Joseph DiTomaso, University of California, Davis

IDENTIFICATION

Other common names: proso millet, proso, millet, hog millet, broom corn millet, broom corn, panic millet

Family: Grass family, Poaceae

Habit: Tall, upright annual grass

Description: Seedlings are upright. The leaves are rolled in the bud, lack auricles and are densely covered in short, stiff hairs. The ligule is a fringe of hairs 0.08–0.16 inch long that are fused at the base. The seed remains attached to the seedling at the root. **Mature plants** reach 1–4.25 feet tall. Stems are upright or occasionally nodding. The collar is pale green or white and partially encloses the stem. The sheath is open and densely hairy. The ligule is similar to that of the seedling. Leaves are 4–12 inches long by 0.25–1 inch wide, with a conspicuous, strongly ridged pale green or white midvein. The leaves are sparsely to densely covered in long hairs (or occasionally are hairless). The roots are fibrous. The **inflorescence** is a 3–12 inches long, terminal, upright to nodding, branching panicle that remains partially enclosed in the leaf. Panicle branches are smooth to rough. Spikelets occur singly at the end of branches and are 0.18–0.21 inch long, smooth and have prominent green veins. Each spikelet produces a single **seed**, which is shiny, smooth, white to reddish-brown or black and 0.13 inch long by 0.1 inch wide. The apparent seed includes a thin, tight covering of fruit tissue.

Similar species: Witchgrass (*Panicum capillare* L.) and fall panicum (*Panicum dichotomiflorum* Michx.) have similar habits and inflorescences to wild-proso millet. Witchgrass is shorter (8–36 inches) and has smaller spikelets (0.08–0.13 inch long) than wild-proso millet. Fall panicum is hairless at maturity, and the seedling leaves have hair only on the underside.

Note: Wild-proso millet populations vary greatly in the degree to which they depart from the domesticated crop and behave as weeds. In general, black seeded populations tend to be weedier than populations with light colored seeds.

MANAGEMENT

Crop rotation is an important component of wild-proso millet management because seed banks of this species are rapidly depleted without substantial annual input of seeds to the soil. Any plants that emerge in alfalfa will be mowed before they can produce seeds, and most of the relatively short-lived seeds will die during the sod phase of the rotation. For example, a Wisconsin study found that four years of alfalfa reduced seedlings emerging in a following corn crop by 80%. Winter wheat is very competitive against wild-proso millet and will usually be harvested before the weed can go to seed. Spring cereals also suppress this weed well since they become well established by the time it emerges, and, in principle, a large proportion of the seeds produced can be captured or destroyed during combine harvest. In contrast, corn, soybeans and dry beans are poor competitors with wild-proso millet. Late spring planting of summer crops, however, can reduce wild-proso millet density relative to earlier planting dates. Crop-like biotypes are more susceptible to competitive stress than are dark-seeded biotypes.

Avoid fall tillage if possible: The relatively large seeds are highly attractive to a wide range of seed predators and are also more susceptible to other forms of mortality when near the soil surface. Even the soil disturbance associated with no-till drilling of cover crops may be sufficient to protect the seeds. Deep tillage in the spring, however, will reduce seedling emergence following a season with heavy seed rain, and since the annual death rate of seeds is high even deep in the soil, relatively few seeds will return to within emergence depth following tillage in subsequent years.

Rotary hoeing can kill a substantial proportion of the first flush of seedlings. Time the hoeing for just before the seedlings emerge, about seven to 10 days after last tillage. Some seedlings will emerge from below the depth of hoeing. Configure tine weeders to bury these just after emergence. The shoot will resprout if broken above the seed, so burying seedlings is preferable to breaking them. Inter-row cultivation of row crops often catches the young wild-proso millet after most have emerged and can provide up to 95% control. Completely covering seedlings with 0.8 inch of soil kills them, so, if possible, hill crop rows while seedlings are still small.

Wild-proso millet is still spreading to new fields in most of its range and preventing new infestations, particularly of the black seeded biotypes, is essential. Till and harvest infested fields last since tillage equipment, combines and forage harvesters are major means of spread. Similarly, clean machinery before moving from an infested to a clean field. Pull up new outbreaks, and if the plants have flowered, remove them from the field.

ECOLOGY

Origin and distribution: The species is native to Eurasia and was one of the first domesticated grain crops. The first black seeded biotype in North America was found in Minnesota and Wisconsin in 1970 and was probably introduced from Europe or Asia. Additional, more crop-like weedy biotypes appear to arise spontaneously by mutation from domesticated proso millet. Wild-proso millet now occurs in most of the United States and southern Canada but is most problematic from southern Ontario through the Midwest to the Northwest. In addition to temperate North America, it occurs in Mediterranean Europe and the Middle East, south Asia, Japan, Australia, New Zealand and South Africa.

Seed weight: Population mean seed weight ranges from 4–7 mg.

Dormancy and germination: Seed dormancy at maturity varies with biotype. In contrast with domesticated proso millet, which had 100% germination at maturity, three wild biotypes had 51–87% germination even though seeds had higher than 99% viability. In another study, only 22% of seeds were capable of germination 10 weeks after appearance of the inflorescence. Biotypes with dark colored seeds take up water more slowly and germinate more slowly than those with white or golden colored seeds. Domesticated proso millet germinates at temperatures of 50–108°F, with the optimum being 68–86°F. The lower threshold for wild-proso millet germination is 45°F. Under field conditions, seedlings begin emerging the week after day/night temperatures exceed 77/49°F. Dormancy is overcome by moist, cold temperatures of 41°F and the absence of light during germination.

Seed longevity: Seeds from light seeded, crop-like populations averaged only 4% survival over the winter, though seeds of one population had 13–40% survival depending on burial depth. None of the crop-like seeds survived a second winter. In contrast, black seeded populations had greater than 70% overwinter survival at 2 inches and most had greater than 90% survival at 8 inches. In Canada, survival of black seeded populations buried at 8 inches for 11 months averaged 40%, and very few seeds survived for 42 months. Another study found an average annual seed mortality of 39%

over a 54 month period. This figure is close to the 42% mortality rate from a one-year study. Seeds survived better in a moderately drained soil than in a well-drained sand or a poorly drained clay loam. Seed predators rapidly remove seeds from the soil surface. Although deep burial at 6–8 inches promotes seed survival relative to more shallow burial at 2 inches, the increase in survival is small relative to many other weed species.

Season of emergence: In Ontario, Wisconsin and Colorado, most seedlings emerge in late May and June, with a few continuing to emerge later in the summer.

Emergence depth: The average depth of emergence is 1–2 inches, and the maximum depth from which seedlings can emerge is 5.4 inches. Emergence from the soil surface is minimal.

Photosynthetic pathway: C_4

Sensitivity to frost: The species tolerates cold better than most C_4 plants but will not tolerate frost.

Drought tolerance: Domesticated proso millet is possibly the most drought tolerant of all cereals, despite a shallow root system, and wild-proso millet is similarly drought tolerant. Young wild-proso millet can tolerate at least two weeks of drought, and many plants survive such droughts even when raked half out of the soil.

Mycorrhiza: No reports are available, but it probably is mycorrhizal based on the mycorrhizal status of other *Panicum* species.

Response to fertility: No information is available on the response of weed biotypes to nutrient applications. Domestic proso millet is less responsive to fertility than most crops. The recommended N application of 36 pounds per acre only resulted in a 6–40% increase in dry weight. The species does best on soils with a pH of 5.8–6.8.

Soil physical requirements: The species is well adapted to medium and fine textured soils but does poorly on course textured soils. It does not tolerate anaerobic or saline soils.

Response to shade: Experiments in Washington and

Illinois found a negative correlation between wild-proso millet seed production and the percentage of light intercepted by sweet corn canopies. Nevertheless, wild-proso millet is moderately shade tolerant. Dry weight was reduced only about 25% with 47% shade. With 90% shade, however, dry weight was reduced about 80%, but this reduction is less than for many other weeds. Plants grew substantially taller but thinner when shaded, which potentially allows them to grow through crop canopies. Even with 90% shade, plants still produced seeds.

Sensitivity to disturbance: Wild-proso millet is highly resistant to physical damage. Crushing young plants with tractor tires or repeated mowing had little effect on survival or subsequent growth. Many plants even survived rotary tilling or disking. In controlled experiments, even complete removal of either roots (below the seed) or shoots had little effect on survival of seedlings at the three- or six-leaf stage, and even at the one- and two-leaf stages, many plants recovered from severe damage. Only a very few plants, however, recovered from complete burial under 0.8 inch of soil.

Time from emergence to reproduction: Maturation is more rapid as day length shortens. Flowering occurs in 2.5–4 weeks with a 10- to 12-hour day length, whereas flowering takes 7.5 weeks with a 16-hour day length. Some seeds will germinate as early as four weeks after appearance of the inflorescence, but even after 10 weeks nearly 80% of seeds were still dormant. Dark seeded biotypes mature in six to seven weeks after emergence, whereas many light seeded biotypes take nine weeks.

Pollination: Wild-proso millet is primarily self-pollinating, with out-crossing rates of less than 10%.

Reproduction: Wild-proso millet plants grown without competition produced 69,000–94,000 seeds per plant. Under competitive conditions, seed production of 420–620 seeds per plant were observed.

Dispersal: Seeds are spread by birds, mammals and farm machinery. Combine harvesters spread the seeds substantial distances within fields and probably between fields as well.

Common natural enemies: *Sphacelotheca destruens*

(head smut) and *Ustilago crameri* (kernel smut) can substantially damage plants. Birds, rodents and insects eat many seeds.

Palatability: Cultivated proso millet is harvested as grain for human and animal consumption or as fodder, and at least the crop-like, light-seed-color weedy biotypes are similarly palatable. Wild-proso millet can, however, cause poisoning of young sheep and goats.

FURTHER READING

Bough, M., J.C. Colosi and P.B. Cavers. 1986. The major weedy biotypes of proso millet (*Panicum miliaceum*) in Canada. *Canadian Journal of Botany* 64: 1188–1198.

Carpenter, J.L., and H.L. Hopen. 1985. A comparison of the biology of wild and cultivated proso millet (*Panicum miliaceum*). *Weed Science* 33: 795–799.

Cavers, P.B., and M. Kane. 1990. Responses of proso millet (*Panicum miliaceum*) seedlings to mechanical damage and/or drought treatments. *Weed Technology* 4: 425–432.

McCanny, S.J., and P.B. Cavers. 1988. Spread of proso millet (*Panicum miliaceum* L.) in Ontario, Canada. II. Dispersal by combines. *Weed Research* 28: 67–72.

Patterson, D.T., A.E. Russell, D.A. Mortensen, R.D. Coffin and E.P. Flint. 1986. Effects of temperature and photoperiod on Texas panicum (*Panicum texanum*) and wild-proso millet (*Panicum miliaceum*). *Weed Science* 34: 876–882.

Witchgrass

Panicum capillare L.

Witchgrass seedling
Scott Morris, Cornell University

Witchgrass hairs on stem and leaf
Scott Morris, Cornell University

Witchgrass plant
Randall Prostak, University of Massachusetts

Witchgrass panicle
Scott Morris, Cornell University

IDENTIFICATION

Other common names: old witch grass, tickle grass, witches' hair, tumble weed grass, fool hay

Family: Grass family, Poaceae

Habit: Sprawling summer annual grass

Description: Seedlings are upright. The seed leaf is lanceolate, parallel to the ground and up to 0.5 inch long by 0.13 inch wide. Subsequent leaves are rolled in the bud, 0.6–1.6 inch long by 0.16–0.24 inch wide, pointed at the tip and lacking auricles. The collar is pale green to white, and the ligule is a 0.04–0.06 inch-long fringe of hairs. Long, silky hairs densely coat the sheath, collar and both leaf surfaces. The hairs are often thickened at the base. **Mature plants** are 8–36 inches tall and usually produce several tillers. Stems are round in cross section, either erect or horizontal, and spreading with upturned tips. The sheath is distinctly veined, green to purplish, and open. Leaf blades are flat, pale green, 2.3–10 inches long by 0.25–0.75 inch wide and have a conspicuous white midvein. The ligule and collar of seedlings and mature plants are similar. Both the sheath and the leaves are densely hairy. The roots are shallow and fibrous. Tillers root only at the base and not at stem joints. The **inflorescence** is a terminal, open, diffusely branching, 8–16 inch-tall and 3–9 inch-wide panicle. The panicle branches are rough, straight to wavy and 4.5–12 inches long. Each branch splits into several smaller branchlets, each of which has a single 0.08–0.13 inch-long, beaked spikelet at the tip. Each spikelet produces a single smooth, shiny, dark brown or gray **seed**. As with all grasses, the seed includes a thin, tight coating of fruit tissue. The seeds are oblong to egg shaped and 0.06–0.13 inch long.

Similar species: Crabgrasses (*Digitaria* spp.), fall panicum (*Panicum dichotomiflorum* Michx.) and wild-proso millet (*Panicum miliaceum* L.) have a similar appearance to witchgrass. Crabgrass seedlings are folded in the bud, lack sheath hairs and have membranous ligules, while witchgrass seedlings are rolled in the bud, have hairy sheaths and have a fringe of hairs as the ligule. The leaves of fall panicum seedlings are hairy only on the underside, while witchgrass leaves are hairy above and below, and fall panicum usually is a larger, more robust mature plant. Wild-proso millet also is a larger plant than witchgrass, reaching up to 4.25 feet in height with spikelets that are 0.18–0.2 inch long.

MANAGEMENT

Since soil disturbance promotes almost 100% of witchgrass seeds near the surface to germinate, spring tillage can make this species highly susceptible to stale or false seedbed management that kills seedlings before planting summer crops. In contrast, late summer or fall fallow will usually prompt little germination. Because seedlings mostly emerge from the top 1 inch of soil and establish relatively slowly, tine weeding and other shallow, in-row weeding methods can be highly effective against this species. Due to their drought tolerance, burying young plants during cultivation may be more effective than partially uprooting them.

Avoid excessive N fertility since this will favor the weed relative to crops, particularly less vigorous crop species. Compared with many other weeds, witchgrass is a poor competitor and can be suppressed by a dense, vigorous crop or cover crop.

ECOLOGY

Origin and distribution: Witchgrass is native to eastern North America from southern Canada to Florida, and a western variety extends to the Pacific. Prior to agriculture, the species persisted on beaches, riverbanks and similarly open, disturbed habitats. It has been widely introduced into temperate areas of the world, including Europe, Asia, Australia, New Zealand and southern South America.

Seed weight: Mean population seed weights vary from 0.15–0.65 mg.

Dormancy and germination: Seeds are dormant when shed from the parent plant and require a period of cold, wet conditions before they are ready for germination. As seeds lose dormancy in spring, they require relatively high temperatures for germination (86–95°F in the day and 59–68°F at night). As spring progresses, the temperature required for germination declines. Light is required for germination, and when present with suitable temperatures, up to 100% of seeds will germinate. Alternating temperatures also are required for high levels of germination. If seeds do not germinate in the spring, they enter secondary dormancy during mid-summer. Since these dormant seeds require another period of chilling, seedling emergence declines substantially during the summer and fall. Dilute solutions of nitrate or ammonium promote germination.

Seed longevity: The seeds can live at least several decades in undisturbed soil, but there are no reports on estimated annual mortality rates. Given its propensity to germinate in light, longevity of seeds near the soil surface in regularly tilled soil is probably much lower than that of seeds buried deeper in soil.

Season of emergence: Witchgrass emerges primarily in late spring and early summer.

Emergence depth: Seeds emerge best from the top 0.5 inch of soil, and none emerge from 2 inches or deeper.

Photosynthetic pathway: C_4

Sensitivity to frost: Plants die from the first frost.

Drought tolerance: Witchgrass is relatively drought tolerant and probably out-competes most crops during dry periods. It is more efficient than C_4 crops like corn at using water for growth and much more efficient than C_3 crops. Young plants have a herring-bone root arrangement that is highly effective at exploiting soil moisture.

Mycorrhiza: Witchgrass was rated as having moderate levels of mycorrhizal infection levels. Witchgrass biomass and P uptake were increased by inoculation with vesicular-arbuscular mycorrhizae at low available P levels.

Response to fertility: Witchgrass is highly responsive to nitrogen. Under high fertility it can accumulate N to the point of becoming toxic to livestock. It is also highly responsive to P but tolerates K and Ca deficiencies well. It is highly salt tolerant, and its ability to tolerate very high pH is indicated by its natural occurrence on alkali flats in the western United States.

Soil physical requirements: The species tolerates a wide variety of soil conditions but does best on sandy to loamy soils. It tolerates compaction.

Response to shade: Witchgrass tolerates moderate shade, but shade slows development and reduces dry matter and tillering. It does not thrive in areas with dense shade.

Sensitivity to disturbance: Witchgrass exists in primarily disturbed and cultivated habitats, resists trampling and sometimes invades overgrazed pastures. Its drought tolerance probably allows survival after partial uprooting during cultivation or hoeing. Because the inflorescence develops slowly, mowing before seed set can eliminate or greatly reduce seed production.

Time from emergence to reproduction: Early emerging witchgrass flowers about 8–13 weeks after emergence, with seeds maturing about 3.5 weeks later. Plants emerging in mid-summer develop more rapidly but are smaller and produce fewer seeds.

Pollination: Witchgrass self-pollinates but probably also cross-pollinates by wind. Under stressful conditions, the flowers often self-pollinate without opening.

Reproduction: A well-developed plant grown without competition in North Dakota produced 11,000 seeds. A very large plant produced 56,000 seeds.

Dispersal: The entire inflorescence breaks off as a unit and rolls as a tumble weed. Witchgrass is a common contaminant of forage seed like timothy and white clover, and contaminated forage and grain seed have been a major source for spreading throughout the world. The seeds float and commonly contaminate surface irrigation water, especially since the species often grows on stream banks. Seeds also disperse on tires, shoes and farm machinery. The seeds pass intact through horses, cattle, sheep and pigs, and are dispersed in feces and manure that is spread on fields.

Common natural enemies: The species is infected by a great variety of fungi and viruses, but their impact in agricultural situations has not been evaluated.

Palatability: No part of the plant makes a desirable food for humans. The plant is unpalatable to livestock except when very young, and it can contain toxic levels of nitrate when growing on highly fertile soils.

FURTHER READING

Clements, D.R., A. DiTommaso, S.J. Darbyshire, P.B. Cavers and A.D. Sartonov. 2004. The biology of Canadian weeds. 127. *Panicum capillare* L. *Canadian Journal of Plant Science* 84: 327–341.

Vengris, J. and R.A. Damon, Jr. 1976. Field growth of fall panicum and witchgrass. *Weed Science* 24: 205–208.

Yellow nutsedge

Cyperus esculentus L.

Sprouting yellow nutsedge tuber with roots and a new shoot
Randall Prostak, University of Massachusetts

Yellow nutsedge tillers
Antonio DiTommaso, Cornell University

Yellow nutsedge inflorescence in fruit
Scott Morris, Cornell University

IDENTIFICATION

Other common names: chufa, rush nut, yellow nut grass, northern nutgrass, coco, coco sedge, edible galingale

Family: Sedge family, Cyperaceae

Habit: Grass-like perennial, spreading by rhizomes and tubers

Description: Seedlings have very similar aboveground characteristics to the more commonly occurring **vegetative shoots**. Newly emerging vegetative shoots have three glossy, light green, grass-like leaves. Leaves are flat or slightly V shaped and set around a very short stem in threes such that the shoot base is triangle shaped. Foliage of the **mature plant** is similar to the young plant with leaves 8–36 inches long by 0.2–0.4 inch wide. Leaves are long, tapered and sharply pointed with a prominent midrib and parallel veins that sometimes give the leaves a ridged appearance. Leaves lack ligules and auricles and have a closed, overlapping sheath that forms a triangular, three-edged, unbranched, stem-like structure at the base. The **inflorescence** develops at the end of a 0.5–3 foot-long triangular, spongy or corky centered, yellow-green stem that is usually no longer than the leaves. Several long, lanceolate, green, roughly horizontal, leaf-like bracts are located just below the inflorescence; they are of equal or lesser length than the inflorescence. The inflorescence consists of multiple stalked, yellowish, bottlebrush-shaped clusters of spikelets. Flowers are arranged into groups of three-sided, 0.1 inch-long, oval shaped, yellow-brown spikelets. The dry fruit contains one light brown seed. Seeds often fail to germinate; instead, extensive **underground** networks of rhizomes and tubers perpetuate yellow nutsedge. Scaly white rhizomes, reaching up to 6.5 feet in all directions, give rise to new tubers at their tips. The tubers are white and scaly when young. Mature tubers are hard, round, nutlike, brown to black in color, and 0.13–0.75 inch in diameter; they eventually shed their scales, enlarge into bulbs and give rise to vegetative shoots and roots.

Similar species: Purple nutsedge (*Cyperus rotundus* L.) has dark green foliage and purple inflorescences. Purple nutsedge foliage tends to be shorter than yellow nutsedge foliage, with tips blunt compared to the acutely elongated tips of yellow nutsedge. Purple nutsedge tubers are produced along the length of the rhizome, not just at the tips.

MANAGEMENT

Although some yellow nutsedge tubers may form as deeply as 18 inches, the vast majority of tubers are found in the top 6 inches of soil. Consequently, this species is sensitive to tillage in late spring and early summer after the tubers have sprouted but before daughter plants or new tubers have formed. Although the tubers can make new sprouts from dormant buds, these will be weaker and more easily controlled by cultivation. Thus, crop rotations that include late spring/early summer tillage help control yellow nutsedge populations. By the same principle, in early-planted row crops, deep cultivation once or twice beginning in early June will substantially reduce a yellow nutsedge population, and in a competitive crop it may provide sufficient control for the remainder of the season. Tillage after harvest of summer harvested crops will help disrupt tuber formation, which occurs primarily in late summer and fall.

The tubers will die if you can dry them out completely. You can do this most effectively by repeatedly turning the soil during dry weather. Solarization can be an effective control measure in climates where soil can be heated in excess of 113°F. However, soil temperatures may not always achieve these levels for sufficient duration throughout the soil profile from which tubers emerge. An integrated approach where fallow tillage precedes solarization can achieve more reliable results.

Since the plants are short (less than 18 inches), a dense planting or other measures that increase the competitiveness of the crop are particularly effective for suppressing yellow nutsedge. Similarly, rotations that include highly competitive crops will help manage yellow nutsedge populations. Although potatoes are relatively competitive, avoid planting them in infested fields because nutsedge rhizomes can penetrate potato tubers and make them unmarketable. Good fertility is more likely to increase the vigor of the crop than the growth of yellow nutsedge.

The large food storage in the tubers allows yellow nutsedge to penetrate even very thick layers of organic mulch materials (e.g., thicker than 6 inches). Allelopathic substances from sweet potatoes, wild radish and rye, however, reduce yellow nutsedge density and vigor. Suppression of yellow nutsedge by rye is most effective if the rye root system is present as well as the straw. The sharp points of the newly emerged shoots easily pierce thin, opaque horticultural plastic film but not clear plastic film, where light induces leaf expansion and blunts the sharpness of the emerging shoot. Thicker plastic mulch, whether opaque or clear, can suppress yellow nutsedge, as can heavier materials such as landscape fabric. Tuber production and patch expansion also are suppressed by plastic film.

One of the most effective methods for managing yellow nutsedge is to occasionally graze swine on the field. Their eradication of tubers will be quicker and more complete if the soil is tilled first. A novel approach using a peanut digger in conjunction with a collection cart has successfully removed substantial numbers of tubers from fields in the southeastern United States. Alternatively, a peanut digger can be used to bring the tubers to the soil surface to dry them out.

ECOLOGY

Origin and distribution: Yellow nutsedge is native to North and South America, southern Europe and Africa. Native populations in North America occur in wetlands, and weedy races in the United States may be introduced. The species occurs widely in the United States and southern Canada with the exception of the Intermountain West and the prairie provinces of Canada.

Seed and tuber weight: Mean seed weights vary from 0.13–0.31 mg. Mean tuber weight varies substantially between locations, ranging from 70–75 mg in Minnesota and Illinois, to 150–160 mg in both Georgia and another site in Illinois, to 700 mg in Maryland.

Dormancy and germination: The tubers are dormant when produced and usually sprout the following spring. Water-soluble chemicals in the skin of the tubers apparently enforce dormancy because washing breaks dormancy. Water percolating through the soil probably acts similarly. Tuber dormancy, however, is also broken by one month of cold treatment (38–50°F). Thus, tubers usually do not sprout until they have overwintered. Temperatures above 54°F are required to initiate tuber sprouting. Cold treatment also reduces the dormancy of seeds. Some seeds will germinate in the dark provided the temperature is higher than 75°F, but germination in light is greater than in dark. In light at 95°F, nearly all seeds will germinate. Nitrate has minimal effect on seed germination. Germination increases with increasing seed mass.

Tuber and seed longevity: Tubers can persist in the soil for three years, but few survive more than two winters in most field situations. Half-life was 4.4–5.8 months. Most seeds near the soil surface germinate or die within three years, but deeply buried seeds last longer. Germination of seeds stored in cold moist conditions was less than for seeds stored in cold dry conditions.

Season of emergence: Dormant tubers sprout in mid-spring and form a non-dormant basal bulb near the soil surface, from which the shoot emerges from late spring to early summer. Later-emerging shoots arise primarily from upturned rhizomes. These thicken to form additional basal bulbs and aboveground shoots. Such shoots will continue to form throughout the growing season. Seed germination is in the spring, but seedlings are extremely rare.

Emergence depth: Percentage emergence is greatest when tubers are in the top 4–8 inches of soil, but emergence from up to 32 inches has been recorded. Seedlings emerge best from seeds very near the soil surface. A few seedlings can emerge from up to 1 inch but none from 1.5 inch. Successful emergence and establishment require long periods of wet soil, so seedling establishment in agricultural fields is rare.

Photosynthetic pathway: C_4

Sensitivity to frost: The foliage is sensitive to frost. Only the tubers overwinter, and even these survive poorly at soil temperatures below 20°F. Tubers at shallow depths of 1–2 inches are most susceptible to winterkilling.

Drought tolerance: Yellow nutsedge can tolerate a few weeks of drought, but growth and tuber production are enhanced by soil moisture conditions optimum for crops.

Mycorrhiza: Yellow nutsedge is readily colonized by mycorrhizae.

Response to fertility: This species is relatively unresponsive to soil fertility.

Soil physical requirements: Yellow nutsedge tolerates a wide range of soil texture and drainage conditions, including sustained flooding, but it does poorly on sandy soils unless the field is irrigated regularly.

Response to shade: This species tolerates shade poorly but can still produce some tubers at low light levels.

Sensitivity to disturbance: Yellow nutsedge plants are most sensitive to disturbance after several leaves have formed. At this point, the original tuber and the newly formed basal bulbs are both low on stored carbohydrates, and any new shoots that sprout will be weak. After that, tubers may be spread around the field by tillage and cultivation implements.

Time from emergence to reproduction: Most reproduction occurs by formation of tubers, and these can begin forming within three weeks of shoot emergence from the previous tuber. Typically, however, long days promote basal bulb formation and vegetative growth, while tuber formation occurs as day length shortens in the fall. Tuber formation of plants that emerged in spring begins in late summer at a day length of 14 hours, but it is delayed by a month or more if emergence did not occur until summer. Flowering occurs from July to September when day length is 12–14 hours. Time from emergence to flowering for plants grown from tubers in Wisconsin was eight weeks. When plants were grown from seeds in New England, flowering occurred in late August and September, approximately 13–16 weeks after emergence. Viable seeds are produced beginning two to three weeks after flowers open.

Pollination: Yellow nutsedge is wind pollinated and is self-sterile.

Reproduction: Most reproduction is vegetative and occurs via both production of tubers and production of new shoots from rhizomes. By the end of the season more than one-third of the total plant dry weight is composed of tubers. A single tuber planted in Minnesota and allowed to grow without competition or management produced 6,900 tubers by the fall and 1,900 daughter plants the following spring in a 34-square-foot patch. A single tuber in eastern Oregon produced 20,000 tubers. No seeds are produced by about 90% of populations because they are composed of a single clone. One seed producing population averaged 1,500 viable seeds per inflorescence. Only 1–17% of flowers contain seeds.

Dispersal: Since seedlings are very rare, dispersal is primarily by transport of the tubers. The low genetic diversity among yellow nutsedge populations is consistent with asexual dispersal by tubers. In natural conditions, transport is by water. Populations probably occasionally establish in low lying fields when flood waters wash in tubers. Most populations in agricultural fields probably establish from tubers brought in on tillage and cultivating machinery. Sources of new populations include root balls of nursery stock and topsoil that is spread after construction. If uprooted nutsedge plants are composted at low temperatures, the resulting compost can potentially spread the weed.

Common natural enemies: Few natural enemies have been reported for this species, and our observations in central New York indicate negligible damage by natural enemies in the field.

Palatability: The foliage is unpalatable. The tubers or "tiger nuts" are highly palatable and have been a foodstuff in Mediterranean countries for centuries. Tiger nuts have higher oil, protein and starch content than most tuber crops, and a milky beverage product of tiger nuts called "horchata de chufa" is considered a health drink. The domestic "chufa" is a variety of yellow nutsedge and is commonly planted to attract wildlife for sport. Swine are fond of the tubers.

Note: Yellow nutsedge tubers and tuber extracts are allelopathic to crops.

FURTHER READING

Johnson, W.C. III, R.F. Davis and B.G. Mullinix Jr. 2007. An integrated system of summer solarization and fallow tillage for *Cyperus esculentus* and nematode management in the southeastern coastal plain. *Crop Protection* 26: 1660–1666.

Johnson, W.C. III, T.R. Way and D.G. Beale. 2015. An undergraduate student project to improve mechanical control of perennial nutsedges with a peanut digger in organic crop production. *Weed Technology* 29: 861–867.

Mulligan, G.A. and B.E. Junkins. 1976. The biology of Canadian weeds. 17. *Cyperus esculentus* L. *Canadian Journal of Plant Science* 56: 339–350.

Stoller, E.W. 1981. Yellow Nutsedge: A Menace in the Corn Belt. U.S. Department of Agriculture Technical Bulletin No. 1642. 16 p.

Stoller, E.W. and L.M. Wax. 1973. Yellow nutsedge shoot emergence and tuber longevity. *Weed Science* 21: 76–81.

Broadleaf Weeds and their Relatives

Annual sowthistles

Annual sowthistle, *Sonchus oleraceus* L.
Spiny sowthistle, *Sonchus asper* (L.) Hill

Annual sowthistle seedlings
Joseph DiTomaso, University of California, Davis

Annual sowthistle rosette
Scott Morris, Cornell University

Annual sowthistle plants
Jack Clark, University of California

Annual sowthistle leaf base
Scott Morris, Cornell University

Annual sowthistle leaf
Scott Morris, Cornell University

Spiny sowthistle rosette
Scott Morris, Cornell University

Spiny sowthistle leaf base
Scott Morris, Cornell University

Spiny sowthistle flower
Scott Morris, Cornell University

IDENTIFICATION

Other common names:

Annual sowthistle: sowthistle, smooth sowthistle, hare's lettuce, hare's colewort, milk thistle

Spiny sowthistle: spiny-leaved sowthistle, sharp-fringed sowthistle, prickly sowthistle, rough milk thistle, rough sowthistle, chaudronnet (Quebec)

Family: Aster family, Asteraceae

Habit: Erect summer or winter annual herbs

Description: Seedlings form alternate-leaved rosettes before bolting. Cotyledons are hairless, have short stalks and are circular to egg shaped. The first true leaves are sparsely hairy, stalked and are circular to spatula shaped. By the four-leaf stage seedlings exude a milky sap when damaged.

Annual sowthistle: Cotyledons are 0.16–0.31 inch long by 0.06–0.16 inch wide. The first true leaves have irregularly toothed edges. Later leaves are oval to egg shaped, irregularly lobed and have winged stalks.

Spiny sowthistle: Cotyledons are 0.14–0.24 inch long by 0.06–0.16 inch wide. The first true leaves have downward facing, spiny teeth along the edges. Subsequent leaves are oval to lanceolate with wavy and spiny edges and short stalks.

Mature plants reach 1–5 feet tall on upright, hairless, hollow stems that branch occasionally towards the top. Leaves are alternate and smooth, and clasp the stem at the base. All parts exude milky sap when damaged. The root system is a short taproot.

Annual sowthistle: Leaves are dull green to blue-green, deeply lobed, 2.4–12 inches long by 0.4–6 inches wide and clasp the stem at the base with a pair of angular lobes. The leaves generally have a triangular to lanceolate lobe at the tip and three or fewer paired lobes on the lower portion of the leaf. Lobes may be reduced or absent on the upper leaves. Leaf edges are irregularly toothed and may have soft spines. Leaves are deeply divided, reaching almost to the midrib. Acute auricles are uneven and slanted.

Spiny sowthistle: Leaves are glossy, green to purplish, oval to lanceolate or occasionally lobed, 2–10 inches long by up to 3 inches wide and clasp the stem at the base with a pair of rounded, spiny lobes. Leaf edges are wavy and covered in sharp, prickly spines. Leaves are not as deeply divided as those of annual sowthistle. Leaves lack stalks but have distinct, large, "rams-horn" auricles clasping the stem.

Flowers are yellow and produced in clusters of one to five flower heads at the ends of stems and branches. The flower head stalks may be smooth or covered in gland-tipped hairs. Each flower head is 0.4–1 inch wide and has up to 250 petals. At the base of each flower head are three to five rows of green, overlapping bracts. **Seeds** are covered with a hard, dry layer of fruit tissue. Seeds are oval to oblong, flattened and have three to five longitudinal ridges on each side. Each seed is attached to a feathery, white pappus up to 0.3 inch long.

Annual sowthistle: Seeds are brown to olive, finely wrinkled, never winged and are 0.08–0.16 inch long by 0.04 inch wide.

Spiny sowthistle: Seeds are light brown, smooth, often winged on the edges and 0.08–0.1 inch long by 0.06 inch wide.

Similar species: Perennial sowthistle (*Sonchus arvensis* L.) and prickly lettuce (*Lactuca serriola* L.) both have milky sap and similar foliage and flowers to spiny annual and annual sowthistle. Perennial sowthistle has larger flowers (1–2 inches wide) than either species and has an extensive, spreading system of perennial storage roots rather than a taproot. Prickly lettuce leaves have prominent spines along the midrib, while the midribs of spiny and annual sowthistle leaves lack such spines. Oxtongue (*Picris*) species have similar flowers and habits to spiny and annual sowthistle. The stems and leaves of oxtongues are covered with short, stiff hairs, while those of spiny and annual sowthistle are smooth.

MANAGEMENT

Annual and spiny sowthistles can act as either spring or winter annuals, making them adapted to both fall- and spring-sown crops. Thus, they are less easily controlled by rotation between crops with different growing seasons than most other weed species. They tend to be particularly abundant and prone to seed production in winter grains. Since germination is favored by light and seedling emergence is limited to the surface 0.4 inch of soil, they are particularly prevalent in no-till systems. A moderately long period in the rosette stage makes sowthistles susceptible to suppression by crops that cast heavy shade and by organic mulch materials. Grain crops planted at high density and narrow rows decrease growth potential of these weeds. The long rosette stage also makes them relatively easy to bury with aggressive tine weeding or by throwing soil into the crop row. Shallow cultivation either in a bare fallow or after planting the crop is effective for flushing out and destroying seedlings. Since the seed bank is relatively short lived, inversion tillage is useful for controlling these species, especially after a season with abundant seed production, provided subsequent inversion tillage does not occur for 30 months. Mowed field edges, cleared stream banks and other non-forested habitats can prevent spread into adjacent fields.

ECOLOGY

Origin and distribution: Annual and spiny sowthistle likely originated near the Mediterranean Sea. Spiny sowthistle has been introduced into North and South America, Africa, Australia, New Zealand and south and east Asia. Annual sowthistle is similarly widespread. Both species occur throughout most of the United States and southern Canada, and spiny sowthistle occurs northward into central Alaska and the Yukon.

Seed weight: Population mean seed weights range from 0.27–0.42 mg for annual sowthistle and 0.3–0.41 mg for spiny sowthistle.

Dormancy and germination: A high percentage of annual sowthistle seeds will germinate immediately after dispersal if they receive light. They emerge considerably better in light than in dark or with only a pulse of light. Germination was good at a fluctuating day/night temperature of 77/50°F, but neither species germinates well at any constant temperature according to one report. However, non-dormant annual sowthistle germinated well at a wide range of constant temperatures from 41–95°F in another report. High moisture levels are required for optimum germination of annual sowthistle. Some germination can occur at moderate salt concentrations and high pH levels greater than 8. Spiny sowthistle germinates well in light immediately after seed collection but acquires a capacity to partially germinate in dark after six months. Cold (43°F), wet conditions decrease spiny sowthistle seed dormancy but germination after 18 weeks of cold stratification is still greater in light than in dark.

Seed longevity: In soil stirred three times a year, annual sowthistle seeds declined by 48–65% per year. For spiny sowthistle, seeds declined by 48–62% per year. Mortality of annual sowthistle seeds was 82% when left on the soil surface for almost eight months through the Australian winter, whereas mortality was 55% for seeds buried 2 inches. In another Australian experiment, annual sowthistle seed mortality was greater than 99% after one year at the soil surface but was 53% when buried 4 inches. Some buried seeds of annual sowthistle lasted for at least five years.

Season of emergence: In cooler climates, most emergence of both species takes place in mid- to late spring, but some seedlings emerge throughout the growing season whenever conditions are favorable for germination. Emergence is greatest at daily maximum and minimum temperatures of 54/39°F to 66/50°F. In warm climates, particularly Mediterranean climates with hot, dry summers, annual sowthistle can behave as a winter annual, with a flush of emergence following planting of fall-sown crops. Because annual sowthistles lack dormancy when exposed to light and they germinate over a wide range of temperatures given sufficient soil moisture, emergence occurs throughout the spring and summer in moist climates, but only during the winter season in Mediterranean climates.

Emergence depth: Emergence of annual sowthistle is best from 0–0.4 inch, with little emergence from 0.8–1.2 inch. Up to 77% of annual sowthistle seeds will germinate on the soil surface. Seeds probably emerge well from the soil surface because the flat shape of the seed allows for good soil contact and water uptake.

Photosynthetic pathway: C_3

Sensitivity to frost: Young plants of both species are highly frost tolerant. Fall emerging individuals overwinter as a rosette of leaves close to the ground and can withstand temperatures several degrees below 0°F. In the Canadian prairies and northward, however, the species overwinter only as seeds. Flower buds are killed by early frost, and flowering plants are partially killed by frost.

Drought tolerance: Annual sowthistle can survive short periods of drought that are sufficiently severe to cause wilting of the leaves. Spiny sowthistle is sensitive to drought stress. Plants watered three times per week grew only half as large as those that were watered daily, and plants watered only once per week failed to reproduce.

Mycorrhiza: Annual sowthistle is strongly mycorrhizal, but spiny sowthistle, although listed as mycorrhizal on disturbed sites in Utah, has a low rate of mycorrhizal infection.

Response to fertility: Annual sowthistle tolerates low N fertility well. It occurs on soils with pH from 6.5–9 and does best at pH over 8. Since P would have very low availability in the high pH conditions in which annual sowthistle thrives, it apparently also tolerates low P. Spiny sowthistle tolerates low fertility well. In one experiment it actually produced more seeds when unfertilized than when given 10-10-10 fertilizer before planting. Nevertheless, another study found that this species had three times the concentration of K and twice the concentration of P and Ca as a barley crop in which it was growing.

Soil physical requirements: Both species occur on a wide range of soil textures from clay to sand. Both species tolerate salt and alkaline soils provided moisture is available.

Response to shade: Annual and spiny sowthistle do not tolerate more than light shade.

Sensitivity to disturbance: Since the leaves of the young plants remain flat on the soil in a rosette for several weeks, they are highly susceptible to burial, as with aggressive tine weeding or when soil is thrown into the crop row. They do not tolerate trampling but can grow new stems if the first stems are broken or cut. Plants cannot regrow from the taproot alone. Plants of annual sowthistle cut at the flowering stage are still able to produce viable seeds.

Time from emergence to reproduction: Annual sowthistle emerging in early spring will flower in June, but overwintering rosettes will begin producing flower stems in March. In warm weather on poor soil, annual sowthistle can flower in as little as six weeks. Spring emerging spiny sowthistle flowers about nine weeks after emergence with the seeds ripening about one week later.

Pollination: Both species self-pollinate but are probably also cross pollinated to some extent by small bees and flies.

Reproduction: A sample of 65 British annual sowthistles produced an average of 140 seeds per head and 44 heads per plant, for a total of 6,100 seeds per plant. A sample of 25 spiny sowthistles averaged 200 seeds per head and 105 heads per plant for a total of 23,000 seeds per plant.

Dispersal: Tufts of hairs help the seeds to disperse long distances on air currents. These hairs also tangle in fur, feathers and clothing, causing the seeds to be carried by animals and humans. Seeds of annual sowthistle have been found adhering to feet and feathers of waterfowl. The seeds of spiny sowthistle (and probably annual sowthistle as well) pass through cattle and are spread with the manure. The seeds also float well and disperse along streams and lakeshores. Annual and spiny sowthistle seeds found in irrigation water had germination rates of 22% and 57%, respectively.

Common natural enemies: Sowthistles are attacked by several species of leaf mining fly larvae. They are susceptible to downy mildew caused by *Bremia lactucae*, which also infects commercial lettuce.

Palatability: European peoples have used sowthistles as a pot herb since ancient times. The content of minerals, vitamins and antioxidants in these species is similar to those of leafy vegetables. The foliage is palatable to livestock, though most animals have difficulty eating immature plants because the leaves lie flat on the ground.

FURTHER READING

Hutchinson, I., J. Colosi and R.A. Lewin. 1984. The biology of Canadian weeds. 63. *Sonchus asper* (L.) Hill and *S. oleraceus* L. *Canadian Journal of Plant Science* 64: 731–744.

Lewin, R.A. 1948. Biological flora of the British Isles. *Sonchus* L. (*Sonchus oleraceus* L. and *S. asper* (L.) Hill). *Journal of Ecology* 36: 203–223.

Widderick, M.J., S.R. Walker, B.M. Sindel and K.L. Bell. 2010. Germination, emergence, and persistence of *Sonchus oleraceus*, a major crop weed in subtropical Australia. *Weed Biology and Management* 10: 102–112.

Bindweeds

Field bindweed, *Convolvulus arvensis* L.
Hedge bindweed, *Calystegia sepium* (L.) R. Br.

Field bindweed (left) and hedge bindweed (right) flower comparison
Antonio DiTommaso, Cornell University

Hedge bindweed climbing on corn
Antonio DiTommaso, Cornell University

Hedge bindweed storage roots
Antonio DiTommaso, Cornell University

Hedge bindweed (left) and field bindweed (right) leaf comparison
Antonio DiTommaso, Cornell University

IDENTIFICATION
Other common names:

Field bindweed: bindweed, European bindweed, lesser bindweed, corn bind, possession bind, bear bind, cornbine, barbine, European glorybind, field morningglory, orchard morningglory, wild morningglory, small-flowered morningglory, creeping jenny, green vine, devil's guts, corn lily, laplove, hedge bells

Hedge bindweed: great bindweed, bracted bindweed, wild morningglory, devil's vine, Rutland beauty, hedge lily

Family: Morningglory family, Convolvulaceae

Habit: Twining perennial herb spreading by thickened horizontal roots

Description: Vegetative sprouts arise from storage roots, lack cotyledons and have normal leaves, although the first leaves that form as the shoot is just pushing out of the soil are often deformed. Seedlings are much less common than vegetative sprouts.

Field bindweed: Cotyledons are 0.35–0.87 inch long by 0.14–0.4 inch wide, long stalked, dark green with many white veins, square to heart- or kidney shaped, with lobes pointing toward the stem and an indented tip. Stems are red near the soil line. First leaves are bell shaped with basal lobes pointing outward.

Hedge bindweed: Cotyledons are 1–2 inches long by 0.5–1 inch wide, have strong veins on the underside, and concave or flat tips and backwards extending, rounded lobes at the base. The stem may be red at the base. First leaves are arrow or heart shaped.

Mature plants run along the ground and twine up other plants. Stems may be hairy or hairless and smooth.

Field bindweed: Leaves are alternate, 1.5–2.5 inches long, spade or arrow shaped, and borne on a long stalk. Lobes are triangular, with rounded ends. Roots can extend over 20 feet deep and run laterally up to 25 feet, forming large clones.

Hedge bindweed: Leaves are alternate, 1.5–6 inches long, heart or arrow shaped, and pointed at the tip. Lobes are squarish when laid flat. Roots extend up to 1 foot deep and run laterally up to 9 feet, forming large clones.

Flowers are white or pink and are petunia shaped in both species.

Field bindweed: Flowers are grouped one to five to a leaf axil and measure 0.5–1 inch across. Leafy bracts at the base of the flower are less than 0.1 inch long.

Hedge bindweed: Flowers are single in leaf axils and measure 1.25–3 inches across. Bracts are heart shaped, 0.25–0.75 inch long and overlap and hide the flower base.

Fruit and seeds: Both field and hedge bindweed have round or egg-shaped capsules with two chambers, each containing one to two seeds. Seeds are gray-brown to black, with two flat sides and one rounded side.

Field bindweed: Capsules are 0.31 inch in diameter. Seeds are 0.12–0.16 inch long and coarsely bumpy.

Hedge bindweed: Capsules are 0.31 inch in diameter and is hidden by remnant flower bracts. Seeds are 0.16–0.20 inch long and slightly rough.

Similar species: Wild buckwheat (*Polygonum convolvulus* L.) has a papery covering extending up the stem from each leaf axil, long tapering leaf tips and long clusters of small, greenish flowers.

A note on development: The seedling develops a taproot from which thickened, permanent lateral roots arise. After growing laterally for 15–30 inches, these bend downward to form secondary taproots. At the bend, a vertical rhizome develops that grows to the soil surface to become a new shoot, and one or more new lateral roots continue the outward spread of the plant. Occasionally, rhizomes and subsequent shoots also develop from buds on the horizontal thickened roots. In undisturbed conditions, most of the thickened horizontal roots are in the top 6 inches of soil. Feeder roots form on both horizontal roots and taproots. In a deep silt loam soil in Kansas, a field bindweed plant spread laterally over 10 feet from the point of germination in seven months and had roots penetrating to nearly 4 feet. By the end of a third growing season, the plant had spread outward more than 16 feet and had roots penetrating as much as 23 feet.

MANAGEMENT

The bindweeds cause severe problems in many crops. The extensive and efficient root system of field bindweed can reduce soil water content to below the wilting point of most crops. Hedge bindweed is more upright in its early growth than field bindweed and is quicker to climb up crop plants. Both species are strong competitors and cause harvesting problems for grain and for both hand harvested and mechanically harvested vegetables.

Winter grains and early planted spring grains suffer less from bindweed competition than do summer crops because their primary period of water use occurs before bindweed is well established. If a fallow period prior to planting suppresses the bindweed in the fall, winter grains will cast enough shade in the spring to effectively suppress field bindweed. Similarly, if your soils do not allow early tillage, use a vigorous fall planted cover crop to compete with bindweed in the spring. Alfalfa can suppress bindweed by dense shading and repeated mowing. Summer planted cover crops also can be effective, but choose species that develop a dense leaf canopy and plant them at high density. In one study, forage soybeans planted July 1 in Minnesota killed every shoot of field bindweed, whereas less-shade-producing covers did not. Sudangrass and German millet were also relatively effective. For vegetable farmers that are not land limited, a sequence of cover crops such as overwintered rye, oats plus peas, then buckwheat or sorghum-sudangrass, and finally overwintered rye plus hairy vetch, with disking before planting each cover crop, will suppress bindweed and prepare the field for a good vegetable crop.

Repeated shallow tillage during fallow periods between crops is also important for controlling bindweed. Root reserves reach a minimum when the shoot is about 12–28 inches long and has four to six fully expanded leaves, and plants are probably most sensitive to cultivation at that stage of development. Full eradication by tillage alone usually requires continuous fallow for two full growing seasons, with shallow tillage every 12–20 days. Usually, such extreme measures are not economically feasible or environmentally desirable. Shorter fallow periods, however, can be worked into the crop rotation either before late planted crops or after early harvested crops. These will not eliminate bindweed but can greatly reduce its density and competitiveness. Note that the amount of soil moisture (affected by rainfall or irrigation) plays an important role in the effectiveness of fallow and repeated tillage. Untimely rainfall or poorly planned irrigation can allow bindweed to persist when managing it with repeated tillage or fallow.

Begin inter-row cultivation of row crops as soon as possible, since tine weeders and rotary hoes are ineffective against young bindweed shoots. A tine weeder can be used, however, to comb larger bindweed shoots out of row crops, but at the cost of some damage to the crop and considerable labor cleaning the weeder. Continue inter-row cultivation as long as you can get through the crop. Use low-pitch half sweeps or vegetable knives set shallowly to cut off bindweed close to the row without damaging crop roots. If bindweed stems are longer than about 8 inches, avoid using rotary tillers, spyders or rolling cultivators, as the vines will spool up on the implement and you will spend more time cleaning the equipment than actually cultivating. Avoid spreading bindweed by scraping off tillage and cultivation implements after working in infested areas. Grazing by sheep, cattle and chickens will suppress aboveground growth and thereby help deplete storage reserves in the roots. Pigs will grub out roots and rhizomes from the plow layer.

Organic mulch is completely useless against these species: We have observed hedge bindweed shoots emerging through an 18-inch thick pile of bark mulch that was waiting to be spread! Synthetic mulch can suppress field bindweed, but it is relatively ineffective against hedge bindweed, since the vines will follow light to the planting holes. This puts the vines in an optimal position for twining up the crop plants and makes hand pulling difficult. Interseeding clover into corn at last cultivation did not control hedge bindweed, but it reduced the height of climbing stems by half and the length of creeping stems by 72%.

Although the bindweeds are difficult to manage, many growers keep them below damaging levels through good crop rotation and persistent cultivation. An integrated program of non-chemical control tactics can provide equivalent control to that obtained using herbicides. We know one vegetable farm that successfully eradicated a bad infestation of hedge bindweed over a three-year period by supplementing mechanical cultivation with consistent hand weeding.

ECOLOGY

Origin and distribution: Field bindweed probably originated in the Mediterranean region of Europe, Asia and North Africa but now occurs widely on those continents and has been introduced into North and South America, Australia and the Pacific Islands. Hedge bindweed is native to Eurasia, Africa and eastern North America, and it has been introduced into New Zealand. Both species occur throughout the United States and southern Canada.

Seed weight: Field bindweed: 8–20 mg; hedge bindweed: 28–34 mg.

Dormancy and germination: Both species have hard

seed coats that prevent absorption of water. Consequently, only 5–25% of freshly collected field bindweed seeds will germinate, and no fresh hedge bindweed seeds will germinate. Damage to the seed coat (scarification) leads to 100% germination in both species. Overwintering in soil tends to soften the spot where the seed attached to the parent plant and increases the germination ability of hard seeds by 30%. Scarified seeds of field bindweed germinate over a wide range of temperatures, from 41–104°F, but percentage germination and speed of germination are greatest at a day/night temperature of 95/68°F. The seeds do not require light for germination. Similarly, 100% of scarified hedge bindweed seeds germinate within one week at day/night temperature regimens ranging from 59/43°F to 77/68°F. Scarified seeds all germinate within about one month. Unscarified hedge bindweed seeds germinated slowly, with 30–60% germinating within six months, depending on temperature. Unscarified seeds germinate best at a day/night temperature of 77/59°F.

Seed longevity: These species form long-term, persistent seed banks with some seeds remaining viable in the soil for at least 20 years. Burial of field bindweed seeds for a single growing season reduced number of intact seeds by 17–40%.

Season of emergence: Shoots from overwintering roots begin emerging in mid-spring and continue emerging throughout the growing season. Seedlings of field bindweed are most abundant in late spring and early summer but continue to emerge throughout the growing season.

Emergence depth: Most roots are found in the top 2 feet of soil, with the highest concentration in the top 6 inches. Accordingly, it is reasonable that emergence from vegetative root tissue occurs primarily from up to 6 inches. Seedlings of field bindweed emerge readily from anywhere in the top 2.4 inches of soil, and 10% of seeds at 4 inches will produce seedlings. Seedlings will usually not emerge from 4.7 inches. Hedge bindweed seeds are larger, and seedlings may emerge from even deeper in the soil.

Photosynthetic pathway: C_3

Sensitivity to frost: The aboveground shoot dies back from frost, but the roots withstand temperatures as low as 21°F.

Drought tolerance: Field bindweed can survive long periods of drought due to its deep root system. When watering frequency was reduced from daily to once per week, shoot and root production by field bindweed seedlings was about halved; shoot production by plants grown from root segments was similarly about halved, but root production of plants grown from root segments was scarcely reduced. With its shallower root system, hedge bindweed may be more susceptible to drought than field bindweed.

Mycorrhiza: Both species are mycorrhizal.

Response to fertility: The limited information available indicates that field bindweed is favored by additions of P and K but tends to be out-competed by N responsive species when N or a balanced nutrient source is used. Weight of twining vines was about twice as great when hedge bindweed was grown in compost and loam as when grown in sand, but productivity of surface runners and belowground parts differed little.

Soil physical requirements: Field bindweed does best on good agricultural soils, but it is also common on sandy soils. It does not tolerate waterlogged soils. In contrast, hedge bindweed is highly tolerant of waterlogged conditions and invades wetlands.

Response to shade: The bindweeds are moderately sensitive to shade. Reducing light by 55% reduced root growth by 35% and shoot growth 60%. In 80% shade, field bindweed shoot dry weight was reduced by 25–30% but stem length about doubled. Under heavy shade (95%), leaf weight decreased 39% but leaf area increased 36%, allowing the plant to continue to grow. With dense shade from a competitive crop, however, field bindweed cannot climb. The 98% shade cast by forage soybeans killed all shoots of field bindweed, and heavy shading cover crops of rye, oats and sudangrass suppressed growth substantially. Both species overcome moderate shade by climbing on crop plants.

Sensitivity to disturbance: These species are extremely resistant to soil disturbance because the root system penetrates below the plow layer, and new shoots can emerge from as deep as 2 feet. Tillage implements fragment roots and vertical rhizomes, resulting in the propagation of these weeds. Root reserves reach a minimum when the shoot is about 12–28 inches long and has four to six fully expanded leaves. Plants are most susceptible to destruction of the shoots at this growth stage. Under favorable growing conditions, clipped seedlings can resprout from the root 19 days after emergence.

Time from emergence to reproduction: Field bindweed plants rarely flower during the first season. In subsequent years, shoots emerge in mid-spring and typically begin flowering six to nine weeks later. Seeds become viable 10–15 days after pollination, but the impermeable seed coat does not fully develop until 30 days after pollination. Lateral storage roots, which are the source of new shoots, form on seedlings within six weeks of emergence.

Pollination: Both species are insect pollinated. Field bindweed sometimes self-pollinates, but some populations cannot self-pollinate. Hedge bindweed has been reported to be unable to self-pollinate, but careful study has shown that it can self-pollinate but requires insects to move pollen within the flower or between flowers of the same clone.

Reproduction: Most reproduction occurs by shoots sprouting up from the spreading root system, and in field bindweed, clones can reach 20 feet in diameter. In tilled fields, reproduction is furthered by fragmentation of roots and rhizomes. In one experiment, 5-inch segments of roots produced 0.4–9.8 shoots per gram of root over the span of 12 weeks. A single 2-inch piece of field bindweed root planted in Saskatchewan produced 25 shoots over the following four months and spread as far as 9.4 feet from the parent plant within 15 months. Seed production requires dry, sunny weather. Capsules normally contain one to four seeds with an average of two. Single plants of field bindweed produce 25–300 seeds per year but determining the scope of an individual is often difficult. Pure stands of field bindweed produce 0.5–180 seeds per square foot.

Dispersal: The seeds pass through the digestive tract of livestock and are dispersed when the animals are moved or when manure is spread. The seeds are more resistant to elevated temperatures during composting than are other weed species and can potentially persist in commercial compost. The seeds commonly contaminate seed grain and beans. We have twice found abundant bindweed seeds in rye sold for cover crop seed. Seeds disperse in surface irrigation water. Migrating birds may disperse field bindweed long distances. Tillage and cultivating machinery spread roots and rhizomes within fields. Crops that are vegetatively propagated (e.g., asparagus crowns or mint roots) can be a source of field bindweed rhizomes.

Common natural enemies: Bindweed gall mite, *Aceria malherbae*, has been established as a potential biocontrol agent and is now widespread in western states. The native Argus tortoise beetle (*Chelymorpha cassidea*) sometimes completely defoliates both bindweed species in the Northeast without damaging the associated grain crops. The native seed feeding beetle *Megacerus discoidus* can cause substantial damage to seeds of hedge bindweed, and the sweet potato plume moth *Oidaematophorus monodactylus* consumes flower buds. *Tyta luctuosa*, a noctuid moth whose caterpillars defoliate field bindweed, has been successfully established in several regions of the United States.

Palatability: People do not eat any part of the plant. Field bindweed shoots and roots are good fodder for cattle.

FURTHER READING

Davis, S., J. Mangold, F. Menalled, N. Orloff, Z. Miller and E. Lehnhoff. 2018. A meta-analysis of field bindweed (*Convolvulus arvensis*) management in annual and perennial systems. *Weed Science* 66: 540–547.

Rask, A.M., C. Andreasen. 2007. Influence of mechanical rhizome cutting, rhizome drying and burial at different developmental stages on the regrowth of *Calystegia sepium*. *Weed Research* 47: 84–93.

Sullivan, P. 2004. Field bindweed control alternatives. ATTRA, National Sustainable Agriculture Information Service.

Weaver, S.E. and W.R. Riley. 1982. The biology of Canadian weeds. 53. *Convolvulus arvensis* L. *Canadian Journal of Plant Science* 62: 461–472.

Canada thistle

Cirsium arvense (L.) Scop.

Canada thistle vegetative tillers
Antonio DiTommaso, Cornell University

Canada thistle flower
Scott Morris, Cornell University

Canada thistle mature seedheads
Antonio DiTommaso, Cornell University

IDENTIFICATION

Other common names: creeping thistle, small-flowered thistle, perennial thistle, green thistle, field thistle, cursed thistle, corn thistle

Family: Aster family, Asteraceae

Habit: Prickly perennial herb spreading by deep thickened storage roots

Description: Cotyledons of the **seedling** are stalkless, fleshy, hairless, oval to rounded and 0.2–0.6 inch long by 0.1–0.2 inch wide. The first true leaves of the **seedling** are egg shaped to lanceolate, hairy and have irregularly lobed, spiny edges. The early leaves form a rosette before bolting. Seedlings begin to develop horizontal storage roots early, and a plant with only two true leaves may have a perennating storage root up to 6 inches long. Vegetative buds develop on the storage root by the third true leaf stage. **Mature plants** are 1–5 feet tall and branched towards the top. Stems are hollow, grooved and sparsely hairy to smooth. Leaves are 2–6 inches long, alternate, oblong and irregularly lobed with sharp spines along the edges. Upper surfaces of the leaves are hairless, dark green and waxy, while lower surfaces are pale green and smooth to hairy. The leaves are stalkless and clasp the stem at the base. The roots are an extensive system of white, fleshy storage roots that can extend to over 15 feet horizontally and vertically and can produce more than 200 new buds annually. Vegetative shoots resemble seedlings but are larger and lack cotyledons. Male and female **flowers** are produced on separate plants in leaf axils and at the end of branches. Male flower heads are round and 0.5 inch wide, while female flower heads are 1 inch wide and flask shaped. The flowers of both sexes are purple-pink, with bases enclosed in overlapping triangular or diamond shaped, scale-like, spineless green bracts. A small leaf is present at the base of each flower stalk. Each female flower head may produce as many as 45 **seeds**. The apparent seeds have a thin dry outer coating of fruit tissue. Seeds are brown, flattened, oblong and 0.1–0.25 inch long. Each seed is weakly attached to a feathery, white brown pappus.

Similar species: Many thistles and thistle-like plants have spiny leaves and similar flower heads to Canada thistle. Nearly all are biennial, however, and have taproots rather than the extensive storage root system of Canada thistle. Bull thistle [*Cirsium vulgare* (Savi) Ten.] leaves have hairy upper surfaces and spiny, winged stems, while Canada thistle leaves are hairless above and the stems lack wings. Sowthistle (*Sonchus*) species may look similar to Canada thistle. Sowthistles have yellow flower heads and bleed milky sap when damaged, while Canada thistle has purple flower heads and clear sap.

MANAGEMENT

Because Canada thistle has a deep root system, the only approach for controlling this weed is to exhaust the storage roots. Food reserves in the roots reach a minimum near the onset of hot weather when the shoots reach about 1 foot tall and then increase as energy flows from the shoots to the storage roots. Consequently, shoots should be removed for the first time by late spring. Repeated removal of the shoots before they attain several leaves will exhaust the storage roots within two years and eliminate the weed. Several studies found a 21-day weeding schedule was optimal. Since buds on the roots will continue to sprout well into the fall, persistence is required. A dense cover crop of sorghum-sudangrass or a mixture of sorghum-sudangrass with compatible species mowed once or twice during the season reduced Canada thistle shoot density and mass to less than 20% of initial values. Competition from pasture species maintained low Canada thistle populations, but populations increased substantially when released from competition. Accordingly, high intensity, low frequency grazing over two to three years provided better control than high frequency grazing for a short duration.

Long fallow periods may not be cost effective unless thistle pressure is severe, but growing crops that allow repeated cultivation close to the row achieves a similar effect for at least part of the season. Both spring and fall sown grain crops tend to promote Canada thistle due to the long period in which the shoots can grow without disturbance, but winter wheat is more competitive than spring grains. In a Danish study, cultivation of spring barley with shovels plus hand cutting of shoots in the rows greatly decreased Canada thistle. Because alfalfa is mowed several times per year over a period of several years, this crop is very useful for managing Canada thistle. Hay that is mowed only once per year is less helpful for managing this weed.

Control of established Canada thistle stands for one year is usually insufficient for long-term control. Generally, because established plants survive many years, multiple control tactics are required for multiple years. An integrated rotation was proposed for central Pennsylvania including three years of alfalfa followed by a three-year sequence of fall brassicas, early spring vegetables and a summer vegetable. On a Maryland organic farm, a program including two years of repeated summer cultivations followed by dense plantings of winter barley for haylage reduced a heavily infested field of Canada thistle by 76%, allowing transition to alfalfa followed by successful establishment of row crops with minimum Canada thistle populations after five years.

The root reserves are sufficient to push the shoot through any amount of loose mulch, and Canada thistle growth will benefit from the soil moisture conserved by the mulching materials. A tough synthetic tarp, however, can prevent shoots from reaching light. Transfer of energy into the dying shoots reduces vigor of the storage roots.

Preventive measures that reduce dispersal are an important part of any control strategy. Canada thistle growing along field margins, fence rows or drainage ditches can easily spread into fields unless eliminated. Because Canada thistle is often a problem in grain crops, thistle stalks, including mature seeds, are often present in baled grain straw. Similarly, manure that includes grain straw bedding is often contaminated with Canada thistle seeds.

ECOLOGY

Origin and distribution: Canada thistle probably originated in southeastern Europe and the eastern Mediterranean. It has spread throughout Europe, North Africa, through central Asia to Japan, and across the northern United States and southern Canada. It is also naturalized in South Africa, New Zealand and southeastern Australia.

Seed weight: Population mean seed weights range from 0.95–1.7 mg.

Dormancy and germination: Most seeds are not dormant when shed from the parent plant, however some investigators report the need for chilling and after-ripening to achieve full germination. Seeds germinate best at warm daytime temperatures (77–86°F) but are inhibited from germination by hot (104°F) temperatures. Light, day/night fluctuation in temperature, and to a lesser extent, nitrate, all increase germination.

Seed longevity: A few seeds may persist in the soil for 17–21 years, but most disappear within the first few years. Seeds persist better when buried over 1 foot, and more poorly when near the soil surface.

Season of emergence: Seedlings emerge primarily in the spring, with some emerging in summer. Shoots from rootstocks begin emerging in mid-spring and emerge continuously until frost.

Emergence depth: Optimum depth for seedling emergence is 0–0.4 inch, but seedlings can emerge from as deep as 2.4 inches. Shoots can emerge from roots several feet deep, but the optimum depth for shoot emergence and plant canopy spread is 4 inches, compared to shallower or deeper depths.

Photosynthetic pathway: C_3

Sensitivity to frost: Shoots tolerate light frost (as in spring) but are killed by hard frost (as in fall). The root system persists over the winter. Seasonally higher concentrations of sucrose and fructans in roots during fall and winter months allow the roots to adapt to cold temperatures. However, root fragments at the soil surface are killed by 14–21°F temperatures.

Drought tolerance: Well developed plants are drought tolerant due to their deep root systems. Shoot establishment and growth from root fragments are reduced by low soil moisture, suggesting that dry conditions can enhance control with tillage practices, and that areas with optimum soil moisture are more prone to proliferation of this weed. Competition for water between shoots and root buds is a primary cause of bud inhibition under dry conditions.

Mycorrhiza: Canada thistle is mycorrhizal and demonstrates a positive growth response to the presence of mycorrhiza.

Response to fertility: Nitrogen increased shoot growth of Canada thistle seedlings but decreased number of root buds initiated. In another pot experiment, nitrogen increased growth of plants derived from root fragments more than that of seedlings. In an Alberta field experiment, nitrogen increased root growth but not shoot growth. In Sweden, soil nitrogen content had little influence on growth or regenerative capacity of Canada thistle roots or shoots. In England, with reduced competition, nitrogen and phosphorus increased Canada thistle shoot stands, while potassium and magnesium had no effect. However, in the presence of pasture species, nitrogen favored the competing pasture grasses and resulted in suppression of Canada thistle. Phosphorus, in contrast, favored Canada thistle both with and without competition.

Soil physical requirements: Canada thistle tolerates a wide range of soil types. It does poorly, however, on wet soils, and is particularly vigorous on well drained, fine textured (clay to silt loam) soils suitable for agricultural production.

Response to shade: The species is shade intolerant, but due to its extensive root reserves, shoots can often grow through competing vegetation. Seedlings will die when light falls below 20% of sunlight. Shoots of established plants suffer when shaded by competing taller species.

Sensitivity to disturbance: Canada thistle is highly resistant to cutting, hoeing and even to deep tillage. The root system commonly extends to depths of 6.6 feet or more, and new shoots quickly reappear after the previous shoots have been removed. Carbohydrate reserves in the root system reach a minimum at about the formation of the flower buds, and the plant is most sensitive to removal of the shoots at that time. Plants arising from root fragments are most sensitive to disturbance when about eight leaves have formed. Plants arising from deeply buried root fragments (8 inches deep) have weaker regrowth potential than plants derived from shallow root fragments (2 inches deep). Root fragments larger than 0.1 inch or 0.05 ounces fresh weight will usually produce new plants, and smaller fragments may also if they are near the soil surface. A single cultivation event increases and disperses the population of propagating rootstock, thereby increasing the abundance of this species, so multiple cultivations or integration with additional suppression tactics are required for control. Shoot emergence from root fragments decreases in early fall, but production of underground shoots continues, suggesting that fall tillage could still be effective for reducing reserves. Seeds are unable to mature and become viable when stems are cut in full bloom.

Time from emergence to reproduction: Plants flower in response to long day lengths of 14–16 hours, generally between late June and August. Thus, plants emerging from rootstocks in spring take about eight to 10 weeks to flower.

Pollination: Canada thistle is pollinated by insects, especially honeybees. Since male and female flowers occur on separate individuals, the species typically cannot self-pollinate. Consequently, many infestations are single clones that cannot produce seeds.

Reproduction: Canada thistle reproduces both vegetatively and by seeds. Seed production varies from 0–5,000 seeds per shoot, depending largely on the thoroughness of pollination. This in turn depends on the proximity of male and female plants. Seeds mainly allow the species to disperse between locations and do little to maintain local populations. Seedlings become perennial from six to 10 weeks after emergence and can thereafter reproduce from sprouts of deep and rapidly spreading roots. In Alberta, by the end of summer, a single plant produced 26 aboveground shoots, 154 underground shoots and 364 feet of roots. Plots initiated with individual root fragments expanded to cover a 161-square-foot area, and roots penetrated to a depth of 4.6 feet in one year and covered 538 square feet to a depth of 7.2 feet in two years. Roots have approximately one to two root buds per 4 inches of root length throughout the year. Root fragments from a 10-year-old stand of Canada thistle had the capacity to produce a similar number of shoots per unit of root, regardless of their location along the 6-foot rooting depth of the stand. Development of underground shoot length is greatest in late fall and winter following death of aerial shoots as the plant becomes primed for shoot emergence when soil warms in spring.

Dispersal: The seeds have a special feathery structure (the pappus) that aids in wind dispersal. The efficiency of wind dispersal is low, but it ensures introduction to new sites and maintains genetic diversity. Canada thistle is common in field borders, which undoubtedly serves as a source of field contamination both by seeds and spreading roots. The seeds also commonly disperse in contaminated crop seed, hay and straw used for livestock or for mulch and compost in gardens. Seeds can blow into irrigation or drainage waters and then be transported long distances while maintaining 52–65% viability.

Common natural enemies: Goldfinches and other birds eat the seeds. Numerous nematodes, insects and fungi have been identified on Canada thistle; a selection of these organisms is discussed below. Four species have successfully established as biocontrol agents on this weed, including *Hadroplontus litura*, *Urophora cardui*, *Larinus carlinae* and *Rhinocyllus conicus*, but their impact is typically limited overall. Larvae of painted lady butterfly (*Cynthia cardui*) can defoliate the plant. Larvae of Canada thistle midge (*Dasypeura gibsoi*) eat the seeds, and larvae of *Orellia ruficauda* attack the flowering heads. None of these organisms is sufficient to control Canada thistle alone. The weevil *Ceutorhynchus litura* has been released in Canada as a biocontrol agent. Larvae mine the leaf veins, stems and root collar, and can greatly reduce a population, but the insect is difficult to get established. The gall fly *Urophora cardui* deposits eggs and produces galls in terminal shoots of Canada thistle but has had little impact on thistle populations despite successful colonization after release in eastern Canada. The shoot-boring weevil, *Apion onopordi*, infested and reproduced in Canada thistle shoots but had little effect on thistle performance. Inoculation with the rust pathogen *Puccinia puntiformis* had no effect on growth of Canada thistle, but in combination with cutting, reduced fertile flower production. The *Phoma macrostoma* fungus is locally systemic in stems, but not roots, of Canada thistle in England and Canada, and causes a bleaching disease. A particularly active strain of this fungus has been registered as a bioherbicide in Canada. *Pseudomonas syringae* pv. *tagetis* also can cause dramatic bleaching symptoms.

Palatability: Canada thistle is unpalatable to humans. Nutritional content of Canada thistle is relatively high, but it has a low palatability for livestock.

FURTHER READING

Anderson, R. 2009. Managing Canada thistle in organic cropping systems. *Small Farmer's Journal* 33: 79–82.

Graglia, E.B. Melander and R.K. Jensen. 2006.

Mechanical and cultural strategies to control *Cirsium arvense* in organic arable cropping systems. *Weed Research* 46: 304–312.

Sciegienka, J.K., E.N. Keren and F.D. Menalled. 2011. Impact of root fragment dimension, weight, burial depth, and water regime on *Cirsium arvense* emergence and growth. *Canadian Journal of Plant Science* 91: 1027–1036.

Tiley, G.E.D. 2010. Biological flora of the British Isles: *Cirsium arvense* (L.) Scop. *Journal of Ecology* 98: 938–983.

Wedryk, S. and J. Cardina. 2012. Smother crop mixtures for Canada thistle (*Cirsium arvense*) suppression in organic transition. *Weed Science* 60: 618–623.

Catchweed bedstraw

Galium aparine L.

False cleavers

Galium spurium L.

Catchweed bedstraw seedling
Antonio DiTommaso, Cornell University

Catchweed bedstraw leaves
Scott Morris, Cornell University

Catchweed bedstraw prickles on stem and fruits
Anurag Agrawal, Cornell University

IDENTIFICATION

Other common names:

Catchweed bedstraw: cleavers, goose grass, spring cleavers, scratch grass, grip grass, catch weed, bedstraw, white hedge, valiant's cleavers

False cleavers: stickwilly, cleavers

Family: Madder family, Rubiaceae

Habit: Lax summer or winter annual herbs, typically scrambling over other plants

Description: The two species are very similar in appearance. The **seedling** stem is green with purple or brown splotches. The first true leaves are elliptical or narrow, spine-tipped and arranged in spoke-like whorls of four or more. Cotyledons are egg shaped with notched tips and long stalks. The upper cotyledon surface is green and hairy. The leaf surface, margins and midvein on the underside have backwards curving, stiff hairs.

Catchweed bedstraw: Cotyledons are egg shaped, 0.24–1.25 inch long by 0.2–0.5 inch wide.

False cleavers: Cotyledons are 0.2–0.4 inch long by 0.08–0.16 inch wide.

Mature plants have four-sided, square, jointed stems that branch near the base and have backward curving bristles clustered near the leaf whorls that cause the weed to cling to neighboring objects. Leaves are in whorls of usually eight, but sometimes six. Leaves have a strong central vein and an indistinct stalk. Leaf edges and midveins are rough, and leaf tips are prickly. Roots are fibrous.

Catchweed bedstraw: Plants form loose, sprawling mats. Stems grow up to 5 feet long. The base of the stem bristles is round. Leaves are green, oval to lanceolate and 0.5–3 inches long by 0.1 inch wide.

False cleavers: Plants are stiffer, rougher and more branched than catchweed bedstraw and have stems up to 6.6 feet long. The base of the stem bristles is flattened. Leaves are linear, yellowish and are 0.5–2.6 inches long by 0.1–0.24 inch wide.

Individually stalked **flowers** are grouped in clusters arising from leaf axils. Flowers are four-petaled. Fruits are split into two identical parts, each containing one seed. The dry fruit (the apparent seed) is gray-brown, a rounded kidney shape and usually covered in short hooked hairs that cling to clothing and fur.

Catchweed bedstraw: Flowers are white and 0.08 inch in diameter. Seeds are 0.08–0.16 inch long.

False cleavers: Flowers are pale yellow to yellow-green but occasionally white and are 0.04–0.06 inch in diameter. Seeds are 0.1 inch long.

Similar species: Smooth bedstraw (*Galium mollugo* L.) is a rhizomatous perennial weed that largely lacks hairs except for the rough leaf margins. It has smaller leaves (0.4–1.1 inch long by 0.08–0.16 inch wide) than catchweed bedstraw. Many other, non-weedy species of bedstraw are present in most regions of North America. Carpetweed (*Mollugo verticillata* L.) has smooth leaves and stems, lacks square stems and has leaves that vary in size and shape within each whorl.

Taxonomic note: These two species are difficult to distinguish, and their weedy behavior is considered similar by farmers and weed professionals. However, there are sufficient differences in flower and fruit size, chromosome numbers and molecular markers to justify classification as separate species. Most of the research that supports the descriptions presented in this chapter was conducted on catchweed bedstraw and will be identified as such, but it should be understood as generally applicable to false cleavers as well. Where research specifically distinguished between these species, this will be noted explicitly in the specific section under discussion.

MANAGEMENT

Deep moldboard plowing (8–10 inches) can help suppress populations of these species. After plowing, seeds will be buried too deeply to emerge during the current season, and since the seeds usually die off quickly, most will be gone before they return to the surface with subsequent tillage. If you wish to avoid annual moldboard plowing, use this practice to bury seeds the spring after years with heavy seed production; then use conservation tillage practices in following years. This will allow extra years for the buried seeds to die. Because the seeds do not survive long in the soil, rotation into a hay crop like alfalfa can also greatly decrease seed populations.

Since catchweed bedstraw responds strongly to P, avoid accumulating excessively high P levels in the soil. If P levels are higher than necessary for good crop production due to regular use of manure or compost as an N source, increase use of legume cover crops for N and decrease applications of compost or manure.

Cultivation with a harrow or tine cultivator can increase germination of both species so that seedlings can be killed during preparation of a seedbed. Because seedlings often emerge from two or more inches, rotary hoeing is relatively ineffective, and tine weeding should be aimed at burial of seedlings just after emergence. In cereal grains a tine weeder can be used to comb these weeds out of the crop just prior to when grain stems elongate. When soil is dry, set the weeder to just graze the soil surface; although much of the plant will be left, the practice can shift the competitive balance in favor of the crop.

Since most seeds are retained on the plant at harvest of spring-planted crops, equipment designed to harvest and destroy weed seeds during crop harvesting operations would be effective at reducing populations of these species.

ECOLOGY

Origin and distribution: Although catchweed bedstraw is native to coastal areas of eastern and western North America to the Aleutian Islands, the weedy races in agricultural fields appear to have been introduced from Europe. It presently occurs in most of the United States and southern Canada. It also is native to Eurasia and has been introduced into temperate areas worldwide. False cleavers is native to Europe and is widespread in Asia, Africa and North America. In North America, it occurs throughout the United States, southern Canada and into Alaska. Molecular analysis revealed that the predominant *Galium* species found in west Canadian fields is false cleavers and that these populations are closely related to each other and to European populations of the same species.

Seed weight: Mean population seed weights of catchweed bedstraw range from 3.7–16.9 mg. Mean population seed weights of false cleavers range from 1.6–2.8 mg.

Dormancy and germination: Germination behavior varies greatly depending on the local race, habitat and time of year the seeds are produced. Many, and sometimes all, of the seeds produced in agricultural fields are capable of germination immediately after falling from the plant. Seeds germinated in darkness just as well as in alternating light/dark conditions. Unlike most weeds, light inhibits germination and exposure to light induces dormancy. In some races, drying leads to dormancy. In others, dry seeds lose dormancy with time. Cold wet conditions will break dormancy, as will nitrates. Optimum temperature for germination of catchweed bedstraw varies considerably with country of origin, from 33–68°F. Catchweed bedstraw seeds produced in agricultural settings in Great Britain germinated at temperatures of 41–59°F, with optimum germination at 48–54°F. False cleavers has a slightly higher optimum temperature requirement of either a constant 72°F or alternating daily temperatures between 50°F–75°F. Germination of catchweed bedstraw was optimum at 40–60% soil moisture holding capacity, whereas false cleavers required a slightly higher optimum of 50–80%.

Seed longevity: Few seeds of catchweed bedstraw last longer than two years in the soil. In two six-year experiments in tilled grain fields, catchweed bedstraw seeds declined at 51% and 65% per year. In France, the annual mortality rate was 41% in untilled soil. Seed survival increases with depth.

Season of emergence: Catchweed bedstraw exhibits a wide range of potential emergence periods, primarily in the fall and spring but sometimes also in summer, and in mid-winter in warmer regions. A dry autumn can delay peak emergence until spring. Seeds from two populations were collected in Turkey. In one population, seeds were collected in June, and seedlings from

these seeds had two peak emergence periods per year in November and in April to June. Seeds from the second population were collected in September, and seedlings from these had one peak emergence period in June to August. Emergence patterns of false cleavers is generally similar to that of catchweed bedstraw, except that false cleavers is more sensitive to dry soil conditions than catchweed bedstraw, which is adapted to a wider window of soil moisture conditions. Once emerged, early growth rate of several populations of both species responded similarly to temperature and moisture conditions.

Emergence depth: Seedlings emerge best from anywhere in the top 2.4 inches of soil, but many can emerge from 4 inches or more. Emergence from a depth of 0.8–2.4 inches was higher than from the soil surface, partly because emergence of both species is enhanced by uniform soil moisture and declines with intermittent drying conditions as occurs at the soil surface.

Photosynthetic pathway: C_3

Sensitivity to frost: Before flowering begins, the plants are very cold hardy. One study showed that they tolerate temperatures of at least -13°F, though another showed severe damage or death at 14°F or lower. The threshold temperature for frost tolerance appears to fluctuate with season, decreasing from 19°F in October to 1°F in December and then increasing to 28°F by April. After flowering begins, the plants are killed by a hard frost.

Drought tolerance: Catchweed bedstraw thrives in moist habitats and is not drought tolerant. One report indicated that false cleavers plants are more drought tolerant than catchweed bedstraw, but both species are rare in areas with low summer rainfall.

Mycorrhiza: Both presence and absence of mycorrhiza have been reported.

Response to fertility: Catchweed bedstraw is highly responsive to both N and P, and its dry matter production continues to increase with added P even at very high levels of P fertility. It responds most, however, when N and P are both applied. Increasing N fertility rates increases this weed's competitiveness with wheat. It grows best at pH from 5.5–8.

Soil physical requirements: Both species thrive in a wide range of soil textures but prefer moist, fertile soils.

Response to shade: Catchweed bedstraw grows best in open conditions but can survive and reproduce in woodland habitats. It scrambles up onto crops by means of the prickles on the stems and leaves, thereby avoiding shade. This weed is highly adaptable to shade conditions by lowering respiration sufficiently to maintain a positive carbon balance. Plants produce more leaves at the expense of root production under shade conditions. Shade of an individual branch leads to death of the shaded branch with little shift in the growth pattern of other branches.

Sensitivity to disturbance: Uprooted plants readily reroot on moist soil. Little regrowth or seed production occurs if stems are removed during or after flowering.

Time from emergence to reproduction: Plants establishing in late summer and fall overwinter and flower from May to September and set seed from June to October. Plants establishing in the spring begin flowering about six weeks later. Generally, as the combination of temperature and day length increase, the number of days between emergence and flowering decreases. Seeds mature about one month after the flower opens.

Pollination: The flower structure normally ensures self-pollination, but a wide variety of insects visit the flowers, and occasional cross pollination may occur.

Reproduction: Plants typically produce 300–400 seeds but under favorable conditions may produce more than 1,500 seeds. When subject to competition from a vigorous grain crop, seed production is as low as nine seeds per plant in one study. Seed retention at spring wheat harvest in western Canada was variable, ranging from 62% to 95%, and 90% of seeds were situated at a height on the plant that could be harvested by combine.

Dispersal: The hooked hairs on the fruit coat help the seeds disperse on fur and clothing. Seeds are a persistent contaminant of crop seed, particularly canola

in the Canadian prairies, and have been spread by humans worldwide. These species also commonly move in straw from infested crops and in combines. The fruits float and disperse along streams and irrigation canals. The seeds pass through livestock without loss of viability and are dispersed in droppings and with the spread of manure.

Common natural enemies: Larvae of the sawfly *Halidamia affinis*, the mirids *Criocoris piceicornis*, *Polymerus nigritus*, and *P. unifasciatus*, and the gall forming aphids *Aphis galiiscabri* and *Galiomum langei* are restricted to *Galium* species. Harvester ants (*Messor barbarus*) consumed 49% of catchweed bedstraw seeds on the soil surface between grain harvest and fall in Spain, but the rate of seed predation in North American fields is unknown.

Palatability: Although catchweed bedstraw has been used as a pot herb, it is also considered to be poisonous and unsuitable for people or livestock. The plant is very coarse, and consumption can cause a low-level poisoning. It is preferred by geese and can be used as a poultry feed. A coffee substitute can be made from seeds.

FURTHER READING

Defelice, M.S. 2002. Catchweed bedstraw or cleavers, *Galium aparine* L.—A very "sticky" subject. *Weed Technology* 16: 467–472.

Malik, N. and W.H. Vanden Born. 1988. The biology of Canadian weeds. 86. *Galium aparine* L and *Galium spurium* L. *Canadian Journal of Plant Science* 68: 481–499.

Taylor, K. 1999. Biological Flora of the British Isles. No. 207. *Galium aparine* L. *Journal of Ecology* 87: 713–730.

Tidemann, B.D., L.M. Hall, K.N. Harker, H.J. Beckie, E.N. Johnson and F.C. Stevenson. 2017. Suitability of wild oat (*Avena fatua*), false cleavers (*Galium spurium*), and volunteer canola (*Brassica napus*) for harvest weed seed control in western Canada. *Weed Science* 65: 769–777.

Wilson, B.J., K.J. Wright and R.C. Butler. 1993. The effect of different frequencies of harrowing in the autumn and spring on winter wheat, and on control of *Stellaria media* (L.) vill., *Galium aparine* L. and *Brassica napus* L. *Weed Research* 33: 501–506.

Chamomiles

Corn chamomile, *Anthemis arvensis* L.
Mayweed chamomile, *Anthemis cotula* L.

Corn chamomile seedling
Joseph DiTomaso, University of California, Davis

Corn chamomile foliage
Joseph DiTomaso, University of California, Davis

Corn chamomile flower
Joseph DiTomaso, University of California, Davis

Mayweed chamomile seedling
Antonio DiTommaso, Cornell University

Mayweed chamomile rosette
Scott Morris, Cornell University

Mayweed chamomile clump of plants
Scott Morris, Cornell University

IDENTIFICATION

Other common names:

Corn chamomile: field chamomile

Mayweed chamomile: mayweed, stinking chamomile, dog fennel, stink-weed, dogs chamomile, dill weed, stinking daisy, hogs fennel, fetid chamomile, stinking mayweed

Family: Aster, Asteraceae

Habit: Upright, branching winter or summer annuals with finely divided leaves

Description: Seedlings have thick, hairless, stalkless, oval shaped cotyledons. The first two leaves are opposite, somewhat thick and divided. Subsequent leaves are thinner, green, lightly hairy, finely divided into leaflets that have smaller lobes, and form a basal rosette. Leaves have a rounded shape; leaflets and lobes are oval or egg shaped.

Corn chamomile: Cotyledons are light green and

0.1–0.2 inch long; in all other ways the seedling resembles mayweed chamomile.

Mayweed chamomile: Short-lived cotyledons are 0.3 inch long with obscure veins. The base of the cotyledons is fused to form a fleshy cup. The seedling stem is green to dull maroon. The first pair of true leaves is opposite with short surface hairs; all subsequent leaves are alternate.

Mature plant stems are upright or leaning; they branch just above the rosette and take root at nodes when in contact with soil. Leaves are club- to egg shaped in outline but are divided two to three times into finger-like leaflets and lobes. Fibrous roots extend from short, broad taproots.

Corn chamomile: Stems are lightly hairy, green, red in strong light and 8–20 inches tall. Leaves are alternate, 0.8–2.4 inches long by 0.4–0.8 inch wide, and leaflet lobes are 0.03 inch wide. Leaflets are less finely divided than mayweed chamomile. Leaves are gray-green and covered in soft hairs.

Mayweed chamomile: Stems are nearly hairless, green, 4–24 inches tall and highly branched. Leaves are alternate, yellow-green, slightly hairy, 0.5–2.5 inches long by 0.25–1.25 inch wide and very finely divided, with narrow leaflets and lobes. Crushed leaves have a strong, unpleasant odor.

Flower heads are daisy like and comprised of 10–16 small ray flowers with 0.25–0.5 inch-long white petals surrounding a central disk of numerous tiny, yellow flowers. Individual flower heads are located on branch ends; one to two rows of small, tapered, hairy green bracts wrap the base of the flower head. **Seeds** are covered with a hard, tight shell of fruit tissue. These seed units are light brown, narrow, four-sided wedges, with 10 longitudinal ribs.

Corn chamomile: Seeds are smooth, 0.06–0.1 inch long and produced from both flower types. Seeds from ray flowers are larger than seeds from disk flowers.

Mayweed chamomile: Seeds are bumpy, 0.06 inch long and produced only from yellow disk flowers.

Similar species: The best way to distinguish the chamomiles from closely related species is by scent, since the entire plant when crushed will exude a distinctive odor at all growth stages. Corn chamomile is lightly chamomile scented, while mayweed chamomile has a strong foul odor. Scentless chamomile [*Tripleurospermum perforatum* (Mérat) M. Lainz, formerly *Matricaria perforata* Mérat] and yellow chamomile (*Anthemis tinctoria* L.) both lack noticeable odors. Scentless chamomile has white flowers, and yellow chamomile has yellow flowers. Pineapple-weed [*Matricaria discoidea* DC., formerly *M. matricarioides* (Less.) Porter] has a petal-less, cone shaped, yellow-green flower head, and its foliage smells like pineapple.

MANAGEMENT

Following sound general weed management principles should keep these species under control. In particular, relieve soil compaction, and if mayweed chamomile is a problem, consider steps to improve soil drainage as well. This will tip the competitive balance in favor of the crops and also improve the success of cultivation. Timely rotary hoeing and tine weeding are important in both row and grain crops. Cultivate row crops shallowly close to the row, and if the crop will tolerate hilling, cover the chamomile seedlings while they are still small. Fertilize the crop for maximum vigor and competitiveness, but avoid over fertilization that will increase other weed species. If the crop rotation allows, use a buckwheat cover crop to suppress corn chamomile emergence and growth before planting a fall crop.

In small grains, high density plantings and closely spaced rows will help suppress these weeds. If necessary, consider cross seeding the grain to achieve a more uniform spacing. Because these weed species are relatively short, long strawed grain varieties help reduce seed production. If you have a severe, persistent problem, set the combine to cut high. Then after harvest, chop the straw and chamomile into a wagon with a forage chopper. Till the field before the chamomile flowers again, and dispose of the chopped material or compost it at high temperature. Mowing will reduce seed production in hay and pasture crops, but plants can flower again below the cutting height, so a second mowing should take place just before secondary flowering begins.

ECOLOGY

Origin and distribution: Both species are native to Europe. They occur throughout Europe, North Africa and the Middle East, and have been introduced into Australia, New Zealand, south Asia and southern South America. Mayweed chamomile occurs throughout southern Canada and the United States, including Alaska and Hawaii. Corn chamomile occurs throughout the eastern and far western United States and southern Canada, and sporadically in the mountain and Great Plains states south to Colorado and Nebraska.

Seed weight: Corn chamomile: 0.5 mg (disk flowers), 1.2 mg (ray flowers); mayweed chamomile: 0.4 mg (disk flowers), the ray flowers are sterile.

Dormancy and germination: The "seeds" of these species and other members of the sunflower family include a shell (pericarp) that surrounds the seed proper. This structure corresponds to the fleshy part of fruits in other families. In most newly produced corn chamomile and mayweed chamomile seeds, this shell is hard and tight around the seed and physically prevents growth of the embryo, even though it allows water uptake. After several months to years, microorganisms in the soil break down this shell and allow seeds to germinate. Smaller seeds germinate more readily than larger seeds because the pericarps of larger seeds completely encase the seeds, whereas that of smaller seeds do not. Light and strong diurnal alternation of temperatures from 68–86°F in the day to 32°F at night promoted maximum germination of mayweed chamomile. Under continuous temperature, mayweed chamomile germinated best at 68°F in one experiment and 86°F in another experiment, but germination was never as high as germination with alternating temperature or when the hard shell was removed. Ammonium and nitrate stimulated germination of corn chamomile and mayweed chamomile. Optimum pH for mayweed chamomile germination was 3–6 and then declined at increasing pH.

Seed longevity: In a seed burial experiment, 89% of corn chamomile seeds survived one year and 47% survived 11 years. In the same study, 63% of mayweed chamomile seeds survived one year and 6% survived 11 years. In soil tilled three times per year, the number of mayweed chamomile seeds declined by 42–51% per year, and only 5% of seeds remained viable after five years.

Season of emergence: Both species emerge mainly in the fall and spring, with some sporadic emergence throughout the summer. In mild climates, these species also emerge in winter.

Emergence depth: A depth of 1.2 inch was considered too deep for emergence, so these species presumably emerge from the upper 1 inch of soil.

Photosynthetic pathway: C_3

Sensitivity to frost: Both species are frost hardy and regularly grow as winter annuals in the northern United States.

Drought tolerance: Corn chamomile is resistant to drought and prefers areas with relatively low summer humidity. Mayweed chamomile is moderately resistant to drought but is most prevalent in low, wet areas with high soil moisture for germination. Growth of mayweed chamomile was similar in wet and dry years and at high and moderately low moisture levels in pot experiments. Root development of mayweed chamomile is optimized at 50°F, and this trait may permit rapid root development during cool, early season conditions and allow greater access to soil moisture during drier conditions later in the season.

Mycorrhiza: Corn chamomile has been reported as mycorrhizal, but other reports indicate that neither this species nor mayweed chamomile is mycorrhizal.

Response to fertility: Corn and mayweed chamomile can grow and flower at low levels of N and P. They greatly increase in size and seed production with N, but competitive ability against cereal grains like barley decreases with increasing N. Corn chamomile is moderately responsive to P. Simultaneously low levels of Ca and K promote early flowering and increase the number of flower heads produced. Corn chamomile grows well on both calcareous soils of pH 7–8 and on moderately acidic soils of pH 5.6–6.6, provided N is readily available. Mayweed chamomile growth was not affected by soil pH in the range of 4.7–6.2.

Soil physical requirements: Corn chamomile usually occurs on well drained soils. It occurs on soil textures ranging from medium clay loam to sand. Within fields, it is frequently most common on headlands, which indicates some tolerance for compaction. Mayweed chamomile occurs on a wide range of soil textures but is most common on clay and clay loam soils. Its affinity for heavy soils is sometimes striking, with population density tapering off quickly at the boundary with lighter textured soils. In contrast with corn chamomile, mayweed chamomile does well on soils with restricted drainage.

Response to shade: In partial shade, mayweed chamomile grows more upright and produces additional leaves to capture light.

Sensitivity to disturbance: Both species grow new shoots quickly if the stems are cut, e.g., during cereal crop harvest. Both species are moderately resistant to trampling. Corn chamomile cannot regenerate if it is completely buried but can regrow if only partially buried. The plants can regenerate roots from the base of the shoot and may therefore reroot following hoeing or cultivation.

Time from emergence to reproduction: Autumn emerging plants resume growth in early spring and begin flowering in late spring. Spring emerging plants flower in mid-summer. Flowering typically continues for two months or longer.

Pollination: Plants of both species are self-incompatible. Flowers are cross pollinated by hover flies and other insects.

Reproduction: Average sized corn chamomile plants produce anywhere from 600–4,200 seeds. Small plants produce fewer than 20 seeds, but large plants can produce more than 18,000 seeds. Up to 10% of the seeds, however, lack an embryo and are non-viable.

Mayweed chamomile flower heads produce 50–75 seeds each. Medium sized plants produce from 550–12,000 seeds. Exceptionally large plants can produce up to 27,000 seeds. Typically, 10–25% of seeds are non-viable, and more than 50% of seeds produced late in the season may be non-viable.

Dispersal: Both species spread as contaminants of forage seed. Plants can become incorporated into grain straw and spread the seeds when the straw is used for bedding or mulch. Seeds also travel in soil clinging to farm implements, shoes and the feet of livestock. Corn chamomile passes through the digestive tract of livestock and is spread with manure and when animals move. Seeds also pass unharmed through some birds. Mayweed chamomile may behave similarly.

Common natural enemies: Corn chamomile is attacked by larvae of the moth *Homoeosoma saxicola*. Mayweed chamomile growth and reproduction can be significantly reduced by various herbivores including aphids, spittlebugs, true bugs (Heteroptera), moths, slugs and snails. Larvae of the beetle *Apion sorbi* attack the base of mayweed chamomile flower heads, distorting them into a spherical shape. Gray mold (*Botrytis cinerea*) can cause extensive damage to mayweed chamomile in wet conditions.

Palatability: Livestock avoid eating both species. Consumption by dairy cattle taints milk.

FURTHER READING

Kay, Q.O.N. 1971a. Biological Flora of the British Isles: *Anthemis cotula* L. *Journal of Ecology* 59: 623–636.

Kay, Q.O.N. 1971b. Biological Flora of the British Isles: *Anthemis arvensis* L. *Journal of Ecology* 59: 637–648.

Kumar, V., D.C. Brainard and R.R. Bellinder. 2008. Suppression of Powell amaranth (*Amaranthus powellii*), shepherd's-purse (*Capsella bursa-pastoris*), and corn chamomile (*Anthemis arvensis*) by buckwheat residues: Role of nitrogen and fungal pathogens. *Weed Science* 56: 271–280.

Common chickweed

Stellaria media (L.) Vill.

Common chickweed seedling
Scott Morris, Cornell University

Common chickweed plant
Scott Morris, Cornell University

Common chickweed flower
Scott Morris, Cornell University

IDENTIFICATION

Other common names: starwort, starweed, bindweed, winter weed, satin flower, tongue grass, chickwhirtles, cluckenweed, mischievous Jack, skirt buttons, cyrillo, white bird's eye, common starwort

Family: Pink family, Caryophyllaceae

Habit: Mat forming winter or summer annual herb

Description: Seedling cotyledons are oval, 0.4 inch long by 0.1 inch wide, pointed at the tip and borne on a 0.4 inch-long hairy stalk. True leaves are opposite, round to teardrop shaped and twice the length of the stalk. Seedlings are light green, upright and about the width of a penny at the four true leaf stage. Branching starts at the five leaf-pair stage. **Mature plants** form dense, light green carpets of foliage with creeping stems up to 20 inches long. Stem tips are angled upward, helping the stems crawl over and onto other plants. Stems branch often, root at the nodes when in contact with the soil and are smooth except for one to two rows of small hairs. Leaves are oppositely arranged and located on 0.25–0.75 inch-long stalks, which may be absent near branch tips. One line of white hairs is present on stalks. Leaves are broadly teardrop to elliptically shaped, pointed at the tip and 0.4–1.25 inch long by 0.6 inch wide. Upper leaves are no wider than 0.4 inch. Shallow, fibrous roots easily break from the shoots. Mats may reach 0.3–3.3 feet in diameter. **Flowers** have five deeply lobed, white petals set inside five larger, green, 0.2 inch long, triangular sepals. The 0.1–0.3 inch diameter flowers are clustered at branch ends. Each flower produces an oval, single chambered capsule containing eight to 10 light yellow-brown to red-brown, irregularly round, flat, 0.06 inch diameter **seeds** with a pebbly surface. The capsule splits to form six teeth when releasing seeds and is retained on the plant.

Similar species: Leaves of mouseear chickweed [*Cerastium fontanum* Baumg. ssp. fontanum] are densely covered in fine hairs at all growth stages. Thymeleaf speedwell (*Veronica serpyllifolia* L.) has rounded, lightly toothed, dark green leaves with notched tips and no hairs on its stems or leaf stalks; its flowers have four white petals with blue streaks. Thymeleaf sandwort (*Arenaria serpyllifolia* L.) has hairy leaves and five non-divided petals. Little starwort (*Stellaria graminea* L.) has flowers similar to common chickweed, but they are borne on long stalks, and plants have long, tapering leaves.

MANAGEMENT

Because common chickweed can grow and set seeds during any season of the year, control throughout all crop rotation phases should be practiced. Otherwise, high densities of seeds may build up in the soil and overwhelm control efforts in crops that are sensitive to common chickweed competition. Thus, for example, flame weeding sweet corn very effectively controls common chickweed and may be worthwhile to prevent competition with subsequent crops, even though common chickweed causes little yield loss in corn. If common chickweed is a problem, sowing density of winter cover crops should be increased. Increased sowing density of both winter and spring grain crops may help reduce yield losses to common chickweed and will also reduce seed production. Because chickweed continues to grow and set seed until it dies, tilling fields immediately after harvest is important in controlling this species. If the species is green and growing, one or two weeks of delay can increase seed production by several fold.

Because common chickweed seedlings are very small and fragile, stirring of the top 1–2 inches of soil two to four times during the first month following tillage is highly effective at removing many of the individuals that will emerge during the season. Since stems root readily in moist soil, cultivation is often ineffective once the plant mat reaches several inches in diameter. Thus, frequent but shallow weeding is more effective at controlling the species. Harrowing in the spring is more effective than fall harrowing for controlling this species in winter wheat. Since most common chickweed seeds that have been incorporated into the soil remain dormant unless exposed to light, flame weeding to create a stale seedbed before planting small seeded or slow growing vegetables is advantageous. Organic mulch materials are effective for suppressing common chickweed due to the species' small seed reserves and prostrate growth form. Common chickweed is highly shade tolerant, however, and can scramble up through loose straw. Thus, plastic mulch, baled straw or a paper barrier under loose straw will be more effective than loose straw alone.

ECOLOGY

Origin and distribution: Common chickweed is native to Europe. It has been widely introduced throughout the world, including parts of the far north and sub-Antarctic islands. In the tropics it occurs primarily at high elevations. It occurs in all 50 states and in all Canadian provinces, but it is generally absent from dry areas.

Seed weight: Seed weight varies between populations, with mean seed weights ranging from 0.36–0.51 mg.

Dormancy and germination: Freshly shed seeds are typically dormant, but some may germinate immediately in some populations. A period of after-ripening at warm soil temperatures (for example, 68°F) breaks dormancy. Thus, most seeds shed in the spring wait until fall to germinate. In addition, a low temperature requirement for after-ripening has been reported. Seeds germinate best at moderate temperatures of 54–68°F and will not germinate at temperatures above 86°F. Once seeds have after-ripened, light, fluctuating temperature and nitrate all promote germination, but light is the most important germination cue. Different populations can possess different after-ripening and environmental requirements for germination, presumably permitting the species to germinate in response to varying conditions.

Seed longevity: Viable seeds of common chickweed have been found beneath medieval ruins in Europe. Deep burial in high moisture, low oxygen conditions is highly favorable for seed survival. Seeds stored under dry, cool laboratory conditions remained viable for 5.5 years. In cool Alaskan conditions, buried common chickweed had a 17% annual mortality rate, but a few seeds were still viable after 24.7 years. Most common chickweed seeds die off quickly, however, in regularly tilled soil. In one study, seed viability declined annually by 19–30% in undisturbed soil but by 34–56% under cultivated conditions. In several other studies in annually tilled fields, common chickweed seeds declined at 33–60% per year. In soil stirred four times per year, seeds declined by 72% per year. Seeds buried in Mississippi lost all viability in 18 months.

Season of emergence: Although common chickweed can emerge any time of year in moist soil at moderate temperatures, most emergence occurs during the spring and fall.

Emergence depth: Optimum depth for emergence is

from just below the surface at 0.4 inch. Seedlings will emerge from seeds as deep as 2 inches, but emergence from deeper is rare.

Photosynthetic pathway: C_3

Sensitivity to frost: Common chickweed can tolerate temperatures down to 7°F, though with some damage. Photosynthesis can occur at temperatures below freezing, and the base temperature for growth of this species was computed to be 26°F. In the Northeast, fall germinating seedlings commonly persist through the winter.

Drought tolerance: The species is drought sensitive, and plants quickly wilt and die during dry conditions.

Mycorrhiza: Most studies have found no mycorrhiza, but one report indicated slight mycorrhizal presence.

Response to fertility: Common chickweed prefers highly fertile soils, and its dense root system is well adapted to access nutrients. Biomass, seed production and seed weight increase with increasing nitrogen. Common chickweed is highly responsive to addition of P when N is adequate. In excessively fertile conditions like manure piles, it concentrates nitrate to toxic levels. Even in agricultural fields, nutrient levels in common chickweed tissues can be high. For example, plants from a Massachusetts onion field had elemental N-P-K levels of 3.4-0.53-7.5 whereas levels in the onions were 2-0.28-2.2. The species grows well on soil with a pH from 5.2–8.2, but growth is severely depressed below pH 5.2.

Soil physical requirements: The species tolerates a wide range of soil types but grows best on moist, fine textured soils.

Response to shade: Common chickweed is shade tolerant. It grows well in partial shade and can flower and set seeds in as little as 1.5% daylight.

Sensitivity to disturbance: Small seedlings are easily killed by cultivation. Larger plants may reroot following cultivation or hoeing if the plant is partially covered with soil. Stems root when in contact with moist soil, which makes large plants difficult to completely remove by pulling. Its low, prostrate growth habit

allows it to survive repeated mowings, as occurs with hay or green manure crops.

Time from emergence to reproduction: Plants usually flower within six weeks of emergence, and seeds set about two weeks later. Since some freshly shed seeds are not dormant or can after-ripen rapidly, the species can complete three to four generations per year. The timing of 50% seed production in Michigan occurs between mid-May and mid-June.

Pollination: Common chickweed usually self-pollinates, and under cold conditions, pollination may occur without opening of the flowers.

Reproduction: Common chickweed flowering is independent of day length, and this species can set seed at any time of the year provided moisture is available and temperatures are above freezing. Flowering and seed set occur continuously as the plant grows. Mature plants typically produce 500–3,000 seeds each, but very large plants may produce 13,000–15,000 seeds.

Dispersal: The seeds have no special adaptations for dispersal, but it has been spread worldwide primarily by humans because of its adaptability to disturbed habitats. Most dispersal probably occurs in soil clinging to shoes, tires, farm implements and animal hooves. Some seeds retain viability when passing through the digestive tracts of cattle, sheep, horses, deer, rabbits, pigs and some birds, including sparrows and gulls. Seeds are also dispersed as a contaminant of pasture seed and other crops.

Common natural enemies: Several species of wildlife graze common chickweed, and several species of insects, nematodes and disease organisms are found associated with this species, but the level of damage is usually trivial.

Palatability: Common chickweed has a mild flavor that makes it palatable as a salad herb. It is nutritious fodder for cows, sheep, horses, pigs and chickens, but it may contain toxic levels of nitrate when grown on nitrogen rich soils.

FURTHER READING
Hill, E.C., K.A. Renner and C.L. Sprague. 2014. Henbit

(*Lamium amplexicaule*), common chickweed (*Stellaria media*), shepherd's-purse (*Capsella bursa-pastoris*), and field pennycress (*Thlaspi arvense*): Fecundity, seed dispersal, dormancy, and emergence. *Weed Science* 62: 97–106.

Mertens, S.K. and J. Jansen. 2002. Weed seed production, crop planting pattern, and mechanical weeding in wheat. *Weed Science* 50: 748–756.

Sobey, D.G. 1981. Biological flora of the British Isles. *Stellaria media* (L) Vill. *Journal of Ecology* 69: 311–335.

Turkington, R., N.C. Kenkel, and G.D. Franko. 1980. The biology of Canadian weeds. 42. *Stellaria media* (L.) Vill. *Canadian Journal of Plant Science* 60: 981–992.

Common cocklebur

Xanthium strumarium L.

Common cocklebur seedling
Antonio DiTommaso, Cornell University

Common cocklebur mature plant
Scott Morris, Cornell University

Common cocklebur streaked stem and immature burs
Scott Morris, Cornell University

IDENTIFICATION

Other common names: clotbur, cocklebur, broad-leaved cocklebur, sheep bur, ditch bur, button bur, noogoora bur, heartleaf cocklebur, rough cocklebur

Family: Aster family, Asteraceae

Habit: Erect summer annual herb

Description: Seedlings have short, thick stems with purple at the base. Seed leaves are lanceolate, unusually large (2 inches by 0.5 inch) and narrower at each end, hairless, thick, dark green above and light green below. The first pair of true leaves is opposite, triangular to egg shaped with a rounded tip and slightly toothed, with three prominent surface veins. All other young leaves are alternately attached, distinctly toothed, egg shaped with a pointed tip and covered in rough, short hairs. The stem becomes green and covered with upward pointing hairs as the plant develops. Fully **mature** stems are green, 1–4 feet tall, highly branched, hairy and flecked with purple-brown to black spots. Ridges are present on the stem. Upright hairs cause leaves to feel abrasive and gritty. Leaves are alternate, 1–6 inches long and oval to triangular or heart-shaped. Edges are wavy, irregularly toothed and slightly lobed into three to five sections: the two lobes near the leaf junction with the stalk give the leaf a heart shaped base, and the ones near the leaf tip may be slightly pointed. The leaf stalk is 0.75–3 inches long. The semi-woody, broad and sturdy taproot can reach 4 feet deep. Heads of either all male or all female green **flowers** appear in leaf axils near branch tips, but both sexes occur on the same plant. Individual flowers are inconspicuous. Male heads are 0.2–0.3 inch across and are subtended by one to three rows of small green bracts. Individual male flowers are round and short-lived, dropping off after shedding pollen. Female flowers are contained within small, green, immature burs. **Fruit** is a woody, brown, egg shaped to elliptical bur. Each bur is approximately 1 inch long by 0.7 inch or less wide, is covered in hard 0.1–0.25 inch spines and has two beaklike projections at the tip. Each bur contains two oblong, pointed, light brown to black **seeds**, covered in a papery, silver-black membrane. Each bur contains one large and one small seed, 0.31–0.60 inch long by 0.2–0.3 inch wide.

Similar species: Four species are sometimes mistaken for common cocklebur: jimsonweed (*Datura stramonium* L.), spiny cocklebur (*Xanthium spinosum* L.), common burdock (*Arctium minus* Bernh.) and great burdock (*Arctium lappa* L.). Jimsonweed can be confused for cocklebur at the seedling stage but can be distinguished by its hairless stems, singly veined leaves and unpleasant odor. Spiny cocklebur has yellow spines at the leaf nodes, more slender leaves and beakless burs. Common burdock and great burdock also produce burs but have larger (20 inches by 16 inches) leaves and burs that break into individual, spined fruits.

MANAGEMENT

Common cocklebur is an extremely competitive weed due to fast emergence and rapid growth supported by the large seed. Populations of one to three plants per 10 square feet can cause soybean yield losses of 52–75%. Common cocklebur emerges faster and in higher numbers, and it is more competitive under tillage than no-tillage conditions, so no-tillage grain production will be less favorable to this weed. Tine weeding and rotary hoeing have limited effectiveness because the seedlings can emerge from deep in the soil. Cultivate row crops early and close to the row, and repeat regularly until the crop is too large to tolerate tractor traffic. If plants set seed before harvest, collect or destroy the seeds during combining.

Because the seeds do not persist well in the soil, rotation to a sod crop for several years will help control this weed. Rotation to a winter grain will also help because the burs do not mature by grain harvest. Disk the field and plant a cover crop after grain harvest to prevent resprouting from the shoot bases and subsequent seed production.

Soil solarization so that mid-day soil temperatures reach 149–156°F will eliminate a high proportion of common cocklebur seeds.

Common cocklebur often establishes first on unmanaged areas like railroad embankments and stream banks, so if you see it there, eradicate it quickly before it can invade your tilled fields.

ECOLOGY

Origin and distribution: Common cocklebur is probably native to North, Central and South America, and possibly also to southern Europe and South Asia. It occurs in Africa and has been introduced into Australia. Its natural habitat is along riverbanks and beaches, and more recently it has colonized agricultural fields, roadsides and other disturbed habitats.

Seed weight: small (upper) seed, 50 mg; large (lower) seed, 60–75 mg. Burs can vary considerably in size. A large proportion of small and intermediate sized burs have either no seeds or only one seed.

Dormancy and germination: Burs usually contain two seeds, and these differ in size and dormancy. The smaller (upper) seed often sits closer to the tip of the bur and is usually dormant. The larger (lower) seed usually germinates the next spring following an after-ripening period. The smaller seed usually does not germinate until summer or the following year; this dormancy is apparently due to germination inhibitors in the seed coat and inability of oxygen to pass through the seed coat. Germination of both types of seeds is promoted by microbial decay or mechanical seed coat damage. The seeds do not require light for germination. A daily temperature fluctuation of 27°F increased germination over that at constant temperatures. Germination was greatest with warm, fluctuating temperatures of 86/68°F to 91/77°F. Under a constant temperature regime, highest germination occurred at 95–104°F, but constant soil temperatures of this magnitude are unlikely to occur under natural field conditions.

Seed longevity: Based on measurements at six to 30 months, common cocklebur seed viability declines at about 50% per year. In Arkansas, no common cocklebur seeds were viable three years after being shed. In Nebraska, some seeds buried in plastic capsules survived for 9 years.

Season of emergence: Most emergence occurs in mid-spring to early summer, with occasional pulses of seedlings appearing later in the summer as well. Emergence is dispersed more evenly from late spring throughout summer under no-till than under tilled conditions. The lowest average soil temperature required for emergence was 63°F, and the minimum day/night soil temperature fluctuation required for emergence was 14°F.

Emergence depth: Common cocklebur emerges well from 0.4–4 inches of soil, and a few seedlings can emerge from as deep as 6 inches. Seeds usually cannot take up enough water to germinate when they are on the soil surface due to poor contact between the large seeds and soil.

Photosynthetic pathway: C_3

Sensitivity to frost: Common cocklebur tolerates only light frost.

Drought tolerance: Common cocklebur is drought tolerant. Roots can extend 7 feet laterally and 4 feet deep, allowing access to water throughout the soil profile. Common cocklebur is an aggressive competitor partially because of its capacity to take up more water than other crops or weeds under similar growth conditions. It can maintain photosynthesis and transpiration under drought stress better than soybeans, but its growth and reproduction are, nonetheless, reduced by prolonged drought.

Mycorrhiza: Common cocklebur is considered a strong mycorrhizal host.

Response to fertility: Common cocklebur is highly responsive to N. In high N conditions, it will store excess N as nitrate and later use it to increase seed production. It grows well at pH 5.2–8.

Soil physical requirements: The species grows in a wide range of soil types from coarse sand to heavy clay. It can tolerate flooded as well as dry soil conditions.

Response to shade: Shade reduces growth and delays flowering of common cocklebur. In the absence of root competition, shade from a soybean leaf canopy reduced cocklebur water uptake, leaf area and weight. Cocklebur has the capacity to maintain functionality of lower leaves within the shade of a crop canopy by adjusting leaf metabolism accordingly. It also has the potential to alter upper leaf growth to increase light capture in response to reduced light levels on lower leaves.

Sensitivity to disturbance: Very rapid growth and ability to emerge from deep layers in the soil make common cocklebur insensitive to tine weeding and rotary hoeing. Small plants regrow quickly from buds at the base of lower leaves if plants are trampled or clipped. Once flowers have been pollinated, the burs will produce mature seeds even if the shoot or branch is severed from the roots.

Time from emergence to reproduction: Plants flower as days shorten, and generally flowering begins in August regardless of age or size. Seed maturation continues until a killing frost. Flowering initiated 16 weeks after emergence in Arkansas and Wisconsin, but 10 weeks after emergence in Quebec. Seeds become viable early in development. A few seeds are viable when burs begin to form, and the majority are viable when burs are fully elongated about three weeks later. In India, time from appearance of first flowers to ripening of first fruit was 23 days.

Pollination: Common cocklebur primarily self-pollinates, but up to 12% of flowers are cross pollinated by wind dispersed pollen.

Reproduction: Vigorous, open grown plants produce 500–5,400 burs, each of which usually contains two seeds. However, in dense stands, plants produced only 71–586 burs per plant. In pure stands in Arkansas, individual plants produced 4,500 burs (9,000 seeds), whereas in Quebec, plants produced 1,300–3,500 seeds. Individual cocklebur plants produced 900 burs (1,800 seeds) when grown with soybeans in North Carolina. The number of burs depends entirely on the size of the plant at the time flowering begins.

Dispersal: The spiny burs cling to animal fur, clothing, grain sacks, etc. They tangle particularly well in sheep's wool and are dispersed when the animals or wool are transported. The burs float well and are readily dispersed in streams, lakes, irrigation ditches and flooded fields.

Common natural enemies: The moth, *Phaneta imbridana*, lays eggs on the bur wall, and the larvae bore in and eat the seeds. The fly, *Euaresta aequalis*, pierces the bur wall and lays its eggs directly on the developing seeds. Together these insects killed from 3–84% of the seeds in New York. Population variation in bur characteristics greatly affects seed mortality, with long spines inhibiting the fly and thick bur walls inhibiting the moth. Two stem beetles burrow down through the stem and overwinter in the root, but they only affect seed set in the branches that they actually damage. The rust *Puccinia xanthii* is common in the United States and affects only cocklebur and ragweed species. It causes deformed leaves, early leaf drop, cracks on the stem, reduced plant size and reduced seed viability.

Spores overwinter on dead plants, which offers some possibility for deliberately spreading the disease using powdered diseased plants.

Palatability: People do not consume common cocklebur, but the seeds have been used in herbal medicine. The cotyledons are highly toxic to livestock. Larger plants have good nutritional value for livestock but are rough and unpalatable. Common cocklebur was unpalatable to grazing sheep.

Note: The pollen of cocklebur produces hay fever symptoms in sensitive individuals, and contact with the stems can also cause dermatitis in some individuals.

FURTHER READING

Bararpour, M.T. and L.R. Oliver. 1998. Effect of tillage and interference on common cocklebur (*Xanthium strumarium*) and sicklepod (*Senna obtusifolia*) population, seed production, and seed bank. *Weed Science* 46: 424–431.

Norsworthy, J.K. and M.J. Oliveira. 2007c. Tillage and soybean canopy effects on common cocklebur (*Xanthium strumarium*) emergence. *Weed Science* 55: 474–480.

Weaver, S.E. and M.J. Lechowicz. 1982. The biology of Canadian weeds. 56. *Xanthium strumarium* L. *Canadian Journal of Plant Science* 63: 211–225.

Common groundsel

Senecio vulgaris L.

Common groundsel seedling
Antonio DiTommaso, Cornell University

Common groundsel seed head and flowers
Scott Morris, Cornell University

IDENTIFICATION

Other common names: staggerwort, stinking Willie, groundsel, grimsel, simson, bird seed, ragwort, chickenweed, common ragwort, garden groundsel, grand mouron, old-man-in-the-spring

Family: Aster family, Asteraceae

Habit: Erect, branched, summer, or occasionally winter, annual herb

Description: Cotyledons of the **seedling** are club to ellipse shaped, measuring 0.1–0.4 inch long by 0.1 inch wide, with a grooved, 0.4 inch-long stalk. First true leaves are purple tinged along their midrib and grooved along their narrow stalk. The stalk may have thin flaps of blade tissue on its edges. Leaves are club to egg shaped and shallowly toothed; teeth point towards the leaf tip. A few hairs may be present along leaf bases, stalks and midribs. **Mature plants** are many-branched and 5–24 inches in height. Hollow, hairless stems will root at leaf-stem junctions, especially if these junctions contact the ground. Leaves are alternate, 6–10 inches long by 0.5–2 inches wide, hairless and club shaped in outline, with a prominent white midvein. Lower leaves are deeply lobed with forward-pointing lobe tips, stalked, wavy edged and purple tinged on their undersides. Upper leaves are irregularly toothed or with opposite lobes. Leaves lack stalks and instead clasp around the stem with blade tissue. All aboveground portions are somewhat succulent and fleshy. The root system consists of a small, inconspicuous taproot and supplemental fibrous roots. Several **flower heads** are grouped together at the ends of branches. Flower heads are 0.4 inch in diameter and made up of many smaller yellow, petal-less flowers. Two sets of small, narrow, green bracts are present directly beneath the flower head. Several spirals of shorter bracts at the top of the flower stalk are blackened on their upper quarter to third. Longer bracts, running from the flower base to the upper reaches of the yellow flowers, make up the second set of black-tipped bracts. The outer coat of the **seed** is actually a thin dry layer of fruit tissue. Seeds are clustered in seedheads. Seeds are gray or red-brown, 0.1–0.2 inch long and cylindrical, with five to 10 longitudinal ridges; small, flat, overlapping hairs develop between these ridges. An easily removed tuft of white bristles is attached to one end of each seed.

Similar species: Other species in the *Senecio* genus can be distinguished from common groundsel by the presence of petals on their flower heads. Mugwort (*Artemisia vulgaris* L.) and common ragweed (*Ambrosia artemisiifolia* L.) seedlings have similar leaves. Mugwort seedling leaves are more triangular, broad and deeply divided; they also have wooly hairs on their undersides. Common ragweed seedlings are very deeply lobed and more closely resemble a young marigold; their cotyledons are round to slightly oval. Pineapple-weed (*Matricaria discoidea* DC.) also has small, yellow flower heads without petals. However, pineapple weed flowers are somewhat acorn-like, and the leaves are very finely divided and emit a pineapple-like scent.

MANAGEMENT

Common groundsel is often present in field crops but is rarely a problem due to its short stature and low competitive ability relative to most field crops. It is a pest in many vegetable and fruit crops due to its rapid reproduction and ability to thrive in all seasons of the year. Clean up field margins, waste ground, edges of parking areas, etc. to prevent seeds from dispersing into fields. Annual moldboard plowing will help control this species, whereas rotary tillage is less effective. Since the seeds are relatively short lived in the soil and must be close to the surface for seedlings to emerge, few seeds will survive long enough to get an opportunity for emergence if the soil column is mixed annually. The fibrous root system and shallow emergence depth make the species highly sensitive to tine weeding. Because young plants form a rosette of leaves before making a flowering stem, the species has little ability to push up through a dense mat of organic mulch material.

Short season crops will help interrupt the life cycle of this rapidly maturing weed. It is almost certain to set seed in a full season crop. If the species is a problem, use precision cultivation tools as long as possible, and then clean up remaining plants by hoeing. Note that to prevent seed set, the plants must be killed before flowers open. Prompt cleanup of fields after harvest is essential for controlling common groundsel.

ECOLOGY

Origin and distribution: Common groundsel is a native of Eurasia, where it is widespread. It has been introduced into North and South America, Australia, New Zealand and South Africa. In North America, its range extends from coast to coast and from Alaska and the Yukon to Mexico. Although it occurs widely in North America, it is primarily a weed of cool, moist regions.

Seed weight: Population mean seed weights range from 0.16–0.25 mg.

Dormancy and germination: Most common groundsel seeds are capable of germination immediately after falling from the parent plant. One exception is seeds collected in early spring, which were mostly dormant. Buried seeds became dormant in late spring and early summer, apparently induced by high temperatures. Six months of burial in the soil is sufficient to break dormancy. Most seeds require light for germination and thus non-dormant seeds do not germinate until exposed to light during tillage. The optimal temperature for germination varies among populations, but generally ranges from 50–68°F. The optimal temperature is also influenced by storage temperature; it has a narrower range of 50–59°F when seed are stored cold whereas it broadens to 50–77°F when seeds are stored warm. Alternating daily temperatures do not increase germination over the optimum constant temperature.

Seed longevity: Common groundsel seed banks are relatively short lived. Over a five-year period, seeds in uncultivated soil declined by 45% per year. In cultivated soil, no seeds were left after five years. In another study, only 6% of deeply buried seeds survived for two years, and no viable seeds shed on the soil surface and subjected to multiple soil disturbances were left after 40 weeks. However, seed buried in porous clay containers for 19 months did not lose viability.

Season of emergence: Peak emergence occurs from early spring to early summer, but some seedlings continue to emerge until late fall. Emergence is likely to cease, however, during hot weather. Flushes of emergence often follow rainfall events or periods when plants are dispersing seeds.

Emergence depth: Most seedlings emerge from within the top 0.8 inch, but a few emerge from as deep as 2 inches.

Photosynthetic pathway: C_3

Sensitivity to frost: Late emerging common groundsel

plants commonly overwinter in cold regions. The weed can survive periods of frozen soil and can survive down to 22°F, but plants wilt and tend to lose leaves. Thus, snow cover is probably necessary for survival during severe cold periods.

Drought tolerance: Common groundsel thrives best in cool to warm, moist conditions. In hot, dry conditions it tends to quickly flower and die. Deliberately drought stressed plants developed a higher ratio of root weight to shoot weight, and the root systems were more vertical and less spreading. The actual depth of soil reached was not greater in drought stressed plants, however, because the total root system was smaller.

Mycorrhiza: Common groundsel is mycorrhizal.

Response to fertility: Common groundsel shows only a minimal growth response to increasing N application rates. The species has a low response to K and competes well at low K levels. Application of a balanced nutrient solution to plants grown on a sandy loam soil increased seed production by 75%. In another experiment, application of N-P-K about doubled production of seed heads in agricultural fields. Nutrient stressed plants produce smaller seeds that have more sporadic germination. Seedlings from nutrient stressed parent plants show increased tolerance of low nutrient conditions.

Soil physical requirements: Common groundsel grows on a wide range of soil types. Generally, the species does best on loose, moist, fertile agricultural soils, but genetically adapted races tolerate more stressful environments like sandy or rocky beaches and the edges of gravel parking areas. The species is salt tolerant.

Response to shade: Common groundsel can survive and grow even under 93% shade, but its growth rate is much reduced relative to full sunlight.

Sensitivity to disturbance: If plants are cut when flowering, 35% of the seeds will still mature. Flower buds form on short shoots in the axils of basal rosette leaves. Normally, these do not develop but rather die after the main stem flowers. If the shoot is cut or trampled, however, the short basal shoots quickly elongate and flower.

Time from emergence to reproduction: In mild, moist conditions, plants flower four to five weeks after emergence and set seed seven to 11 days later. Overwintering plants require several months, depending on how long freezing and near freezing temperatures last. In regions with mild winters, the species can flower in any month of the year, but even moderately hot conditions such as day/night temperatures of 84/72°F can inhibit seed set. An increase in daylength from eight to 13 hours can decrease time to flowering and seed production. Populations subjected to a heavily weeded garden habitat exhibited a shorter vegetative period before flowering, whereas populations from field margins and coastal areas had a longer vegetative phase.

Pollination: Common groundsel mostly self-pollinates but occasionally is cross pollinated by insects, particularly hover flies.

Reproduction: Typically, common groundsel seed production ranges from 1,100–1,800 seeds per plant but can average as high as 38,000 per plant. Seed production in a pure stand was 8,000–13,000 seeds per plant but was 900–2,200 in competition with wheat. Since the species can complete three generations per year even in regions with long, cold winters like central New York, neglected populations can increase rapidly.

Dispersal: The seeds have a plume that assists with wind dispersal. Seeds also disperse in irrigation water, in contaminated crop and cover crop seed, and on vehicles and farm machinery. The seeds occur in manure.

Common natural enemies: A rust (*Puccinia lagenophorae*) damages leaves and can reduce seed production by half or more. Rust infection also reduced plant recovery from freezing temperatures. Rust only occurred on plants fertilized with P. Infection is more severe in summer and autumn than in spring, probably because of lower inoculum levels in spring.

Palatability: Common groundsel is not consumed as food by people. It contains toxic alkaloids that can damage the liver and lungs. Consumption of 25–50% of bodyweight over several weeks causes poisoning in livestock. Cattle and horses are more susceptible than sheep and goats. Chickens can also be poisoned. The onset of symptoms from poisoning is typically delayed by several weeks. Since the toxic alkaloids are contained in the flowers and more flowers are produced under good moisture and fertility conditions, these conditions can also increase toxicity potential to livestock. This species has value as a food source for birds and invertebrates, and therefore may promote biodiversity if retained at low population levels in agricultural fields.

FURTHER READING

Holm, L., J. Doll, E. Holm, J. Pancho and J. Herberger. 1997. *Senecio vulgaris* L. In *World Weeds. Natural Histories and Distribution.* pp. 740–750. Wiley and Sons: New York, New York.

Müller-Schärer, H. and J. Frantzen. 1996. An emerging system management approach for biological weed control in crops: *Senecio vulgaris* as a research model. *Weed Research* 36: 483–491.

Popay, A.I. and E.H. Roberts. 1970b. Ecology of *Capsella-bursa-pastoris* (L.) Medik. and *Senecio vulgaris* L. in relation to germination behavior. *Journal of Ecology* 58: 123–139.

Robinson, D.E., J.T. O'Donovan, M.P. Sharma, D.J. Doohan and S. Figueroa. 2003. The biology of Canadian weeds. 123. *Senecio vulgaris* L. *Canadian Journal of Plant Science* 83: 629–644.

Common lambsquarters

Chenopodium album L.

Common lambsquarters seedling
Antonio DiTommaso, Cornell University

Common lambsquarters plant
Antonio DiTommaso, Cornell University

Common lambsquarters plant in flower
Scott Morris, Cornell University

IDENTIFICATION

Other common names: white pigweed, white goosefoot, pigweed, fat hen, mealweed, frost blite, frost bite, bacon weed, lamb's quarters, goosefoot, netseed lamb's quarters, wild spinach, green pigweed

Family: Goosefoot family, Chenopodiaceae

Habit: Tall, erect, summer annual herb

Description: Seedlings have thumb- to long ellipse-shaped, 0.16–0.6 inch long by 0.04 inch wide cotyledons. Upper sides of cotyledons are dull green; seedling stems and undersides of cotyledons are maroon-green. The first two leaves are egg shaped and opposite; all subsequent leaves are alternate, although the first few leaves may appear opposite. Young leaves have a silvery or pink, grainy or mealy coating. The diamond-to-egg-shaped young leaves arise from a silvery to light green, mealy stem; leaf edges are either entirely or lightly toothed. Stems of **mature plants** are upright, pyramidal, highly branching, hairless, ridged, maroon-speckled and can reach 3–5 feet tall. Leaves are alternate, 1.2–4 inches long by less than 3.5 inches wide, green, and diamond or egg to triangle shaped, with mealy undersides. Lower leaves are broader, longer and usually toothed and stalked. Upper leaves are sometimes lanceolate, stalkless and toothless. The taproot is short and branched. Individual **flowers** are very small, inconspicuous and silvery to green; they are clumped together in dense clusters located in stem-leaf junctions and at branch tips. **Seeds** are round, black and 0.06 inch wide, or they can be brown and slightly larger with a flatter oval shape. Papery flower tissue closely coats most of the seed surface at dispersal.

Similar species: Halberdleaf orach (*Atriplex patula* L.) is similar to common lambsquarters and has a white mealy bloom on young growth. Halberdleaf orach is shorter and bushier than common lambsquarters, and its leaves are generally more lanceolate, less toothed and often have a pair of narrow lobes near the base of the leaf blade. Common lambsquarters seedlings are sometimes confused with several young pigweed species (*Amaranthus* spp.). Young, true leaves of common lambsquarters seedlings have a grainy or fuzzy, silver to pink bloom when they first emerge, but pigweed seedlings do not. Mature common lambsquarters leaves generally have toothed, sculpted edges while pigweeds have generally oval shaped, untoothed leaves. Other *Chenopodium* spp. do not have mealy young tissues, diamond- to egg-shaped leaves and light colored, succulent looking stems.

MANAGEMENT

Common lambsquarters germinates readily in response to tillage and cultivation. More than most weed species, a one- to two-week lag between initial and final seedbed preparation is effective at flushing out and destroying seedlings. A short fallow in the spring with repeated surface cultivation causes even greater depletion of the surface seed bank. The key is to use progressively shallower tillage, with the final seedbed prepared to a depth of no more than 1.5 inch. The species rarely emerges from deeper than that, and deeper tillage will raise seeds into the near-surface zone favorable for germination. Continue killing seedlings after crop planting by tine weeding with blind cultivation. Once the crop is large enough to tolerate inter-row cultivation, hill up slightly when the lambsquarters seedlings are still in the early seed-leaf stage. Even crops like cabbage and tomatoes that are not normally hilled will tolerate 1 inch of soil against the base, and this is sufficient to bury the tiny seedlings. Continue hilling with subsequent cultivations if the crop will tolerate it.

In spring and fall planted grains, a dense, uniform and vigorous stand is critical for maximizing the crop's initial competitive advantage. Light harrowing just as the lambsquarters seedlings begin to emerge can substantially reduce density of the weed, but any substantial reduction in stand density of the grain is likely to prove counterproductive.

Straw mulch and other mulch materials are highly effective for suppressing this species since its small seeds provide few resources for pushing the seed leaves up out of the mulch mat. Because the seed leaves stay together in a vertical position until they reach the light, however, a few seedlings will usually penetrate at least 2 inches of loose straw, so either use a deeper mulch layer or compact the mulch after application.

Common lambsquarters is highly responsive to N fertility. Avoid excess fertilization, and, in particular, avoid heavy fertilization before the crop is well established. On the other hand, incorporation of a legume cover crop can enhance seed bank decline of this species by triggering fatal germination of buried seeds.

Because common lambsquarters is a prolific producer of long-lived seeds, consistent efforts to limit seed production will greatly assist long-term management. Since most seeds remain on the plant long after they mature, they can be captured or destroyed during combine harvest. Clean up fields promptly after harvest if this weed is present. If possible, remove plants that have flowered, as they can continue to form seeds even after mowing or light tillage that leaves the flowering stalks on the soil surface. Hand rogue at least the larger plants out of intensive vegetable systems. Many of the seeds remain on the plant until early winter, so fall cleanup after harvest can reduce lambsquarters density the following year. In a long-term vegetable crop rotation, seed banks of common lambsquarters tended to decrease with deep plowing to 14–16 inches but increased with rotary tillage or shallow plowing to 6–7 inches. Despite the potential longevity of common lambsquarters seed banks, the species can be virtually eliminated from a field through proper crop rotation coupled with other good management practices.

ECOLOGY

Origin and distribution: Common lambsquarters was introduced from Europe. It is present in all 50 states and in all Canadian provinces and territories except for Nunavut. It has been introduced throughout the world and is widespread from 70° N to 50° S except in extreme deserts.

Seed weight: 0.5–0.7 mg depending on the population and type of seed; the same plant may produce several types of seeds with different size and dormancy characteristics.

Dormancy and germination: Common lambsquarters frequently produces two or more visually distinct types of seeds on the same plant, and these differ in their dormancy properties. Many seeds are usually dormant immediately after falling from the parent plant, but some are not. A period of cold, wet conditions breaks dormancy, and warm weather in early spring promotes subsequent emergence. Seeds produced under short day-length conditions are less dormant than those produced during long days. Common lambsquarters germination is substantially increased by white light but is inhibited by light depleted in red wavelengths, such as light that has passed through a plant leaf canopy. Germination is increased by the presence of nitrate and by large day/night temperature fluctuations. The germination promoting effects of light, nitrate and fluctuating temperature act together such that more seeds germinate with two of these cues than with one, and maximum germination is usually reached when the seeds receive all three cues. All of these cues tend to occur during or shortly after tillage. Seeds from plants grown in high nitrate conditions have higher N concentrations and higher germination rates than seeds from unfertilized soil. Seeds germinate best with daytime temperatures of 64–77°F. The process of germination begins at temperatures as low as 43°F, but relatively few seeds germinate below 55°F or above 91°F. A few weeks of exposure to temperatures over 59°F causes some seeds to enter secondary dormancy, and thus, warm weather in summer induces dormancy in a substantial fraction of seeds.

Seed longevity: Common lambsquarters seeds can remain viable in the soil for many decades. As an extreme example, viable seeds have been recovered from under medieval ruins in Europe. Most seeds, however, do not last so long. In undisturbed soil, mortality rates of 8–35% per year have been observed, with most rates near the lower end of the range. In New York, annual seed mortality was variable, with a more typical average rate of 21% in one experiment but an unusually high rate of 78% in another. In long-term conventional and organic plots in the mid-Atlantic states, annual seed mortality averaged 51%. In annually tilled fields, decline in the common lambsquarters seed bank has varied from 14–42% per year, with most loss rates nearer the higher value. In soil stirred six or more times per year, seed loss was 31–52% per year.

Season of emergence: Common lambsquarters emerges throughout the growing season but with a strong peak in spring. It is categorized as an early emerging species that begins emergence in early spring, but it has one of the longest emergence durations of all weeds studied.

Emergence depth: Optimum depth for emergence is near the soil surface, 0.1–0.2 inch, and few seedlings emerge from deeper than 1.2 inch.

Photosynthetic pathway: C_3

Sensitivity to frost: Common lambsquarters is sensitive to frost.

Drought tolerance: The species is relatively drought tolerant.

Mycorrhiza: The species is non-mycorrhizal.

Response to fertility: Common lambsquarters is a heavy feeder of plant nutrients. It has a strong growth response to increasing N applications up to at least 480 pounds per acre and a moderate response to P up to about 46 pounds per acre of P_2O_5. Plants grew substantially larger with 5,830 pounds per acre of composted chicken manure as compared to 2,920 pounds per acre. The species concentrates N in excess of its needs, and increasing N can favor this weed relative to crops. Common lambsquarters also has a strong growth response to increasing K, and it is strongly competitive when K levels are high. The species tolerates a wide range of soil pH but tends to grow poorly on very acidic soils.

Soil physical requirements: Common lambsquarters is regularly found on all soil textures from sand to clay and peat, but it grows most vigorously on fine textured soils. It emerges best in a moderately rough seedbed. It tolerates some soil compaction and waterlogging but with reduced emergence and growth. The species tolerates salinity.

Response to shade: Common lambsquarters is intolerant of heavy shade, especially shortly after emergence. Plants react to shade by growing taller and allocating a larger proportion of tissue to stem and leaves and a smaller proportion to the inflorescence. Nevertheless, plants will still flower in dense shade, and shading that begins after the plant flowers has no effect on seed production.

Sensitivity to disturbance: Young plants are unable to survive mowing or trampling, and older plants re-grow poorly. They dry out quickly when uprooted.

Time from emergence to reproduction: Plants flower more quickly with short day lengths than with long days. Flowering occurred five to six weeks after emergence in Canada and 81 days after emergence in Wisconsin. Some seeds become viable within two weeks of flowering, and most are viable within three weeks. Early maturation may be triggered by protracted drought. Mature seeds usually remain on the plant for weeks to months before being released.

Pollination: Common lambsquarters is self pollinated or wind pollinated.

Reproduction: The plants set seeds over a relatively short period as they mature. A small plant (about 1 foot) can produce several hundred seeds, and large plants can produce more than 100,000 seeds. Plants growing at low density with cabbage, onions or without a crop produced 30,000–370,000 seeds per plant, plants in soybeans produced 30,000–175,000 seeds per plant, plants in corn produced 8,000–117,000 seeds per plant, and plants in rapeseed produced 175–2,250 seeds per plant. Within each of these crops, most variation was due to differences in growing conditions between years. Plant size at maturity decreases as the date of emergence is delayed, and seed production is closely correlated with total plant weight.

Dispersal: Because seeds of common lambsquarters often reach high densities in soil, they are easily spread between sites by soil clinging to large animals, shoes, tires and machinery used for farm operations and road construction. Seeds have been spread in ship ballasts. The seeds survive well in the digestive tracts of cows, sheep and pigs, and manure is commonly contaminated with common lambsquarters seeds. Some seeds will pass through chickens, sparrows and ducks. Seeds also disperse in streams and irrigation water.

Common natural enemies: In wet weather, damping-off fungi can kill large numbers of lambsquarters seedlings, particularly if they are shaded by a crop or mulch. Leaf miner damage (curved tracks on the leaves) is commonly observed but is rarely severe enough to check the weed's growth.

Palatability: Young lambsquarters are highly palatable and can be used as a salad green or pot herb. The foliage is high in vitamin C, carotenoids and essential minerals. Seeds and dried flower heads can be ground and added to soups and breads. The cleaned seeds can be cooked as a grain or ground for flour. Plants can, however, contain oxalic acid and nitrate and can be toxic to sheep and pigs if large amounts are consumed rapidly.

FURTHER READING

Bassett, I.J. and C.W. Crompton. 1978. The biology of Canadian weeds. 32. *Chenopodium album* L. *Canadian Journal of Plant Science* 58: 1061–1072.

Henson, I.E. 1970. Effects of light, potassium nitrate and temperature on the germination of *Chenopodium album* L. *Weed Research* 10: 27–39.

Mohler, C.L., B.A. Caldwell, C.A. Marschner, S. Cordeau, Q. Maqsood, M.R. Ryan, A. DiTommaso. 2018. Weed seed bank and weed biomass dynamics in a long-term organic vegetable cropping systems experiment. *Weed Science* 66: 611–626.

Williams, J.T. 1963. Ecological Flora of the British Isles. *Chenopodium album* L. *Journal of Ecology* 51: 711–725.

Common milkweed

Asclepias syriaca L.

Common milkweed follicles
Antonio DiTommaso, Cornell University

Common milkweed dehiscent follicles
Scott Morris, Cornell University

Common milkweed stand in flower
Randall Prostak, University of Massachusetts

IDENTIFICATION

Other common names: silkweed, cotton weed, Virginia silk, wild cotton, silky milkweed, common silkweed, showy milkweed, swallow wort

Family: Milkweed family, Asclepiadaceae

Habit: Erect, unbranched, perennial herb spreading by deep thickened storage roots

Description: Seedlings have light green, smooth stems. Cotyledons are 0.25–0.5 inch long, largely untapered, dull green with prominent veins, long stalks and round tips. True leaves are opposite, waxy, pointy tipped and dark green with a prominent, white midvein. Vegetative sprouts arising from underground roots are far more robust and common than seedlings; sprout stems are capped with a folded clump of leaves that unfolds as the stem elongates. **Mature plants** reach 3–5 feet tall on a single hollow, hairy, unbranched stem; the stem is green, turning red with maturity. Leaves are opposite and borne upon 0.4 inch-long stalks. The gently tapered, 3–8 inches-long, elliptical leaves are green and hairless on the top, lighter green and hairy on their undersides. Leaf midveins are prominent and white. Secondary veins do not reach the edge of the leaves. The mature plant has a large underground system of thick, white, horizontal storage roots. In deep, well-drained soil, roots can penetrate to 8 feet. The entire plant exudes a white, milky sap when cut. Globes of fragrant, stalked, purple-pink to white **flowers** grow from upper leaf axils and stem tips. **Seeds** develop in 3–5 inch-long, teardrop-shaped, bumpy, hairy, spiny, gray-green seedpods; pods tend to grow in pairs. Upon seed maturity, pods split open, shedding up to 200 seeds each, and turn grayish brown. Pod interiors are glossy yellow. Seeds are large, 0.25–0.5 inch, brown, oval with one flat end; the center of the seed is raised and surrounded by a thin, papery margin. Long, silky hairs are attached to the flattened end.

Similar species: Hemp dogbane (*Apocynum cannabinum* L.) differs from common milkweed by its green-white flowers, branching habit in its upper third, smaller leaves measuring 2–4.5 inches long, and long, narrow, curved seedpods.

MANAGEMENT

Common milkweed does well in grain and early planted corn because the shoots emerge after tillage and planting. In contrast, soil preparation for summer planted crops eliminates the first flush of shoots and forces the plant to use additional root reserves to regenerate shoots. Shoots can emerge from roots well below the plow layer, so a single deep moldboard plowing will not control this weed. Consequently, repeated shallow cultivations that cut shoots before they can replenish the roots are required. For heavy infestations, a tilled fallow period from late spring through summer will help get the weed under control. Carbohydrate storage in the roots reaches a minimum in July to September, so mid-summer fallow may be most effective at exhausting the roots. Buds on the roots become dormant in late summer, however, so fallowing at this time is not effective.

Rotation with alfalfa helps reduce vigor of common milkweed populations due to frequent mowing. Hay crops that are only mowed once per year do not help control this weed. Seedling establishment of common milkweed was highest in spring wheat, intermediate in soybeans and lowest in corn, reflecting the duration of competition throughout the season. Generally, any crop that establishes a leaf canopy before common milkweed emergence and maintains a competitive canopy through the season will greatly suppress this weed.

Common milkweed also serves several beneficial functions within agroecosystems. It hosts aphids that provide a food source for parasitic wasps, which, in turn, attack and control the European corn borer (*Ostrinia nubilalis*). It also contains compounds that support reproduction and survival of the monarch butterfly (*Danaus plexippus*) during their migration to overwintering sites in Mexico. Consequently, maintenance of low milkweed densities in crop fields may contribute to a balanced landscape management strategy that realizes the ecological benefits of common milkweed while avoiding its agricultural liabilities.

ECOLOGY

Origin and distribution: Common milkweed is native to eastern North America and occurs from Georgia to Oklahoma, northward to southern Canada. It has been introduced into Europe and Japan, and the Willamette Valley of Oregon.

Seed weight: Mean seed weights for various populations range from 3.5–7.4 mg.

Dormancy and germination: Seeds are dormant when shed from the mother plant and require one to nine weeks of cold (41–48°F), wet conditions or about one year of dry storage to break dormancy. The longer the exposure to cold conditions, the lower the temperature required for optimum germination. This ensures that seeds will not germinate when shed in the fall but will be ready for germination the spring following dispersal. Seeds germinate at constant temperatures from 59–95°F, but germination is best at fluctuating day/night temperatures from 68/50°F to 95/68°F. Seed germination percentage increases in the presence of nitrate. Light has little influence on germination of cold stratified seeds. Optimum pH for germination is 4–8.

Seed longevity: Seeds survive in the soil for up to nine years, but most seeds will probably germinate or die within the first three years following production.

Season of emergence: Seedlings emerge primarily in the spring. Optimum temperature for emergence was 81°F with minimal emergence at temperatures below 59°F. Shoots from overwintering rootstocks emerge from mid-spring throughout the summer, but peak emergence occurs in late spring and early summer.

Emergence depth: Seeds germinate poorly on the soil surface, but seedlings emerge well from 0.2–1.6 inch. A few seedlings can emerge from 2.4–2.8 inches. New shoots can emerge from rootstocks buried as deep as 3.5 feet in the soil, but most emerge from roots in the top 12 inches.

Photosynthetic pathway: C_3

Sensitivity to frost: Common milkweed seedlings and shoots normally do not emerge until after the last frost, and the shoots usually begin to die back prior to the first frost in the fall, so frost has little consequence for this species. This species was entirely killed when present at the first frost in the fall.

Drought tolerance: Full grown plants are drought tolerant because of their deep taproot, but seedlings are more drought sensitive than other weed species.

Mycorrhiza: Common milkweed had a low mycorrhizal status in prairie settings. Low levels of infection by one mycorrhizal species increased milkweed biomass, but infection by another species decreased biomass.

Response to fertility: Common milkweed grows well on soils with relatively low nitrogen, but plant size responds to balanced fertilization with N, P and K. More biomass is allocated to roots than to shoots under low fertility conditions than under high fertility conditions. Reproductive pod and seed output is also increased by a balanced fertilizer. It tolerates pH from 4–10 but does best on slightly alkaline soils.

Soil physical requirements: The species does not tolerate wet soils. It thrives on soils with a wide range of textures provided drainage is adequate. It appears to tolerate compaction given its common occurrence on headlands and roadsides. It tolerates salt and can grow at substantially higher salt concentrations than sorghum. It is absent from boron deficient soils.

Response to shade: Common milkweed grows best in 30–100% of full sunlight. Mid-summer shading of 75% had little effect on pod and seed production. Shade from a dense annual weed leaf canopy in the absence of soil competition substantially reduced reproduction of common milkweed roots.

Sensitivity to disturbance: The roots of mature plants can penetrate 7–12 feet into the soil, and root fragments from as deep as 4–5 feet are capable of resprouting. This regeneration capacity protects them from damage during tillage and cultivation. Seedling roots develop buds about three weeks after emergence that are capable of regenerating a new shoot if the original is destroyed. Plants in the three-leaf stage are capable of 65% resprouting when clipped at ground level, and plants in the four-leaf stage or greater are capable of 100% resprouting with multiple shoots. Thus, only occasional mowing or tillage will not prevent the development of a large colony.

Time from emergence to reproduction: Seedlings do not flower the first year. Shoots from older plants flower roughly six to eight weeks after emergence. Seeds mature about six weeks after flowering. In Missouri, plants reached peak pod production in late July/ early August, and pods dried and opened by the end of September.

Pollination: Common milkweed is self-sterile and cross pollinated by insects, mostly bees and wasps. Bumblebees are primarily responsible for pollination by day, but moths pollinate at night. The majority of pollinated flowers abort, with very few producing mature seedpods.

Reproduction: Stalks typically produce four to six pods, each containing 100–425 seeds. The number of pods depends on the level of pollination. Normally, vegetative reproduction does not begin until the second or third year of life. An undisturbed plant in southern Ontario produced a colony of 56 shoots during its fourth growing season.

Dispersal: Seeds are dispersed primarily by wind. Each seed has a tuft of fine, silky hairs that provide buoyancy in the air. Seeds often cling to the pods for several months after the pods open, which prolongs the period of dispersal into late winter. Root growth can spread the plant up to 10 feet per year. Root fragments are dispersed within fields and occasionally between fields on tillage implements. Root segments as small as 1 inch can produce shoots.

Common natural enemies: Common milkweed flowers can be destroyed by the mid-summer herbivores *Tetraopes tetrophthalmus* and *Diabrotica cristata*, while pods are destroyed in late summer by *T. tetrophthalmus*, *Rhysseratus lineaticollis* and *Oncopeltus fasciatus*. It hosts several specialized insects including caterpillars of the monarch butterfly (*Danaus plexippus*), the milkweed longhorned beetle (*Tetraopes teraophtalmus*) and the small milkweed bug (*Lygaeus kalmii*), but they usually do little to control the weed. Virus diseases cause yellowing or mottling, clumping of stems, and deformed stems and leaves.

Palatability: Milky sap contains a poison that causes nausea and potential heart damage. Nevertheless, young shoots, buds and flowers can be used as a pot herb provided the water is changed repeatedly during cooking to remove the bitter sap. Common milkweed is poisonous to livestock when 1–2% of body weight is consumed, but toxicity decreases as plants mature. Dried milkweed in hay remains toxic.

Note: Common milkweed appears to contain

allelopathic compounds that inhibit the growth of selected crops and weeds.

FURTHER READING

Bhowmik, P.D. 1994. Biology and control of common milkweed (*Asclepias syriaca*). *Reviews of Weed Science* 6: 227–250.

DiTommaso, A., K.M. Averill, M.P. Hoffman, J.R. Fuchsberg, and J.E. Losey. 2016. Integrating insect, resistance, and floral resource management in weed control decision-making. *Weed Science* 64: 743–756.

Evetts, L.L. and O.C. Burnside. 1975. Effect of competition on growth of common milkweed. *Weed Science* 23: 1–3.

Common purslane

Portulaca oleracea L.

Common purslane seedling
Scott Morris, Cornell University

Common purslane flowers
Antonio DiTommaso, Cornell University

Stand of common purslane
Antonio DiTommaso, Cornell University

IDENTIFICATION

Other common names: pusley, purslane, pursley, wild portulaca, low pigweed, common portulaca, wild portulac, little hooweed

Family: Purslane family, Portulacaceae

Habit: Succulent, prostrate, taprooted summer annual herb

Description: Young **seedling** stems begin upright, reaching 0.5 inch in height, then become prostrate. Stems and cotyledons are maroon tinged and fleshy. Cotyledons are 0.08–0.4 inch long by 0.04–0.08 inch wide and oblong. First true leaves are 0.5–1 inch long, fleshy, smooth and green with maroon hued undersides. The first six to 10 true leaves are opposite. **Mature plants** are succulent, crawling and grow prostrate in 2–3 foot diameter circles. Stems are highly branched, round, fleshy and turn maroon with age. Leaves are club shaped, round-tipped, fleshy, green, hairless and smooth. Leaves may be alternate or opposite and sometimes have red margins. The root system consists of a thick taproot with secondary fibrous roots. Yellow, five-petaled **flowers**, 0.13–0.4 inch in diameter, are present individually in leaf axils and branch junctions, and in clusters on terminal branch ends. Flowers open only in sunny weather. **Seeds** are black, flat, round to kidney shaped and small (less than 0.04 inch in diameter), with a bumpy surface. Seeds develop in 0.25 inch diameter globe-shaped capsules that split horizontally to release the seeds upon maturity.

Similar species: Horse purslane (*Trianthema portulacastrum* L.) is also a prostrate succulent, but it has stalked leaves and pinkish purple flowers. Common purslane is sometimes confused with prostrate pigweed (*Amaranthus blitoides* S. Watson), prostrate knotweed (*Polygonum aviculare* L.) and various spurges (*Euphorbia* spp.). Prostrate pigweed has non-fleshy leaves, which distinguish it from common purslane. Prostrate knotweed can be distinguished by the presence of papery appendages (ocreae) wrapping the stem above each leaf. Spurges release a milky, white sap when cut.

MANAGEMENT

Because newly emerged common purslane seedlings are very small and fragile, stirring the top 1–2 inches of soil with tillage implements two to four times after the soil warms is highly effective at removing most of the individuals that will emerge during the season.

Cultivation of large plants is often ineffective because the plants reroot or set seed without rerooting. Thus, cultivation of plants beyond the cotyledon stage should aim at burial rather than breakage or uprooting. Although stem fragments can develop new root systems, studies have shown that this requires several weeks and that the resulting individuals are less vigorous and produce fewer seeds than new seedlings that germinate in response to the cultivation. Flame weeding kills seedlings, but large plants recover unless unusually long exposure times are used.

Organic mulch materials are highly effective for suppressing common purslane. Because the species is susceptible to rotting under continuously shady, humid conditions, and because of its prostrate growth form, less mulch is required to control common purslane than for most other annual weeds. Due to its shade intolerance and prostrate growth form, dense planting helps control this weed in crops that will tolerate high density and that grow quickly. Rapid coverage of the ground is critical, however, because even small, dying plants can set seed.

Common purslane requires high nutrient levels for rapid growth. In particular, it is more responsive to P than some vegetable crops like lettuce. Continuous reliance on manure derived compost as an N source for vegetable production leads to P accumulation, and this appears to foster problems with common purslane.

Because the seeds are highly persistent in the soil, removal of plants that escape control before they set seed is useful for long-term control. Since flowers can self-pollinate without opening, plants can set seed while still appearing to be immature.

ECOLOGY

Origin and distribution: Common purslane probably originated in South America and spread into North America with Native American agriculture. It may, however, have originated in North Africa and spread to eastern North America with European colonization. It currently occurs throughout the United States and southern Canada, and has a worldwide distribution, except at latitudes above 60° N.

Seed weight: 0.08–0.15 mg.

Dormancy and germination: Some seeds are capable of germination immediately after falling from the capsule, although dormancy develops as seeds mature. Light is usually required for germination of freshly shed seeds, but germination in the dark becomes more likely as the seeds age. Initially, the species germinates best when soil temperatures exceed 86°F. This temperature requirement lowers to 68°F, however, as spring progresses. Germination of aged seeds is better at alternating than at constant temperatures. Light filtered through a leaf canopy inhibits germination. Nitrate promotes germination of seeds in the dark. Germination is not affected across a pH range of 5–9.

Seed longevity: Common purslane seeds have germinated after 30–40 years of burial. In several annually tilled soils in New Zealand seeds declined by 29% per year. On three soil types in a tilled small grain-fallow rotation in Saskatchewan, the common purslane seed bank declined by 17–27% per year. Seed decline in a Mississippi soil was approximately 60% per year in untilled soil and 76–87% per year in tilled soil. Seeds of common purslane remained viable after soil was solarized with a daily maximum of 149°Fm and 30% survived continuous cooking of moist soil at 140°F.

Season of emergence: Most seedlings begin emerging in late spring when maximum air temperature exceeds 86°F and rain is adequate, and they continue to emerge during the heat of the summer.

Emergence depth: Common purslane emerges best when the seeds are at the soil surface, provided the soil is moist. Emergence declines rapidly for seeds placed deeper, and seedlings do not emerge from seeds deeper than 1 inch in the soil.

Photosynthetic pathway: C_4

Sensitivity to frost: Common purslane is very sensitive to frost and sometimes even dies from chilling prior to the first frost.

Drought tolerance: This species is extremely drought tolerant due partially to water storage in the succulent leaves and stems. It also has small stomates and a waxy outer layer on the leaves, both of which reduce water loss from the leaves. Under extreme drought conditions, the plants shed their leaves and grow new ones when water again becomes available.

Mycorrhiza: Common purslane is not mycorrhizal.

Response to fertility: Common purslane is most problematic on highly fertile soils. The species shows a linear increase in growth up to N levels of 133 pounds per acre. It responds to N, P and K but responds most to P. The species shows a strong response to available phosphorus, and competition for phosphorus is the apparent cause for competitive suppression of lettuce by this species. The species is found on soils with a pH of 5.6–7.8.

Soil physical requirements: Common purslane grows best on sandy soils but tolerates a wide range of soil textures on well-drained soils. It establishes best on finely prepared seedbeds. It is relatively tolerant of saline soils, partially because of its capacity to produce antioxidant compounds.

Response to shade: Common purslane grows slowly in shade, and since the plant has a prostrate growth form, it is a poor competitor for light. Under moist shady conditions, the species is prone to fungal diseases. Even dying plants can set seed, however.

Sensitivity to disturbance: Common purslane germinates in response to environmental cues that are associated with tillage (light, warm soil, nitrate). Soil disturbance that commonly occurs in vegetable fields during summer results in a flush of new seedlings. Because it is succulent, common purslane does not die easily following uprooting. It may then reroot or continue to mature and set seeds even without contact with the soil. The plant fragments easily, and the pieces can develop new root systems from stem fragments, but only stem fragments with nodes produce new leaves, and the presence of leaves on the fragment improves survival and new leaf growth.

Time from emergence to reproduction: Plants flower four to eight weeks after emergence and set seeds seven to 16 days later. Plants emerging in mid-summer can flower within two to three weeks of emergence. Consequently, the species can complete two or more generations per year. Seeds mature on main branches first and subsequently on secondary branches, giving a sequential production of seeds from individual plants throughout the season.

Pollination: The species is primarily self pollinated. Flowers open in the morning on bright, hot days, and if conditions are not suitable for opening, the flowers can successfully self-pollinate and turn to mature capsules without opening.

Reproduction: Flower buds form in the axils of leaves. The capsules look very much like the flower buds. Consequently, plants commonly appear to go from a pre-reproductive state to seed shed overnight. One foot diameter plants produce about 7,000 seeds, but a large plant may produce over 100,000 seeds. Undisturbed common purslane does not spread vegetatively, but plant fragments created during cultivation can reroot.

Dispersal: The seeds have no apparent adaptations that aid dispersal, and they fall directly onto the ground around the capsule. The principal mode of dispersal is probably in soil clinging to shoes, tires and machinery. Seeds have been recovered from cattle manure, deer feces and bird feces, and occasional dispersal to new sites by animals may occur.

Common natural enemies: A sawfly, *Sofus pilicornis*, mines the leaves. A white rust, *Albugo portulacae*, commonly attacks common purslane and can destroy plants under favorable conditions.

Palatability: Common purslane is edible as a salad vegetable or pot herb. The digestibility, protein content and mineral content is higher than that of most other weed species. The plant is particularly rich in omega-3 fatty acids, vitamin E and vitamin C. An upright growing cultivar has been developed. However, livestock fed large amounts of common purslane can develop nitrate and oxalate poisoning.

FURTHER READING

Mitich, L.W. 1997. Common purslane (*Portulaca oleracea*). *Weed Technology* 11: 394–397.

Miyanishi, K. and P.B. Cavers. 1980. The biology of Canadian weeds. 40. *Portulaca oleracea* L. *Canadian Journal of Plant Science* 60: 953–963.

Proctor, C.A., R.E. Gaussoin and Z.J. Reicher. 2011. Vegetative reproduction potential of common purslane (*Portulaca oleracea*). *Weed Technology* 25: 694–697.

Common ragweed

Ambrosia artemisiifolia L.

Common ragweed seedling
Antonio DiTommaso, Cornell University

Common ragweed young plant
Antonio DiTommaso, Cornell University

IDENTIFICATION

Other common names: Roman wormwood, hog weed, bitter weed, wild tansy, mayweed, hay fever weed, blackweed, Roman wormweed, annual ragweed, short ragweed, small ragweed

Family: Aster family, Asteraceae

Habit: Erect, branched, summer annual herb

Description: Seedling stems are purple-spotted to purple. Cotyledons are dark green, paddle shaped, thick and 0.2–0.4 inch long. Purple flecking is possible along the edges. Young leaves are green, deeply divided into lobes with slightly pointed to rounded tips; leaves are hairy above and densely hairy below. The first four pairs of true leaves are opposite, but later leaves are alternate. **Mature plants** are 1–3 feet tall, erect, branched and are with a shrub-like appearance. Stems are densely covered with rough, 0.1 inch hairs. Leaves are 2–4 inches long, with or without hairs on all surfaces. Leaves are fern like, deeply divided into many lobes or composed of many smaller, strongly divided and lobed leaflets. Leaf stalks are longer on the lower part of the plant. Roots are fibrous and shallow, often close to the soil surface. **Flowers** are green, 0.1 inch wide and arranged in clusters of male and female flowers on each plant. Male flowers are produced in spikes at branch tips of upper branches, whereas female flowers occur singly or in clusters in lower leaf axils. Plants can flower even when 2 inches tall if cut or mowed. **Fruits** are top shaped, woody, brown with spots or streaks, 0.14 inch long by 0.1 inch wide. Ridges on the side extend over the top, resulting in a crown-like appearance. A longer beak rises out from the center of the crown. **Seeds** are brown, 0.1 inch long by 0.08 inch wide and disperse in the hard fruits.

Similar species: Seedlings of giant ragweed (*Ambrosia trifida* L.) and corn chamomile (*Anthemis arvensis* L.) are similar to common ragweed. Giant ragweed cotyledons are larger, measuring 0.75–1.75 inches long by 0.5 inches wide, and young leaves are either unlobed or, more commonly, are divided into three lobes. The first leaves of the chamomile species are more finely lobed or divided, have longer stalks and lack dense hairs on the underside. Western ragweed (*Ambrosia psilostachya* DC.) is similar to common ragweed, however western ragweed is a perennial with a taproot. The leaves of western ragweed are lanceolate and less finely divided than those of common ragweed. Mugwort (*Artemisia vulgaris* L.) has similar foliage to that of common ragweed but has silvery-white hairs on the undersides of the leaves. Also, mugwort is a perennial with branching rhizomes.

MANAGEMENT

Common ragweed seeds germinate primarily in the spring. If the seeds are near the soil surface, delaying tillage as long as possible allows more to germinate before planting and decreases density in the crop. Similarly, if the seeds are mixed through the soil profile, a lag between tillage and seedbed preparation allows many to be eliminated before planting. Thus, rotations that include late planted crops tend to decrease the species, provided weed control in the crop is adequate. Since common ragweed does not set seed until late in the summer, early harvested crops like winter grains or short cycle, spring planted vegetables provide an opportunity to prevent seed production and break the life cycle. Because the seeds are highly persistent in the soil, however, you may need several years of good management before the population is well controlled.

Tine weeders and other in-row cultivators will kill many seedlings before they emerge but fewer after they emerge because the seedlings grow quickly. After emergence, aim to bury the fast growing seedlings before they begin to elongate. In a year when prolonged rains prevented tine weeding of soybeans until four weeks after planting, we controlled a dense stand of common ragweed by aggressive tine weeding followed immediately by inter-row cultivation that threw about 1 inch of soil into the row. At that time, the second leaves with three leaflets were starting to emerge on the soybeans.

Straw mulch is less effective for suppressing common ragweed than for other annual weeds but can still be useful if applied in especially heavy layers or if a paper barrier is used under the straw. An early-planted winter annual cover crop that establishes a complete leaf canopy by early spring will suppress ragweed establishment, whereas a late-planted cover crop with an incomplete canopy will permit establishment of a competitive ragweed population in no-till planted crops. Also, interseeded red clover in wheat can moderately suppress common ragweed. We have found that mowing the clover at a height of about 5 inches after common ragweed has flowered but before seeds form effectively prevents seed set because the clover rebounds from mowing more rapidly than the ragweed.

Because the seeds are so persistent in the soil, prompt cleanup of fields after harvest is recommended. Rogue large individuals out of vegetable crops. Capture the seeds when combine harvesting soybeans. These can be ground and fed to livestock. If seeds are produced, leave them on the soil surface until spring to maximize seed predation.

ECOLOGY

Origin and distribution: Common ragweed is native to central and eastern United States and southern Canada, but in forested regions it was probably rare and restricted to disturbed, open areas prior to the advent of Native American agriculture. The species has been introduced to locations in Europe, Asia and South America.

Seed weight: Population mean seed weights with the fruit coat ranged from 1.7–4.6 mg, with most means greater than 4.4 mg. One population had a mean without the fruit coat of 2.4 mg. Seeds harvested in France ranged from 1.2–7.7 mg and had similar germination rates regardless of seed weight.

Dormancy and germination: Common ragweed seeds are dormant when shed in fall. The seeds require several weeks of chilling at 41°F for germination. Light and alternating temperatures in the range of 50–86°F increase germination of seeds after chilling. Optimum alternating day/night temperatures required for germination were 77/68°F to 95/86°F. Seeds that do not germinate under dark, low-temperature, early spring conditions (as would occur when buried in soil) and are subsequently exposed to increasingly warm late spring soil temperatures become dormant and again require a chilling period for germination. Seeds chilled in the absence of oxygen remain dormant, suggesting that seeds buried at soil depths that restrict oxygen would remain dormant despite cold winter temperatures. A range of concentrations of N-P-K nutrients had no effect on germination. The presence of established plants can reduce germination and emergence of common ragweed, presumably through impact on the temperature and light environment of seeds.

Seed longevity: Seeds can persist for up to 40 years in the soil, but most die earlier. Viability was 85% after 20 years of burial. In shorter experiments covering two to three years, the annual seed mortality rate was 7% and 12%.

Season of emergence: Emergence occurs mainly in spring with declining emergence rates in early summer. If a seedbed is prepared in early spring, most emergence will have occurred by late spring. It is typically one of the earliest emerging weeds and has a relatively short duration of emergence.

Emergence depth: Seedlings emerge best from seeds at or very near the soil surface. Nearly all seedlings emerge from the top inch of soil, though a few can establish from larger seeds placed as deeply as 3.1 inches, and no seedlings can emerge from 4 inches.

Photosynthetic pathway: C_3

Sensitivity to frost: Common ragweed is killed by frost.

Drought tolerance: The species is drought tolerant. Extreme drought limits above and belowground growth, but the ratio of roots to shoots is not affected. Photosynthesis declines with high water stress, but remarkably rapid recovery is observed when water stress is relieved. Despite tolerance to short-term drought conditions, common ragweed generally thrives best in environments with high annual rainfall, and specifically with high October rainfall that presumably favors seed development and maturation.

Mycorrhiza: Common ragweed is considered a strong mycorrhizal host.

Response to fertility: Common ragweed is highly tolerant of infertile soils, but it also responds to fertility. On infertile sites, plants less than 4 inches tall may mature, whereas on highly fertile sites, plants may reach 6 feet. The species is known to be a strong accumulator of N, P and K and many micronutrients, particularly zinc.

Soil physical requirements: Common ragweed tolerates a wide range of soil textures and drainage conditions. It is highly tolerant of extreme soil compaction. Common ragweed tolerates moderate soil salinity by decreasing allocation to root growth relative to shoot growth. Salt tolerant populations occur on roadsides that receive deicing salt during the winter.

Response to shade: Common ragweed tolerates moderate shade, but photosynthesis progressively declines and growth is stunted as shade increases. Light reduction of 62% only reduced its growth rate under cool conditions (63/45°F) but not under warm conditions of 73/55°F to 84/66°F. It is one of the few annual weeds that regularly mature (always at a small size) in species-diverse hay meadows and goldenrod dominated fields.

Sensitivity to disturbance: Seedlings are fragile and easily broken, and they quickly dry when uprooted. However, because common ragweed rapidly produces a long taproot, it is more difficult to kill in the seedling stage than are many other small seeded annuals. Medium sized to large plants will reroot readily in moist soil. Common ragweed has greater capacity for regrowth following repeated cuttings than other upright annual broadleaf weeds. Repeated mowing of the plants at 8, 16 or 32 inches causes some mortality and reduces size at the end of the season but does not prevent reproduction. One mowing in mid-summer may reduce biomass substantially, but plants generally have a high capacity for regrowth by lateral stems that can produce viable seeds. No-till management of cover crops with a roller-crimper does not kill established common ragweed plants.

Time from emergence to reproduction: Common ragweed flowers two to five months after emergence, generally from August to October in response to decreasing day length. Flowering is initiated when day length decreases to 14 hours per day or less. Genotypes from northerly locations flower earlier than those from southerly locations.

Pollination: Common ragweed has male and female flowers on the same plant and can be either self- or cross pollinated by wind.

Reproduction: Common ragweed seeds mature as the plant senesces. Typical sized plants produce 3,000–60,000 seeds, but small, highly stressed individuals may produce as few as one or two seeds. Plants in Quebec produced up to 18,000 seeds without competition and 3,000 seeds when growing with crops.

Dispersal: Common ragweed seeds are each tightly encased in a woody fruit that is the dispersal unit. These have no adaptations for dispersal. However, because they often reach high densities in soil, they are easily spread from one site to another in soil clinging to boots, tires, tillage equipment, etc. Common ragweed is one of the most prevalent species along field margins; from there it can migrate into fields. Seeds seem to be heat tolerant and may be spread in compost. The seeds survive well in the digestive tracts of cows, sheep, horses, etc., and manure is commonly contaminated with common ragweed seeds. Horses are particularly fond of common ragweed, and horse manure is especially likely to be contaminated. Birds also sometimes disperse common ragweed. Long distance dispersal can take place in contaminated crop seed, birdseed or hay. Hay transported from North America to feed horses during World War I is speculated to have contributed to the spread of this species in France.

Common natural enemies: Two native chrysomelid beetles, *Ophraella communa* and *Zygogramma suturalis*, feed on the foliage. *Ophraella communa* is most adapted for use as a biological control agent in subtropical climates. White rust (*Albugo tragopogi*) was isolated from local populations in Quebec and severely reduced growth, pollination and reproduction of 14% of common ragweed plants tested.

Palatability: The seeds are edible but so small and difficult to extract from the hard fruit coat that their use as a grain is impractical. Palatability to sheep was lower than that of many annual weeds despite having similar protein and nutrient levels.

Note: The pollen is extremely allergenic and causes hay fever symptoms in 10% of people in the United States. Common ragweed plants can produce over a billion pollen grains per plant when growing with crops. Some people also develop allergic skin reactions on contact with the foliage and pollen.

FURTHER READING

Bassett, I.J. and C.W. Crompton. 1975. The biology of Canadian weeds. 11. *Ambrosia artemisiifolia* L. and *A. psilostachya* DC. *Canadian Journal of Plant Science* 55: 463–476.

Bazzaz, F.A. 1974. Ecophysiology of *Ambrosia artemisiifolia*: A successional dominant. *Ecology* 55: 112–119.

Buttenschøn, R.M., S. Waldispühl and C. Bohren. 2010. Guidelines for management of common ragweed, *Ambrosia artemisiifolia*. http://euphresco.org.

Mutch, D.R., T.E. Martin and K.R. Kosola. 2003. Red clover (*Trifolium pratense*) suppression of common ragweed (*Ambrosia artemisiifolia*) in winter wheat (*Triticum aestivum*). *Weed Technology* 17: 181–185.

Common sunflower

Helianthus annuus L.

Common sunflower seedlings
Anita Dille, Kansas State University

Common sunflower vegetative plant
Anita Dille, Kansas State University

Common sunflower plants in flower
Anita Dille, Kansas State University

IDENTIFICATION

Other common names: sunflower, wild sunflower

Family: Aster family, Asteraceae

Habit: Tall, upright, branched summer annual herb

Taxonomic note: The species *Helianthus annuus* has been classified into a cultivated subspecies (ssp. *macrocarpus*), a wild subspecies (ssp. *lenticularis*) and a weedy subspecies (ssp. *annuus*). Since weedy populations are often more genetically related to nearby wild populations than to each other, the validity of the weedy forms as a taxonomic entity is in doubt.

Description: Seedling stems below the cotyledons are green to purplish. Cotyledons are oblong, hairless, 0.5–1.5 inch long by 0.25–0.5 inch wide and fused at the base. The first one to three pairs of true leaves are opposite, and all subsequent leaves are alternate. Early leaves are dull green above, light green below, toothed on edges and coated with rough, stiff hairs on both surfaces. Leaves are oval to lanceolate with a tapered and rounded tip. **Mature plants** are typically 2–10 feet tall. The stems are erect, branch occasionally towards the top and are densely covered in coarse, spreading white hairs. The leaves are 4–12 inches long, egg shaped to triangular or heart shaped, stalked, toothed and conspicuously three-veined. Both surfaces of the leaves are covered with stiff white hairs. Upper leaves have shorter stalks than lower leaves and may be lanceolate. The root system is a taproot with branching and spreading fibrous roots. One to 12 **flower heads** occur at the end of stems and branches. The flower heads are long stalked, 3–15 inches wide and consist of 20–40 yellow petal-like ray flowers surrounding numerous red- to purple-brown disk flowers. The yellow ray flowers are 0.6–1.6 inches long, whereas the disk flowers are 0.2–0.3 inch long. Beneath the flower head are two to three rows of overlapping, hairy, green, tapered oval bracts. **Seeds** are encased in a hard, dry fruit known as an achene. These units (hereafter called seeds) are 0.13–1 inch long by 0.1–0.6 inch wide, oval and flattened, and are tipped with two to four bracts that detach at maturity. The seeds are white, gray, brown or black, and are often mottled or streaked.

Similar species: Several species of *Helianthus* are similar to common sunflower. Prairie sunflower (*H. petiolaris* Nutt.) is an annual species with smaller, 1–2 inch-wide flowers and narrower, 2–6 inch-long by 0.5–3 inch-wide triangular to lanceolate leaves. The rhizomatous perennial Texas blueweed (*H. ciliaris* DC.) is shorter than common sunflower, 28 inches tall and has lanceolate, blue-green leaves. Jerusalem artichoke (*H. tuberosus* L.) is a rhizomatous perennial that produces knobby, irregular tubers at the rhizome tips. Jerusalem artichoke leaves are narrower, 1.6–4.7 inches wide, than common sunflower, and the flower heads are smaller, 2.5–3.5 inches across.

MANAGEMENT

Sunflower is a very early emerging spring species, so under a winter wheat crop, for example, few seedlings will successfully emerge or be a problem. The dense canopy of winter wheat and spring canola during the period when common sunflower usually emerges suppresses emergence. Also, winter grains are harvested before most common sunflower seeds are mature, so these crops will tend to reduce the population. On the other hand, in no-till summer crops, it will be up before corn, soybeans or grain sorghum are planted, so it needs to be managed at that time, or it will become a serious problem in those crops.

For bad infestations, capture the seeds in the combine and separate them from the crop afterward. Common sunflower seed has a high oil content, and captured seed could be a potential fuel source. Alternatively, the seeds can be ground to destroy viablility and fed to livestock. Grazing cattle on infested grain stubble can reduce seed production. Avoiding fall tillage is critical for managing common sunflower since incorporating seeds in soil protects them from seed predators. The relatively large, oily seeds are highly favored by quail and small mammals, and these animals frequently destroy most surface seeds over the winter. This greatly reduces the density of seedlings emerging the next spring. Although seeds can potentially survive for many years in the soil, few actually do so. Consequently, using the methods above to greatly reduce input to the seed bank for even a single year can bring a severe infestation down to manageable levels. Hand roguing escaped sunflowers is effective for preventing an increase of low level infestations in row crops.

The seedlings are large, fast growing and can emerge from deep in the soil. Consequently, rotary hoes are relatively ineffective against this weed. Tine weeding should be aimed at breaking or burying the seedlings. If common sunflower plants between rows get large, avoid using shovels on flexible S-shanks.

These will tend to walk around the strong taproots and leave the plants in place. Instead, use sharp sweeps fixed at a shallow angle, and run about 1.5–2 inches deep to sever the shoot from the root. New shoots cannot sprout from root tissue alone. The large nutrient storage in the seeds allows seedlings to penetrate even thick layers of mulch.

Common sunflower hybridizes freely with the domesticated sunflower, allowing potential introduction of genes for disease and insect resistance from cultivars. This may increase survival, reproduction and competitive ability of the weed.

ECOLOGY

Origin and distribution: Common sunflower is native to western North America. It now occurs throughout the United States, including Hawaii and Alaska, and in much of Canada and Mexico. It has been introduced into the Caribbean, South America the Middle East and parts of southern Asia.

Seed weight: Mean seed weights for North American populations range from 6.5–8.7 mg, although seeds produced late in the season (October) can average as little as 4.3 mg.

Dormancy and germination: Viability of common sunflower seeds is sometimes as low as 72% but can be as high as 94–99%. Seeds are dormant when mature. After a few months of dry after-ripening or overwinter burial in cold, moist soil, most seeds become capable of germination, but 26–42% may still remain dormant. Shallow burial of the seeds over the winter greatly increases germination the following spring. Some seeds will germinate at 39°F, but germination is quicker and more complete at 68–77°F. Light is not required for germination. Germination is not reduced by pH as low as 4 or as high as 10, and some seeds will germinate in water containing 10,000 parts per million salt.

Seed longevity: A few seeds survived at least 17 years when buried at 8 inches and left undisturbed. Soil burial in the autumn after seed dispersal greatly facilitates formation of a persistent seed bank. Seed mortality rates vary greatly from year to year. Average mortality rates for seeds buried for two to four years range from 26–47% per year. Mortality of seeds in the soil is low through the first winter, but most seeds on the soil surface do not survive the winter, and very few seeds persist on the soil surface for more than two years.

Season of emergence: In the northern Great Plains, seedlings emerge from late April to early June, with most emerging in May. In Kansas they emerge from late March through early May. Seeds from a Kansas population tested at sites across the Midwest showed most emergence during a roughly two-week period with peak emergence from late March to early May. The timing of emergence depended on soil temperature and moisture, and it varied substantially between years at a given site. Warmer winters prolong the period over which seedlings emerge relative to cooler winters.

Emergence depth: Seedlings can emerge from depths of at least 4 inches.

Photosynthetic pathway: C_3

Sensitivity to frost: Seedlings in the cotyledon stage can survive temperatures as low as 23°F, and mature plants can survive down to 28°F.

Drought tolerance: Common sunflower is moderately drought tolerant. The main taproot can reach a depth of 10 feet, which allows established plants to extract water from deep in the soil where few other plant roots can reach.

Mycorrhiza: Wild common sunflower has high mycorrhizal infection rates. In domestic sunflower, high mycorrhizal infection improves P uptake and growth. Production of domestic sunflower can be lower following non-mycorrhizal crops such as canola, and the weed may respond similarly.

Response to fertility: USDA classifies common sunflower as having a low fertility requirement, but in Argentina it is found only on fertile soils that are high in organic matter and available phosphorus. It grows on soils with a pH between 5.5–7.8. Domesticated sunflower is highly responsive to N, with yields often continuing to increase with application rates in excess of 161 pounds per acre.

Soil physical requirements: Common sunflower can grow on soil of fine, medium and coarse textures. It can survive in poor soil, shallow soil over limestone and mild waterlogging. It has medium tolerance to salinity.

Response to shade: Common sunflower is intolerant of shade.

Sensitivity to disturbance: No information available.

Time from emergence to reproduction: Flowering begins nine to 17 weeks after emergence, and continues for four to 13 weeks, depending on the population. In the central Great Plains, common sunflower flowers from July to October, and seeds mostly disperse from September through October, though some may continue to disperse through the winter.

Pollination: Sunflowers are self-incompatible and are insect pollinated.

Reproduction: Three populations growing without crop competition averaged 38 heads per plant, 136 viable seeds per head and 5,300 seeds per plant. In another study, a typical plant growing with minimal competition produced 7,200 seeds, but some of these were immature whereas others had already shattered. An Argentine biotype grown in competition with domestic sunflower averaged 34 heads per plant, 190 seeds per head and 6,500 seeds per plant.

Dispersal: The seeds are dispersed locally by birds and small mammals. They also disperse in surface irrigation water. Combine harvesters contribute to expansion of patches of common sunflower and may also occasionally move seeds to previously un-colonized fields. The species is common along roadsides, and seeds are probably moved to new locations when soil is removed from road ditches. Seeds probably also move in soil clinging to tires and machinery.

Common natural enemies: Insects, birds and small mammals eat many sunflower seeds. In one study, 42% of seeds left on the soil surface were eaten in the first 10 days. Red sunflower seed weevil (*Smicronyx fuivus*), gray sunflower seed weevil (*S. sordidus*), sunflower moth (*Homeosoma electellum*), banded sunflower moth (*Cochylis hospes*) and sunflower budworm (*Sulemia helianthana*) damaged 44–58% of heads but less than 2% of seeds in some common sunflower populations in Kansas. Many of the diseases and insects of domestic sunflower probably also attack common sunflower.

Palatability: Seeds are edible for humans and livestock. Common sunflower is acceptable forage at all stages of development but should not be used as the sole feed due to a high Ca/P ratio.

FURTHER READING

Alexander, H.M., C.L. Cummings, L. Kahn and A.A. Snow. 2001. Seed size variation and predation of seeds produced by wild and crop-wild sunflowers. *American Journal of Botany* 88: 623–627.

Burton, M.G., D.A. Mortensen, D.B. Marx and J.L. Lindquist. 2004. Factors affecting the realized niche of common sunflower (*Helianthus annuus*) in ridge-tillage corn. *Weed Science* 52: 779–787.

Cummings, C.L. and H.M. Alexander. 2002. Population ecology of wild sunflowers: effects of seed density and post-dispersal seed predators. *Oecologia* 130: 274–280.

Robinson, R.G. 1978. Control by tillage and persistence of volunteer sunflower and annual weeds. *Agronomy Journal* 70: 1053–1056.

Dandelion

Taraxacum officinale F.H. Wigg.

Dandelion seedling
Antonio DiTommaso, Cornell University

Dandelion rosette
Scott Morris, Cornell University

Dandelion plant in flower
Antonio DiTommaso, Cornell University

Dandelion mature seeds on seed head
Antonio DiTommaso, Cornell University

IDENTIFICATION

Other common names: common dandelion, blowball, cankerwort, faceclock, pee-a-bed, wet-a-bed, lion's tooth, Irish daisy

Family: Aster family, Asteraceae

Habit: Rosette forming, taprooted perennial herb with leafless flowering stalks

Description: Cotyledons are 0.2–0.4 inch long by 0.08–0.14 inch wide, yellow-green, hairless, untoothed, flat, and oval to upside down teardrop shaped. The first true leaves of the **seedling** are alternate, hairless, spatula to upside down teardrop shaped, and gray-green on their lower surface; edges are wavy, irregularly toothed and lobed at the third leaf stage. Later emerging leaves of the young plants are widely toothed and deeply lobed, with lobe tips pointed towards the center of the rosette. Scattered flat hairs may be present on both sides of the leaf. **Mature plants** form a basal rosette of leaves. Leaves are stalkless, narrow and deeply divided toward the base, 2–15.5 inches long, deeply lobed. Lobes are triangular, often oppositely paired and point toward the crown. Scattered hairs are occasionally present on the midvein and leaf underside. Large, fleshy taproots may branch and can reach depths up to 6.5 feet. The entire plant exudes milky white sap when cut. Single **flowers** develop on 2–30 inch-long, hollow, sparsely hairy, leafless, white-green to reddish tinged stalks. The bright yellow flower head is 1–2 inches in diameter, 0.5–1 inch tall and set directly above two rows of long, green, narrow bracts. The outer bract row is curled back to touch the stem, and the inner row encases the immature flower head. The flower head consists of 100–300 yellow petaled flowers, each with five points at the tip. **Seeds** are gray-yellow to light brown, 0.25 inch long, cylindrical and attached to a feathery pappus by a 0.4 inch-long stalk. The seedhead is gray-white and spherical.

Similar species: Redseed dandelion [*Taraxacum laevigatum* (Willd.) DC.] is very similar to dandelion, but in contrast to that species, the terminal lobe of the leaf is about the same size as the lateral lobes, and the seeds are red to purple brown. Chicory (*Cichorium intybus* L.) can be distinguished from dandelion by the prominent, coarse hairs on its older leaves and by its lobes pointing forward, backward and perpendicular to the leaf axis. Common catsear (*Hypochaeris radicata* L.) has similar yellow flowers, but its stems are tall and branching with scattered leaves and multiple flowers. Annual and perennial sowthistle (*Sonchus* spp.) have waxy, blue-green leaves with large auricles at the leaf base.

MANAGEMENT

Avoid high K levels in the soil to avoid problems with dandelion. Dandelion competes poorly with forage grasses and grains for K, so when K is in relatively short supply, the grasses win. In contrast, since grasses and grains show a greater response to N and P than dandelion, high N and P levels will favor the grasses. Excessive liming also promotes dandelion.

In pastures, dandelion is more likely to be a problem under continuous stocking than under intensive rotational grazing. With rotational grazing, the leaves will grow upright in response to competition from the grasses, which makes them more susceptible to grazing when the animals are present. In contrast, under continuous stocking, the leaves will be prostrate and hard

for the animals to bite, and growth rates are likely to be higher due to reduced competition from the grasses. Dandelion can establish in alfalfa after it is mowed and tends to increase during the alfalfa phase of a crop rotation. Alfalfa can promote dandelion emergence and establishment, especially in small gaps in the alfalfa stand. Mowing alfalfa in the flowering stage is more effective at suppressing dandelion than mowing in the vegetative or bud stage. Seeding forages in narrow row spacing reduces subsequent dandelion density and growth. Although dandelion makes nutritious forage, it slows the drying of hay by up to a full day. Generally, removal of leaves or the uppermost root from established plants has little effect on survival in lawns or hay.

Moldboard plowing is the most effective tillage regimen for controlling severe dandelion infestations: Plowing puts the root crown deep in the soil from which resprouting is difficult, and inversion of the roots decreases the likelihood of sprouts arising from small roots. Conversely, dandelion density often increases with adoption of reduced tillage practices. Resprouting of roots following tillage is less in May than at other times, probably because root reserves are reduced by flowering at that time.

Frequent tine weeding is effective for removing new seedlings that arise following the spring burst of seed production. Flame weeding will kill small seedlings, but it is ineffective against larger plants because the tissues that produce new leaves are protected by slight burial in the soil and by a whorl of newly forming leaves.

Because of its prostrate growth form, dandelion emerging from seeds is one of the most sensitive weed species to mulching with straw. Seedlings have a difficult time emerging through a loose mat of straw. Long leaves emerging from root sprouts have no trouble, however, in emerging through even a thick mat of loose straw. Matted straw from bales will prevent emergence of root sprouts, but in rainy weather, the straw will begin to rot and provide a favorable habitat for establishment of seedlings.

Dandelion often poses little problem for field crop production in humid regions, even though the bright yellow flowers make the field appear very weedy. It is short in height and a poor competitor for nutrients, so if the crop is more than a foot tall and not water stressed, dandelions may cause little yield loss. Moreover, the flowers provide nectar and pollen for beneficial insects that help control pests. In drought years and in dry regions, however, a dandelion infestation may decrease yield substantially.

ECOLOGY

Origin and distribution: Dandelion probably originated in Europe but spread through Eurasia and North America prior to human agriculture. It occurs throughout the United States and Canada, up to nearly 65° N, and is considered both native and introduced throughout North America. European settlers introduced dandelion very early during the colonization of New England, perhaps deliberately for its medicinal properties. The species now inhabits all regions of the world, including tropical and sub-Antarctic regions.

Seed weight: 0.34 mg (seeds produced in summer) to 0.54 mg (seeds produced in spring).

Dormancy and germination: Newly dispersed dandelion seeds are capable of immediate germination if they receive the right cues. A few freshly dispersed seeds will germinate regardless of conditions, but light, fluctuating temperatures and nitrate all increase germination. Light is the most effective stimulus for germination, but 100% germination requires light plus one of the other factors. If seeds experience a period of cold and dark, as occurs when buried in the soil during winter, they subsequently require light for germination. Seeds will germinate at temperatures from 41–95°F, but percentage germination and speed of germination are reduced at temperatures above 68°F. Germination is highest at alternating day/night temperatures of 59/41°F and 77/59°F, and is reduced at 95/77°F.

Seed longevity: Seeds can last one to five years in the soil, but in normal agricultural conditions only a few survive to the next season. The overwhelming majority of seedlings come from recently dispersed seeds. In one experiment, a majority of seeds were consumed by ground beetles within two to three weeks after shedding, but the 2–4% of viable seeds that remained were sufficient to maintain high soil populations. In another experiment recording the fate of dandelion seeds shed in spring, 29–48% of seeds became nonviable, 35–44% were consumed by ground beetles, 5–25% were consumed as seedlings by slugs and only 2–13% survived as seedlings.

Season of emergence: Some seedling emergence occurs throughout the growing season, but the bulk of new seedlings appear in early to mid-summer, shortly after the burst of seed production in late spring. Many established rosettes overwinter, even in cold climates, and shoot establishment from rootstocks occurs in early spring.

Emergence depth: Emergence is not affected by depth between 0–0.8 inches but declines for seeds at progressively greater depths. No emergence occurs from seeds below 3 inches. Emergence from small root fragments can occur from 2–4 inches of soil.

Photosynthetic pathway: C_3

Sensitivity to frost: Dandelion is very frost hardy and shoots of at least some plants overwinter throughout its range. In warmer parts of the United States, plants continue to grow throughout the winter.

Drought tolerance: Young plants are highly sensitive to drought, but well-established plants are very drought resistant.

Mycorrhiza: Dandelion forms mycorrhizal associations.

Response to fertility: In competition with grasses, increased N and P favored grasses relative to dandelion, even when N was applied at very high rates. Without competition, dandelion growth increases with P fertility but does not respond to N. Dandelion has a higher requirement for K than many grasses and thus competes poorly for this element when it is in moderate to short supply. Dandelion tolerates soil with a pH as low as 4, but establishment and growth increases up to a pH of 8. In one study, a soil with pH of 7.1 produced 45% more fresh weight and had 67% more plants than a soil with pH of 6.2. Below pH of 5.2, growth is poor.

Soil physical requirements: Dandelion occurs on a wide range of soils from wet, fine textured meadow soils to rocky ridge tops. In vegetated areas, seedlings establish best on ridges, but on bare soil, seedlings establish best in hollows.

Response to shade: Dandelion is shade tolerant and will grow even under competitive crops like potatoes.

Sensitivity to disturbance: Dandelion has a great capacity for resprouting from damaged roots or following the removal of leaves. It does not normally reproduce vegetatively, but root fragments as small as 0.05 inch diameter by 0.25 inch length can establish new plants. The likelihood of establishment increases with the length and diameter of the fragment. Root fragments that are moved out of their original vertical position are less likely to resprout. Fragments from near the root crown are more likely to sprout than are fragments from near the root tip. Root fragments derived from plants at peak flowering in mid-spring have lower survival than those derived from plants in summer. Flowers that are cut from the stalk will set seed, but the seeds are not viable.

Time from emergence to reproduction: Some plants that establish in the spring may flower in the fall, but most do not flower until the following spring. As with most perennial herbs that do not spread by roots or rhizomes, flowering is more related to achievement of a minimum size rather than to the age of the plant. Dandelions that have formed a strong taproot and 20 leaves (some of which may have subsequently died) are usually ready to flower. Peak flowering begins in mid-spring and continues for two to six weeks, but it can also continue throughout the year with a secondary peak in fall. A few seeds may become viable as early as three days after the flowers wither and close, but most are not viable until six to eight days after flowering in the summer or about 10 days after flowering in the spring or fall.

Pollination: Dandelions in North America have three sets of chromosomes rather than a multiple of two like most species. Consequently, they are incapable of sexual reproduction except by extremely rare genetic accidents. The embryos develop into seeds without fertilization. However, insect visitors may be needed to trigger seed set.

Note on evolution: Why does dandelion produce pollen if the seeds develop without fertilization? Common dandelion almost certainly arose by hybridization of two related species that were insect-pollinated. The hybridization left a chromosome arrangement that makes sex essentially impossible, but due to some fluke, the hybrid was able to develop seeds asexually. Since both parent species produced pollen, the resulting dandelion does too. Although pollen is not needed for seed production, its manufacture cannot be selected out of the population because without sex, evolution proceeds extremely slowly.

Reproduction: The flower bud forms near ground level in about one week. The shoot grows up in 48 hours once the bud is formed. The flowers typically open on two to three successive days. They are open only in morning during summer but through most of the day in spring and fall. They close or remain closed if the weather is cold or wet. The shoot lies horizontal while the seeds mature. Presumably this is an adaptation to reduce seed losses to grazing, but it also protects the developing seed head from mowing. Roughly two days later, the stalk returns to an upright position and the seed head opens for dispersal. The length of the flower stem is longer in summer when the temperature is higher and day length is longer. Under good conditions, a dandelion plant will produce 50–150 seed heads per year with an average of 250 seeds per head.

Dispersal: Seeds are wind dispersed by means of umbrella-like fluff. Updrafts are most important for long distance dispersal, and a model of dandelion seed transport determined that 99.5% of seeds would land within 11 yards, 0.05% would travel more than 108 yards, and 0.01% would travel greater than 1,083 yards. However, dispersed dandelion seeds were one of the most likely of weeds tested to be intercepted by standing vegetation, and many seeds may not become incorporated into the soil. Seeds also disperse in the feces of cattle, horses and birds, in soil on shoes, tires and farm implements, and via water in irrigation ditches.

Common natural enemies: Sparrows and finches eat the seeds while they are on the stalk, and rodents and sparrows eat seeds after dispersal. Deer, ducks, geese and grouse eat the leaves. The cyprinid wasp, *Phanacis taraxaci*, forms galls on the leaves. The weevil, *Ceutorhynchus punctiger*, attacks buds, seeds and leaves, and shows some potential as a biological control agent. The aphid, *Aphis knowltoni*, feeds on the roots. Several fungi attack dandelion in North America, including species of powdery mildew (*Sphaerotheca*), and *Puccinia taraxaci* and *Synchytrium* *tarazaci*, both of which make tiny growths on the leaves. The latter caused partial stunting of the plants. *Phoma exigua* and *P. herbarum* may prove useful as biocontrol agents.

Palatability: Leaves of dandelion are commonly marketed for use as a salad green and as a cooked green. Leaves are least bitter in early spring. It is one of the most nutritious of all vegetables, containing high concentrations of minerals and vitamins, including a higher concentration of beta carotene than carrots. The flowers can be used to make wine. Dandelion foliage has good forage quality, but palatability to grazing lambs was lower than alfalfa. Chickens relish both the roots and leaves.

Note: Incorporation of dandelion tissues into soil inhibits growth of tomato crown and root rot (*Fusarium oxysporum* f.sp. *radicis-lycoperici*).

FURTHER READING

Hacault, K.M. and R.M. Van Acker. 2006. Emergence timing and control of dandelion (*Taraxacum officinale*) in spring wheat. *Weed Science* 54: 172–181.

Mann, H. and P.B. Cavers. 1979. The regenerative capacity of root cuttings of *Taraxacum officinale* under natural conditions. *Canadian Journal of Botany* 57: 1783–1791.

Stewart-Wade, S.M., S. Neumann, L.L. Collins and G.J. Boland. 2002. The biology of Canadian weeds. 117. *Taraxacum officinale* G. H. Weber ex Wiggers. *Canadian Journal of Plant Science* 82: 825–853.

Tilman, E.A., D. Tilman, M.J. Crawley and E.A. Johnston. 1999. Biological weed control via nutrient competition: potassium limitation of dandelions. *Ecological Applications* 9: 103–111.

Deadnettles

Henbit, *Lamium amplexicaule* L.
Purple deadnettle, *Lamium purpureum* L.

Henbit seedling
Scott Morris, Cornell University

Henbit plant
Scott Morris, Cornell University

Henbit flower
Scott Morris, Cornell University

Purple deadnettle seedling
Scott Morris, Cornell University

Purple deadnettle plants
Scott Morris, Cornell University

IDENTIFICATION

Other common names:

Henbit: bee nettle, blind nettle, giraffe head, dead nettle, henbit dead nettle, henbit nettle

Purple deadnettle: red deadnettle

Family: Mint family, Lamiaceae

Habit: Henbit is a winter annual or summer annual, sprawling with upward curving stems that root at the nodes. Purple deadnettle is a sprawling winter annual with stems rooting at the nodes.

Description: Seedlings have stalked cotyledons with two lobes at the base and a flat to shallowly indented tip. Young leaves are opposite with large, rounded teeth and distinct stalks.

Henbit: Cotyledons are round to thumb shaped and 0.13–0.5 inch long by 0.1–0.2 inch wide. Seedling stems are purple. The upper leaf surfaces and prominent veins on the underside of the leaf are covered with fuzzy hairs. Young leaves are round in outline and have two to four teeth per side.

Purple deadnettle: Cotyledons are oval shaped, red-stalked, hairless and 0.47 inch long by 0.43 inch wide. Fuzzy hairs are present on the blades and edges of young leaves. Young leaves are round to broadly egg or heart shaped, and they have two to four teeth per side with one large lobe at the tip.

Mature plants have square, green to purple-tinged, weak stems that branch in a prostrate or leaning manner at the base with tips curving upright. Leaves are opposite and heavily cross-veined, giving the surface a wrinkled appearance. The distance between leaves shortens near stem tips, especially flowering tips. The fibrous root system is shallow and supplemented by roots growing from stem-leaf junctions in contact with the ground.

Henbit: Stems have downward pointing hairs and purple tinges; they can reach 4–16 inches tall. Lower leaves are stalked and similar in appearance to seedling leaves. Upper, stalk-less, fan shaped leaves encircle the stem almost completely. Leaves are broadly round-toothed to lobed, softly hairy, strongly veined, round tipped and 0.5–2 inch long (lower) or broad (upper).

Purple deadnettle: Stems are hairless to lightly hairy, streaked purple and hollow; they can reach 4–20 inches tall. Lower leaves are stalked and similar in appearance to seedling leaves. Upper, small-stalked leaves are maroon, angle down to the ground and do not encircle the stem. Leaves are shallowly lobed, lightly hairy, pointy tipped and 0.4–0.5 inch long.

Flowers are whorled in circles set just above upper leaf pairs.

Henbit: Six to 10 pink to dark purple flowers are present per leaf pair. Flowers near stem tips are 0.5–0.75 inch long, with lightly hairy petals fused into a flared tube. Flowers on lower whorls never fully open; they are small, inconspicuous and self-pollinating.

Purple deadnettle: Three to six flowers are present per leaf pair. Flowers have 0.5–0.7 inch-long, fused petals that are partially wrapped by green structures and are hidden from view by leaf pairs.

Fruit and seeds: Four brown, egg- or thumb-shaped, 0.06 inch-long, nut-like seeds are produced per flower. Nutlets have two flat sides, one round side and white spots.

Henbit: Nutlets are long and tapered to a flat tip.

Purple deadnettle: Nutlets are bumpy, broader and rounder than henbit nutlets.

Similar species: Young Persian speedwell (*Veronica persica* Poir.) has similar leaves but is distinguishable by its round stems and blue or white flowers. Ground ivy (*Glechoma hederacea* L.) also has blue flowers and leaves that are long stalked, triangularly toothed and have a shiny surface. Ground ivy, however, is a strongly aromatic, mat-forming perennial with creeping stems. Spotted deadnettle (*Lamium maculatum* L.) has white spots on the leaves, and healall (*Prunella vulgaris* L.) has larger, round tipped, largely untoothed, egg-shaped or lanceolate leaves.

MANAGEMENT

Tillage in late fall is effective for killing purple deadnettle and fall emerging henbit to prevent seed production the next spring. Unfortunately, soils are often too wet for tillage that time of year, and fall tillage leaves soil exposed to winter erosion. Flame weeding is an effective alternative. If the stand is dense, reduce ground speed to ensure that sufficient heat reaches the base of the plant. In areas where spring germinating henbit is a problem, tillage for late spring planted crops will destroy mid- to late-emerging cohorts, but early emerging plants may already have set seed. Thus, tillage or flame weeding in late April may be necessary in addition to tillage in May. In vegetable systems, an early planted crop of transplanted lettuce may become weedy with spring henbit, but the tillage before and after planting can be effective for breaking the weed's life cycle.

Cultivation of cool season vegetable crops to control henbit and purple deadnettle will be more successful if these weeds are small. The spreading, fibrous root system of plants with several nodes holds soil well, and in cool weather the plants are slow to dry and can easily reroot. They are especially difficult to kill after stems begin to root at the nodes. For controlling purple deadnettle and fall emerging henbit in winter grains, till and plant midway in the planting window to allow a large proportion of seedlings to emerge first but still produce a competitive crop. Seed at a heavy rate to allow aggressive tine weeding after crop establishment, and try to rotary hoe or tine weed before crop emergence as well. Plant spring grains as early as possible to get the crop established before peak henbit emergence. Weed as for the winter grains. If planting of spring grains has to be delayed, consider planting a later crop species instead so that tillage will kill off the henbit after it emerges.

Management of winter cover crops is critical for effective long-term control of both species in a vegetable rotation. If fall rye or forage radish cover crops establish quickly and produce a high amount of ground cover, then henbit and purple deadnettle will be severely suppressed. However, fall cover crops that do not establish well and leave significant gaps in the leaf canopy will allow either of these weeds to establish and set seed before cover crop termination in spring. One alternative is to shift to an earlier planted cover crop of a winter hardy clover. Use buckwheat as a nurse crop so that the clover does not become weedy with summer annuals, and mow it before seeds form. The lack of late summer/fall tillage and the low light under the clover will reduce henbit and purple deadnettle emergence, and the clover will compete strongly with plants that do emerge. Use a high seeding rate of clover to ensure a dense, competitive stand. In regions with hot, sunny summers, solarization can effectively destroy the surface henbit seed bank to prepare the soil for fall crops.

Purple deadnettle is particularly adapted to the moist, protected soil environment of fields with incomplete crop residue cover left untilled over winter. However, complete ground cover by wheat straw reduced henbit density by 77% relative to a no-straw control following minimum tillage and completely suppressed henbit in an organic no-till crop of winter fava beans.

ECOLOGY

Origin and distribution: Both species are native to the Mediterranean region of Eurasia and North Africa. Henbit has spread throughout Europe and much of Asia, and it has been introduced into Australia, New Zealand, North America and temperate and mountainous parts of South America. It occurs throughout the United States and southern Canada. Purple deadnettle is now a weed across most of Europe, North America and New Zealand. It is most common in the eastern half of the United States and along the Pacific Coast, including Alaska. It occurs only sporadically in the High Plains and Intermountain West.

Seed weight: Henbit population mean seed weight ranges from 0.5–0.6 mg. Purple deadnettle mean seed weight ranges from 0.65–0.92 mg.

Dormancy and germination: Henbit seeds are dormant when shed from the parent plant. After exposure to high summer temperatures for several weeks, seeds from Kentucky populations germinate well at any temperature regimen from 41°F to 86/68°F day/night. Both dormant seeds and seeds that have lost dormancy through warm summer temperatures become conditionally dormant during exposure to winter temperatures (e.g., 41°F) and then will only germinate well at temperatures below 68/50°F. Consequently, seeds germinate in hot weather in the late summer or in cool weather in the fall or early spring, but not in the warm weather of late spring or summer. This ensures that the plants will have time to grow during cool seasons of the year. In the cold climate of Sweden, dormancy was broken by a long (24 weeks) period of warm, dry conditions, and emergence in the field occurred almost entirely in the spring. Another study showed that non-dormant seeds of henbit germinated best at 68°F.

Purple deadnettle seeds are also dormant when shed in the spring. Hot summer temperatures break dormancy of seeds from Kentucky by late spring or early summer, but germination is inhibited by hot summer temperatures of 86/59°F or higher. When temperatures cool in the fall to 68/50°F or lower, the seeds can then germinate. A few weeks of cold winter temperatures (e.g., 41°F) induce a secondary dormancy so that the seeds will not germinate in the same mild temperatures in the spring that allowed germination in the fall. This secondary dormancy is again broken by hot temperatures in late spring or early summer, and the cycle repeats. This dormancy cycle causes purple deadnettle to act as a strict winter annual, at least in the central United States. In Sweden, warm moist or dry conditions break dormancy as in the central United States. However, most emergence occurs in the spring, implying that cold temperatures during the winter do not induce dormancy. In Sweden, seedlings emerging in the fall often die without reproducing. Dormancy cycles in the northern United States and Canada require study.

Germination of non-dormant seeds of both species is generally stimulated by light, although one study found that light inhibited germination of henbit.

Seed longevity: Viable seeds of both species have been found beneath medieval ruins in Europe, but such sites are favorable for long-term seed survival. Roughly 20–50% of henbit seeds survived 20 years under sod. Purple deadnettle seeds in annually tilled winter wheat and oilseed crops declined by 39% and 60% per year. In another experiment, purple deadnettle seed density declined at an average of only 20% per year, with the decline faster in spring wheat than in winter wheat. Average seed loss over a five-year period from soil stirred three times per year was 43–60% per year for henbit and 43–53% per year for purple deadnettle.

Season of emergence: Henbit emerges in mid-spring and in early autumn, but emergence is primarily in autumn in the South and primarily in spring in the prairie provinces of Canada. In Michigan, emergence occurs from spring to fall, with peak emergence in late summer. Purple deadnettle emerges only in autumn in Kentucky, throughout the growing season with a peak in autumn in England, and primarily in the spring in Sweden.

Emergence depth: Most seedlings of henbit and purple deadnettle emerge from the top 1 inch of soil, though a few can emerge from as deep as 2.5 inches. One study sampled the effective seed bank of these species by taking cores only 0.5 inch deep, indicating that the researchers thought most seedlings emerged only from that shallow surface layer of the soil.

Photosynthetic pathway: C_3

Sensitivity to frost: Both species are frost tolerant. Henbit can tolerate temperatures down to 22°F but suffers some damage at that temperature. Fall emerging plants of both species can overwinter in the northern United States, but snow cover likely improves survival substantially.

Drought tolerance: Henbit is drought-intolerant.

Mycorrhiza: Both species are mycorrhizal.

Response to fertility: Henbit response to N was intermediate among 24 weeds and crops studied and was roughly similar to wheat. Of the same 24 species, it was among the group of species most responsive to P fertility. In another experiment, however, four years of manure or compost either with or without additional sulphate of ammonia had negligible effects on late spring henbit density in vegetable crops. Eight years of application of various N fertilizers decreased henbit relative to plots that received no N, probably because increased vigor of fertilized winter grains suppressed the low growing henbit. On a sandy soil, henbit was insensitive to pH from 5 to around 7, but below pH 5 density decreased, although its proportion of the weeds present increased.

Soil physical requirements: Henbit is often found in light, sandy soils but occurs on a range of rich agricultural soils. Purple deadnettle is usually found in rich, loamy or sandy-loam soils.

Response to shade: Henbit is shade intolerant. Purple deadnettle partially avoids shade through increased stem elongation in environments with competition from other plants.

Sensitivity to disturbance: Fragments of purple deadnettle reroot readily after shallow tillage or hoeing.

Time from emergence to reproduction: In warm climates, fall germinating henbit flowers in late winter to early spring. In the northern United States and southern Canada, fall germinating plants flower from April to June. Spring germinating plants emerging in early April in Wisconsin began flowering 36 days later, but some spring and summer emerging plants in the northern United States do not flower until September. Purple deadnettle flowers from late March/early April to mid-May in Ohio. A relatively short time is required for viable seed to be produced after flowering, because peak shedding of henbit seeds occurred in mid-May, within a month from when flowering began.

Pollination: Henbit forms closed self-pollinating flowers on the lower nodes and sometimes on the whole plant, but it often produces open, cross-pollinated flowers at the upper nodes. Pollination is by solitary bees and honeybees. Purple deadnettle self-pollinates, but it is also pollinated by bumblebees and other bee species.

Reproduction: A typical henbit plant produces between 200–2,000 seeds, but a particularly large plant produced 60,000 seeds. Purple deadnettle produces about 600 seeds per plant but can produce 27,000 seeds if grown without competition. Purple deadnettle spreads as a mat of stems that root at the nodes; if fragmented by tillage, rooted stem sections can re-establish.

Dispersal: Both species form persistent seed banks and probably commonly disperse in soil clinging to farm machinery and the bodies of cattle. Purple deadnettle seeds have been found in soil adhering to an automobile traveling in rural areas and also in commercial topsoil. Seeds of both species have specialized bodies that provide a food source attractive to non-seed-eating ants. These presumably encourage dispersal by ants.

Common natural enemies: Both species are favored food sources for a variety of slug species.

Palatability: These species are sometimes eaten as salad or pot herbs, particularly in Japan. Henbit has digestibility, crude protein and mineral content similar to or higher than common forage grasses. Grazing on henbit can cause mild neurological problems in sheep, cattle and horses, but the condition is rare and reversible. Purple deadnettle has no reported toxic properties for ruminants.

FURTHER READING

Baskin, J.M., C.C. Baskin and J.C. Parr. 1986. Field emergence of *Lamium amplexicaule* L. and *L. purpureum* L. in relation to the annual seed dormancy cycle. *Weed Research* 26: 185–190.

Defelice, M.S. 2005. Henbit and the deadnettles, Lamium spp.--archangels or demons? *Weed Technology* 19: 768–774.

Hill, E.C., K.A. Renner and C.L. Sprague. 2014. Henbit (*Lamium amplexicaule*), common chickweed (*Stellaria media*), shepherd's purse (*Capsella bursa-pastoris*), and field pennycress (*Thlaspi arvense*): fecundity, seed dispersal, dormancy, and emergence. *Weed Science* 62: 97–106.

Holm, L., J. Doll, E. Holm, J. Pancho and J. Herberger. 1997. *World Weeds: Natural Histories and Distribution*. Wiley: New York.

Lawley, Y.E., R.R. Weil and J.R. Teasdale. 2011. Forage radish cover crop suppresses winter annual weeds in fall and before corn planting. *Agronomy Journal* 103: 137–144.

Docks

Broadleaf dock, *Rumex obtusifolius* L.
Curly dock, *Rumex crispus* L.

Broadleaf dock seedling
Antonio DiTommaso, Cornell University

Broadleaf dock rosette
Antonio DiTommaso, Cornell University

Curly dock plant (left) and broadleaf dock plant (right)
Scott Morris, Cornell University

Curly dock seedling
Scott Morris, Cornell University

Curly dock rosette
Antonio DiTommaso, Cornell University

Curly dock fruit
Scott Morris, Cornell University

IDENTIFICATION

Other common names:

Broadleaf dock: bitter dock, blunt-leaved dock, red-veined dock, broad-leaved dock, celery seed

Curly dock: yellow dock, sour dock, narrow-leaved dock, curled dock

Family: Buckwheat family, Polygonaceae

Habit: Erect, taprooted, perennial herbs with basal rosettes of large leaves

Description: Young leaves of **seedlings** develop basal rosettes. Leaf edges unfurl outwards from underneath the leaf.

Broadleaf dock: Cotyledons are egg shaped, 0.4–0.7 inch long by 0.14–0.25 inch wide and gritty on both sides. First true leaves are ovate to oblong shaped. All subsequent young leaves are oblong and wavy edged; they have heart-shaped bases, white hairs beneath their stalks and midveins, and possible red flecking on the veins and blade.

Curly dock: Thumb shaped cotyledons are hairless, gritty to the touch, and 0.3–0.6 inch long by 0.1–0.2 inch wide with fleshy, possibly red-flecked, dull green upper sides and pale green undersides. Cotyledon leaf stalks are ridged. Stems are green with a red base. The young true leaves are egg shaped, smooth and flat along edges, hairless and possibly red-flecked.

Mature plants form vigorous basal rosettes. All leaves have a transparent, thin membrane (ocrea) covering the leaf-stem juncture. Flower stalks are chestnut to red-brown with smaller and more tapered leaves than those in the rosette. The underground part of both species consists of a large, fleshy, yellow taproot that is surmounted by about 2 inches of underground stem containing many buds. The root and underground stem are difficult to distinguish but behave differently when fragmented by tillage (see the section "Response to disturbance").

Broadleaf dock: Leaves are 12 inches long by 6 inches wide, hairless, tapered and heart shaped at the base; they are flatter, less wavy, broader and less crinkled than curly dock. Lower leaf stalks may have red veins. The taproot is usually highly branched.

Curly dock: Rosette leaves are curly, wavy, crinkly edged, elongated and tapered at the base. Rosette leaves are 6–12 inches long and up to 2.5 inches wide. Stem leaves are up to 6 inches long and 1 inch wide. Leaves are alternate, hairless, shiny, green and increasingly ruddy as the season progresses. The taproot is often unbranched or only minimally branched.

Flowers are green-red, inconspicuous, small, without petals and arranged into spiral clusters at the ends of closely spaced stalk branches. Individual flowers hang from the cluster on small stalks. The inflorescence turns red-brown as the seeds mature.

Broadleaf dock: Stalks are hairless, 3–4 feet tall and occasionally present in small groups. Flowers are 0.07–0.1 inch wide. Clusters make up two thirds of the stem length.

Curly dock: Stalks are jointed, reddish-brown, hairless and up to 5 feet tall. Flowers are 0.25 inch wide. Clusters are 6–24 inches long.

Fruit and seeds: Both species have heart-like, winged, brown fruits surrounding three-sided, teardrop-shaped, glossy, reddish brown seeds.

Broadleaf dock: Wings on fruits are reddish brown, 0.13–0.25 inch long, toothed and spiny. Seeds are 0.08 inch long when removed from fruit.

Curly dock: Wings on fruits are corky or papery, dark brown, 0.13–0.25 inch long and entire. Seeds are 0.07 inch long when removed from fruit.

Similar species: Common burdock (*Arctium minus* Bernh.) and great burdock (*Arctium lappa* L.) have young seedlings similar to the docks. Seedlings of burdocks have smooth, shiny cotyledons with downy undersides. Young true leaves have leaf stalks that encase the stem, whereas dock stems extend around the leaf stalk. Docks have ocreae, whereas burdocks do not.

MANAGEMENT

Both species proliferate in poorly managed pastures. They establish well in bare, trampled areas where livestock congregate. With intensive rotational grazing, small dock plants are eaten and trampled back before they get well established, and the rapid regrowth of the grass will tend to out-compete the dock. Broadleaf dock had fewer plants and smaller taproots when ryegrass-white clover swards were cut four times per season (simulating grazing) than when cut twice per season (for silage). Other experiments have shown similar decreases in vigor of dock as cutting frequency increases. Although most livestock will not eat dock unless it is very young, deer and some breeds of sheep will eat even large plants. High seeding rates reduce dock infestations in newly established crops of both grasses and alfalfa. Similarly, undersowing to increase the density of thinning grass stands can help suppress dock.

In both forage crops and annual crops dock problems increase with over fertilization. Moderate fertility rates are the key: You want enough nitrogen for the crop to grow vigorously and to compete with the dock for light, but excess N will favor the dock relative to the crop.

Tillage is effective for managing dock. The worst problems in annual crops come when the field is rotated from a perennial forage to an annual crop. If the dock becomes old and dense in the forage phase of the rotation, special measures are needed to get it under control. One way to do this is to plow the field in mid-summer and work the dock roots to the surface where they can dry out using a spring tooth harrow. Turn the soil and roots periodically with the harrow to ensure a good kill. Another successful approach is to rototill the field to a depth of about 4 inches. This will chew up the underground vertical stem from which most new shoots arise. The rotary tillage should be slow to pulverize the underground stems to the maximum degree. Simply severing the neck from the root can actually increase regeneration. Follow the initial rotary tillage with faster passes using other implements to kill new sprouts. This approach can be used to spot-treat heavily infested areas of a pasture or forage field. Successive tillage operations should occur when the leaves are 2 inches long. Moldboard plowing will reduce dock populations, but some plants will recover even from burial at the bottom of the plow layer. A mechanical dock puller that can remove 600 plants per hour has been developed in Germany.

Since both dock species have very long growing seasons, winter cover crops help to suppress dock populations through competition. Prevention of new infestations (see the section "Dispersal") and early eradication of isolated plants can prevent much grief, particularly on farms with perennial forage and pasture.

ECOLOGY

Origin and distribution: Both species are native to Europe. Both species have been introduced widely in Asia, Africa, Australia and North and South America. Broadleaf dock occurs as far north as Greenland, and curly dock occurs in Iceland and the interior of Alaska. In North America, broadleaf dock is most common east of the Mississippi River and in the Pacific Northwest, with populations more scattered in the Great Plains and Inter-Mountain states. Curly dock, however, occurs throughout the United States and the agricultural zones of Canada.

Seed weight: Mean population seed weights of broadleaf dock range from 0.7–2.8 mg. Mean population seed weights of curly dock range from 1–3 mg with most near 1.4 mg.

Dormancy and germination: For both species, seed dormancy characteristics apparently depend on growth conditions and vary greatly between individuals and position of the seed on the seed stalk. Broadleaf dock seeds are dormant when shed but gradually lose dormancy during succeeding months. Germination occurs at 50–95°F with maximum germination at 68–77°F. Temperature alternation and light increased germination, whereas exposure to light filtered through leaves decreased germination. Exposure to high nitrate levels reduces germination at optimal light levels. Seeds from early bolting plants are more dormant than those from late bolting plants.

Curly dock seeds are dormant when the dry fruit hull surrounding the seed restricts germination, but they also may be non-dormant when shed at maturity in summer. Exposure of moist seeds to low temperatures (32–59°F) for one or more weeks breaks dormancy. Exposure to temperature alternations and light promote germination. Seeds germinate in response to a wide range of alternating day/night temperatures, ranging from 68/50°F to 95/68°F. High N along with a deficiency of P and K fertilization of the mother plant reduces seed germination. Secondary dormancy can be induced by several days of warm temperatures (up to 86°F) or dry conditions. Although most research has been conducted on each of these species separately, an extensive review suggests both species have similar dormancy and germination behavior.

Seed longevity: Seeds of broadleaf dock have been shown to survive at least 40 years under favorable conditions, and seeds of curly dock can remain viable in the soil for as long as 80 years. In pastures, seed densities of several thousand per square foot are common. Nevertheless, seed mortality can be high; in one experiment in Ontario, less than 15% of curly dock seeds and 1% of broadleaf dock seeds survived more than one year. In a series of five-year experiments in which the top 3 inches of soil was stirred three times per year, seeds of curly dock declined by an average of 37–47% per year, and seeds of broadleaf dock declined by 49–56% per year.

Season of emergence: Seedlings emerge throughout the growing season, but notable flushes of emergence occur during the spring and fall when soil temperature is above 59°F and daily fluctuations are greatest. Overwintering rosettes resume growth in early spring.

Emergence depth: Emergence of dock seedlings is best at or near the soil surface, but a few seedlings can emerge from as deep as 2–3 inches. Emergence from fragments of belowground stem can occur from a depth of 6 inches if soil is not waterlogged, but emergence is best from 0–3 inches.

Photosynthetic pathway: C_3

Sensitivity to frost: Established plants of both species are very frost tolerant and commonly overwinter with a rosette of small reddish leaves, though curly dock sometimes dies completely back to the ground. Seedlings of broadleaf dock are frost sensitive.

Drought tolerance: Mature plants of broadleaf dock are drought tolerant, but seedlings can die from prolonged drought conditions. Curly dock is very drought tolerant.

Mycorrhiza: Most reports describe the absence of mycorrhiza from both species. However, one study reported the presence of mycorrhiza.

Response to fertility: Both species respond strongly to N, and their high abundance is an indicator of excessive N fertility. Curly dock responds most to N and P together, and seed production is enhanced most by balanced N, P and K. Curly dock responds more to P than to K. Curly dock foliage contains high concentrations of Zn, but whether it also requires high Zn levels in the soil is unclear.

Soil physical requirements: These species are present on almost all soil types but less so on peat or very acidic soils. Curly dock is an indicator of compaction. It tolerates waterlogged soils and several weeks of complete immersion in flood water.

Response to shade: Neither species establishes well in closed communities, such as a well-managed, vigorous pasture. However, regrowth after a period of intense competition and shading was surprisingly high.

Sensitivity to disturbance: Mowing has little effect on aboveground growth of both species but reduces belowground growth. Leaves are quickly regenerated, and mowing induces production of new flowering stalks. Broadleaf dock was better adapted to frequently cut grasslands than curly dock because of greater investment in root biomass. Docks tolerate extreme trampling and often become dominant in areas where cattle congregate. New plants establish readily from fragments of the short, vertical underground stem that sits on top of the roots. New plants will also form from fragments of the true root, but apparently only in the spring.

Time from emergence to reproduction: Broadleaf dock rarely flowers during the first season. Curly dock can flower during either the first or second season, depending on conditions. In subsequent years, both species make new flower stalks repeatedly from spring until hard frost in late autumn. Curly dock that begins regrowth in early February to March will begin flowering in April to early May. Seeds mature six to 18 days after flowers open. The majority of plants become capable of regeneration from rootstock at 38–51 days after planting.

Pollination: Both species are wind and bee pollinated, but both have a high degree of self-fertility.

Reproduction: Seed production is proportional to the size of the plant and varies from 100 to over 60,000 seeds. About half of curly dock plants die after flowering, and a typical plant lives about three years. Both species are relatively short-lived perennials with less than 2% of broadleaf dock and 4–24% of curly dock surviving for more than four years. Vegetative reproduction is limited to expansion of the diameter of old clumps unless the plant is broken by tillage. Flowering plants allocate substantially less biomass to roots than non-flowering plants. Even after tillage, most new individuals arise from seeds rather than from fragments of roots or underground stalks.

Dispersal: The seeds blow short distances in the wind. They also disperse by floating on water. The wings of the fruit tend to cling to clothing and the fur of animals. Much long distance dispersal probably occurs, however, by passage through livestock, with the seeds being deposited in new locations when the animals are moved or the manure is spread. The seeds in manure often come from infested pastures. Seeds may also be present in small grain, forage and cover crop seed. Birds also disperse the seeds.

Common natural enemies: Deer apparently prefer dock over other forage. Host specific natural enemies normally cause little damage to these species. In an experiment, simultaneous application of the rust *Uromyces rumicis* and the beetle *Gastrophysa viridula* greatly reduced leaf area and subsequent regrowth of broadleaf dock. Curly dock was more susceptible than broadleaf dock to *U. rumicis* infection.

Palatability: Leaves of broadleaf dock were formerly cooked and eaten like spinach, and also fed to pigs. Cattle, sheep, horses, rabbits and chickens all reject the foliage and may be poisoned if forced to eat it due to lack of other forage. Curly dock has low digestibility and palatability and was completely rejected by grazing lambs.

Note: Extracts and residue of broadleaf dock inhibit growth of a wide range of crops, including white clover, sunflowers, wheat, corn and soybeans.

FURTHER READING

Cavers, P.B. and J.L. Harper. 1964. *Rumex obtusifolius* L. and *R. crispus* L. *Journal of Ecology* 51: 737–766.

Niggli, U., J. Nösberger and J. Lehmann. 1993. Effects of nitrogen fertilization and cutting frequency on the competitive ability and the regrowth capacity of *Rumex obtusifolius* L. in several grass swards. *Weed Research* 33: 131–137.

Pye, A., L. Anderson and H. Fogelfors. 2011. Intense fragmentation and deep burial reduce emergence of *Rumex crispus* L. *Acta Agriculturae Scandinavica, Section B—Soil and Plant Science* 61(5): 431–437.

Zaller, J.G. 2004. Ecology and non-chemical control of *Rumex crispus* and *R. obtusifolius* (Polygonaceae): a review. *Weed Research* 44: 414–432.

Field pennycress

Thlaspi arvense L.

Field pennycress seedling
Antonio DiTommaso, Cornell University

Field pennycress plant
Scott Morris, Cornell University

Field pennycress fruit
Scott Morris, Cornell University

IDENTIFICATION

Other common names: bastardcress, fanweed, stinkweed, mithridate mustard, frenchweed, French weed, bastardweed, dish mustard, field thlaspi

Family: Mustard family, Brassicaceae

Habit: Erect, branched, summer or winter annual herb

Description: Cotyledons are hairless, unequally sized, 0.2–0.4 inch long by 0.1–0.3 inch wide, oval shaped and bluish-green in color. The first leaves on the **seedling** are opposite and oval shaped, with wavy margins; all subsequent leaves are alternate. Young leaves form a basal rosette. **Mature plants** reach 20–32 inches tall. Basal leaves are long stalked (0.5–2 inches); the pale green blades are hairless, 1–2.5 inches long, egg shaped and widest above the middle; they wither at maturity. Stem leaves are stalkless, dark green, irregularly toothed, lanceolate to linear, 0.75–2 inches long and tapered; they persist to maturity. Pointed auricles are present at the base of stem leaves. An extensive, fibrous root system extends from a thin taproot. All plant parts exude an unpleasant odor when crushed. **Flowers** are inconspicuous, 0.13 inch wide, have four white petals and are clustered at stalk ends. The stalk continues to elongate at maturity, producing new flowers at its tip and maturing seeds below. Seedpods are round, 0.5 inch wide and flat with winged, papery edges. Seedpods have a central line and a notched tip. **Seeds** are light to dark brown or purplish, 0.06 inch long and egg shaped. Seeds have one straight side, one round side and distinctive fingerprint-like surface ridges.

Similar species: Shepherd's-purse [*Capsella bursa-pastoris* (L.) Medik.] seedpods are more heart or triangular shaped, and the flower stalks are mostly leafless and unbranched. Thoroughwort pennycress [*Microthlaspi perfoliatum* (L.) F.K. Mey.] has smaller seeds contained in 0.25 inch-long seedpods and has rounded auricles.

MANAGEMENT

Delaying planting of both spring and fall seeded crops allows more field pennycress seeds to germinate before tillage. Where feasible, rotation with summer planted crops reduces field pennycress populations, but only if the weed is prevented from setting seed in the spring before planting. Early spring cultivation during the fallow year of grain-fallow rotations will prevent seed production and will improve good control. Otherwise, field pennycress produces seeds exceptionally early in the spring, and substantial seed input can occur in fallow fields while other fields are being prepared and planted with crops. Some winter wheat varieties are substantially more competitive against this weed than others (Holm et al. 1997), and the same probably applies for other grains. Tall, leafy, small grain varieties are likely to be more competitive since this species does not tolerate shade. Banding N in grain crops decreases the density of field pennycress. A substantial fraction of field pennycress seedlings can be selectively removed from many crops with well-timed tine weeding, but usually many will survive. Unlike most annual weeds, field pennycress populations are not reduced by rotation into alfalfa. Fall germination and early spring seed production allows field pennycress to complete its life cycle while avoiding competition with alfalfa. Also, mowing causes branching and rapid regrowth, with little reduction in final plant size or seed production. Smooth brome and crested wheatgrass, however, effectively suppress field pennycress.

Use dense winter cover crops of rye or spelt to suppress field pennycress in vegetable crops, but be sure to incorporate these in the spring before the weed goes to seed. Rework the seedbed before planting to flush seeds out of the soil and kill seedlings. Tine weed or hoe around crops until they are well established to eliminate early flushes.

ECOLOGY

Origin and distribution: Field pennycress originated in Eurasia and is widespread from Japan to Spain. It is present throughout the United States and in Canada as far north as the Yukon, but it is most problematic in the northern Great Plains region. It has also been introduced into Australia, New Zealand and Argentina.

Seed weight: 0.74–1.5 mg.

Dormancy and germination: Various strains of the species show differing germination behavior. In most field-grown populations, seeds are dormant when shed from the parent plant and require an additional month or more of after-ripening. Seeds of various strains respond differently to cold. For some non-dormant strains, cold temperatures, for example 36°F, induce deep dormancy that is only broken by exposure of the seeds to several months of warmer temperatures

(43–95°F). In other dormant strains, several weeks of cold temperatures break dormancy. Regardless of strain, during the transition between dormancy and non-dormancy, germination is promoted by red light but inhibited by light filtered through a crop canopy. Nitrate and day/night temperature fluctuations also promote germination of non-dormant seeds. All of these factors probably contribute to the flush of germination that commonly follows fertilization and tillage. Non-dormant seeds are capable of optimum germination at a wide range of day/night temperatures, from 59/43°F to 95/68°F. Germination of this species requires relatively higher soil moisture compared to crops.

Seed longevity: The seeds can persist 17–30 years in undisturbed soil. In a five-year experiment, the number of field pennycress seeds declined by 50% per year in soil stirred four times per year, but only 10% per year in undisturbed soil.

Season of emergence: Most emergence occurs in the spring; relatively few individuals emerge in mid-summer, and a second, usually smaller, peak in emergence occurs in the fall. In Wisconsin, field pennycress is among the earliest weeds to emerge in spring.

Emergence depth: Most seedlings of field pennycress emerge from the top 0.8 inch of soil. The emergence requirements of a shallow seed depth combined with high soil moisture may explain why non-dormant seeds capable of germinating at high temperatures do not emerge in mid-summer when surface soil is usually dry.

Photosynthetic pathway: C_3

Sensitivity to frost: The species is very cold hardy, and cold acclimated plants can survive 7°F with little damage. Fall germinating individuals overwinter and flower the following spring.

Drought tolerance: The root system of field pennycress has a higher density and total length than most grain crops and other weeds, which makes the species a good competitor for moisture. Relative to other species, however, field pennycress requires more water per unit of growth.

Mycorrhiza: This species is non-mycorrhizal.

Response to fertility: Plant size increases steadily with increasing N application rate up to 143 pounds per acre. Field pennycress has been categorized as intermediate in N response relative to other weed species. The species also continues to respond to phosphorus fertilization up to very high application rates.

Soil physical requirements: Field pennycress occurs on all types of soil suitable for crop production.

Response to shade: The species is intolerant of shade and usually cannot push through a crop canopy.

Sensitivity to disturbance: Removal of the shoot tip increases branching but has little effect on mature plant weight or seed production. Seeds in green pods will continue to mature even after being plowed under.

Time from emergence to reproduction: Plants establishing in the spring flower five to seven weeks after emergence, whereas overwintering plants flower a month earlier. The seeds require about two weeks to fully mature; however, immature seeds become viable as little as six days after pollination.

Pollination: Self-pollination is common, but 10–20% of flowers are cross pollinated by insects.

Reproduction: Spring emerging plants grown in the field in Saskatchewan with little competition produced an average of 14,000 seeds per plant, whereas fall emerging plants produced an average of 9,400 seeds per plant. Field pennycress averaged 300 seeds per plant in high density monocultures.

Dispersal: Seeds are dispersed in soil on tires, field machinery, shoes and combines. It is commonly found in manure and can survive digestion in the guts of transported livestock. Viable seeds have been recovered from droppings of several species of wild birds. It is also a common contaminant of grain and forage seed, which is a major way it has moved between continents. The seeds disperse short distances with wind and longer distances in irrigation water.

Common natural enemies: Field pennycress is susceptible to the pathogens and insects that commonly attack crops in the mustard family, but there is no record that damage levels are detrimental to growth or reproduction. Carabid beetle activity was highly correlated with field pennycress seed predation from the soil surface; losses from fields in late summer averaged 23% per week.

Palatability: Consumption by cattle gives meat and milk an unpleasant flavor and in large quantities can cause poisoning. Field pennycress seeds can contaminate canola and lower the quality of canola oil. The species is cultivated in parts of Europe for the tender young shoots, which are eaten in salads or cooked like spinach.

Note: Field pennycress is currently being domesticated as a winter annual oilseed crop that would be complementary with a corn-soybean rotation in the upper Midwest. It also has potential as a winter hardy cover crop that can provide early spring pollinator services.

FURTHER READING

Holm, L., J. Doll, E. Holm, J. Pancho and J. Herberger. 1997. *World Weeds: Natural Histories and Distribution*. John Wiley & Sons: New York, NY.

Mitich, L.W. 1996. Field pennycress (*Thlaspi arvense* L.): The stinkweed. *Weed Technology* 10: 675–678.

Warwick, S.I., A. Francis and D.J. Susko. 2002. The biology of Canadian weeds. 9. *Thlaspi arvense* L. (updated). *Canadian Journal of Plant Science* 82: 803–823.

Flixweed

Descurainia sophia (L.) Webb ex Prantl

Flixweed foliage
Joseph DiTomaso, University of California, Davis

Flixweed plant in flower
Joseph DiTomaso, University of California, Davis

Flixweed seed pods
Joseph DiTomaso, University of California, Davis

IDENTIFICATION

Other common names: herb sophia, tansy mustard

Family: Mustard family, Brassicaceae

Habit: Slender, erect, branching winter annual or biennial herb

Description: Seedlings have light green cotyledons that are 0.2–0.3 inch long by 0.04–0.06 inch wide, oval to club shaped, and rough due to surface hairs. The seedling stem is short and sparsely hairy. The first true leaves are opposite on the stem, usually tri-lobed, stalked and hairy. All subsequent leaves are alternate and subdivided into leaflets. Leaflets have many narrow, closely spaced lobes. Seedling leaves are 0.13–0.3 inch long and develop into a basal rosette that withers during flowering. Stem and leaves have star shaped hairs. The **mature plant** bolts from the rosette; stem height ranges from 8–32 inches. The upper half of the hairy stem has branches up to 15 inches long. Stem leaves are 0.75–4 inches long by 0.5–2 inches wide, narrowly oval, pinnate, covered in star-shaped hairs, and narrower when located higher on the stem. Leaflets are narrow and oval to linear in shape with pointy tipped lobes. The taproot is large and sometimes branches. **Flowers** are green-yellow to yellow, 0.13 inch wide, upward facing, individually stalked, four-petaled and clustered into a triangular group at the stem tip. Flower stalks are 0.5 inch long and are attached perpendicular to the stem. The stem continuously lengthens as new flowers develop at the tip and as seeds ripen lower on the stem. **Fruit** are long, skinny pods divided into two chambers by a papery, translucent, internal membrane. Pods are green, bluntly round tipped, 0.5–1.25 inch long by 0.04 inch wide, straight or slightly curved; they turn brown at maturity. **Seeds** are thumb to egg shaped, smooth, dull orange and 0.02–0.05 inch long by 0.01–0.02 inch wide. Each chamber has a row of 10–20 seeds. Seeds are sticky when wet.

Similar species: Pinnate tansymustard [*Descurainia pinnata* (Walter) Britton] has less finely divided, lacy looking leaves and shorter fruit at maturity; its seeds are in two rows per pod chamber, and the seedling stem is maroon. Smallflowered bittercress (*Cardamine parviflora* L.) seedlings have similar rosettes, but their leaves are not subdivided into smaller leaf units with central branches.

MANAGEMENT

If feasible, rotate with summer planted crops to reduce flixweed populations. In grain-fallow rotations, good control of flixweed in the fallow years is essential. First tillage with a flat bladed implement should occur before flowers drop. Otherwise, disk lightly to chop up the plants, but, in any case, do not let filled pods form. Following grain-grain rotations, harrow or disk the field shallowly several times between harvest and winter, and then again in the spring. Plow and plant two weeks later. Tine weed the grain before emergence and again as soon as the crop will tolerate it. If further flushes of the weed occur, continue tine weeding until grain stalks begin to elongate. Consider overseeding with a clover at the last tine weeding. This will not compete with the crop but will compete with newly emerging flixweed in the fall and allow the soil to rest from excessive tillage. Flixweed pods often do not shatter until grain harvest so consider trapping chaff to prevent seed return. Alternatively, capture the flixweed seed with the grain in the combine and separate the two later with a grain cleaner.

Although flixweed tolerates crop competition better than most annual weeds, a vigorous crop is still an essential component of management. Ensure that planting density is sufficient to compensate for stand losses that may result from aggressive tine weeding. Except during drought years, the crop is likely to respond to good soil fertility more than flixweed.

ECOLOGY

Origin and distribution: Flixweed is native to Europe and occurs from North Africa to Scandinavia and across Asia to northern India and China. In North America, it occurs throughout the United States, except in parts of the Southeast, and northward in Canada and Alaska to the Arctic. It has also been introduced into South America and New Zealand.

Seed weight: 0.12 mg.

Dormancy and germination: Nearly all flixweed seeds are dormant when shed from the parent plant. Germination is limited in the first fall and spring after seed shed but becomes more plentiful in subsequent falls and springs. Dormancy is broken by exposure to summer-like soil temperatures, with optimum day/night temperatures of 77/59°F. However, seeds will not germinate at this temperature but will wait for cooler temperatures of 59/43°F to 68/50°F. This ensures that the main peak in germination occurs during cool weather in the fall, but after the seeds have experienced summer heat. Light is a primary requirement for germination; few seeds will germinate in dark regardless of temperature. Although some research has shown that a few days of cold (39°F), wet and darkness can break dormancy, other research shows that extended periods of such conditions for longer than five days can induce secondary dormancy to such an extent that spring germination is limited. This ensures that the seeds will not germinate during extended periods of cold in winter or early spring. Seeds germinate best with alternating day/night temperatures. Nitrate breaks dormancy.

Seed longevity: Flixweed forms a persistent seed bank that declines slowly, even with tillage. Nearly all seeds remained viable after 30.5 months burial in Sweden. Viable seeds in undisturbed soil in Alaska declined by 25% per year. Based on the number of seeds surviving after six years in a spring grain-fallow rotation in Saskatchewan, the rate of decline in tilled soil was 23% in clay, 33% in loam and 32% in sandy loam.

Season of emergence: Flixweed emerges primarily in autumn and early spring, with higher emergence in fall than in spring.

Emergence depth: As would be expected for a species with small seeds, flixweed seeds must be near the soil surface for seedling emergence.

Photosynthetic pathway: C_3

Sensitivity to frost: Flixweed will survive well at -13°F. It commonly overwinters as a rosette of small leaves; snow cover protects it from cold temperatures and drying winds in far northern latitudes.

Drought tolerance: Flixweed requires moist soil, and it inefficiently uses soil moisture. It is probably abundant in dry regions because of its ability to grow during cool, moist periods of the year.

Mycorrhiza: Flixweed is not mycorrhizal.

Response to fertility: Flixweed shoot growth is relatively unresponsive to N, particularly when growing with a competitive crop. However, roots increase greatly with increasing N fertility, which may increase its competitiveness for water in high N soils. It is relatively unresponsive to P. The extensive taproot system of flixweed facilitates the extraction of nutrients from deep layers of the soil.

Soil physical requirements: The species grows on a wide range of soils from clay to sand, but it is considered an indicator of sandy or stony soils. It tolerates a wide range of pH and can thrive on calcareous sites that are low in humus.

Response to shade: Flixweed can grow in semi-shade. It responds to shade by reduced allocation to roots and greater allocation to leaf area, which is typical of shade-tolerant plants.

Sensitivity to disturbance: Close mowing reduces seed production.

Time from emergence to reproduction: Overwintering seedlings flower about four weeks after resuming growth in spring. Plants emerging in late April in Saskatchewan flowered about one month later, but plants that were not exposed to cold temperatures required several months to reach maturity. Seeds matured in mid-summer, approximately six weeks after the beginning of flowering.

Pollination: Flixweed normally self-pollinates.

Reproduction: Average seed production has been estimated to range from 4,000–76,000 seeds per plant, but an exceptionally large plant produced 700,000 seeds. In a competitive wheat crop, seed production was reduced to 150–250 seeds per plant.

Dispersal: The seeds produce a sticky substance that attaches them to bird feathers. Most seeds die in the rumen, but some can pass through and are spread in manure. Seeds are also spread as contaminants of grain and forage seed. The seeds survive in water for several years and disperse in irrigation water.

Common natural enemies: Flea beetles, cabbage seedpod weevils and beet leaf hoppers use flixweed as an alternative host, but they have little adverse effect on this species.

Palatability: The seeds have a high oil and protein content. Native Americans parched seeds and cooked them into a porridge, made them into bread or used them to thicken soup. Young plants can be eaten raw but have a strong odor and bitter taste. Bitter taste and odor can be removed by boiling for three to four minutes with two changes of water. Flixweed is low in palatability for most livestock, but it sometimes forms an important part of the diet of cattle in the Intermountain West. Consumption of mature plants by cattle can cause poisoning.

FURTHER READING

Best, K.F. 1977. The biology of Canadian weeds. 22 *Descurainia sophia* (L.) Webb. *Canadian Journal of Plant Science* 57: 499–507.

Blackshaw, R.E. 2004. Application method of nitrogen fertilizer affects weed growth and competition with winter wheat. *Weed Biology and Management* 4: 103–113.

Landau, C.A., B.J. Schutte, A.O. Mesbah and S.V. Angadi. 2017. Flixweed (*Descurainia sophia*) shade tolerance and possibilities for flixweed management using rapeseed seeding rate. *Weed Technology* 31: 477–486.

Mitich, L.W. 1996. Flixweed (*Descurainia sophia*). *Weed Technology* 10: 974–977.

Galinsogas

Hairy galinsoga, *Galinsoga quadriradiata* Cav. = *G. ciliata* (Raf.) S.F. Blake
Smallflower galinsoga, *Galinsoga parviflora* Cav.

Hairy galinsoga seedling
Scott Morris, Cornell University

Hairy galinsoga mature plant
Scott Morris, Cornell University

Hairy galinsoga flower
Scott Morris, Cornell University

Smallflower galinsoga plant
Joseph DiTomaso, University of California, Davis

Smallflower galinsoga seed head
Joseph DiTomaso, University of California, Davis

IDENTIFICATION
Other common names:

Hairy galinsoga: French weed, common quickweed, shaggy soldier, ciliate galinsoga, fringed quickweed, quickweed, Peruvian daisy, shaggy galinsoga

Smallflower galinsoga: gallant soldier

Family: Aster family, Asteraceae

Habit: Highly branched, summer annual herbs

Description: Seedlings have 0.4 inch-long, stalked cotyledons and opposite young leaves.

Hairy galinsoga: Cotyledons are hairless and square or egg shaped, with flattened, slightly indented tips. Seedling stems are short, green and sometimes turn maroon with time. Young leaves are triangular or egg shaped and light green, with a pointed tip. Three prominent veins are red tinged above and hairy below. Dense hairs are present on the stem, stalks, toothed leaf edges and upper leaf surface; hairs located on leaf edges point toward the leaf tip.

Smallflower galinsoga: Cotyledons are round to lima bean shaped, with indented tips and a fringe of tiny hairs along the margin. Young leaves are paddle shaped to broadly oval, slightly cupped and sometimes toothed, with a tapered tip. Very young leaves are grooved along one to three veins.

Mature plants are upright and much-branched. Branching occurs in even pairs from axils of the opposite leaves.

Hairy galinsoga: Densely hairy stems reach 4–28 inches in height. Coarse hairs are present on leaf stalks, the entire upper leaf surface and veins on the lower leaf surface. The broad leaves are 1–3 inches long by 0.5–2 inches wide, egg shaped to triangular, with coarsely toothed, hairy edges. Roots are shallow and fibrous.

Smallflower galinsoga: Irregularly hairy stems are 12–24 inches tall. Hairs are concentrated near stem-leaf joints and on leaf stalks. Leaf stalks are 0.5 inch long, thin and may be absent on younger leaves. Leaves are egg shaped to lance-egg shaped with a pointed tip, 0.3–4.25 inches long by 0.1–2.75 inches wide, light green and finely to coarsely toothed. A small taproot may be present.

Flower heads of both species are clustered at branch ends; they have four to five small, usually white, 0.1 inch-long ray flowers. Petals have three rounded teeth at their tip. The center of the flower head is a mounded group of distinct yellow disk flowers that each produce one seed.

Hairy galinsoga: Flower heads are 0.25 inch in diameter, with occasional pink petals. Petaled ray flowers do not produce seeds.

Smallflower galinsoga: Flower heads are 0.13 inch in diameter. Petals are only white, and the petaled ray flowers produce seeds. There are 15–50 yellow disk flowers per ray flower.

Fruit and seeds: The apparent seeds are covered with a tight hairy coating of fruit tissue. These four-sided seeds are topped with a crown of papery brown scales.

Hairy galinsoga: Seeds are 0.1 inch long and torpedo shaped.

Smallflower galinsoga: Seeds are 0.06 inch long and cylinder or oval shaped. Slightly wedged seeds are derived from the ray flowers.

Similar species: Seedlings can be confused for Virginia copperleaf (*Acalypha virginica* L.), but the two can be distinguished by looking for the notched, not broadly indented, tip of the Virginia copperleaf cotyledon. Unlike *Galinsoga* spp., all but the first true leaves of Virginia copperleaf are alternate.

MANAGEMENT

Hairy and smallflower galinsoga are particularly problematic in low-growing vegetable and specialty crops. Since their seeds do not persist for more than a couple of years in the soil, one of the best tactics for managing these weeds is to rotate fields into a sod crop periodically. Three or four years in sod is usually sufficient to nearly or even completely eradicate an infestation. Reduction of the population will be more complete if galinsoga is prevented from going to seed during the establishment of the sod, for example by use of a nurse crop or mowing. A few years of aggressive control to avoid seed production can also greatly reduce populations. For example, a competitive summer annual cover crop can reduce galinsoga seed production by more than 98%. After the seed bank has been depleted, avoid reintroducing galinsoga in soil clinging to shoes and machinery.

Controlling the galinsogas requires regular attention. Because these species go to seed so rapidly, clean up fields immediately after harvest to reduce population growth. During summer fallow periods, either keep the soil clean cultivated at three- to four-week intervals to flush out seedlings, or plant with a competitive cover crop. If time and the season permit, try to work in a tilled fallow period before planting. Since the seeds can only emerge from the top 0.4 inch of soil, a thorough cultivation at shallow depth can effectively deplete the surface seed bank without bringing up new seeds.

Two alternatives are available for handling very heavy seedings that result from an occasional year with poor control. First, since the seeds mostly germinate near the soil surface, do not till between seed production and when the first flowers appear on spring germinating plants. Most of the seeds will have germinated by then and will be destroyed when a seedbed is prepared for an early summer crop. Second, moldboard plow the seeds under as deeply as you can. For the next few years use relatively shallow tillage while the deeply buried seeds die off.

ECOLOGY

Origin and distribution: Both species originated in Central America. Both species occur in moist areas of the United States, primarily in the Northeast and northern states of the Midwest, with scattered occurrence elsewhere in the moister parts of the United States. Both species are widespread in temperate and tropical North and South America, and they have been introduced into various parts of Europe, Asia, Africa and Australia. They are rare in dry regions of the world, even in irrigated crops.

Seed weight: Mean population seed weights of hairy galinsoga range from 0.19–0.23 mg. Mean population seed weights of smallflower galinsoga range from 0.17–0.27 mg.

Dormancy and germination: Freshly shed seeds of both species will germinate immediately if exposed to light and warm temperatures. However, seeds of hairy galinsoga shed in summer were dormant until late fall in the Czech Republic, and some lots of smallflower galinsoga harvested in summer months were dormant. Germination approaches 100% in light at constant temperatures ranging from 54–97°F or at fluctuating day/night temperatures ranging from 59/50°F to 95/68°F, but it decreases at cooler temperatures of 59/43°F or lower. Both species show markedly reduced germination in the dark, which allows them to persist until the following season when buried in the soil. Nitrate has no effect on germination.

Seed longevity: Studies of seed longevity are limited. In India, soil seed populations declined by more than 99% in cropland from winter to summer due to a combination of emergence and mortality. In Mexico, viability of seeds from disk flowers of smallflower galinsoga declined by approximately 70% in the first year after burial, whereas those from ray flowers declined by approximately 50%. The general absence of

galinsoga from farms where row crops are rotated with several years of sod indicates that few seeds survive in the soil for more than three years.

Season of emergence: Emergence begins in early spring, peaks in late spring and early summer but continues until frost, particularly following soil disturbance.

Emergence depth: Only seeds at or very near the soil surface produce seedlings. Most galinsoga seeds produce seedlings when positioned on the soil surface, but only half emerge from 0.1 inch, and none emerge from 0.4 inch.

Photosynthetic pathway: C_3

Sensitivity to frost: Both species of galinsoga die when exposed to even mild frost.

Drought tolerance: Both species are drought sensitive. Hairy galinsoga is somewhat less sensitive to drought than tomatoes.

Mycorrhiza: Smallflower and hairy galinsoga are mycorrhizal.

Response to fertility: Both species thrive when N, P and K levels are all high, whereas low levels of any of these nutrients will reduce growth and flowering. In the year following a corn crop, hairy galinsoga showed a much stronger response to P with no added N than to N with no added P. At reduced levels of N and K, smallflower galinsoga increases root growth, but hairy galinsoga does not.

Soil physical requirements: Both species occur on a variety of soil types but prefer damp, rich soil.

Response to shade: Both species grow best in full sunlight and show markedly reduced growth when shaded. Hairy galinsoga exhibits strong shade avoidance responses when grown in competition, including increased stem length and leaf area, delayed flowering and seed production, and greatly reduced biomass and seed production.

Sensitivity to disturbance: Plants of all sizes usually wilt quickly when uprooted unless the weather is unusually wet. Stem fragments buried in moist soil can apparently produce roots, thereby facilitating their persistence in cultivated vegetable crops. Well-rooted plants tend to break near the base when pulled or hoed and then quickly regenerate.

Time from emergence to reproduction: Plants flower until the first killing frost. Both species first flower 24–60 days after emergence. Flowering time of hairy galinsoga was delayed from 34 days in the absence of competition to 46 days under competitive conditions, but always occurred when there were 10 leaves on plants. Seeds are released eight to 14 days after flowering.

Pollination: Both species readily self-pollinate but are also cross pollinated by insects.

Reproduction: Once a plant begins to flower, it will continue to grow and flower until frost. Consequently, early emerging plants that are not killed by human intervention can produce many seeds. Medium sized plants of hairy galinsoga about 18 inches in diameter averaged 40,000 seeds per plant. Plants grown in the absence of competition produced approximately 25,000 seeds within 67 days, whereas those grown under moderate competition within a broccoli crop produced no more than 4,000 seeds per plant during the same time period. A plant of smallflower galinsoga in Japan produced 400,000 seeds. Since plants mature rapidly and seeds can germinate immediately upon dispersal, both species are capable of completing three to four generations per year in the Northeast.

Dispersal: Seeds disperse short distances by wind, but the fine hairs attached to the seed are much reduced relative to most wind dispersed species in the aster (composite) family. The seeds are hairy and stick readily but not persistently to fur and clothing.

Common natural enemies: Both species are commonly attacked by a diversity of aphid and leafhopper species. They are also susceptible to galinsoga mosaic virus.

Palatability: Leaves, stems and flowers of hairy galinsoga are cooked in soups in southern Mexico. Smallflower galinsoga contains high concentrations of protein and minerals and is consumed as a leafy vegetable in several areas of the world.

FURTHER READING

Damalas, C.A. 2008. Distribution, biology, and agricultural importance of *Galinsoga parviflora* (Asteraceae). *Weed Biology and Management* 8: 147–153.

Kumar, V., D.C. Brainard and R.R. Bellinder. 2009. Effects of spring-sown cover crops on establishment and growth of hairy galinsoga (*Galinsoga ciliata*) and four vegetable crops. *HortScience* 44: 730–736.

Warwick, S.I. and R.D. Sweet. 1983. The biology of Canadian weeds: 58. *Galinsoga parviflora* and *G. quadriradiata* (= *G. ciliata*). *Canadian Journal of Plant Science* 63: 695–709.

Giant ragweed

Ambrosia trifida L.

Giant ragweed seedling
Antonio DiTommaso, Cornell University

Young plant of giant ragweed
Antonio DiTommaso, Cornell University

Giant ragweed male flower
Antonio DiTommaso, Cornell University

IDENTIFICATION

Other common names: great ragweed, buffalo weed, kinghead, crown weed, wild hemp, horse weed, bitterweed, tall ambrosia, tall ragweed

Family: Aster family, Asteraceae

Habit: Tall, branched, summer annual herb

Description: Seedling cotyledons are round to paddle shaped, 0.75–1.75 inch long by 0.25–0.5 inch wide. The stem is shiny and green with purple spots. The first two true leaves are lanceolate to oval shaped with widely spaced, shallow teeth, and they may have two basal lobes. Subsequent young leaves are opposite, roughly hairy and divided into three deep lobes. **Mature plants** are typically 3–12 feet tall, though they may reach up to 20 feet in height. Stems are single or branching and are covered in short, rough hairs. Leaves are opposite, up to 12 inches long and 8 inches wide, usually divided into three to five lobes, toothed along the edges and hairy on all surfaces. Leaf stalks are 0.4–2.75 inches long and sometimes winged. Uppermost leaves are lanceolate. Roots are fibrous, occasionally with a short taproot. **Flowers** are green, 0.125 inch in diameter and arranged in separate clusters of male and female flowers on a single plant. Male clusters are arranged in dense, 3–8 inch-long spikes at the ends of stems and branches. Female flowers are located in clusters of leafy bracts below the male spikes and in the upper leaf axils. Each female flower produces one thick-walled fruit encasing a single seed. Fruits are 0.2–0.6 inch long, ridged, brown to grey/black, with a blunt beak at the apex. Ridges terminate in five to eight blunt spines that form a crown around the beak. **Seeds** are brown and oval to egg shaped.

Similar species: Common ragweed (*Ambrosia artemisiifolia* L.) leaves are much more finely dissected than those of giant ragweed. Young giant ragweed plants may resemble sunflowers (*Helianthus* spp.). Sunflower leaves are not lobed and become alternate as the plant grows, while giant ragweed leaves are lobed and opposite.

MANAGEMENT

Giant ragweed is a rapid growing, competitive species that can cause substantial yield losses even at low densities. It is increasingly a problem in field crops and is associated most strongly with soybean production and minimum tillage. Efforts should be made to control it in fencerows and field borders since problematic populations in fields are often associated with its presence in border habitats.

Rotation with hay or pasture will give the relatively short-lived seed bank a chance to decline in density. Inclusion of crops or cover crops that provide early spring leaf canopy or residue cover that lowers soil temperature delays initiation of giant ragweed emergence. Rotation with cereal grains provides an opportunity after harvest to kill giant ragweed before it can go to seed. Also, since the majority of seeds are retained on plants after typical grain harvest dates, seeds can be captured or destroyed during combine harvesting. On vegetable farms, it can be eliminated by tillage after short season spring crops, which prevents reproduction, or by tillage before summer planted crops, which kills seedlings after the bulk of emergence has already occurred.

Because seeds on the soil surface are subject to high rates of seed predation, if possible, avoid fall tillage after harvest of corn, soybeans and other late harvested crops. If a cover crop is required, interseed it during the last cultivation, or broadcast it on the soil surface before or after harvest. Refuges for rodents and large invertebrate seed predators (e.g., grassed drainage ways, hedgerows) can potentially increase predation of giant ragweed seeds. However, every effort should be made to prevent reproduction when earthworm populations are high, because they can facilitate seed survival by burying seeds in their burrows within a few days.

Giant ragweed tends to emerge early in the spring, but seeds on the soil surface may not get sufficient moisture for germination. Shallow tillage early in spring to cover the seeds and promote emergence followed by a later tillage before planting soybeans or dry beans can further eliminate a substantial proportion of the previous year's seed production.

Due to its large seed size, giant ragweed seeds can emerge from below the planting depth of crops and grow very quickly, which makes rotary hoeing largely ineffective. The window for tine weeding is very narrow and corresponds to the stage when the seed leaves are just unfolding. A stiff tined implement can break and bury a substantial portion of the seedlings at this stage. Use a belly mounted cultivator or a three-point hitch mounted cultivator with a good guidance system to cultivate as close as possible to the crop row. Low pitch sweeps run very shallow are most effective for cutting off the giant ragweed without damaging the crop. Hilling up in the crop row is often ineffective since the weed usually grows as fast as the crop. If you need to hill up to control other species, you may need to make two passes with different machines or modify your cultivator. One possible configuration is to run shallow sweeps in front and a large sweep with a hilling attachment on the center shank in the rear. Because giant ragweed quickly emerges above the canopy of soybeans and dry beans, a front mounted mower can greatly reduce seed production and competition with the crop.

ECOLOGY

Origin and distribution: Giant ragweed is native to stream banks and floodplains of North America. It currently occurs throughout most of the United States, southern Canada and into Mexico, although its range has probably increased since European settlement. It is most common along tributaries of the Mississippi River north of the Ohio River. It has been introduced into Europe and Asia.

Seed weight: The large seeds of giant ragweed vary substantially in size and shape within plants, between plants and between locations and years. Average seed weights from 17–45 mg (including the fruit coat) have been reported.

Dormancy and germination: Less than 5% of freshly produced giant ragweed seeds will germinate. Stratification at 39°F is required for germination. A minimum of six weeks of stratification was required to alleviate dormancy. Excising embryos released them from dormancy, suggesting that dormancy is partially imposed by the seed coat and associated structures. Nitrate can reduce but not eliminate the stratification requirement. Light has little influence on germination. Seeds that had been exposed to natural winter conditions in Illinois germinated at temperatures ranging from 46–97°F, but germination was greatest at 50–75°F. Optimum germination occurs at soil moisture content near field capacity.

Seed longevity: Giant ragweed seed banks deplete rapidly. In the Duvel long-term experiment, most seeds were lost in the first year, but a few seeds survived 21 years. In the absence of seed production, 96% of the seeds in the soil were depleted in two years. In Illinois, one study showed that 7% of seeds produced in the fall survived to the spring, while another study determined that 5–14% survived for one year. The large seeds of giant ragweed suffer very high rates of seed predation. A study in Ohio showed about 40% overwinter loss of seeds due to rodents. Viable seeds declined to 8–34% after one year, and few seeds lasted more than four years unless they remain deeply buried. However, earthworms can facilitate burial and persistence of giant ragweed seeds in no-till fields.

Season of emergence: In the absence of soil disturbance, most seedlings emerge in early spring. Giant ragweed is typically one of the first annual species to emerge. It also produces flushes of additional emergence following tillage or cultivation during spring and summer. Emergence began in late March in Ohio and lasted only a month for natural populations but lasted throughout the spring months for agricultural populations. Emergence similarly occurred throughout all spring months in Minnesota agricultural fields and was enhanced by colder overwinter temperatures, which presumably facilitated loss of dormancy. The trend toward longer emergence periods is positively associated with the increasing presence and difficulty of controlling this species in agricultural production.

Emergence depth: This species emerges best from the top 0.5–2 inches, but a substantial percentage of seedlings can emerge from 4 inches. None emerge, however, from 8 inches. Plant survival and vigor following emergence declines with increasing burial depth of the seed. Shallowly buried seeds (0.2 inch) and seeds on the soil surface have poor germination.

Photosynthetic pathway: C_3

Sensitivity to frost: Giant ragweed is damaged but not killed by moderate frost. However, giant ragweed often matures and begins senescing before the first frost.

Drought tolerance: Giant ragweed is not well adapted to drought. Well established plants can survive several weeks of dry weather, but the species is absent from non-irrigated land in regions with long summer droughts. Presence of this weed was associated with high rainfall and moderate temperatures in October, which probably facilitates production of viable seeds.

Mycorrhiza: Although there are no reports of mycorrhizal associations with giant ragweed, it is likely to be mycorrhizal because the closely related species, common ragweed, has been described as a strong mycorrhizal host (see the Common ragweed chapter).

Response to fertility: The species is usually found on highly fertile soils but shows only a moderate response to additional fertilization. Giant ragweed can accumulate up to 100 pounds per acre of nitrogen and could be highly competitive with nitrogen-requiring crops.

Soil physical requirements: Giant ragweed will grow on a range of soil types, but most typically the species occurs on silty lowland soils.

Response to shade: In pure stands, plants that emerge 10–30 days later than the early emerging individuals are severely reduced in size, which indicates that the species can be suppressed by shade. The very rapid growth rate and tall stature of giant ragweed, however, often allows it to overtop crops before they can cast significant shade.

Sensitivity to disturbance: Plants re-grow well after cutting, even when cut close to the ground. Plants cut during combine harvesting of cereal grains send up side shoots that produce seeds. Cutting at 2–4 inches reduced giant ragweed growth more than seed production, and it took repeated cuts to have a substantial impact on seed production.

Time from emergence to reproduction: Plants typically emerge in early spring, flower in mid-summer and mature seeds in late summer. Giant ragweed flowers in response to shortening day length and usually flowers two to three weeks earlier than common ragweed. First viable seeds were produced approximately three weeks after pollination and fertilization.

Pollination: Giant ragweed is wind pollinated. It will self-pollinate, but the female flowers are receptive before the male flowers release pollen, so cross pollination is normal. In high-density environments, male flower production declines in favor of female flower production. In greenhouse tests, plants produced by self-pollination were less vigorous than plants produced by cross-pollination.

Reproduction: Individual plants in an Ohio corn field produced an average of 150 and 240 seeds in successive years, but less than half of these were viable. At a low density, individual plants produced from 1,300–3,600 seeds, with viability ranging from 59–77%.

Dispersal: Giant ragweed is a common inhabitant of non-crop areas and field edges, and its presence in these areas is positively associated with its presence in crop fields. The difficulty of managing this weed is most strongly associated with its occurrence near waterways. Seeds were identified on the surface of irrigation water in Nebraska. Giant ragweed seeds float for several hours to a few days, which probably allows them to disperse along stream bottom lands during flood events. The species probably also occasionally moves with soil clinging to tires, machinery and animals, but the low rate of seed production and consequent low seed density in the soil probably makes such events rare. Occasional movement in combine harvesters seems likely. Low ability to disperse out of valleys may be the reason giant ragweed is less common on upland farms.

Common natural enemies: The host specific fungus *Puccinia xanthii* f. sp. *Ambrosia-trifidae* attacks giant ragweed leaves and reduces seed production and seed size. Ten to 25% of seeds are killed by insects (fruit fly, beetles, a moth) while still on the plant, and taller plants appeared to be most susceptible. Rodents consume a large proportion of seeds on the soil surface during fall and winter, and insects kill many seeds that are on the soil surface during the summer. Several families of stalk boring insects can be prevalent in giant ragweed and may interfere with translocated herbicide activity.

Palatability: The seeds were gathered and eaten by Native Americans in the Mississippi Valley. Dehulled seeds contain 47% protein and 38% fat. The foliage is high quality forage for livestock. However, it was found to be unpalatable to sheep despite its nutritional value.

Note: The pollen of giant ragweed causes severe hay fever symptoms in sensitive individuals.

FURTHER READING

Abul-Fatih, H.A. and F.A. Bazzaz. 1979. The biology of *Ambrosia trifida* L. II. Germination, emergence, growth and survival. *New Phytologist* 83: 817–827.

Bassett, I.J. and C.W. Crompton. 1982. The biology of Canadian weeds. 55. *Ambrosia trifida* L. *Canadian Journal of Plant Science* 62: 1003–1010.

Harrison, S.K., E.E. Regnier and J.T. Schmoll. 2003. Postdispersal predation of giant ragweed (*Ambrosia trifida*) seed in no-tillage corn. *Weed Science* 51: 955–964.

Regnier, E.E., S.K. Harrison, M.M. Loux, C. Holloman, R. Venkatesh, F. Diekmann, R. Taylor, R.A. Ford, D.E. Stoltenberg, R.G. Hartzler, A.S. Davis, B.J. Schutte, J. Cardina, K.J. Mahoney and W.G. Johnson. 2016. Certified crop advisors' perceptions of giant ragweed (Ambrosia trifida) distribution, herbicide resistance, and management in the corn belt. *Weed Science* 64: 361–377.

Hemp sesbania

Sesbania herbacea (Mill.) McVaugh = S. *exaltata* (Raf.) Rydb. ex A.W. Hill = S. *macrocarpa* Muhl. ex Raf.

Hemp sesbania seedling
Jack Clark, University of California

Hemp sesbania plant
John Gwaltney, SoutheasternFlora.com

Hemp sesbania flower
John Gwaltney, SoutheasternFlora.com

Hemp sesbania fruit
John Gwaltney, SoutheasternFlora.com

IDENTIFICATION

Other common names: coffeeweed

Family: Legume family, Fabaceae

Habit: A tall, erect species, hemp sesbania behaves as an annual herb in most of the United States because it is killed by frost, but in perennially warm climates it is a semi-woody perennial.

Description: Cotyledons are fleshy, oblong to lanceolate, hairless and 0.4–1 inch long by 0.12–0.15 inch wide. The first true leaf of the **seedling** is simple, lanceolate and 0.8 inch long by 0.2 inch wide. All subsequent leaves are divided, with four to eight pairs of short-stalked, oppositely arranged, oblong to lanceolate leaflets with pointed tips. Young leaves are 0.8–2 inches long, while individual leaflets are 0.2–0.6 inch long by 0.06–0.12 inch wide. Leaf arrangement is alternate. **Mature plants** are typically 3–6 feet tall but occasionally reach over 10 feet, with smooth, green and generally unbranched stems that become woody with age. The leaves are fern-like, alternate, 4–12 inches long and divided into 20–70 oppositely arranged leaflets. Each leaflet is 0.75–2.5 inches long by 0.1–0.3

inch wide, hairless above and hairless to sparsely hairy below. Leaflet shape is similar to that of the seedling. The taproot is large and branched. The pea-like **flowers** occur in clusters of two to six on 0.7–3.8 inch-long stalks growing from the leaf axils. Flowers are yellow to yellow-orange, often spotted with purple and 0.4–0.6 inch long. Each flower is replaced by a four-sided, curved and jointed seedpod. The seedpods are 6–8 inches long by 0.1–0.2 inch wide with a 0.2–0.4 inch-long beak at the tip. Pods contain 30–40 **seeds**. Seeds are oblong, 0.1–0.3 inch long by 0.08–0.1 inch wide, glossy and solid brown to mottled tan and brown.

Similar species: Partridgepea [*Chamaecrista fasciculata* (Michx.) Greene] has similar foliage to hemp sesbania. Partridgepea leaves, however, have a distinctive orange gland at the base and have fewer leaflets (16–30) than those of hemp sesbania.

MANAGEMENT

Hemp sesbania is an important weed of rice and summer row crops. However, because its diffuse canopy of small leaflets allows passage of light and the profusion of N-fixing nodules on its root system minimize competition for N with the crop, this weed is often less competitive than its rapid growth and great stature would indicate.

Hemp sesbania grows best at hot, mid-summer day/night temperatures of 86/77°F or above. Soybean growth is slowed less by cooler, 77/59°F day/night temperatures, than hemp sesbania. Hence, establishing soybeans relatively early in the growing season helps give the crop a competitive head start. Similarly, corn and spring cereals that can germinate and grow at cooler temperatures than are suitable for hemp sesbania can competitively suppress it. Since the seeds survive well in undisturbed soil, rotation with sod crops like alfalfa will only moderately reduce the seed bank.

Since some hemp sesbania seedlings will emerge from below crop planting depths, tine weeders and rotary hoes are only partially effective. Cultivate shallowly close to the row. Plants do not usually emerge above the soybean canopy until late July or early August, but most competition occurs four to 10 weeks after soybean emergence. Nevertheless, if possible, mow off plants just above the crop canopy shortly after the weed begins to flower to reduce seed production. The tall height, tough stems and superficial root system probably make this weed susceptible to mechanical weed pullers in soybeans, cotton and some other crops. Yield loss in soybeans can be reduced by high crop density (e.g., 209,000 plants per acre). High density planting in rice probably will also help suppress the weed. Irrigation of soybeans with unmanaged hemp sesbania in Arkansas increased growth and light interception by the weed at the expense of the crop but still substantially increased crop yield.

Since hemp sesbania supports nitrogen fixing symbiotic bacteria, it is relatively independent of soil N for growth. Hence any N deficiency will competitively favor this weed over the crop. It produces root exudates that inhibit soybean nodulation, so N fertilization sometimes helps soybeans compete with this weed.

ECOLOGY

Origin and distribution: Hemp sesbania is native to the Southeast, southern Midwest and southwest United States; it occurs south to Central America and sporadically northward to southern Ontario. It has been introduced into Russia.

Seed weight: Population mean seed weights range from 6–15 mg.

Dormancy and germination: Hemp sesbania seeds have physical dormancy maintained by an impermeable seed coat and at maturity, few seeds will germinate. Permeable seeds germinate well at temperatures from 58–104°F, but germination is negligible at 41°F. Light neither promotes nor inhibits germination. Seeds have to take up 58–65% of their weight in water to germinate, and two or more cycles of partial hydration followed by dehydration greatly decrease germinability, probably by killing the seeds. However, seeds will germinate under greater moisture stress than soybeans.

Seed longevity: Hemp sesbania seeds persist well in the soil seed bank, with 18% surviving 5.5 years of undisturbed burial, resulting in a computed 27% annual rate of mortality per year. Nevertheless, when a hemp sesbania seed bank was tilled annually to depths of 2–6 inches, emergence declined by an average of 65% per year over a four-year period, with slightly faster declines with deeper tillage. Overwinter flooding of fields causes only minor deterioration of seeds.

Season of emergence: In a study in Mississippi, seedlings emerged primarily from May through July, with often two distinct periods of emergence. A series of soybean experiments planted mid-May to early June found that most hemp sesbania emerged simultaneously with the crop, and in another experiment 98% of seeds emerged within a six-week period.

Emergence depth: Hemp sesbania emerges best from 0.4–1.2 inch, but a few seedlings can emerge from depths up to 5 inches. Under favorable conditions, seeds in the top 1.2 inch of soil can emerge in as little as two days, giving the weed a head start over many crops.

Photosynthetic pathway: C_3

Sensitivity to frost: Hemp sesbania does not tolerate frost.

Drought tolerance: The species grows best in moist conditions but can tolerate drought once it is well established.

Mycorrhiza: Hemp sesbania is mycorrhizal.

Response to fertility: Hemp sesbania roots form an association with nitrogen fixing rhizobial bacteria and are not responsive to N fertility. It tolerates pH down to 4.5 and also tolerates alkaline conditions.

Soil physical requirements: Hemp sesbania is most troublesome on clay and heavy loam soils, and it grows poorly on sandy soils. **Hemp sesbania tolerates flooding and waterlogged soils,** so it can be a problem in both rainfed and paddy rice conditions. **It also tolerates salinity and has been used as a green manure for reclaiming saline soils.**

Response to shade: It is highly intolerant of shade, but if given the opportunity, it grows 10 feet or more in height and thus can potentially overtop the tallest of crops.

Sensitivity to disturbance: Killing plants at initial pod set, defined as 75% blossom drop when pods were less than 3 inches, reduced production of viable seeds by 94%.

Time from emergence to flowering: Plants **flower from June to October, with flowers appearing** about six to seven weeks after emergence. The time from initial pod set to seed maturity is about six weeks. Thus, plants emerging in May mature by mid-August. In California, seed production continues from June to October, but seed production from a given flush of emergence occurs more or less simultaneously.

Pollination: Flowers are visited by bees but, as with related species, may be self pollinated as well.

Reproduction: Widely spaced plants produced 21,500 seeds per plant. Plants spaced 3 feet produced 2,100–9,400 seeds per plant.

Dispersal: Legume seeds with hard seed coats survive digestion in cattle, so many seeds that are eaten are likely to disperse in feces or when manure is spread. Seeds tolerate flooding well and probably move in flood water. The seeds can contaminate crop seed and may occasionally be dispersed during disposal of cotton gin trash.

Common natural enemies: The anthracnose fungi *Colletotrichum truncatum* is being developed as a biocontrol agent.

Palatability: Hemp sesbania has better digestibility and higher crude protein than common southern forage grasses but has a higher than optimal Ca:P ratio after flowering. Although other species of *Sesbania* cause irritation of the digestive tract and diarrhea in domestic animals, hemp sesbania has low toxicity. Nevertheless, 2–3% hemp sesbania seeds in the diet of chickens will reduce food intake and egg production, and 7% creates low level mortality.

FURTHER READING

Lovelace, M.L. and L.R. Oliver. 2000. Effects of interference and tillage on hemp sesbania and pitted morningglory emergence, growth, and seed production. 2000. *Proceedings of the Southern Weed Science Society* 53: 202.

Norsworthy, J.K. and L.R. Oliver. 2002b. Effect of irrigation, soybean (*Glycine max*) density and glyphosate on hemp sesbania (*Sesbania exaltata*) and pitted morningglory (*Ipomoea lacunosa*). *Weed Technology* 16: 7–17.

Woon, C.K. 1987. Effect of two row spacings and hemp sesbania competition on sunflower. *Journal of Agronomy and Crop Science* 159: 15–20.

Horsenettle

Solanum carolinense L.

Horsenettle with flowers
Antonio DiTommaso, Cornell University

Horsenettle fruit
Scott Morris, Cornell University

Horsenettle leaf spines
Scott Morris, Cornell University

IDENTIFICATION

Other common names: ball nettle, bull nettle, apple-of-Sodom, wild tomato, sand brier, devil's tomato, devil's potato, Carolina horse nettle

Family: Nightshade family, Solanaceae

Habit: Thorny, branched perennial herb spreading by deep horizontal roots

Description: Seedlings have oval to oblong-shaped, 0.5 inch-long cotyledons. The cotyledons have a shiny green surface, light green undersides and hairy margins. The stem of the seedling is purplish and has short, stiff hairs. The first pair of true leaves has untoothed, smooth edges and scattered, star shaped hairs on its upper surface. Later young leaves have wavy, lobed edges, star-shaped hairs and sharp, stiff, curved prickles on both surfaces. **Mature plants** arise from thick, deep, horizontal storage roots. In deep, well-drained soil, horsenettle roots can penetrate to 8 feet. Semi-woody stems with prickly hairs reach 1–3 feet. Stems are erect, angled at leaf nodes and branch moderately. Leaves are alternate and 2.5–4.5 inches long by 1–3 inches wide; they have an egg-shaped outline with two to five shallow lobes on each edge.

Star shaped hairs are present on the stems and on both leaf surfaces. Large, 0.25–0.5 inch, sturdy, yellow or white, curved, painful spines and prickles are present on stems, leaf stalks, leaf veins and flower stalks. Small clusters of one to five, white to purple, potato-like **flowers** sit atop long, sturdy, leafless stalks. Flowers are star shaped with a central cone of yellow, pollen filled anthers. Five hairy green sepals occur at the base of the five fused, 0.75–1 inch-diameter flower petals. Immature, green **berry fruits** turn yellow and shrivel as they mature; mature berries are 0.5–0.75 inch across and hold 40–170 flat, round seeds. **Seeds** are 0.1 inch across, smooth, shiny, and range from pale to dark yellow or orange.

Similar species: Buffalobur (*Solanum rostratum* Dunal) cotyledons have a clear central vein when viewed from below, and young leaves have deep, round lobes that indent nearly to the midvein. Clammy groundcherry (*Physalis heterophylla* Nees) has densely hairy stems, single yellow or greenish flowers arising in leaf-stem junctures and berries covered in a thin, papery membrane; it does not have prickly or spiny leaves.

MANAGEMENT

Summer tillage is a key management strategy. Since

the species emerges relatively late, it is likely to increase during a sequence of spring planted, long season crops like corn and soybeans. On grain farms, rotation into a small grain crop allows midseason tillage following harvest. If possible, allow the plants to resprout and then till again before planting a winter grain or fall cover crop. This will not eliminate the weed but will reduce density. On vegetable farms, a series of short season crops, for example spring spinach, summer lettuce and a fall brassica, allows for repeated tillage during times when horsenettle is most sensitive to disturbance. When spring planted full season crops are unavoidable, cultivate more deeply than usual (e.g., 4 inches). If horsenettle is not yet well established in a field, repeated hoeing in the row to supplement cultivation can eliminate the weed in a couple of years. Unlike most perennials, only a few supplemental hoeings per season will be required because this weed is slow to resprout and relatively slow to begin replenishing root reserves.

Horsenettle is common in overgrazed pastures. If the infestation is severe, renovate the pasture through deep tillage and reseeding to develop a dense stand of palatable species. Horsenettle sprouts back vigorously following mowing early in the season. Recovery from mowing later in the season is slower. Unfortunately, horsenettle responds to repeated mowing by producing a rosette of leaves close to the ground that largely escapes further damage. Mowing is effective at preventing flowering and fruit production, thereby reducing the likelihood of livestock poisoning. Because horsenettle is relatively slow to resprout from roots after elimination of the shoots, intensive rotational grazing is effective for controlling this species. If possible, grazing episodes should be timed to catch the weed while stems are still soft and easily damaged by trampling.

ECOLOGY

Origin and distribution: Horsenettle is native to the southeastern United States but has been spread northward into southern Canada and westward through most of the United States, but it is uncommon between the western Corn Belt and the far west. It has been introduced into Japan, South Asia, Australia, New Zealand and parts of Latin America.

Seed weight: 1.1–1.9 mg.

Dormancy and germination: Many seeds are dormant in the fall but lose dormancy after a few months of storage. Seeds do not need cold to break dormancy. The optimum temperature for germination is 68–86°F. Germination is increased by nitrate and fluctuating temperatures but not by light.

Seed longevity: Seeds remain viable for at least three years when buried at 3–5 inches.

Season of emergence: Seedlings begin emerging in mid-May. Shoots sprouting from roots begin emerging in mid-spring and continue emerging through summer. The temperature range most suitable for sprouting from roots is 59–86°F. Horsenettle emergence is reduced where horsenettle previously grew, suggesting that populations are self-regulating.

Emergence depth: Seedlings emerge well from 0.5–2 inches, and some seedlings can emerge from 3–4 inches, but no seedlings emerge from depths of 5–6 inches. Emergence is low for seeds near the soil surface (0.25 inch or less). All root cuttings emerged when buried 12 inches or less.

Photosynthetic pathway: C_3

Sensitivity to frost: Shoots are killed by the first frost. Only 10% of seedlings established in fall survive winter. Freezing also kills roots, but roots below the frost line survive.

Drought tolerance: Horsenettle thrives in hot weather and is drought resistant due to deep penetration of the root system. Root segments on the soil surface do not tolerate three days of drying.

Mycorrhiza: Horsenettle exhibited moderate levels of mycorrhizal infection in a prairie habitat.

Response to fertility: Horsenettle tolerates infertile soil. It is among the few early colonizers of extremely poor, eroded soil of abandoned fields in the Carolina Piedmont region. Horsenettle's vegetative and fruit biomass is doubled by supplemental nitrogen fertilizer. Root length is doubled by a balanced fertilizer when plants are grown from seed but not when grown from root segments. This suggests that root segments have sufficient nutrients for establishment without reliance on soil nutrients.

Soil physical requirements: Horsenettle grows on a wide range of soil types, from sand and gravel to clay. It establishes best and grows most rapidly, however, on coarse textured soils.

Response to shade: Horsenettle tolerates moderate shade, but growth and fruit production are suppressed by heavy shade.

Sensitivity to disturbance: Starch stored in the roots reaches a minimum from one to two months after stems emerge and when flowering is initiated. Thus, destroying the shoots at that time is particularly effective for decreasing vigor of the plants. Mowing only temporarily reduced aboveground shoots for a month but often reduced flowering and fruiting for the duration of the summer. Disturbance of established plants, even if they survive, will reduce their capacity to sprout and produce shoots in the subsequent year. New plants can establish from root sections as short as 0.4–0.8 inch, and 6-inch sections can produce shoots when buried 18 inches deep. Root segments require about one month to produce new shoots. Most shoots sprout from taproots rather than from lateral root segments. Seedlings become capable of regenerating from the root following shoot removal when 15–20 days old.

Time from emergence to reproduction: Shoots begin flowering about five to eight weeks after emergence, peak in early summer and continue until the fall. Berries and seeds begin maturing one to three months after flowering.

Pollination: Although all flowers have both male and female parts, flowers near the top of the plant function as males, whereas those near the base function as females. Individuals are normally self-incompatible, but flowers can self-pollinate when cross pollen is scarce. The flowers are pollinated by bumblebees and carpenter bees. Feeding damage by larvae of a host specific moth, *Frumenta nundinella*, can cause fruits to develop without pollination.

Reproduction: Most populations are maintained primarily by vegetative reproduction, but seeds are important for spread of the species. The taproot can grow 4 feet and horizontal roots can spread more than 3 feet per year. Tillage increases vegetative reproduction by cutting up and moving roots. Horsenettle berries contain 13–160 seeds (average of 86), and single shoots produce up to 5,000 seeds.

Dispersal: Horsenettle seeds pass through cattle unharmed and are dispersed in manure. Probably many species of mammals eat the fruits and subsequently disperse the seeds in their feces. Root fragments can be spread by harvesting and tillage equipment in agricultural fields or by earth-moving equipment in non-agricultural areas.

Common natural enemies: Eggplant lacebug (*Gargaphia solani*) causes leaf yellowing and early leaf drop, and potato bud weevil (*Anthonomus nigrinus*) can destroy a substantial proportion of the flowers. First generation larvae of the host specific moth, *Frumenta nundinella*, can cause substantial damage to leaves and flowers. Second generation larvae can heavily damage fruits and seeds. Populations of the moth are normally kept in check, however, by a parasitoid wasp. A downy mildew caused by *Erysiphe cichoracearum* can infect foliage in fall.

Palatability: Both leaves and fruit of horsenettle are toxic to people and livestock, and immature berries are particularly poisonous. Toxicity increases in autumn. Toxicity is unaffected by fertility and moisture conditions, suggesting the importance of toxic compounds as a deterrent to insects and pathogens under all growing conditions.

FURTHER READING

Bassett, I.J. and D.B. Munro. 1986. The biology of Canadian weeds. 78. *Solanum carolinense* L. *Canadian Journal of Plant Science* 66: 977–991.

Wehtje, G., J.W. Wilcut, T.V. Hicks and G.R. Sims. 1987. Reproductive biology and control of *Solanum dimidiatum* and *Solanum carolinense*. *Weed Science* 35: 356–359.

Horseweed

Conyza canadensis (L.) Cronquist = *Erigeron canadensis* L.

Horseweed seedling
Antonio DiTommaso, Cornell University

Horseweed plant
Antonio DiTommaso, Cornell University

Horseweed mature plant
Joseph Neal, North Carolina State University

IDENTIFICATION

Other common names: blood stanch, butterweed, Canada fleabane, colt's tail, fireweed, hogweed, mare's tail, pride weed, flea wort, mule tail, fleabane, bitterweed, Canadian fleabane, stickweed

Family: Aster family, Asteraceae

Habit: Erect, winter or summer annual herb with densely leafy, essentially unbranched stalks

Description: Short hairs give **seedlings** a fuzzy appearance. Cotyledons are small, 0.1 inch long and egg shaped. Young leaves are oval, toothed, hairy and arranged in a basal rosette. Leaves at the rosette base have distinct stalks, and upper leaves taper to the rosette base. Toothed leaf edges become deeper with age. A short taproot is supplemented by a fibrous root system. Seedlings emerging in the spring form basal rosettes, yet produce only a few leaves, and soon after emergence these plants begin to bolt. **Mature plants** bolt on a single stem that can reach 1–6 feet in height. The basal rosette frequently disintegrates when plants mature. Stems are covered with bristly, short hairs and many closely spaced, hairy, alternately arranged, downward angled, baseball-bat-shaped leaves measuring 4 inches long by 0.4 inch wide. Toothing, when present, is widely spaced and shallow. The leaf tip has two small, uneven indentations located across from one another. The upper stem branches produce a panicle of small, 0.2 inch diameter, white, daisy-like **flower heads**. These consist of 25–50 white to pink-tinged, one-petaled flowers along the edge surrounding 7–12 small, yellow, knobby disk flowers. At the base of each flower head are one to two rows of small, green bracts. The outer coat of the **seed** is a thin dry layer of fruit tissue. Seeds are brown, torpedo shaped and attached to a pappus of bristly hairs. Seed length, not including the pappus, is less than 0.1 inch.

Similar species: Hairy fleabane [*Conyza bonariensis* (L.) Cronquist] is shorter, branched from the base and has densely hairy gray-green leaves. Its ray flowers are greenish-yellow rather than white, and its disc flowers are inconspicuous and white rather than yellow. Annual fleabane [*Erigeron annuus* (L.) Pers.] and rough fleabane (*Erigeron strigosus* Muhl. ex Willd.) are similar to horseweed at all stages. Annual fleabane flower heads are two to three times larger in diameter than those of horseweed. The stem has ridges and branches, and the leaves are more distinctly toothed. Rough fleabane leaves are paddle shaped and have hairs that lie parallel to the leaf surface.

MANAGEMENT

The keys to controlling horseweed are tillage, crop competition and management of the species in areas around fields. Horseweed rosettes are easily destroyed by spring tillage, which is why the species is most frequently found in overwintering crops like strawberries and winter or no-till grain crops. Moreover, plow tillage that buries the previous year's seeds will permanently eliminate nearly all of them since most will die before subsequent tillage returns them to the soil surface environment that they need for establishment. Since new seedlings arise at the soil surface and are slow to establish, they are easily killed by tine weeding in field crops and by precision weeding tools and hoeing in vegetable crops. However, plants become difficult to control with tine weeding in late spring because of rapid growth at this time.

Horseweed is sensitive to crop competition, so steps that increase crop competitiveness can greatly reduce horseweed problems (see Chapter 3 for methods). Spring germinating horseweed can be a problem in weakly competitive crops like carrots and onions. To avoid problems in these crops, precede them with two years of highly competitive or short season crops in which horseweed is suppressed and prevented from going to seed. A rye or forage radish cover crop that establishes a leaf canopy quickly in fall can effectively suppress establishment of this species.

Since horseweed reproduces prolifically and the seeds blow easily into surrounding fields, cleaning up waste ground to prevent seed production is critical. Plant stream sides, banks of irrigation ditches, edges of equipment yards and other potentially weedy habitats with competitive species that suppress horseweed. Mowing will usually not completely eliminate seed production because new sprouts will grow from dormant buds. Mowing greatly reduces seed production, however, and lowers the launching point so that seeds cannot blow as far. If horseweed has not yet gone to seed at harvest, be sure to clean up the field afterward with disking or other appropriate tillage.

Even low rates of organic mulch effectively suppress horseweed if placed before seedlings begin to bolt, probably because energy reserves in the seed are negligible and the seedlings are tiny and weak. Horseweed is especially sensitive to rye straw, which releases toxins that poison the seedlings in addition to casting shade. Consequently, straw mulch can be very effective for suppressing horseweed in crops where spring tillage is not an option, like strawberries and garlic.

ECOLOGY

Origin and distribution: Horseweed is native to North America but is probably much more widespread now than prior to European settlement. It occurs throughout southern Canada, the entire United States, Mexico and much of Latin America. It is widely naturalized in Europe, temperate and tropical Asia, Australia, New Zealand and southern Africa.

Seed weight: 0.032–0.072 mg.

Dormancy and germination: Most seeds are not dormant when dispersed from the parent plant. Chilling is not required for germination. Light is the primary requirement for germination of most populations. Seeds germinate best at day/night temperatures of 68/50°F to 86/68°F. No seeds germinated at 54/43°F. The lowest continuous temperature giving maximum germination was 63°F. The base temperature for germination ranged from 49°F for a population from Ontario to 57°F for a population from Spain.

Seed longevity: Horseweed seeds are generally short-lived in the soil. Horseweed soil seed populations declined by 76% in 10 months from October to August, but 6% of the seeds remained after two years of good weed control. Horseweed is present in seed banks of pastures where the plant has not been present above ground for many years, and it was the most abundant species in the seed bank after 12 years of a tillage experiment, suggesting that a portion of seeds may last for several years in the soil. However, these observations could be complicated by seeds that blew in from elsewhere, and long-distance seed dispersal is probably more important for survival of this species than seed longevity (see the section "Dispersal").

Season of emergence: Many plants emerge in the fall and overwinter as a rosette of leaves. Other plants emerge in the spring, and a few seedlings may emerge throughout the summer. Emergence was reported as greatest in fall in cooler climates such as Canada and the upper midwestern United States, and it was greatest in spring in milder climates such as Indiana and southern Sweden. The lack of seed dormancy means that microsite conditions play an important role in determining whether horseweed functions as a winter or summer annual species. In Tennessee emergence occurred during months when daytime temperatures were between 50–60°F, leading to peak emergence in the fall at some sites and in the spring at other sites.

Emergence depth: Seedling emergence declines very rapidly with depth of burial. The majority of the seedlings emerge from seeds at the soil surface. In one experiment, only 4% as many seeds emerged from 0.1 inch as emerged from seeds placed on the soil surface, and no seedlings emerged from greater depths. In another experiment, 10% emerged from 0.4 inch.

Photosynthetic pathway: C_3

Sensitivity to frost: Winter annual horseweed plants tolerate frost well and overwinter as a rosette of leaves, even in very cold climates. During sunny winter weather, leaf temperature may be as much as 18°F above air temperature, and leaves photosynthesize actively. Plant survival in Iowa and Minnesota ranged from 59–91%, with higher survival rates associated with larger rosettes before winter. However, in southern Indiana, only plants with small rosettes survived winter, and their survival rate was only 24%. Likewise, in Sweden few rosettes were found to survive winter. The principle cause of winter mortality is frost heaving, but the specific interactions between soil conditions and rosette size remain unresolved. Possibly, rosettes survive better in regions with continuously cold conditions and a snowpack than in regions with alternating freezing and thawing temperatures.

Drought tolerance: Horseweed tolerates drought well.

Mycorrhiza: Horseweed is mycorrhizal.

Response to fertility: Horseweed commonly colonizes badly eroded, abandoned farmland, which indicates a tolerance of low soil fertility and limited response to applied nitrogen.

Soil physical requirements: Horseweed occurs most commonly on sandy, stony or loam soils with good drainage. It does not tolerate flooding. Seeds will germinate in salty environments.

Response to shade: Horseweed growth is stunted by shading from crops or other weeds.

Sensitivity to disturbance: Overwintering rosettes of horseweed are easily killed by spring tillage, and the species is generally rare or absent from spring tilled fields. Slow development of the seedling makes the species easily controlled by tine weeding, raking or hoeing. Mowing prevents or delays seed production.

Time from emergence to reproduction: Flower stalks of overwintering plants begin elongating in mid-spring and flower about two to three months later. Seeds mature 10 days to three weeks after pollination. Seed production continues for several weeks. Spring emerging plants rapidly pass through a rosette phase resulting in shorter time to bolting and flowering than fall emerging plants.

Pollination: Pollen is released before the flowers fully open, which promotes extensive self-pollination. Insects regularly visit the flowers, however, and the outcrossing rate for an Ontario population was estimated at 4%.

Reproduction: Flower heads typically produce 35–70 seeds each. The number of flowers produced by a plant depends on plant height. Thus, fall emerging plants 16 inches tall produced about 2,000 seeds whereas plants 5 feet tall produced about 230,000 seeds. Uncrowded plants on fallow ground produce about 200,000–400,000 seeds each. On the other hand, spring emerging, glyphosate-resistant plants growing within a crop canopy had a maximum production of only 10,000–60,000 seeds per plant. Plants emerging earliest, either in the fall or spring, had the highest survival rate and produced the most seeds.

Dispersal: The tiny, plumed seeds stayed aloft longer than those of 19 species tested in a controlled seed drop experiment. Although the majority of seeds remain within 330 feet of the source plant, dispersal distances of up to 1,640 feet were documented. Some seeds that ascend into the atmosphere by turbulence could be carried 75 miles or more. Seeds float well on water and probably disperse long distances in streams and irrigation canals. The plumed seeds are also caught by and carried on vehicles.

Common natural enemies: Aster yellows mycoplasma can decrease seed production by over 50%.

Palatability: The species is not eaten by people. It is little used by livestock and is irritating to horses.

Note: Horseweed can produce allelopathic compounds that inhibit forage species including orchardgrass, white clover and alfalfa under controlled environments. However, numerous crops have shown no adverse influences when seeded into terminated stands of horseweed under field conditions.

FURTHER READING

Buhler, D.D. and M.D.K. Owen. 1997. Emergence and survival of horseweed (*Conyza canadensis*). *Weed Science* 45: 98–101.

Davis, V.M. and W.G. Johnson. 2008. Glyphosate-resistant horseweed (*Conyza canadensis*) emergence, survival, and fecundity in no-till soybean. *Weed Science* 56: 231–236.

Weaver, S.E. 2001. The biology of Canadian weeds. 115. *Conyza canadensis* L. Cronquist. *Canadian Journal of Plant Science* 81: 867–875.

Jimsonweed

Datura stramonium L.

Jimsonweed seedling
Antonio DiTommaso, Cornell University

Jimsonweed flower
Randall Prostak, University of Massachusetts

Jimsonweed plant with immature fruit
Antonio DiTommaso, Cornell University

IDENTIFICATION

Other common names: stramonium, Jamestown weed, thorn apple, mad apple, stinkwort, angel's trumpet, devil's trumpet, dewtry, whiteman's weed, purple thornapple

Family: Nightshade family, Solanaceae

Habit: Erect, branched, summer annual herb

Description: The **seedling** is large, with cotyledons that are 1.2–1.6 inch long by 0.2 inch wide, lanceolate, hairless and thick. Cotyledons have an obvious vein, and the petioles have hairs on the upper surface. The seed coat often remains on the cotyledon tips after emergence. The seedling stem gradually turns purple, starting at the base. The first pair of true leaves is opposite, egg to triangle shaped with non-toothed, entire edges, and strongly veined on the underside; the first true leaves appear gray-green due to very small, scattered hairs on the puckered leaf surface. Other young leaves show edge irregularity and have flat hairs only above veins and on the leaf stalk. **Mature plants** reach 1–6 feet in height and branch in their upper portions. Stems are purple, 1–2 inches thick and have either no hairs or very small, inconspicuous hairs.

Leaves are alternate, oval to triangular in outline, and irregularly toothed, giving them an overall oak-leaf appearance. Leaves are dark green, hairless, strongly veined and 2–6 inches long by 1–4 inches wide. The taproot system is shallow, broad, thick and highly divided. The entire plant has a strong, unpleasant odor. Trumpet-shaped, white or purple, solitary **flowers** sit on stalks arising from leaf-stem junctures on upper plant portions. Flowers have five green, ridged, 1.4–1.8 inch-long sepals covering the base of the five fused petals. The petals are 2–4 inches long and 1–2 inches wide; each petal tip has a pointed, thin projection that extends 0.2–0.3 inch beyond the rim of the flower. Each plant produces up to 50 egg-shaped, 0.8–1.2 inch-long by 0.75 inch-wide, green **fruit** capsules; the entire fruit is covered with 0.13–0.4 inch-long prickles and spines. The capsule turns brown and splits into four chambers. Capsules of vigorous plants contain approximately 600–700 seeds. **Seeds** are kidney shaped, flat, black, wrinkled and 0.13–0.14 inch long.

Similar species: Seedlings of common cocklebur (*Xanthium strumarium* L.) can be differentiated from those of jimsonweed by their lack of odor and larger, 2 inch-long by 0.4 inch-wide cotyledons. Young common cocklebur leaves are covered with small, rigid hairs, and older leaves do not resemble oak leaves.

MANAGEMENT

Because this weed is extraordinarily responsive to nutrients, avoiding over-fertilization, particularly with P and N, is crucial for long-term control. Keep this species controlled around manure and compost storage areas to avoid spreading seeds into fields. Avoid incorporating seeds into the soil after crop harvest, if possible, because jimsonweed is particularly susceptible to loss of viability when the seeds are left exposed on the soil surface.

Jimsonweed is relatively intolerant of crop competition. Its growth form changes when subject to competition, with the plant becoming less branched and sloughing the lower leaves. Even though plants commonly emerge above the canopy of mid-sized crops like soybeans, heavy competition during early growth greatly decreases total productivity and seed set. Thus, steps that increase the competitive ability of the crop are highly desirable. Since most of the jimsonweed plant lies above the canopy of most crops, topping the plants with a mower should greatly decrease both competition with the crop and weed seed production.

Because seedlings commonly emerge from below the working depth of rotary hoes and tine weeders, these implements are relatively ineffective for controlling jimsonweed. Tine weeding should focus on maximizing lateral movement of soil to bury seedlings. In crops that tolerate hilling, bury young seedlings as early as crop growth will permit, since jimsonweed makes rapid vertical growth.

No-till and cover crop surface residue can reduce jimsonweed emergence. Both conditions reduce soil temperature and light at the soil surface, thus depriving jimsonweed seeds of the relatively warm temperatures and light they require for germination.

ECOLOGY

Origin and distribution: The origin of jimsonweed is uncertain but most likely is in tropical South or Central America, although it may have originated in Asia. It now occurs throughout the tropical and temperate regions of the world. It is widespread in southeastern Canada and throughout the United States except for the coldest parts of the Midwest and far west.

Seed weight: Most population mean seed weights range from 5.6–11.7 mg, but mean weights of 1.8–2.4 mg have been observed for populations near the northern edge of the species range. Seed germination generally increases as seed weight increases, with weights over 6 mg per seed providing greatest germination.

Dormancy and germination: Variable conditions have been reported to break seed dormancy. Seeds are kept dormant by a combination of a hard seed coat and germination inhibitor that eventually washes out of the seed. Seeds buried at 8 inches had no dormancy when exhumed, whereas seeds buried at 2 inches maintained dormancy for one year before losing dormancy after exposure to cold winter temperatures. A large day/night temperature fluctuation of 90/54°F effectively overcame dormancy whereas a smaller fluctuation of 90/81°F did not. Seeds buried in soil at 41–50°F became sensitized to germinate when exposed even briefly to light. In most situations, light is a critical factor in triggering germination. Re-burial of photosensitized seeds can induce dormancy, probably due to volatile compounds in the soil that increase in proportion to depth. Base temperature for germination ranged from 46–52°F. Seeds germinate best at 68–95°F with alternating daily temperatures being most stimulatory. Germination is reduced by low moisture conditions more than that of many other weed species.

Seed longevity: In one experiment, over 91% of seeds buried at 22 inches survived more than 39 years. Seeds did not lose viability when buried 2–8 inches for 22 months in Israel. Most seeds remained viable after burial at 8 inches for 17 years at one location, but all lost viability after three years at another location in Nebraska. First year assessment of seed mortality in these experiments indicated an annual loss rate of 14% and 6%. Based on an experiment in which the seeds were buried at 4 inches, we calculated an average loss per year of about 50%. Annual tillage without allowing seed replacement virtually eliminated a jimsonweed seed bank in six years. Burial of jimsonweed seeds by fall tillage promotes seed survival. The wide range in longevity reported for this species may be related to differences in the duration of hard seed coat integrity.

Season of emergence: Most seedlings emerge from mid-spring to early summer, but some seedlings continue to appear throughout the summer following rain.

Jimsonweed has been categorized as "middle-emerging" relative to other weeds.

Emergence depth: Seedlings emerge well from 0.4–2 inches; a few can emerge from as deep as 4 inches, but none emerge from 6 inches. Seeds on the soil surface have reduced germination. Lower emergence from 1.6 inch in clay than in sandy soils was related to poorer gas exchange in clay soils.

Photosynthetic pathway: C_3

Sensitivity to frost: Jimsonweed is frost sensitive. Seeds in immature capsules do not continue to mature after frost.

Drought tolerance: Jimsonweed can survive in sandy pastures and similar dry sites but thrives best on fertile soil and high rainfall. This species was more competitive with crops under above-average rainfall than under drought conditions. It has a higher rate of water loss and greater physiological sensitivity to water stress than other weeds.

Mycorrhiza: No studies have reported on the mycorrhizal status of jimsonweed.

Response to fertility: Jimsonweed growth responds strongly to nitrogen. It accumulates higher concentrations of N than most crops. Alkaloid content also increases with N. It also shows a strong growth response to P and K. In one study, it was the most responsive to P out of 10 warm-season weeds. The species is commonly found in nutrient-rich sites like barnyards and around manure piles. Jimsonweed tolerates soil pH as low as 4.7, but growth is reduced below pH 5.4.

Soil physical requirements: Jimsonweed does best on good quality agricultural soils but can tolerate a wide range of soil conditions.

Response to shade: Light shade (25%) stimulates growth of jimsonweed, thereby prompting the plant to grow through partial crop canopies. Moderately heavy shade (75%), however, substantially reduces growth.

Sensitivity to disturbance: Cut or trampled plants regenerate from buds near the base of the stem.

Immature capsules will continue to ripen on cut branches or uprooted plants.

Time from emergence to reproduction: Flowering begins about five to nine weeks after emergence, with the time to flowering less for later emerging plants and for populations from more northerly latitudes. Each flower is open for only one day. Jimsonweed is indeterminate, so it continues to flower as plants continue growth into late summer. Seeds mature and are capable of germination about one month after fertilization, but capsules usually do not open until about seven weeks after fertilization. In Delaware, plants emerging later than early July failed to produce mature capsules.

Pollination: Jimsonweed usually self-pollinates, but occasionally plants may be cross pollinated by insects. Inbreeding depression of seed production can occur, suggesting the need for outcrossing to maintain vigorous populations.

Reproduction: Vigorous plants at low densities produce up to 50 capsules and 30,000 seeds per plant. In contrast, plants stressed by competition may only produce 1,300–1,500 seeds.

Dispersal: Seeds are dispersed by combines and in soil clinging to tillage implements, tires, shoes and livestock. Seeds can contaminate grain and cover crop seed. Several introductions through contaminated soybean seed have been reported. The seeds and capsules float well and disperse along streams and irrigation ditches. The spines on the capsules are not effective for dispersing the seed.

Common natural enemies: Three-lined potato beetles (*Lema trivittata* and *L. trilineata*) destroy seedlings and can cause severe defoliation of larger plants. Greater damage to foliage from *L. trilineata* and other herbivores was observed in inbred than in outcrossed populations.

Palatability: Both leaves and seeds are highly toxic to humans and livestock due to alkaloids and sometimes nitrate. Many people have died from grain contaminated with jimsonweed seeds. Since the foliage is extremely unpalatable and poisoning causes loss of appetite, damage to grazing livestock is usually limited.

FURTHER READING

Pawlek, J.A., D.S. Murray and B.S. Smith. 1990. Influence of capsule age on germination of non-dormant jimsonweed (*Datura stramonium*) seed. *Weed Technology* 4: 31–34.

Weaver, S.E. and S.L. Warwick. 1984. The biology of Canadian Weeds. 64. *Datura stramonium* L. *Canadian Journal of Plant Science* 64: 979–991.

Zhang, J., M.L. Salas, N.R. Jordan and S.C. Weller. 1999. Biorational approaches to managing *Datura stramonium*. *Weed Science* 47: 750–756.

Kochia

Bassia scoparia (L.) A.J. Scott = *Kochia scoparia* (L.) Schrad.

Kochia older seedling
Anita Dille, Kansas State University

Kochia plant
Wale Osipitan

Kochia flowering stem
Anita Dille, Kansas State University

IDENTIFICATION

Other common names: summer cypress, belvedere, burning bush, fireball, Mexican fireweed, belvedere, mock cypress, red belvedere, belvedere cypress, red belvedere, broom cypress

Family: Goosefoot family, Chenopodiaceae

Habit: Tall, much branched, taprooted summer annual herb

Description: Cotyledons of kochia **seedlings** are stalkless, elliptical, 0.18 inch long by 0.1 inch wide and softly hairy, with dull green upper surfaces and sometimes bold pink or magenta on the underside. Stems are hairy and green to reddish in color. Young leaves are lanceolate to wide-thumb shaped, pointed at the tip, red-tinged underneath, gray-green above due to the presence of numerous soft hairs, and initially have a whorled leaf arrangement like a basal rosette. **Mature plants** are highly branching, 1–6.5 feet tall, pyramidal to round, and bushy, with a soft, airy texture. Plants look blue to gray-green during the growing season, turning red-green or red-purple in the fall. Stems are red tinged and more or less hairy. Alternate leaves are narrow, 1–2 inches long by 0.13–0.25 inch

wide, and are attached by short stalks. Soft hairs are present on the untoothed leaf edges, leaf undersides and typically on upper leaf surfaces. Leaves decrease in size higher up on stems. The root system is a taproot with branched fibrous roots. The stem of the mature, dry plant breaks free from the root to become a tumbleweed in the fall. Green, petal-less, 0.13 inch-long **flowers** occur singly or in clusters of two to six in leaf axils of the upper stems or in short spikes with long, white hairs and a 0.13–0.5 inch-long, leaf-like bract below each flower cluster. **Seeds** are contained in papery bladders derived from the green portion of the flowers. Seeds are egg shaped with many irregularities and no bigger than 0.1 inch, with a gritty, bumpy, dull, grooved surface; they range in color from transparent brown with yellow spots to dark red-brown or black.

Similar species: Russian-thistle (*Salsola tragus* L.) has narrower, needle-like leaves at all stages compared to kochia. Forage kochia [*Bassia prostrata* (L.) A.J. Scott] is a perennial, semi-evergreen species with 1–5 inch-long, linear leaves. Common lambsquarters (*Chenopodium album* L.) has broader, diamond-shaped leaves and is a hairless species with a white or pink-dusted leaf surface, especially when the plant is young.

MANAGEMENT

Kochia relies primarily on prolific seed production and abundant seedling emergence for population growth, suggesting that management of these life stages will have the greatest impact on controlling this species. Because kochia seeds survive for a short time in the soil, preventing seed production for just a single year will largely control even severe infestations. Two or three years of vigilance can essentially eradicate the population. Consequently, rotating fields into a hay crop for a few years should be highly effective for managing kochia. Winter cereals are good competitive crops when they establish a vigorous leaf canopy by the time kochia emerges in spring. Avoid uncompetitive crops such as flax or beans. Growing a fall-established triticale-legume cover crop mix in a wheat-fallow system in Kansas effectively suppressed kochia without adversely affecting the following wheat crop. Yellow sweetclover significantly reduced kochia population when grown as a green manure fallow in Alberta.

Kochia is more abundant in no-till than in conventional-tillage systems, in part because of its ability to germinate in cool soil associated with crop residue at the soil surface. Management including carefully targeted tillage operations were most effective for reducing kochia populations. Because seeds need to be within the top inch or less of soil to successfully emerge and because seeds buried more deeply often germinate and die, inversion tillage is an effective control measure. Because the peak of emergence occurs early in the spring, delaying tillage and planting will decrease the current year's infestation, even if tillage is shallow. Cool-season crops like wheat, however, will compete better with kochia, a warm-season weed, when planted during the cool weather of spring.

Because kochia emerges only from the top inch of soil, tine weeding is effective against this weed in both small grains and row crops. For example, harrowing of spring wheat and barley with either a tine weeder or a spike-tooth harrow at the crop's three-leaf stage gave 66% and 62% control of kochia, respectively. Spike tooth harrowing at both the three- and five-leaf stage increased control to 82%, but a second tine weeding at the five-leaf stage did not improve control. None of the weeding treatments affected yield of either crop despite about 19% damage when harrowing was done twice. Pre-emergence tine weeding or rotary hoeing would probably provide additional control. Due to its

small seed reserves, kochia starts slowly, but it grows rapidly once established. Consequently, burying in-row kochia seedlings with soil as soon as the crop will tolerate it could be effective.

In the northern Great Plains, seeds mature after the time of small grain harvest. This indicates that many of the early maturing seeds could be captured during combine harvesting of grain and, with prompt cleanup of the field after harvest, could prevent further seed production that year. Plants remaining after grain harvest can produce many seeds, so destroy kochia re-growth before the first killing frost. Avoid irrigation practices that increase soil salinity, as this will favor kochia relative to crops. After a field has been cleaned up, a snow fence can prevent the majority of mature plants from rolling in from adjacent infested areas.

ECOLOGY

Origin and distribution: Kochia is native to eastern Europe and western Asia. It occurs throughout southern Canada and the United States except parts of the Southeast, but it is particularly a problem in the Great Plains and Intermountain West. In addition to North America, it has spread widely in Europe, China and Japan, and has been introduced into Africa, Argentina, Australia and New Zealand.

Seed weight: Population mean seed weights vary from 0.2–0.85 mg.

Dormancy and germination: Kochia seeds have little dormancy and germinate greater than 75% at temperatures from 41–95°F, but germination is inhibited at 104°F. Alternating temperature does not enhance germination. Germination is very rapid, with seedlings commonly breaking through the seed coat within 24 hours at 68–77°F. Light is not required for germination. Kochia seeds can germinate in soil that is too dry for establishment of drought tolerant crops, and kochia can also germinate in solutions of up to 10,000 parts per million salt. High salt conditions reduce germination less at high temperatures than at low ones. Kochia can germinate at a wide range of pH from 2–12.

Seed longevity: Most kochia seeds either germinate or die in their first year. In undisturbed soil, up to 3% can survive two to three years when buried at 12 inches, but survival is much poorer near the soil surface. In the western United States and Canada, seeds buried in fall had good viability in the early spring when emergence typically occurs, but they suffered high mortality over summer months and had low viability by fall. Seed placement from the soil surface to a depth of 4 inches had little effect on seed mortality in these experiments.

Season of emergence: Most seedlings emerge early in spring, but some emerge later following rains. In Colorado, for example, 80% of seedlings emerged between April 11 and June 20, with emergence commencing when the average daily air temperature was 49°F, a finding consistent with research showing emergence commencing when the soil temperature reached 50°F. Populations from several Great Plains states had different emergence patterns, suggesting different emergence biotypes that would require location-specific management tactics. Post-emergence herbicides have selected for late emerging biotypes.

Emergence depth: Seedlings emerge best from the top 0.4 inch, and few emerge from deeper than 0.8 inch. Seedlings establish better from seeds on the soil surface than at 0.1 inch.

Photosynthetic pathway: C_4

Sensitivity to frost: Seedlings in spring have tolerated a nighttime temperature of 9°F, but six-week-old plants that were cold acclimated for three weeks did not survive 18 hours at temperatures of 22°F or lower.

Drought tolerance: Kochia is very drought tolerant, and historically infestations have tended to increase during drought periods to the exclusion of other weed species. Drought tolerance is partially due to an extensive root system; under favorable moisture conditions the roots penetrate 7 feet, and under drought conditions they can penetrate up to 16 feet. Kochia is less drought tolerant under high-salt soil conditions.

Mycorrhiza: Kochia is not mycorrhizal.

Response to fertility: Kochia is highly responsive to N, and its productivity continues to increase up to very high N application rates (480 pounds per acre). Plants can absorb up to 90% of available soil nitrogen. Kochia only responds to P when soil levels are very low, although very low soil N fertility increases the range of P application rates to which kochia responds. High soil P levels will suppress growth.

Soil physical requirements: Kochia is most common in dry pastures, rangelands and cropland with alkaline soils. It does not occur on highly acidic soils. It tolerates a wide range of soil types and is well adapted to saline soils.

Response to shade: Compared with cropped sunflowers, kochia is relatively intolerant of shade.

Sensitivity to disturbance: Kochia regrows profusely after cutting by producing branches from axillary buds. For example, when grown for hay it is typically mowed four times during the growing season. Plants that are cut off by a combine produce many seeds that remain in place since the truncated plant does not break off and tumble.

Time from emergence to reproduction: Kochia plants flower 57–109 days after emergence. It flowers in response to decreasing day length, with shorter light periods and longer times from emergence required for more southerly populations (e.g., New Mexico as compared to South Dakota).

Pollination: Kochia commonly self-pollinates but has a moderate rate of outcrossing by wind and bee dispersed pollen.

Reproduction: Kochia typically produces 10,000–30,000 seeds per plant, with the potential for up to 100,000 seeds per plant. Greater than 99.9% of seeds are retained on plants at wheat harvest, suggesting the potential for removal and destruction of seeds at harvest. However, plants re-growing following combine harvest of grain can produce 2,600–4,000 seeds each.

Dispersal: Mature kochia break off near the base of the stem and roll as tumbleweeds, dispersing seeds as they bounce along over long distances if unobstructed. As a result, kochia has the highest rate of spread of any introduced weed in the western United States. Some kochia seeds can pass through the rumen of cattle unharmed and can be expected to move about with the cattle and be spread with manure. The seeds also disperse in irrigation water.

Common natural enemies: Kochia appears to be free of damaging diseases and insect herbivores in North America, although grasshoppers do eat it.

Palatability: The palatability and nutritional value of kochia for livestock is better than that of some grasses, such as bromegrass, but less than that of alfalfa. Plants contain up to 25% protein, and biotypes have been investigated as forage species on arid saline soils, which are not suitable for common forages. Kochia can be toxic to livestock if it composes more than 50% of the diet for several weeks, particularly if it is fed fresh. Note that "forage kochia" [*Bassia prostrata* (L.) A.J. Scott] is a perennial shrub suitable for forage in western rangeland and should not be confused with the weedy annual kochia described here.

Note: Kochia pollen is an important allergen and a common cause of allergic sensitization.

FURTHER READING

Eberlein, C.V. and Z.Q. Fore. 1984. Kochia biology. *Weeds Today* 15: 5–7.

Friesen, L.F., H.J. Beckie, S.I. Warwick and R.C. Van Acker. 2009. The biology of Canadian weeds. 138. *Kochia scoparia* (L.) Schrad. *Canadian Journal of Plant Science* 89: 141–167.

Moore, J., J. Dodd, S. Lloyd, C. Hanson, T. Grice and J. Thorpe. 2003. Kochia (*Bassia scoparia*). *Weed Management Guide.*

Petrosino, J.S., J.A. Dille, J.D. Holman and K.L. Roozeboom. 2015. Kochia suppression with cover crops in southwestern Kansas. *Crop, Forage & Turfgrass Management* 1(1): 1–8.

Schwinghammer, T.D. and R.C. Van Acker. 2008. Emergence timing and persistence of kochia (*Kochia scoparia*). *Weed Science* 56: 37–41.

Morningglories

Ivyleaf morningglory, *Ipomoea hederacea* Jacq.
Pitted morningglory, *Ipomoea lacunosa* L.
Tall morningglory, *Ipomoea purpurea* (L.) Roth

Ivyleaf morningglory seedling
Scott Morris, Cornell University

Ivyleaf morningglory climbing on a fence
Antonio DiTommaso, Cornell University

Ivyleaf morningglory flower
Scott Morris, Cornell University

Pitted morningglory seedling
Scott Morris, Cornell University

Pitted morningglory foliage
Antonio DiTommaso, Cornell University

Tall morningglory seedling
Antonio DiTommaso, Cornell University

Tall morningglory flower and leaves
Antonio DiTommaso, Cornell University

Blade portion of one cotyledon of (from left to right) ivyleaf, pitted and tall morningglory
Antonio DiTommaso, Cornell University

Ivyleaf, pitted and tall morningglory leaf shape
Antonio DiTommaso, Cornell University

IDENTIFICATION

Other common names:

Ivyleaf morningglory: common morningglory, ivy-leafed morningglory, entireleaf morningglory

Pitted morningglory: small white-flowered morningglory, white morningglory

Tall morningglory: common morningglory, purple morningglory

Family: Morningglory family, Convolvulaceae

Habit: Twining summer annual herbs

Description: Seedling stems are green to maroon and ridged. Cotyledons are two lobed, hairless, symmetrical and green.

Ivyleaf morningglory: Cotyledons are deeply notched at the tip and slightly notched at the base, forming a butterfly shape. Each 0.6–1.75 inch-long by 0.6–1.5 inch-wide cotyledon fits within a square to rectangular or trapezoidal outline. The first true leaf is unlobed, heart shaped and hairless. All other true leaves are three lobed, with upright hairs on both blade surfaces, stems and leaf stalks.

Pitted morningglory: Cotyledons are split more than 75% of their length into a V and have narrower and pointier lobes than tall and ivyleaf morningglories. Fully expanded cotyledons are 1.9 inch

long by 2.6 inches wide at the broadest point. All young leaves are heart shaped and tapered to the tip. Leaves are nearly or completely hairless.

Tall: Cotyledons are green, 0.6–1 inch long by 0.6–1 inch wide, butterfly shaped and fit into a square outline. All young leaves are heart shaped and covered in small hairs that lie parallel to the leaf surface.

Mature plants trail and climb by twining, branching stems that can reach 12 feet, although 6.5 feet is more common. Leaves are alternate. Roots are coarsely branched.

Ivyleaf morningglory: Stems are hairy. Leaves are 2–5 inches long by 2–5 inches wide, tri-lobed and covered in upright hairs; leaves generally have a heart-shaped base and tapered lobes.

Pitted morningglory: Stems are smooth. Leaves are nearly hairless, up to 3.7 inches long by 3.1 inches wide, heart shaped and tapered to a narrow, pointy tip. A small taproot is present.

Tall morningglory: Stems are hairy. Leaves are heart shaped with a blunt tip and basal lobes that may overlap slightly. Leaves are 2–5 inches long and nearly that wide and are covered with appressed hairs.

Flower petals are fused into a funnel shaped flower with many stamens and three pistils. Clusters of flowers form in leaf axils.

Ivyleaf morningglory: Petals are 1.1–2 inches long, initially off-white to light blue but turning pink-purple after opening. Clusters of one to three form in each leaf axil. Thin, hairy or bristly, green, 0.5–1 inch-long sepals curve backwards at the tip.

Pitted morningglory: Petals are 0.6–0.8 inch long and characteristically white to pinkish. Otherwise, the flowers resemble those of ivyleaf morningglory.

Tall morningglory: Petals are 1.75–2.75 inches long and light colored when young, turning purple-pink or blue as they open and mature. Clusters of three or more form in each leaf axil. Narrow, lanceolate sepals are 0.5 inch long and clasp the base of the fused petals. Otherwise, the flowers resemble those of ivyleaf morningglory.

Fruit and seeds: Flowers produce egg-shaped capsules that remain partially concealed by the remaining sepals. Capsules contain two to four (usually three) compartments. Seeds are dark brown to black and wedge shaped. Seeds have two flat and one round side.

Ivyleaf morningglory: Each capsule produces four to six seeds. Seeds are large, almost 0.25 inch wide and covered in minute hairs.

Pitted morningglory: Each capsule produces four, 0.2 inch-wide, smooth seeds.

Tall morningglory: Each capsule produces four to six 0.2 inch-diameter seeds, similar to those of ivyleaf morningglory.

Similar species: Several other morningglory species exist, with smallflower morningglory [*Jacquemontia tamnifolia* (L.) Griseb.] being of great importance in the southern United States. It can be distinguished from the morningglories described above by the dense appressed hairs on the foliage and by its tight clusters of small, bright blue flowers. Bindweeds, such as field bindweed (*Convolvulus arvensis* L.), are often confused for true morningglories. Bindweeds are perennials, whereas the weedy morningglories discussed here all arise from seeds. Wild buckwheat (*Polygonum convolvulus* L.) is a twining annual that has leaves resembling pitted morningglory, but wild buckwheat has small green flowers and sheaths extending up the stem from the base of leaves.

MANAGEMENT

Morningglories are difficult weeds to control, and their vining habit makes them aggressive competitors. Early planting dates for corn, soybeans and cotton allow the crop to become competitive before most of the morningglories emerge. If morningglories emerge and vine before the crop canopy closes, substantial crop yield losses can result. Tine weed aggressively to break or bury newly emerged seedlings. You may need to do this more frequently than for smaller seeded, and therefore slower growing, weeds. Because many emerge from deep layers of the soil, relatively few can be uprooted, and rotary hoeing is likely to be less effective than tine weeding. To avoid crop damage, cultivate frequently enough to kill morningglories before inter-row plants have attached to crops. If possible, use a belly mounted cultivator or accurate guidance system with shallow pitched half sweeps to get close to the rows. Shorter season crop varieties may allow harvest before the vines interfere severely with harvesting equipment. Consider setting the combine to capture morningglory seeds with the grain, and then clean them out later or use equipment to capture or destroy seeds in the chaff. Flame weed in well-established cotton while morningglory seedlings are still small.

Winter wheat suppresses morningglory establishment by keeping the soil cool and undisturbed during the peak germination period. Seedlings that do emerge are suppressed by shade. A dense, vigorous stand of spring grain has a similar effect, provided the grain is planted early. Morningglories are favored by short annual crop rotations like alternating peanuts and cotton, so diversifying rotations to include an established sod during the warm periods of the year will suppress morningglory populations.

Although some seeds will likely survive solarization, even a few days at 140–158°F in moist soil eliminates a large proportion of pitted morningglory seeds. Both plastic and straw mulch are ineffective for controlling morningglories. The vines find their way through the planting holes in plastic—they are essentially led to the crop by light. Large seed reserves allow the seedlings to emerge through 6 inches or more of straw.

ECOLOGY

Origin and distribution: Ivyleaf morningglory probably originated in tropical America. Authorities disagree about whether tall and pitted are native to the Southeast and southern Midwest or were introduced from the American tropics. Ivyleaf morningglory is now widely established east of the Rocky Mountains and in the Southwest. Pitted morningglory occurs from Massachusetts to Iowa and south to Florida and Texas, and also in California. Tall morningglory is native to tropical America but is now established in most of the United States. All three species are most problematic in warm, humid regions, and their occurrence tends to be scattered in the northern and western states.

Seed weight: Ivyleaf morningglory, 27–35 mg; pitted morningglory, 19–25 mg; tall morningglory, 19–25 mg.

Dormancy and germination: Usually, a high percentage of morningglory seeds have a hard seed coat that prevents them from absorbing water and thereby maintains dormancy. A substantial proportion of recently matured seeds of all three species will germinate, but germination declines after the seed coat hardens. Seeds require exposure to moist conditions or high relative humidity (as occurs in spring) to become sensitized, followed by high temperatures (as occurs in early summer) to break dormancy. However, dry conditions (as occurs in late summer) will desensitize seeds so dormancy is not broken by high temperatures. Seeds germinate best at warm temperatures (59–95°F). Moistened seeds of pitted morningglory germinated best at these warmer temperatures, but seeds with broken seed coats germinated equally well at day/night temperatures from 59/43°F to 95/68°F. Pitted morningglory germination was more tolerant of drought than several other weed species including ivyleaf morningglory. Pitted morningglory germinated best at pH 6–8, whereas tall morningglory germinated well at a pH range of 5–7. Light generally has little effect on germination of morningglory species, but it did promote germination of tall morningglory seeds that had been buried in the soil. Seeds of tall morningglory germinate poorly in soil with low oxygen. This is not directly due to excess carbon dioxide or a lack of oxygen but occurs because seeds in a low oxygen environment produce volatile organic compounds that enforce dormancy. Venting of volatiles during tillage would promote germination. Flooding also greatly reduced germination of tall morningglory.

Seed longevity: Morningglory seeds can have a relatively long life in soil because of their hard seed coats. Ivyleaf mornigglory seeds can persist after burial for up to 17 years. However, more typically, 36% and 70% of ivyleaf seeds disappeared between late autumn and the following August, and mortality was little affected by depth of burial. Pitted morningglory seeds buried at 22 inches had 31% survival after 39 years. In contrast, in another experiment, only 13% of pitted morningglory seeds remained viable after burial for 5.5 years. When a sowing of pitted morningglory was tilled annually to 6 inches and no seed return was allowed, the number of seedlings emerging declined by 58% per year.

Season of emergence: All three species have some emergence throughout the warm months of the year, but peak emergence tends to occur in early summer. Studies in Illinois showed flushes of ivyleaf emergence primarily in June and July following rainfall. Out of 23 summer annual weeds tested, ivyleaf morningglory was the latest emerging species in spring but had the longest emergence duration once it began emerging.

Emergence depth: Generally, all three species emerge best from the top 1–2 inches of soil, but substantial numbers can emerge from 4 inches. A few individuals can even emerge from 6 inches. Tall morningglory seeds at 0.5 inch emerged faster and plants became more competitive with crops than seeds emerging from 2 inches.

Photosynthetic pathway: C_3

Sensitivity to frost: The first frost usually kills morningglory plants.

Drought tolerance: Ivyleaf morningglory does not compete effectively for soil moisture. Tall morningglory is more susceptible to drought than cotton. Tall morningglory responds to drought by allocating more resources to leaves while maintaining overall biomass compared to control plants, but reproduction is reduced.

Mycorrhiza: Tall morningglory is mycorrhizal. Data are

lacking for the other two species.

Response to fertility: Ivyleaf is a poor competitor for N and grows poorly without N fertilization. Tall morningglory responds to low N by allocating more resources to roots than control plants, but overall growth and reproduction are severely reduced. Tall morningglory is more responsive to soil P and K than corn or soybeans but less responsive than forage grasses. It grows poorly on soils with a pH below about 5.3.

Soil physical requirements: Pitted morningglory tolerates poor soil drainage.

Response to shade: Large seeds and a viny growth habit often allow morningglories to escape competition for light by climbing up competing plants. Light interception by the crop leaf canopy must exceed 90% for suppression of ivyleaf morningglory. Cotton competitively suppressed newly emerging ivyleaf once the crop leaf canopy closed, but plants still produced a few seeds. Rapid canopy closure in narrow-row soybeans suppressed pitted morningglory growth. Tall morningglory has longer internodes, longer stems, thinner stems and larger leaves when growing in shade. These shade-responsive traits are enhanced when support is available, allowing the species to climb and access light within a crop canopy, even when it emerges after the crop. But tall morningglory is unable to flower in shade.

Sensitivity to disturbance: Ivyleaf morningglory has a high potential to recover from removal of the growing tip when vines are greater than 4 inches long.

Time from emergence to reproduction: Ivyleaf morningglory flowers four to seven weeks after emergence. Tall morningglory flowers six to eight weeks after emergence. Seeds of both species mature about four weeks after flowers open, and plants continue to flower and set seeds until they are killed by frost. Pitted morningglory tends to flower as days shorten to 13 hours in late August, regardless of whether they emerge in May or late June.

Pollination: Ivyleaf morningglory is primarily self pollinated, and flower structure indicates that pitted is likely self pollinated as well. Tall morningglory has a high rate of cross-pollination (70%), mostly by bumblebees and small butterflies. The prevalence of self-pollination in ivyleaf morningglory flowers is explained by the close proximity of anthers and stigma, whereas this distance is highly variable in the outcrossing tall morningglory. Also, the larger and more conspicuous flowers of tall morningglory attract more bumblebees and provide greater potential pollinator energy gain than do ivyleaf flowers. Self-pollination incurs no penalty; outcrossed and self-pollinated populations of tall morningglory perform similarly in response to drought or nitrogen stress.

Reproduction: Plants emerging mid-summer and growing without competition often produce about 5,000–7,000 seeds, but under good conditions plants can produce as many as 11,000 (ivyleaf), 16,000 (pitted) and 26,000 (tall) seeds. Crop competition, however, greatly decreases seed production. For example, pitted morningglory competing with soybeans produced only 5% and 0.4% as many seeds as plants growing without competition in two successive years. A study of ivyleaf morningglory showed that plants emerging in early July produced the most seeds, with earlier and later emerging individuals producing substantially fewer seeds.

Dispersal: Given the similarity of morningglory seeds to seeds of field bindweed and their hard, impermeable seed coat, the seeds probably pass through livestock and are spread with manure. Since deer frequently feed on morningglory, they may spread the seeds as well. These species probably also disperse when seeds in contaminated feed grain pass through livestock.

Common natural enemies: Ivyleaf morningglory can be heavily damaged by white rust (*Albugo ipomoeae-panduratae*) and orange rust (*Coleosporium ipomoea*) but produce thousands of seeds per plant anyway. Tall and pitted morningglory are less damaged by rusts. Tall morningglory (and probably other species) can be heavily attacked by tortoise beetles (*Deloyala guttata* and *Metriona bicolor*), the sweet potato flea beetle (*Chaetocnema confine*) and the corn earworm (*Heliothis zea*).

Palatability: Ivyleaf and tall morningglory have better digestibility and higher crude protein than warm season forage grasses, but they are deficient in P.

FURTHER READING

Elmore, C.D., H.R. Hurst and D.F. Austin. 1990. Biology and control of morningglories (*Ipomoea* spp.). *Reviews of Weed Science* 5: 83–114.

Gomes, L.F., J.M. Chandler and C.E. Vaughan. 1978. Aspects of germination, emergence, and seed production of three *Ipomoea* taxa. *Weed Science* 26: 245–248.

Leon, R.G., D.L. Wright and J.J. Marois. 2015. Weed seed banks are more dynamic in a sod-based, than in a conventional, peanut-cotton rotation. *Weed Science* 63: 877–887.

Oliveira, M.L. and J.K. Norsworthy. 2006. Pitted morningglory (*Ipomoea lacunosa*) germination and emergence as affected by environmental factors and seeding depth. *Weed Science* 54: 910–916.

Nightshades

Black nightshade, *Solanum nigrum* L.
Eastern black nightshade, *Solanum ptycanthum* Dunal
Hairy nightshade, *Solanum physalifolium* Rusby = *S. sarachoides* auct. non Sendtn.

Black nightshade seedlings
Jack Clark, University of California

Black nightshade plant
Jack Clark, University of California

Black nightshade ripe fruit
Antonio DiTommaso, Cornell University

Eastern black nightshade young plant
Antonio DiTommaso, Cornell University

Eastern black nightshade underside of leaves
Randall Prostak, University of Massachusetts

Eastern black nightshade flowers
Randall Prostak, University of Massachusetts

Hairy nightshade seedling
Jack Clark, University of California

Hairy nightshade plant
Joseph DiTomaso, University of California, Davis

Hairy nightshade ripe fruits
Joseph DiTomaso, University of California, Davis

IDENTIFICATION
Other common names:

Black nightshade: deadly nightshade, poison berry, garden nightshade, hound's berry, garden huckleberry

Eastern black nightshade: deadly nightshade, poison berry, garden nightshade, West Indian nightshade

Hairy nightshade: No other common names located

Family: Nightshade family, Solanaceae

Habit: Erect, branched, summer annual herbs

Description: Seedlings have egg-shaped to lanceolate cotyledons. The stem of the seedling is hairy. Young leaves are alternate.

Black nightshade: Cotyledons are 0.1–0.4 inch long by 0.1 inch wide and pointy tipped with small, sticky hairs on the leaf edges, below the midvein and scattered on the leaf surface. The first leaf is spade shaped, and subsequent leaves are oval or egg shaped. Young, light green leaves are wavy and hairy on all surfaces and edges, with prominently veined, purple tinged undersides.

Eastern black nightshade: Cotyledons are 0.4–0.5 inch long by 0.1–0.2 inch wide on 0.1–0.2 inch-long stalks and have green upper surfaces and green to maroon, prominently veined undersides. The stem of the seedling is green with a red hue. Stems and stalks above the cotyledons are purplish, with absent or inconspicuous hairs. Young leaves are egg shaped and nearly hairless, with wavy edges and purple-tinged undersides.

Hairy nightshade: Cotyledons are egg shaped to lanceolate, 0.17–0.4 inch long and green on both surfaces, with hairs on leaf edges and stalks. Young leaves are oval to elliptical, wavy edged and hairy.

Mature plants have round, ridged, woody, green stems. Leaves are stalked and alternate on the stem. When present, toothing is coarse, irregularly spaced and rounded. Fibrous roots branch from the shallow taproot.

Black nightshade: Branching begins low on the stem, and stems can reach 1.75–3 feet tall. Dark green leaves are 1–5 inches long by 0.5–2.5 inches wide, oval to egg shaped, with wavy sides that taper to a tip. Edges have purple undersides. Lower leaves have teeth, but upper leaves do not. Sparse, non-sticky hairs are typically present on stems, leaf undersides and leaf stalks.

Eastern black nightshade: Stems are hairless to lightly hairy and can reach 1–3.3 feet. Leaves are 1–3 inches long by 0.4–2.25 inches wide and elliptical to triangular egg shaped. Leaves are nearly hairless. Edges are toothed, wavy and red-purple underneath.

Hairy nightshade: Stems are slender, often coated in sticky, spreading hairs and typically reach 1–2 feet but sometimes 3 feet in height. Leaves are egg shaped to lanceolate, 0.75–3 inches long by 0.25–2.25 inches wide, with fine, spreading hairs on the undersides of large veins and sometimes on leaf surfaces. Leaf margins are lightly toothed or untoothed.

Flowers are star shaped, individually stalked and made up of five fused petals and a cone of yellow anthers; flowers have five small, green sepals at the flower base and are clustered.

Black nightshade: Each cluster has five to 10, white to pale blue, drooping, 0.25–0.4 inch-wide flowers. Flower stalks arise from an unbranched central axis.

Eastern black nightshade: Each cluster has four to seven drooping, white to light purple, 0.25–0.4 inch flowers. Flower stalks arise from a common point.

Hairy nightshade: Each cluster has four to five, white, 0.3–0.5 inch-wide flowers. The flower stalks generally arise from an unbranched central axis but occasionally can arise from a common point.

Fruit and seeds: Berries are green when immature. Seeds are flat.

Black nightshade: Dull, blue-black berries are 0.4–0.6 inch wide and each contains 15–60 yellow to light brown (sometimes white), pointed, oval, 0.06 inch-long seeds. Berries fall off the plant before the first frost. Berries are not enclosed by sepals.

Eastern black nightshade: Shiny, purple-black berries are 0.4 inch wide, and each contains 50–110 round, yellow to white, dull, 0.06–0.08 inch-long seeds. Berries are retained after the first frost and are not enclosed by sepals.

Hairy nightshade: Yellow-green to olive or slightly purple-green berries are 0.2–0.4 inch wide, and each contains 10–36 round or oval, 0.07–0.09 inch-diameter, tan seeds. Berries are partially enclosed by sepals.

Similar species: Silverleaf nightshade (*Solanum elaeagnifolium* Cav.) is a perennial with large, showy purplish flowers, thorny stems and gray-green leaves, and it spreads by storage roots. American black nightshade (*Solanum americanum* Mill.) has wide cotyledons, hairless stems and leaves that have no or only shallow toothing; its habit is upright, but its branches are closely clumped, not spreading. Leaves and flowers of horsenettle (*Solanum carolinense* L.) are similar, but it is a perennial with large, sharp spines on stems and leaves, and it spreads by underground storage roots. Bittersweet nightshade (*Solanum dulcamara* L.) is a larger, vining perennial with red berries.

MANAGEMENT

Black nightshades not only compete with crops and reduce crop yields but also interfere with harvest and lower crop quality. The berries stain and cause soil to stick to beans, thereby causing much crop loss. They are impossible to separate from peas, leading to downgrading of the crop. The berries also add moisture to a range of combine harvested crops, leading to growth of molds. Normal tillage and cultivation practices are usually sufficient to keep light infestations under control. When growing dry beans, food grade soybeans or peas, however, eradicate any newly established populations and attack established populations with all measures available. Avoid excessive N as this will favor these weeds relative to most crops. Use good cropping practices, relatively high seeding rates and narrow row spacing to get dense, uniform stands that will suppress nightshade. Soil disturbance promotes nightshade emergence, and when practical, no-till systems can reduce nightshade densities. These species form moderately persistent seed banks, but they germinate throughout spring and summer in response to tillage. Consequently, the seeds are relatively easy to flush out of the soil with tilled fallow periods interspersed into the crop rotation or with short cycle vegetable crops in which the weeds do not have time to produce mature fruits. Sod crops set back populations of these weeds since they do not tolerate repeated mowing or competition from the perennial forages. In pastures these nightshades can be controlled by avoiding overgrazing. Intensive rotational grazing is especially effective since the repeated grazing episodes will destroy successive flushes of seedlings.

These weeds harbor many diseases and insect and nematode pests of solanaceous crops like potatoes, tomatoes and peppers. If they are allowed to grow during break periods between these crops, they will cancel many of the benefits of crop rotation for disease and pest control. Consequently, escapes should be hand rogued out of break crops and removed from the field.

Solarization to temperatures of 122–131°F for at least two days completely inhibits germination of these species from the upper soil layer.

ECOLOGY

Origin and distribution: Black nightshade was apparently introduced from Europe. It occurs on the East Coast and West Coast but is primarily a problem on the West Coast. It also occurs in Asia, Africa, Australia, and South and Central America. Eastern black nightshade is native to eastern North America and is the most common member of the black nightshade group east of the Rocky Mountains. Hairy nightshade is a native of southern South America that has been introduced into North America, Europe, Africa, Australia and New Zealand. In North America it occurs throughout most of southern Canada and the United States except the Deep South. It is most common, however, in the prairie region and the Pacific Northwest.

Seed weight: Mean population seed weights for black nightshade range from 0.80–1.3 mg, whereas mean seed weight for eastern black nightshade is 0.43 mg.

Dormancy and germination: The percentage of fresh black nightshade seeds that are dormant varies with the population, and some populations have little or no dormancy once they are separated from the berry. As little as one week of wet storage at 40–59°F is sufficient to break dormancy. Black nightshade seeds germinate best with fluctuating temperatures with an amplitude of 9–27°F and including a high temperature of 77–86°F. Constant temperatures may result in poor germination. However, some studies have found good germination at constant temperatures of 77–86°F provided seeds received a pretreatment with wet, cool conditions to reduce dormancy. The base temperature for germination is 46–50°F. Hot summer temperatures (e.g., 93–100°F) induce secondary dormancy. Nitrate and light promote germination of black nightshade. Light is a requirement for germination at constant temperature but is less important at alternating temperatures.

Eastern black nightshade germinates best in light at temperatures from 77–86°F or with alternating temperatures including a high temperature of 86°F. Several days' exposure to full sunlight, however, completely inhibits germination. Nitrate increases germination of seeds that have been exposed to light. The base temperature for emergence is 52–55°F. Eastern black nightshade requires several more days to emerge than other weeds and tomatoes. Germination is optimum at pH 5–8.

Hairy nightshade seeds are usually dormant initially but lose dormancy after 1–6 months, under a wide range of conditions. Seeds germinate well at 77–92°F but poorly at 50–73°F and at 97°F. Fluctuating temperatures increase germination of seeds that have not fully after-ripened by lowering the base temperature from 70°F to 54°F. Seeds of black and hairy nightshade enter a secondary dormancy in late summer/early fall.

Given their seed germination characteristics, stimulation of emergence of these nightshade species by tillage is not surprising.

Seed longevity: Black nightshade seeds can survive up to 39 years in undisturbed soil. Experiments in annually tilled soil, however, showed numbers of black nightshade seeds declined by 37% per year. Under similar tilled conditions, average seed losses over a five-year period were 28% and 45% per year. Hairy nightshade appears to be less persistent than black nightshade, but at least a few seeds last up to five years in the soil. All hairy nightshade seeds survived one year in experiments at two locations in Nebraska, but survival dropped to 11% and lower after two and subsequent years at one location, whereas it remained at 65% after 17 years at the other location.

Season of emergence: The black nightshades begin emerging in mid-spring and continue through summer whenever moisture is adequate. Eastern black nightshade was classified as a late-emerging species in comparison to seven other common weeds in the mid-Atlantic states, but it was classified as a middle-emerging weed with a longer emergence period than most other species in Nebraska. Emergence of nightshades tends to peak in spring when air temperature reaches 68°F or when mean soil temperatures reaches approximately 59°F with an amplitude of 18°F.

Emergence depth: Black nightshade emerges well from seeds within the top 1.6 inch of soil. A few seedlings can emerge from seeds at 2.4–3.1 inches but none from seeds at 4 inches. Eastern black nightshade emerged better from the soil surface than from a 0.6 inch depth, probably due to its smaller seeds.

Photosynthetic pathway: C_3

Sensitivity to frost: Black nightshade plants are sensitive to even light frost, but hairy and eastern black nightshades are more tolerant to light frost. The seeds of black and eastern black nightshades do not tolerate cold temperatures such as are found during winter on the Canadian prairies, but hairy nightshade seeds can tolerate such conditions.

Drought tolerance: All three species require moist soil for good growth. They wilt quickly when water stressed but may recover again if the drought period is short. Black and eastern black nightshades are generally absent from regions with dry climates except in irrigated fields, but hairy nightshade tolerates drier conditions.

Mycorrhiza: Black nightshade is mycorrhizal.

Response to fertility: These species are highly responsive to N and P fertility and will store excess N as nitrate. For example, black nightshade accumulated more than twice as much nitrate at the flowering stage as tomatoes and peppers, and the uptake of major nutrients by these crops was reduced when competing with eastern black nightshade. Hairy nightshade plants increased in size up to nitrogen application rates of 430 pounds per acre with a relative response that was similar to wheat. The species showed the strongest response to increasing P levels of 22 weeds and two crops tested. The black nightshades grow best at a pH of 6–6.5.

Soil physical requirements: These species will grow on a wide range of soil textures from sandy loam to clay, but they do best on fertile soils with good moisture holding capacity. They do not tolerate poor drainage.

Response to shade: Eastern black nightshade tolerates only light shade. Shade levels of 60%, 80% and 94% relative to full sunlight reduced growth by 48%, 83% and 98%, and berry production by similar amounts. Plants within soybean rows were only one third as large as plants between the rows. However, even the smallest plants subjected to the most shade produced some seeds to replenish the seed bank. Black nightshade is more shade tolerant than eastern black. For example, moderate shade of 50–60% had little effect on plant biomass and increased the leaf area of black nightshade, but greater than 75% shade reduced all measures of plant size. In hot dry conditions, partial shade appears to promote growth and survival of black nightshade.

Sensitivity to disturbance: Plants resprout vigorously following severe and repeated cutting.

Time from emergence to reproduction: Black nightshade plants emerging in spring require seven to nine weeks before flowering. In mid-summer, flowers begin opening five to six weeks after emergence. Eastern black nightshade flowered five to 10 weeks after emergence. Nightshade berries mature four to five weeks after pollination, although maximum seed viability of eastern black nightshade was not reached until six to eight weeks after flowering. Fruit production in all species continues until frost.

Pollination: All three species commonly self-pollinate but also are cross pollinated by insects.

Reproduction: Eastern black nightshade plants growing outside in Ontario produced 50–100 berries and at least 2,500–5,000 seeds each. In Minnesota, eastern black nightshade without competition produced 2,700–7,400 berries containing 252,000–825,000 seeds per plant. However, maximum seed production with soybean competition was 2,500 seeds per plant. In Illinois, individual plants produced 6,000 berries with 500,000 seeds under the most favorable conditions. In California, black nightshade averaged 60 seeds per berry and 60,000 seeds per plant, whereas hairy nightshade averaged 20 seeds per berry and 16,000 seeds per plant. Black nightshade berries fall off the plant when mature, but eastern black nightshade plants remain upright with some berries intact on the plant after frost has killed the plants.

Dispersal: Seeds of all three species are spread by birds, rodents and livestock that eat the berries. The seeds pass unharmed through the bird and mammal digestive tracts. The berries also disperse by water as well as with hay, straw or crop seed.

Common natural enemies: The nightshades are attacked by Colorado potato beetles (*Leptinotarsa decemlineata*), tobacco and potato flea beetles (*Epitrix hirtipennis* and *E. cucumeris*) and spinach leaf miners (*Pegomya hyoscyami*). They also host bean aphids (*Aphis fabae*), which can cause leaf curling and death of nightshade shoot tips. Like domesticated species in the nightshade family, these weedy nightshades are attacked by many fungal and viral diseases.

Palatability: Berries and foliage can be poisonous to humans and livestock, and immature berries are particularly toxic. Nightshade plants mixed with forage remains poisonous and can cause intestinal dysfunction and many other symptoms in livestock. Toxicity is associated with alkaloids and high accumulations of nitrate. However, concentration of the toxic principles apparently varies greatly and is reduced in boiled leaves and ripe berries. The nutritional value of leaves and berries is generally high, although it can vary depending on soil fertility, age and species. In the United States, the berries of these weeds are sometimes used to make jam. In Africa, Central America and southeast

Asia, people eat the fruit of black nightshade and use the leaves as a cooked vegetable. Domesticated varieties of black nightshade are grown for the fruit and leaves in many countries and are called "garden huckleberry" in the United States. Leaves of plants grown in shade were considered more palatable and less bitter than leaves grown in full sun.

FURTHER READING

Bassett, I.J. and D.B. Munro. 1985. The biology of Canadian weeds. 67. *Solanum ptycanthum* Dun., *S. nigrum* L. and *S. sarrachoides* Sendt. *Canadian Journal of Plant Science* 65: 401–414.

DeFelice, M.S. 2003. The black nightshades, *Solanum nigrum* L. et al. – poison, poultice, and pie. *Weed Technology* 17: 421–427.

Peachey, R.E., R.D. William and C. Mallory-Smith. 2004. Effect of no till or conventional planting and cover crops residues on weed emergence in vegetable row crop. *Weed Technology* 18: 1023–1030.

Rich, A.M. and K.A. Renner. 2007. Row spacing and seeding rate effects on eastern black nightshade (*Solanum ptycanthum*) and soybean. *Weed Technology* 21: 124–130.

Stoller, E.W. and R.A. Myers. 1989. Effects of shading and soybean *Glycine max* (L.) interference on *Solanum ptycanthum* (Dun.) (eastern black nightshade) growth and development. *Weed Research* 29: 307–316.

Palmer amaranth

Amaranthus palmeri S. Watson

Palmer amaranth foliage and petioles
Joseph DiTomaso, University of California, Davis

Palmer amaranth plant
Joseph DiTomaso, University of California, Davis

Palmer amaranth inflorescence
Mike Stanyard

Palmer amaranth inflorescence, closeup
Joseph DiTomaso, University of California, Davis

IDENTIFICATION

Other common names: carelessweed, Palmer pigweed

Family: Pigweed family, Amaranthaceae

Habit: Tall, erect annual herb

Description: Seedling stems are usually red, sometimes green and sometimes slightly hairy. Cotyledons are lanceolate, hairy and red to green in color. Young leaves are alternate, egg shaped and roughly five times longer (0.4–0.5 inch) than wide, with notched tips.

The coarse, red stem of **mature plants** will reach 1.5–6.5 feet tall with long, thin branches spaced far apart on the stem. Upper stems and branches are greener, less hairy and have narrower, pointier leaves than those located lower on the plant. Leaves are alternate, hairless, egg shaped to lanceolate, 2–8 inches long by 0.5–2.5 inches wide, with conspicuous white veins on the undersides and long leaf stalks. The relatively shallow taproot is reddish near the soil surface. Plants are either male or female, with individual, inconspicuous, green **flowers** grouped in spikes; spikes are soft or lightly spiny, thin, 0.3–0.8 inch wide, often nodding and largely unbranched. The longest spike is located on the main stem; smaller spikes are present at branch tips and in upper leaf-stem joints. Terminal spikes of male plants are yellow and 0.5–1.5 feet, with thin, triangular, green bracts around individual flowers. Female spikes are similar, with thick, stiff, notched bracts with a small spine located in the notch. **Fruits** are brown, triangular or pyramid shaped sacs, with three to five spines at their tip; sacs rupture at maturity, releasing one glossy, round, dark red-brown to black, 0.04–0.06 inch-long **seed**.

Similar species: Palmer amaranth seedlings are similar to seedlings of redroot pigweed (*Amaranthus retroflexus* L.), smooth pigweed (*Amaranthus hybridus* L.), Powell amaranth (*Amaranthus powellii* S. Watson) and waterhemp [*Amaranthus tuberculatus* (Moq.) Sauer]. Mature Palmer amaranth has thinner stems than redroot pigweed, and its combination of non-wavy, egg-shaped, round-tipped leaves with very long leaf stalks and white veined undersides sets the mature plant apart. Other *Amaranthus* spp. have crowded, highly branching spikes, different from the long, nodding, thin, lightly branching terminal spikes of Palmer amaranth. Only waterhemp is like Palmer amaranth in having separate male and female plants.

MANAGEMENT

Since Palmer amaranth has a relatively short-lived seed bank, rotation of land into several years of a perennial sod should reduce population levels substantially. It will rarely be a problem in winter grain crops and early planted spring grains since these crops do most of their growth before this weed emerges in late spring. Since they are harvested before it matures, tillage after grain harvest interrupts the weed's life cycle.

Since Palmer amaranth seeds must be near the soil surface for successful emergence, inversion tillage reduces densities of the weed relative to reduced tillage systems. Nevertheless, because the species often produces such a prodigious number of seeds, even in competitive crops, annual tillage will still not eliminate all seeds that can emerge to infest the crop. High seed retention on plants at harvest suggests the potential utility of seed collection and destruction during combining. Removal of harvested residue and chaff from the field reduced subsequent Palmer amaranth populations by up to 70%.

Palmer amaranth is among the fastest growing of all weeds. Its photosynthetic rate is very high, even for a C_4 species, and its leaves track the sun, which ensures a high photosynthetic rate throughout the day. When plants are small, they can double in size every two to three days, and plants can reach 4 inches tall two weeks after the seeds begin germination and over 6.6 feet by maturity. Root growth is also five-fold greater than that of soybeans. This gives the weed an advantage in competition for water and nutrients. The extremely rapid emergence and growth of Palmer amaranth means that seedlings are only briefly susceptible to rotary hoeing and tine weeding, and good timing of operations is particularly important when managing this weed. In corn and sorghum, two pre-emergence blind cultivations may be useful, one about three days after planting and the second just before crop emergence. Similarly, tine weeding or rotary hoeing at close intervals (e.g., four days) after crop emergence will also reduce populations. This will require slowing tine weeders down when the crop is very small, but early control is critical: The rapid growth means that any early emerging seedlings that escape will be too big to bury by the time row crops are large enough to hill.

High corn density and irrigation improved corn competitiveness against Palmer amaranth in Kansas. Increasing the population of drilled soybeans suppressed Palmer amaranth emergence in Arkansas. Since soybeans will germinate and grow at soil temperatures too low for germination and growth of Palmer amaranth, early planting can give the soybeans a competitive head start against this weed. Alternatively, a period of fallow with repeated shallow tillage in late spring can flush a large proportion of the seeds out of the surface seed bank, resulting in less weed pressure in the crop. The long season of emergence, rapid growth and exceptional seed production potential of Palmer amaranth makes all of the tactics discussed above only partially effective. In addition, development of glyphosate resistant populations appears to have facilitated the rapid development of taller, more aggressive plants that are better adapted to compete in various crop canopy structures. Good control may require hand pulling plants before they go to seed.

Black plastic mulch is effective for suppressing most Palmer amaranth in vegetable crops, but plants that emerge through planting holes can overwhelm crops. In bell pepper production, Palmer amaranth plants should be removed from planting holes within five weeks of transplanting the crop. Cover crop mulches can contribute to integrated management of this small seeded species. Residue from a rye cover crop suppressed a Palmer amaranth population by 57–58% in Arkansas. Even a relatively light layer of rye mulch (980–1,960 pounds per acre) can reduce Palmer amaranth density by over 70%, but rye mulch biomass generally needs to exceed 8,900 pounds per acre for greater than 80% Palmer amaranth suppression. This level of rye requires early fall planting and high residual soil nitrogen. Two years of a rye cover crop following a single inversion tillage to bury surface seeds controlled Palmer amaranth by 92% compared to 69% by inversion tillage alone. Incorporation of winter brassica cover crops, including turnips, garden cress, oilseed rape and Indian mustard suppressed Palmer amaranth in transplanted bell peppers by over 40%.

ECOLOGY

Origin and distribution: Palmer amaranth is native to the canyons and desert washes of the Southwest. In the past 50 years it has become a problem weed in much of the Southeast and the southern Midwest. It now occurs sporadically northward to Wisconsin, Ontario and Massachusetts and can successfully compete with crops and complete its life cycle throughout Illinois. It has also been introduced to Europe, Asia and Australia.

Seed weight: Mean population seed weights vary from 0.44–0.49 mg.

Dormancy and germination: Some newly matured seeds can germinate immediately, but most are dormant. Seeds produced from spring emerging plants have higher germinability (40–70%) than those produced by fall emerging plants (15–20%). Palmer amaranth has a minimum temperature threshold for germination of 63°F, and in field conditions seedlings begin emerging when temperatures near the soil surface reach 64°F. It germinates best at 86–99°F and can germinate in day/night temperature regimens as hot as 113/104°F. Fluctuating temperatures provide moderate promotion of germination. After seeds have been buried or left on the soil surface for six months, light increases percentage germination. Reduced light levels and reduced day/night temperature amplitude under a closed soybean canopy reduced Palmer amaranth emergence up to 76%. At optimal temperatures, germination is very rapid, with most seeds germinating in less than one day. Application of poultry litter has been observed to increase emergence, possibly indicating that germination is promoted by nitrate as are other *Amaranthus* species.

Seed longevity: Palmer amaranth seeds are relatively short lived. In one study, seed survival was 44–61% after one year, and only 9–22% of seeds survived for three years. In another experiment, seed survival after one year at a 6 inch depth was 20%. A single year of good weed control reduced seeds in the top 2 inches of soil by 80–99%. However, despite 98% reduction of the seed bank after six years, 7 million seeds per acre were still present.

Season of emergence: In South Carolina, seedlings emerged primarily from mid-May to mid-July and required ample rainfall. In California, optimum emergence occurred from May to September.

Emergence depth: Palmer amaranth emerges best from the top 0.5 inch of soil. Although a very small percentage of seedlings emerged from 3 inches when planted in pots, emergence in the field did not occur from below 1.5 inch.

Photosynthetic pathway: C_4

Sensitivity to frost: Palmer amaranth does not tolerate freezing temperatures.

Drought tolerance: Palmer amaranth is highly drought tolerant. The leaves can maintain photosynthesis despite substantial water stress, and the plant forms a deep taproot. Plants can germinate, grow to substantial size and complete their life cycle following a single rain. Rapid growth and high seed production is the primary response of this species to hot, droughty environments.

Mycorrhiza: Palmer amaranth appears to lack mycorrhizae.

Response to fertility: Growth was more responsive

to N than to P and K, and N deficiency reduced all Palmer amaranth growth characteristics, particularly under high light intensity. Application of 164 pounds per acre of P_2O_5 at planting and a double application of 58 pounds per acre of N at three and six weeks after planting had little effect on the growth of dense, pure stands of Palmer amaranth, but in another study, increased density and growth of the weed was observed when rice received poultry litter. Palmer amaranth populations with a history of high nitrogen fertilization and glyphosate resistance have developed higher nitrogen-use efficiency than populations with a low nitrogen history and glyphosate sensitivity, suggesting co-evolution of traits for nitrogen-use efficiency and glyphosate resistance.

Soil physical requirements: Palmer amaranth roots penetrate even highly compacted soil layers, and the species maintains high productivity on compacted soil.

Response to shade: Plant height increased as rapidly in 87% shade as in full sunlight, with shaded plants producing fewer, thinner leaves and fewer branches. Time to flowering was doubled in plants growing in 88% shade. Palmer amaranth is difficult to shade with a summer annual crop because it grows so fast and tall that it can even overtop crops like corn.

Sensitivity to disturbance: Cutting Palmer amaranth stems at 6 inches above soil level prevented yield loss of cotton, but plants re-grew and produced 116,000 seeds per plant. When cut at 0 or 1 inch, seed production was reduced, but plants still added 690 and 28,000 seeds per plant, respectively, to the seed bank. Hail damage favors Palmer amaranth relative to corn.

Time from emergence to reproduction: Plants emerging from March through June mostly flowered in five to eight weeks, whereas plants emerging in July or later when day lengths were declining flowered in three to four weeks. High plant densities can accelerate initiation of flowering by 10–20 days. Plants continue to flower over a period of 40 days after flower initiation in Arkansas. Some viable seeds were produced as early as two to three weeks after the beginning of flowering.

Pollination: Flowers are normally wind pollinated.

Since male and female flowers occur on separate plants, self-fertilization is theoretically impossible, but apparently some seeds can develop without pollination.

Reproduction: Without competition, female plants emerging in spring or summer produce on average 200,000 to over 600,000 seeds per plant. Palmer amaranth grown in cotton produced 310,000–435,000 seeds in Georgia. Plant establishment three weeks later reduced seed production in cotton by 76%. When grown at low density in competition with corn, plants produced over 35,000 seeds per plant. Plants grown with soybeans across five Midwestern states produced 13,000–60,000 seeds per plant. Plants retain 95–100% of seeds at soybean maturity and lose only 3% of seeds during the month after soybean maturity.

Dispersal: Seeds disperse in moving water, on mowers and cotton pickers, in mud clinging to tractor tires and by animals. Seeds pass intact through the guts of some birds or are regurgitated after being retained for several days, so dispersal by birds seems likely. Cotton gin trash can be contaminated with Palmer amaranth, and gin trash composting procedures are frequently inadequate for killing weed seeds. Plants allowed to grow along field margins also contribute significantly to seed flux into fields. Pollen can disperse at least 980 feet, allowing for the rapid spread of genetic traits adapted to agricultural practices.

Common natural enemies: The pigweed flea beetle, *Disonycha glabrata*, can cause substantial damage to most pigweed species. Seed consumption by fire ants, ground beetles and rodents accounted for high seed losses from the soil surface, especially during summer months.

Palatability: Native Americans ate both the cooked seeds and cooked leaves. Plants are considered good forage for livestock at all stages of growth. **Heavy consumption for five to 10 days, however, can cause swelling of tissues around the kidneys (perirenal edema) in pigs, cattle and sheep.**

Note: Incorporation of 9–13 tons per acre dry weight of Palmer amaranth residue substantially reduces growth of grain sorghum and several vegetable species.

FURTHER READING

Davis, A.S., B.J. Schutte, A.G. Hager and B.G. Young. 2015. Palmer amaranth (*Amaranthus palmeri*) damage niche in Illinois soybean is seed limited. *Weed Science* 63: 658–668.

DeVore, J.D., J.K. Norsworthy, and K.R. Brye. 2013. Influence of deep tillage, a rye cover crop, and various soybean production systems on Palmer amaranth emergence in soybean. *Weed Technology* 27: 263–270.

Ehlringer, J. 1983. Ecophysiology of *Amaranthus palmeri*: a sonorant desert summer annual. *Oecologia* (Berlin) 57: 107–112.

Keeley, P.E., C.H. Carter and R.J. Thullen. 1987. Influence of planting date on growth of Palmer amaranth (*Amarantbus palmeri*). *Weed Science* 35: 199–204.

Ward, S.M., T.M. Webster and L.E. Steckel. 2013. Palmer amaranth (*Amaranthus palmeri*): a review. *Weed Technology* 27: 12–27.

Perennial sowthistle

Sonchus arvensis L.

Young shoots of perennial sowthistle
Antonio DiTommaso, Cornell University

Perennial sowthistle leaves clasping the stem
Randall Prostak, University of Massachusetts

Perennial sowthistle flowering plant
Randall Prostak, University of Massachusetts

IDENTIFICATION

Other common names: corn sow thistle, field sowthistle, creeping sow thistle, gutweed, milk thistle, field milk thistle, swine thistle, tree sow thistle, dindle, marsh sowthistle

Family: Aster family, Asteraceae

Habit: Erect, mostly unbranched, perennial herb arising from a basal rosette and spreading by shallow, thickened storage roots

Description: Seedlings have short-lived, slightly fleshy, 0.2–0.3 inch-long by 0.05–0.2 inch-wide, round or oval cotyledons with a small notch at the tip. The first true leaves are paddle shaped, irregularly toothed along the margins and dull blue-green. Young leaves form a rosette and are lanceolate, alternate and hairless, with prickly teeth along the margins and leaf stalks. Teeth point towards the base of the leaf. All parts exude a white sap when injured. **Mature plants** are 2–5 feet tall, with hollow stems that may branch at the top. Stems are hairless towards the base, with sparse to dense gland-tipped hairs towards the top, or occasionally the top is hairless. Leaves are lanceolate, 2–16 inches long by 1–4 inches wide, alternate, hairless and irregularly lobed. Leaf margins are toothed and prickly. Leaves are larger and more densely arranged on the lower stem. Lower leaves have two to six triangular to lanceolate lobes or are occasionally unlobed. Lobes on the upper leaves are reduced or absent. Leaf bases have a pair of rounded lobes that clasp the stem. The root system is extensive and can reach up to 10 feet in depth and spread over 9 feet horizontally. Roots are yellow-white, thick, brittle and fleshy, with buds that can produce vegetative shoots from up to 20 inches below ground. Small clusters of **flower heads** develop at the end of upper stems and branches. Each bright yellow, dandelion-like flower head is 1–2 inches wide. At the base of each flower head are overlapping, narrow bracts. Each bract is 0.5–1 inch long, with many short, yellow gland-tipped hairs (occasionally hairless). The apparent seed includes a thin, tight coat of fruit tissue. These **seeds** are red-brown to dark brown, rectangular, 0.1–0.14 inch long by 0.04–0.06 inch wide and have five to 12 longitudinal ridges. Each seed is attached to a feathery, white 0.4–0.6 inch-long pappus.

Similar species: Prickly lettuce (*Lactuca serriola* L.), annual sowthistle (*Sonchus oleraceus* L.) and spiny sowthistle [*Sonchus asper* (L.) Hill] are similar to perennial sowthistle. Prickly lettuce seedlings are hairy and leaves of mature plants have a distinctive row of sharp spines along the underside of the midvein. Annual and spiny sowthistle flower heads are only 0.75 inch or smaller, and the seeds have fewer ridges than those of perennial sowthistle. Spiny sowthistle leaves have larger, spinier teeth than perennial sowthistle leaves, while annual sowthistle leaves lack prickles. Annual sowthistle leaves usually have a large triangular terminal lobe that is not found in the other two sowthistle species. All three species are annuals and have taproots rather than the extensive root system of perennial sowthistle.

MANAGEMENT

Because perennial sowthistle thrives best in wet soils, improving soil drainage can improve control. In particular, improving soil drainage will improve the ability of your crops to competitively suppress the weed.

As with most creeping perennial weeds, good control involves fragmenting the root system to weaken subsequent regrowth, repeatedly killing the shoots and strong competition from crops. In Scandinavian studies, weight and carbohydrate storage in the roots reached a minimum in late May, and this corresponded to the five- to seven-leaf stage of development. Burial of the plant at that stage reduced the number of shoots produced the rest of the season relative to burial earlier or later. Subsequently turning new shoots under two to three times when they reached four to six leaves killed the storage roots completely. Performing the operations at the four-leaf stage eliminated the plants in 74 days or less but required one to three additional operations, whereas waiting until the six-leaf stage required up to 84 days but only required one to two additional operations. The four- to six-leaf stage was reached by 3 inch root fragments in about two to three weeks, and this corresponded to the point at which the root fragments were most depleted. Similar to the tillage experiments, cutting the shoots off just below soil level when they reached the six-leaf stage eliminated the storage roots in three operations, whereas cutting the shoots at four leaves required four operations. Waiting until the plants had six leaves caused shrinkage of the storage roots but not complete death in one experiment but produced the most rapid elimination of the plants (42 days) in another experiment.

In general, a well-planned tillage fallow period can eliminate the weed in time to plant a winter grain or late season vegetable crop. Spring plowing is more effective than autumn plowing for control of perennial sowthistle. Fallow tillage in the fall is less effective than tillage in the spring because the buds become dormant and do not sprout readily in the fall. Nevertheless, tillage following grain harvest can be a useful control tactic because it kills top growth that is feeding the storage roots and because breaking up the storage roots makes the plants less vigorous in the spring. Two years of hay mowed three times per season also substantially reduces perennial sowthistle density.

As with most perennial weeds, cutting up the roots with tillage implements greatly decreases the vigor of the subsequent shoots. Moreover, burying the small fragments 8–12 inches deep also reduces the number and subsequent vigor of the shoots. Thus, some European organic growers control the weed by chopping the storage roots into small pieces with a disk or field cultivator and then moldboard plowing to bury the fragments. When the shoots from deeply buried small fragments were allowed to grow unchecked, they produced large storage roots by the end of the season so that chopping up and burying the roots was essentially futile if no further actions were taken. When fields were planted with barley, however, the combination of fragmentation, deep burial and crop competition resulted in a substantial decrease in weight of storage roots. A short tilled fallow period could probably be substituted for the deep tillage in this control strategy.

ECOLOGY

Origin and distribution: Perennial sowthistle is native to Europe and western Asia and is most common in northwestern Europe. It has been introduced into North and South America, Australia and New Zealand. It is distributed throughout the northern United States and southern Canada but only occurs sporadically in the southern and southwestern states.

Seed weight: Mean population seed weights range from 0.38–0.69 mg.

Dormancy and germination: Seeds have little or no innate dormancy and will readily germinate immediately after dispersal. Seeds germinate best at 77–86°F. Few seeds will germinate at constant temperature outside this range, but fluctuating temperatures with the high above 86°F increase germination. Many perennial sowthistle seeds will germinate in the dark, but exposure to light increases the germination. Seeds can tolerate wetting for five days and subsequent drying without losing viability.

Seed longevity: Few seeds last in the soil longer than five years. Most seeds (80%) germinate within the first year. Seeds apparently survive better in clay soils than in sandy loam. In soil worked three times a year, mortality of seeds was 48–65% per year.

Season of emergence: Seedlings emerge primarily in late spring through mid-summer. Shoots begin emerging from rootstocks as soon as the soil warms, which is late April in many of the areas where the weed is a major pest.

Emergence depth: Seedlings emerge best from the top 0.2 inch of soil, though a very few can emerge from as deep as 1.2 inch. Root fragments can produce shoots from anywhere in the plowed horizon, but emergence from fragments at the soil surface or deeper than 8 inches is reduced. Shoots from small, deeply buried fragments are weaker and may not emerge at all.

Photosynthetic pathway: C_3

Sensitivity to frost: Shoots die back after frost. Overwintering roots survive soil temperatures as low as 3°F without damage but cannot survive -4°F.

Drought tolerance: Perennial sowthistle seedlings are highly sensitive to drying and generally only establish in wet spots or in areas where crop residues or cover of other plants keep the soil moist. Plant growth is best in saturated soil and is progressively reduced at field capacity and lower soil moisture levels. Well established plants, however, often have some deep roots that help the plant survive dry periods.

Mycorrhiza: There have been two reports of the presence of mycorrhiza on this species and one report of their absence.

Response to fertility: Perennial sowthistle does best on neutral to slightly alkaline soils. Plant growth was

optimum at pH 6.2 and 7.2 but 30% lower at pH 5.2. Nitrogen had little influence on shoot emergence but increased the mass of thickened roots by fall. The species accumulates higher N, P, K and Mg concentrations than winter wheat and higher K and Ca concentrations than spring barley. It can achieve K tissue concentrations of 5%.

Soil physical requirements: Perennial sowthistle is most common on loam or clay soils, particularly in areas with high precipitation, and it is relatively rare on dry, sandy and gravelly soils. Compaction reduces its growth and ability to produce new shoots. The species tolerates moderate salinity, but only in wet soils.

Response to shade: Perennial sowthistle is sensitive to shade. Plant weight and reproduction are reduced in shade, and leaf growth increases at the expense of stems and roots as shade increases.

Sensitivity to disturbance: Ability of young plants to re-establish if the root is fragmented develops within a few weeks of seedling emergence. Basal stems and new roots growing from established plants have the same capacity for vegetative reproduction. Most of the thickened roots from which new plants arise lie within the top 8 inch plow layer where they can be broken up by tillage implements. This creates more, but weaker, sprouts. Deep burial of root fragments by inversion tillage reduces the number and vigor of subsequent sprouts. Thickening of new roots into overwintering storage roots begins when the shoot has five to seven leaves. Buds on storage roots become dormant in late summer and fall, so tillage late in the season does not induce a flush of new shoots. Bud dormancy is enhanced primarily by shortening day length but maximally in combination with temperatures below 62°F. Bud dormancy is overcome by exposure to 36–40°F for a month or more.

Time from emergence to reproduction: Most plants do not flower during their first year unless conditions are highly favorable. Flowering begins when the shoot has 12–15 leaves, which is in early July in the northern United States. Flowering continues until late summer. Initiation of flowering is delayed several weeks by soil moisture below field capacity and by shading. Seeds develop about 10 days after the flowers open.

Pollination: Perennial sowthistle is pollinated by bees, flies and blister beetles. The species does not self-pollinate, and thus populations consisting of a single clone do not produce seeds.

Reproduction: Perennial sowthistles produce an average of about 30 seeds per head. A particularly large, isolated shoot reportedly produced 62 seed heads and nearly 10,000 seeds, but in our experience, five to 20 seed heads is more typical of plants growing in agricultural fields. Seeds can mature on cut stems once the flowers are pollinated. Spreading roots are the primary means of vegetative reproduction, and they enable plants to spread rapidly. The edges of clones of perennial sowthistle in North Dakota spread outward at a rate of 1.6–9.2 feet per year.

Dispersal: Tufts of hairs help wind disperse seeds moderate distances, but seeds probably rarely disperse farther than the adjacent field by wind. Hooked cells on the hairs cause the seeds to cling to fur and clothing. The species also spreads in contaminated seed and hay, and in combines. Since storage roots are mostly in the plow layer, root fragments probably disperse in soil clinging to tillage machinery.

Common natural enemies: Perennial sowthistle is susceptible to several nematodes, including root-knot nematodes (*Meloidogyne incognita*) and cyst nematodes (*Heterodera sonchophila*). It is also susceptible to *Pseudomonas solanacearum* wilt, but plants typically recover from wilting symptoms in the evening. Although several insect species have been explored for biological control of perennial sowthistle, none have demonstrated significant impact on populations of this species.

Palatability: The leaves have sometimes been eaten as a salad or pot herb. Perennial sowthistle is acceptable quality forage for most livestock but not lambs.

FURTHER READING

Brandsæter, L.O., K. Mangerud, M. Helgheim and T.W. Berge. 2017. Control of perennial weeds in spring cereals through stubble cultivation and moldboard ploughing during autumn or spring. *Crop Protection* 98: 16–23.

Lemna, W.K. and C.G. Messersmith. 1990. The biology of Canadian weeds. 94. *Sonchus arvensis* L. *Canadian Journal of Plant Science* 70: 509–532.

Håkansson, S. and B. Wallgren. 1972b. Experiments with *Sonchus arvensis* L. III. The development from reproductive roots cut into different lengths and planted at different depths, with and without competition from barley. *Swedish Journal of Agricultural Research* 2: 15–26.

Vanhala, P., T. Lötjönen, T. Hurme and J. Salonen. 2006. Managing *Sonchus arvensis* using mechanical and cultural methods. *Agriculture and Food Science* 15: 444–458.

Pigweeds

Powell amaranth, *Amaranthus powellii* S. Watson
Redroot pigweed, *Amaranthus retroflexus* L.
Smooth pigweed, *Amaranthus hybridus* L.

Powell amaranth seedling
Antonio DiTommaso, Cornell University

Powell amaranth older seedling
Antonio DiTommaso, Cornell University

Powell amaranth developing inflorescence
Antonio DiTommaso, Cornell University

Redroot pigweed seedling
Joseph DiTomaso, University of California, Davis

Redroot pigweed leaves
Antonio DiTommaso, Cornell University

Hairy stem of redroot pigweed
Antonio DiTommaso, Cornell University

Smooth pigweed seedlings
Antonio DiTommaso, Cornell University

Smooth pigweed inflorescence
Scott Morris, Cornell University

IDENTIFICATION
Other common names:

Powell amaranth: rough pigweed, amaranth pigweed, green amaranth, green pigweed, careless weed

Redroot pigweed: rough pigweed, green amaranth, pigweed, wild beet, amaranth pigweed, red root, careless weed, redroot amaranth, common amaranth

Smooth pigweed: green amaranth, pigweed, wild beet, spleen amaranth, rough pigweed, amaranth pigweed, red amaranth, careless weed, prince's feather, slender pigweed, smooth amaranthus

Family: Amaranth family, Amaranthaceae

Habit: Erect, often branched, summer annual herbs

Description: Seedlings have reddish-pink stem bases and oval shaped true leaves.

Powell amaranth: Stems are nearly hairless and red tinged. Cotyledons are 0.5 inch long and lanceolate, with red or purple undersides. Young leaves are green with red-purple undersides, slightly notched, not wavy along the edges, with a prominent midvein. Very few hairs are present on stems, stalks and leaves.

Redroot pigweed: Stems are hairy and pale green. Cotyledons are 0.5 inch long, lanceolate (four to five times longer than wide) and with red or purple undersides. Young leaves are green, wavy edged and notched with a prominent central vein. A central vein is prominent on the upper surface near the blade base. Small, tough hairs are present on leaf stalks.

Smooth pigweed: Stems are hairy and reddish-purple. Cotyledons are dull green with red-purple undersides, 0.13–0.5 inch long and less than 0.13 inch wide. Young leaves are dark green, slightly wavy-edged and slightly notched with a prominent midvein. Hair is dense on the stems.

Mature plants have a shallow, sometimes red taproot. Leaves are green and alternately arranged.

Powell amaranth: Stems are nearly hairless and can reach 5 feet tall. Leaves are 0.6–3 inches long by 0.2–1.6 inch wide, diamond shaped, pointy tipped, non-ruffled and shiny green (sometimes red-tinged) with white veins on the blade undersides.

Redroot pigweed: Upper stems are coated with curly hairs and can reach 6 feet tall. Leaves are 1.5–6 inches long by 2.5 inches wide, oval to diamond shaped, wavy-edged, and dull green with white central veins on the blade undersides.

Smooth pigweed: Upper stems are coated with short hairs and can reach 6.5 feet tall. Leaves are up to 6 inches long, oval to egg shaped, wavy-edged, long stalked and dark green with strong veins and light green to magenta-tinged undersides.

Flowers are small, greenish (turning brown upon maturity) and clustered into long groups located at branch ends and in leaf-stem joints. Clusters on branch ends are larger than those in leaf axils. Individual flowers are either male or female.

Powell amaranth: Flowers are clustered into thin, stiffly upright, mostly unbranched spikes. Branch end clusters can reach 10 inches long by 0.4–0.8 inch wide.

Redroot pigweed: Flowers are tightly clustered into stiff, branching panicles. Branch end clusters can reach 2–8 inches long by less than 1 inch wide.

Smooth pigweed: Flowers are clustered into nodding, branching spikes. Branch end clusters can reach 6 inches long; those in axils can reach 0.5–3 inches long. Clusters are softer than those of the other species.

Fruit and seeds: Seeds develop singly in small, bladder-like fruits (utricles). Seeds are glossy and dark brown to black, no larger than 0.05 inch, oval to ellipse shaped, flattened and notched at the narrow end.

Powell amaranth: Utricles do not easily rupture.

Instead, both seed and sac often fall off the plant as an intact, light tan, 0.1 inch-long unit.

Redroot pigweed: The utricles rupture around the middle, dispersing seeds.

Smooth pigweed: The utricles rupture around the middle, dispersing seeds.

Similar species: Common lambsquarters (*Chenopodium album* L.) seedlings have opposite, fuzzy or grainy looking, white to pinkish young leaves. Palmer amaranth (*Amaranthus palmeri* S. Watson) leaves are hairless with white veined undersides and long stalks. It has a soft, 1–1.5 foot-long inflorescence atop its main stem. Waterhemp [*Amaranthus tuberculatus* (Moq.) Sauer] has tapering, thumb-shaped leaves, long, coarsely branched inflorescences and stems that can reach 8 feet high. Tumble pigweed (*Amaranthus albus* L.) stems are pale green to white; spiny amaranth (*Amaranthus spinosus* L.) stems are grooved, hairless and spiny; and prostrate pigweed (*Amaranthus blitoides* S. Watson) is distinguished by its red, flexible, ground-hugging stems. Livid amaranth (*Amaranthus blitum* L.) has a matting habit, green to red stems and deeply notched, 1.5 inch-long leaves.

MANAGEMENT

Pigweed germination is generally responsive to tillage and cultivation after the soil warms. Thus, a one- to two-week lag between initial and final seedbed preparation in late spring or in summer helps flush out and destroy seedlings. The final seedbed preparation should be to a depth of no more than 1.5 inch to avoid raising seeds to near the surface. Tine weed or rotary hoe when seedlings are first emerging. Once the crop is large enough to tolerate inter-row cultivation, hill up slightly (less than 1 inch is usually sufficient) before true leaves of pigweed seedlings appear. Even crops like cabbage and squash that are not normally hilled will tolerate this amount of soil against the base. When growing a crop that tolerates heavy hilling, like corn or potatoes, pile the soil up as high as possible to kill any remaining seedlings at later cultivations.

For small grain crops, a dense, uniform and vigorous stand is important for maximizing the crop's competitive advantage. Harrowing spring grains at between the three-leaf stage and stem elongation is often well timed to eliminate many newly emerged and pigweed seedlings at the white thread stage.

Straw mulch and other mulch materials are highly effective for suppressing these species since their small seeds provide minimal resources for pushing the cotyledons up out of the mulch. Because the seed-leaves stay together in a vertical position until they reach the light, however, some seedlings will usually penetrate 3,600 pounds per acre of loose straw, so either use a heavier mulch layer or compact the mulch after application since dense mulches are more difficult for pigweed to penetrate.

Pigweeds are highly responsive to N and P fertility, so avoiding excess fertilization is critical to management. A legume cover crop mulch, particularly if soil coverage is incomplete, also can stimulate pigweed emergence. If possible, apply fertility amendments after the crop is established.

Pigweeds are only moderately persistent in the seed bank so rotating with sod crops and use of bare fallow can decrease pigweed seed density. Also, a few years of good control will dramatically reduce pigweed seed bank density, and incorporation of a legume cover crop can accelerate seed bank decline.

Pigweeds are prolific seed producers, so clean up fields promptly after harvest if these weeds are present. If possible, remove plants that have flowered, as they can continue to form seeds even after mowing or light tillage that leaves fragments of flowering stalks on the soil surface. Hand rogue at least the larger plants out of the crop if this is economically feasible. Many seeds remain on plants until soybean harvest, which provides an opportunity to capture or destroy seeds during combining.

ECOLOGY

Origin and distribution: Redroot pigweed is native to eastern and central North America and Powell amaranth to the mountains of western North America. Smooth pigweed is native from eastern North America through Mexico to South America. All three species occur throughout most of the United States, though Powell amaranth is absent from inland parts of the Southeast and smooth pigweed from parts of the Rocky Mountain states. All three species have been introduced further northward in Canada, and redroot pigweed occurs in Alaska. Redroot pigweed has been introduced throughout Europe and Asia, to Australia and New Zealand, and most of Africa. Smooth pigweed has been introduced into Africa, south and east Asia, Australia and New Zealand.

Seed weight: Powell amaranth, 0.40–0.54 mg. Mean seed weight for various redroot pigweed populations ranges from 0.25–0.48 mg with cooler and drier locations having larger seeds. Smooth pigweed, 0.33–0.46 mg.

Dormancy and germination: Germination is stimulated by high soil temperatures (86–104°F). Higher temperatures are required to stimulate germination in younger seeds relative to older ones. A period of burial in the soil increases germinability and decreases the minimum temperature for germination to 68°F or lower. Germination increases with exposure to light, and redroot and smooth pigweed are sensitive to the equivalent of 0.01 seconds of sunlight. Nitrate also stimulates germination of these species, possibly by making seeds more sensitive to light. Seeds of redroot pigweed produced under long day length, cool temperatures or nutrient stress are more dormant, as are Powell amaranth seeds produced in competitive environments. Low soil moisture conditions induce secondary dormancy in pigweed seeds. The first seeds produced on a plant are less dormant than seeds produced at the end of the growing season. This allows a second generation to emerge and reproduce within a growing season in moderate to warm climates, but it prevents extensive germination of seeds too late to successfully produce mature plants.

Seed longevity: When buried in containers at the bottom of the plow layer or below it, a few redroot pigweed seeds have survived several decades, and smooth pigweed sustained only 12% annual mortality in one report. More typically, seed survival of redroot pigweed and Powell amaranth is poor near the soil surface, but even deep in undisturbed soil the annual survival rate is substantially lower than that of many other annual broadleaf weeds. A natural population of redroot pigweed seeds in Mississippi disappeared completely in three years, both with and without annual spring tillage, but this is probably an extreme case. In Michigan, the annual mortality of buried redroot pigweed seeds was 41–81%. In another study with annual tillage, a redroot pigweed seed bank declined 36% per

year. In New York, the annual mortality rate of Powell amaranth seeds buried 6 inches ranged from 45–88%. In a seven-year experiment with monthly tillage to 10 inches, the number of Powell amaranth emerging declined by 41% per year. In Maryland and Pennsylvania, mortality of buried smooth pigweed seeds was 39% after one year and 71% after two years.

Season of emergence: These species emerge mainly in late spring and early summer but continue to emerge throughout the growing season, particularly after soil disturbance. Redroot pigweed and smooth pigweed are classified as late emerging weeds with a relatively long (approximately two month) emergence duration.

Emergence depth: Optimal depth for emergence is 0.2–0.8 inch. One greenhouse study of redroot pigweed found an optimum emergence depth of 0.2 inch in typical medium textured soils, whereas another found that optimum emergence extended to 0.8 inch. A field study of Powell amaranth found reduced emergence at 1.2 inch relative to 0.2–0.8 inch and no emergence from 2 inches.

Photosynthetic pathway: C_4

Sensitivity to frost: These species are killed by frost.

Drought tolerance: These species are moderately drought tolerant. Redroot pigweed produces at least double the root length of other weed and crop species tested during the first month of growth, partially by growing thinner roots. A fast-growing root system would allow this small seeded species to more quickly access soil moisture throughout the soil profile in competition with other species. In addition, the C_4 photosynthetic pathway of these species is unusual for broadleaf weeds and allows the amaranths to thrive under greater heat and drought conditions than many other species.

Mycorrhiza: These species are basically non-mycorrhizal, but mycorrhizal associations do occur occasionally.

Response to fertility: Redroot pigweed is highly responsive to N, P and K and will increase in growth with N application rates up to 480 pounds per acre. Not surprisingly, redroot pigweed becomes more competitive against crops as N application rate increases. Redroot pigweed emergence can be reduced by compost or manure applications, but growth of seedlings that did emerge was increased by compost. Powell amaranth had the highest response to varying compost rates of seven crops and weeds tested, and smooth pigweed is probably also highly responsive to compost and other organic fertilizers. Smooth pigweed growth, however, was unresponsive to P over a range that nearly doubled the growth of lettuce, but it interfered with lettuce anyway by concentrating P in its tissues. Redroot pigweed does poorly on soils with pH below 5.2, and Powell amaranth grows more poorly at pH 4.8 compared to 6 or 7.3.

Soil physical requirements: All three species thrive on a wide range of soil textures from sand to clay and muck. A comparison of redroot pigweed growth on different soil types, however, found best growth on sandy loam and poorest growth on silty clay. Most redroot pigweed populations are intolerant of salt, but adapted populations are moderately tolerant.

Response to shade: Growth of these species is substantially reduced by low light. In partial shade, plants are less branched and allocate more energy to stem growth, and this sometimes helps them to grow out from under a competing crop canopy.

Sensitivity to disturbance: Newly emerged seedlings are tiny, fragile and easily broken or buried. Plants resprout if cut above the seed leaf node. Plants of moderate size or larger will reroot readily in moist soil.

Time from emergence to reproduction: All three species flower in response to shortening days. Redroot pigweed emerging in early summer (long day conditions) flowers in about 6.5 weeks, with seeds maturing seven to eight weeks later. Similarly, in Wisconsin, smooth and redroot pigweed emerging in May flowered in seven and eight weeks, respectively. In contrast, plants emerging under shorter days in late summer can flower and produce seeds very rapidly with as few as three to four leaves at the time of flowering. For example, under short day conditions, flowering occurred in three weeks with seeds maturing three weeks later. Although competition can greatly influence the number of leaves at flowering, the number of

days to flowering was unaffected under a wide range of competitive environments under field conditions.

Pollination: These species are primarily self pollinated but, since the flowers are either male or female, wind or gravity is needed to move the pollen.

Reproduction: Plants continue to flower and produce mature seeds until frost. To detect mature seeds, rub a portion of the flower cluster between your fingers and look for hard black seeds. Plants grown in favorable conditions typically produce 25,000–120,000 seeds, though plants with over 1 million seeds have been reported for redroot pigweed. Plants emerging in a well-established crop, however, may produce only a few dozen seeds.

Dispersal: Because seeds of these species often reach high densities in soil, they are easily spread from one site to another in soil clinging to boots, tires and tillage machinery. They are also picked up and dispersed by combines. Wind usually only blows the seeds a few feet, but some seeds remain on the inflorescence into the winter and can blow longer distances on crusted snow. A substantial percentage of seeds survives passage through the digestive tracts of ruminants and rabbits and are dispersed with their droppings. Cow manure is commonly contaminated with pigweed seeds, and spreading the manure disperses these seeds. Both seeds and bits of the inflorescence float and disperse in irrigation water.

Common natural enemies: Larvae of the micromoth *Coleophora lineapulvella* can greatly reduce seed production. Mice, carabid beetles and crickets consume many seeds after dispersal. European corn borer (*Ostrinia nubilalis*) sometimes causes substantial damage. The tarnished plant bug (*Lygus lineolaris*) is often seen on pigweed seed heads and can reduce seed production of grain amaranth in the Midwest by 80% or more.

Palatability: The young foliage can be used in salads or as a pot herb. The foliage may contain high levels of nitrate or oxalate, but these can be removed by cooking and draining off the water. The seeds can be cooked as grain or ground into flour. Smooth pigweed and Powell amaranth are believed to be the wild progenitors of the grain amaranths *A. cruentus* and *A. hypochondriacus*, respectively. All three of the weedy species discussed here make nutritious and highly digestible forage. Consumption of large quantities of fresh pigweed over several days, however, can cause poisoning of pigs, cattle and sheep.

FURTHER READING

Brainard, D.C., A. DiTommaso and C.L. Mohler. 2006. Intraspecific variation in germination response to ammonium nitrate of Powell amaranth (*Amaranthus powellii*) seeds originating from organic and conventional vegetable farms. *Weed Science* 54: 435–442.

Costea, M., S.E. Weaver and F.J. Tardif. 2004. The biology of Canadian weeds. 44. *Amaranthus retroflexus* L., *A. powellii* S.Watson and *A. hybridus* L (updated). *Canadian Journal of Plant Science* 84: 631–668.

Holm, L.G., D.L. Plucknett, J.V. Pancho and J.P. Herberger. 1977. *The World's Worst Weeds: Distribution and Biology*. The University Press of Hawaii: Honolulu.

Little, N.G., C.L. Mohler, Q.M. Ketterings and A. DiTommaso. 2015. Effects of organic nutrient amendments on weed and crop growth. *Weed Science* 63: 710–722.

Weaver, S.E. and E.L. McWilliams. 1980. The biology of Canadian weeds. 44. *Amaranthus retroflexus* L., *A. powellii* S. Wats. and *A. hybridus* L. *Canadian Journal of Plant Science* 60: 1215–1234.

Plantains

Blackseed plantain, *Plantago rugelii* Dcne.
Broadleaf plantain, *Plantago major* L.
Buckhorn plantain, *Plantago lanceolata* L.

Blackseed plantain rosette
Scott Morris, Cornell University

Red base of the leaf stems of blackseed plantain
Scott Morris, Cornell University

Broadleaf plantain seedlings
Joseph DiTomaso, University of California, Davis

Broadleaf plantain rosette
Antonio DiTommaso, Cornell University

Broadleaf plantain (left) and blackseed plantain (right) leaves
Scott Morris, Cornell University

Broadleaf plantain (top) and blackseed plantain (bottom) spikes
Scott Morris, Cornell University

Buckhorn plantain seedling
Jack Clark, University of California

Buckhorn plantain rosette
Scott Morris, Cornell University

Buckhorn plantain flower
Scott Morris, Cornell University

IDENTIFICATION

Other common names:

Blackseed plantain: black-seeded plantain, pale plantain, purple-stemmed plantain, red-stalked plantain, Rugel's plantain

Broadleaf plantain: common plantain, dooryard plantain, Englishman's foot, greater plantain, greater ribwort, rat tail plantain, ripplegrass, round-leafed plantain, soldier's herb, waybread, white man's foot

Buckhorn plantain: black jacks, English plantain, lamb's tongue, narrowleaf plantain, narrow-leaved plantain, ribgrass, ribleaf, ribwort plantain

Family: Plantain family, Plantaginaceae

Habit: Perennial herbs with a basal rosette and a leafless flowering stalk

Description: Seedlings form alternate-leaved rosettes. True leaves have three to five prominent, parallel veins.

Blackseed plantain: Cotyledons are spatula shaped, three-veined and up to 0.4 inch long and 0.08 inch wide. Young leaves are long stalked, dull green, oval to egg shaped and sparsely hairy. Leaf stalks have upturned edges and are usually pale pink to purplish red at the base.

Broadleaf plantain: Cotyledons are spatula shaped,

0.06–0.28 inch long by 0.02–0.04 inch wide and three-veined. Young leaves are long stalked, oval to egg shaped and sparsely hairy. Leaf stalks have upturned edges and are green or occasionally pinkish at the base.

Buckhorn plantain: Cotyledons are grass-like, 0.67–1.4 inch long by 0.04–0.06 inch wide, hairless, without a stalk and tapered at the base. The first true leaves are long, narrow and hairy along the margins; later leaves are sparsely hairy, with denser hairs at the base.

Mature plants form a rosette. Roots are fibrous with a short taproot.

Blackseed plantain: Leaves are 0.8–8.7 inches long by 0.5–5.6 inches wide, oval to egg shaped, pale green, hairless and tapered at the tip. The leaves have five to nine veins. Leaf stalks are generally as long as the leaves and are usually pink to purplish red at the base. Leaf edges are wavy and smooth to occasionally toothed.

Broadleaf plantain: Leaves are 0.8–8 inches long by 0.4–4 inches wide, oval to egg shaped, pale to dull green, leathery and sparsely hairy. The leaves have three to seven veins. Leaf stalks are generally as long as the leaves and are green or occasionally pink at the base. Leaf edges are wavy and smooth to occasionally toothed.

Buckhorn plantain: Leaves are lanceolate, 3.1–12

inches long by 0.4–1.6 inch wide and have three to seven veins. The leaves have silky, white to brownish hairs at the leaf base and are otherwise hairless to sparsely hairy. Leaf stalks are short and inconspicuous or absent. Leaf edges are generally smooth but may have scattered, small teeth.

Inflorescences, produced at the end of upright and leafless stalks, are spikes of clustered, inconspicuous flowers.

Blackseed plantain: Plants produce from one to nine flower stalks that can reach up to 20 inches tall. Flower spikes are cylindrical, up to 12 inches long and 0.23 inch wide, and are composed of up to 300 or more flowers. The flowers are pale green to whitish and 0.08–0.12 inch wide. Flowers are replaced by cylindrical, 0.16–0.24 inch-long capsules that split below the middle to release four to nine dark brown, glossy, 0.06–0.08 inch-long seeds. The seed surface is smooth.

Broadleaf plantain: Plants produce from one to 30 flower stalks that can reach up to 20 inches tall. The cylindrical flower spikes are up to 10 inches long and 0.24 inch wide. Each spike is composed of up to 400 or more pale green to whitish, 0.08–0.11 inch-wide flowers. Seeds are produced in oval, 0.16–0.2 inch-long capsules that turn from green to brown as they mature. Ripe capsules split horizontally at the middle to release six to 30, irregular, glossy, brown, 0.02–0.06 inch-long seeds. The seed surface has a fine, web-like pattern visible under magnification.

Buckhorn plantain: Plants produce from one to 30 flower stalks, which are up to 18 inches tall. The flower spike is conical to cylindrical and 0.8–3.2 inches long by 0.16–0.24 inch wide. Individual flowers are 0.08–0.12 inch wide and brown to yellow-brown. The stamens protrude from the flowers on up to 0.16 inch-long stalks. Seeds are produced in brown, oval, 0.12–0.16 inch-long capsules. Each capsule contains one to two oblong, glossy, dark brown to black, 0.08 inch-long seeds.

Similar species: Hoary plantain (*Plantago media* L.) is hairier and has much shorter leaf stalks than broadleaf plantain or blackseed plantain; the leaves are hairier and wider than those of buckhorn plantain (0.6–4.3 inches). Bracted plantain (*Plantago aristata* Michx.) has narrow leaves similar to buckhorn plantain, however the leaves and flowering stalks of bracted plantain are much hairier than those of buckhorn plantain, and the flowering spikes have long (up to 0.8 inch), hairy bracts throughout. Paleseed plantain (*Plantago virginica* L.) leaves and stalks are densely covered in soft, wooly hairs unlike broadleaf, blackseed or buckhorn plantain.

MANAGEMENT

These species are primarily weeds of perennial forages, pastures and lawns. Generally, broadleaf plantain inhabits more highly disturbed habitats, has a shorter lifespan, earlier flowering, produces more and smaller seeds, and functions more like an annual plant than the other species. Blackseed and buckhorn plantain inhabit less disturbed sites, have a longer lifespan and lower investment in seed production, and behave more as perennials. Buckhorn plantain thrives in hay meadows and is particularly common in those with red clover. Except in seed production fields, this is scarcely a problem since the species is sufficiently productive and nutritious to have warranted development as a forage crop, with several commercially available named varieties. Buckhorn plantain is less common in pastures, but it is a useful component of a mixed species pasture because it continues to produce during hot, dry periods when productivity of C_3 grasses and legumes declines. Broadleaf and blackseed plantain thrive in overgrazed pastures but tend to be out-competed in vigorous hay meadows. Using an intensive rotational grazing system helps suppress broadleaf and blackseed plantain in pastures by shading the plants during the growth phase of the forage and forcing a more upright growth form that is easier for livestock to eat. Fertility and good soil aeration will help favor forage grasses and legumes relative to plantains. Sowing uncontaminated forage seed is critical for preventing outbreaks of all three species.

Although broadleaf and blackseed plantain are commonly found in grain and vegetable crops, they rarely compete significantly with crops due to their relatively slow development and prostrate growth habit. Dense populations usually indicate soil problems. Shallowly moldboard plow with complete inversion (skim plow) to bury large plants, but then adopt a reduced tillage regimen with cover crops to improve soil structure. Avoid rotary tillage. Use a tine weeder to regularly harrow out seedlings and rotate into crops that allow hilling up to cover seedlings in the row.

ECOLOGY

Origin and distribution: Blackseed plantain is native to the eastern half of the United States and southern Canada. Broadleaf and buckhorn plantain are natives of Europe that have been spread widely throughout the world, including isolated oceanic islands. They tend to be uncommon in the lowland tropics. They occur throughout the United States. In Canada, broadleaf plantain occurs northward to above the Arctic Circle, whereas buckhorn plantain occurs primarily in southeastern and southwestern Canada.

Seed weight: Mean population seed weights for blackseed plantain range from 0.35–0.7 mg. Mean population seed weights for broadleaf plantain range from 0.06–0.34 mg. Buckhorn mean seed weights range from 0.8–2.9 mg.

Dormancy and germination: Usually, a substantial proportion of fresh seeds of all three species are capable of immediate germination, though most seeds of broadleaf plantain produced in midsummer are dormant. Dormant seeds lose dormancy over the winter. Optimal temperature for germination of broadleaf plantain is 77–86°F; however, exposure too cold for several weeks lowers the minimum temperature required for an optimum response to 59°F. Buckhorn plantain had a similar response to temperature, but the minimum temperatures giving maximum germination were lower than those for broadleaf plantain. Nitrate stimulates germination of broadleaf and buckhorn plantain, and fluctuating temperatures promote germination of broadleaf plantain. Light promotes germination of all three species, but there can be complex interactions between light, temperature and nitrate. For example, dormancy of broadleaf and blackseed plantain was relieved by prechilling at 41°F for two weeks in combination with light, nitrate and alternating day/night temperatures of 86/68°F, whereas either light or nitrate was sufficient for germination of buckhorn plantain. Cold stratification and fluctuating temperature partially relieved the light requirement of broadleaf plantain.

Germination of cold-stratified buckhorn plantain seeds was minimally increased by light or by light that passed through a leaf canopy.

Seed longevity: In an experiment that compared all three species, blackseed, broadleaf and buckhorn plantain survived 21, 30 and 16 years, respectively. Broadleaf plantain can remain viable in soil for 50–60 years. In one experiment, 68–87% of seeds of broadleaf plantain were viable after three years, and in other experiments, seeds maintained up to 10% viability after 40 years of burial in undisturbed soil. Buckhorn plantain seeds had 34–76% viability after three to four years and 8% viability after 10 years of burial. In various experiments in cultivated soil, broadleaf plantain seeds declined by 33–43% per year and buckhorn plantain seeds declined by 60–67% per year.

Season of emergence: Emergence from mature plants occurs in mid-spring, whereas emergence of seedlings occurs throughout the growing season, with a peak in late spring. On average, buckhorn emerged earlier than broadleaf plantain, probably because it has a lower temperature minimum for germination.

Emergence depth: Broadleaf plantain seedlings emerge only from the top 0.2 inch of soil. Buckhorn plantain emerges best from the top 0.8 inch of soil, with declining emergence to 2.4 inches. No seedlings emerge from 3.1 inches. Information on emergence depth is unavailable for blackseed plantain, but the small seeds of this species probably restrict emergence to the top 1 inch or less of the soil.

Photosynthetic pathway: C_3

Sensitivity to frost: All three species are highly frost tolerant. Even in northern areas they often overwinter as rosettes of small leaves, though in open sites, leaves may be lost for the winter.

Drought tolerance: Buckhorn plantain is relatively drought tolerant and commonly occurs in droughty sites like the gravel of railroad embankments. Blackseed plantain is probably the least drought tolerant of the three species, with broadleaf plantain having intermediate drought tolerance. Mature plants of all three species have dense, deep root systems to

32 inches, which allow them to extract soil moisture during drought.

Mycorrhiza: Broadleaf and buckhorn plantain are mycorrhizal. All three plantain species were highly colonized in prairie habitats.

Response to fertility: Broadleaf responds more to increased nutrient levels than does blackseed plantain. But, increased nutrient levels delayed flowering of blackseed more than that of broadleaf plantain. Broadleaf plantain commonly occurs in cracks in concrete pavement, indicating that it tolerates very high pH (higher than 8) and low N and P. Broadleaf and buckhorn plantain were associated with fields with a pH of 6.5–7.8. Buckhorn plantain growth, reproductive output and seed weight were increased by increased nutrient availability. Buckhorn plantain responds to P but tolerates low levels of K. Buckhorn is more responsive to P than is broadleaf plantain. It prefers neutral to calcareous soils and is excluded from soil with a pH below 4.5. Unlike most non-legumes, buckhorn plantain has bacteria associated with its roots that fix substantial amounts of N.

Soil physical requirements: All three species regularly occur on a wide range of soil types from clay to sand. Buckhorn plantain is most common on dry soils. All three species are highly tolerant of soil compaction. Broadleaf plantain tolerates long periods of waterlogging. The presence of buckhorn plantain on seashores and the abundance of both buckhorn and broadleaf plantain along roadways receiving de-icing salts indicate that both species are salt tolerant.

Response to shade: All three species tolerate some shading, but blackseed and buckhorn appear to be more shade tolerant than broadleaf plantain. Buckhorn plantain seedlings are able to grow moderately well at light levels similar to those in a dense grassland, whereas broadleaf and blackseed require periodic mowing or grazing of taller vegetation. In shaded conditions buckhorn plantain leaves become longer and more vertical, and seed weight is increased.

Sensitivity to disturbance: Since the stem lies just below ground and the leaves are fibrous and resistant to tearing, these species are highly tolerant of traffic by people, animals and tires. Even the flowering stalks are flexible and difficult to crush. Because of its upright habit, buckhorn plantain is somewhat less tolerant of traffic than the other two species. None of the species can grow back out of the soil when completely buried by tillage, but the extensive fibrous root systems of larger plants tend to help them survive and reroot after cultivation. Consequently, shallow cultivation or hoeing that cuts the plant off just below the stem will be more effective than deeper cultivation that leaves more roots attached. Removal of buckhorn plantain cotyledons results in reduced early plant growth but no adverse effect on reproduction.

Time from emergence to reproduction: Without competition, broadleaf plantain can flower within six to 10 weeks of emergence. Blackseed plantain is slower. Development of both species is slowed by competition and cool temperatures, and consequently, many plants remain vegetative their first season when shaded by taller plants. Older plants begin flowering in mid-June and continue flowering through fall. Seeds develop two to three weeks after the flowers open. Time to flowering in buckhorn plantain varies greatly with conditions and populations. A 13-hour day length is required to initiate flowering in broadleaf and blackseed, whereas a 16-hour day length is required to induce flowering in buckhorn plantain.

Pollination: These species are wind pollinated. Some plants of buckhorn plantain produce only flowers of one sex, thereby ensuring cross pollination. Other plants of buckhorn, and all plants of broadleaf and blackseed plantain, can self-pollinate.

Reproduction: In a pasture in Ontario, broadleaf and blackseed plantain plants produced an average of 565 and 662 seeds per year. In favorable conditions, blackseed plantain can produce 5,000 seeds per plant per year, and broadleaf plantain can produce 15,000–36,000 seeds per plant per year, depending on location. Seeds of broadleaf and blackseed plantain often remain in the capsules through the winter. In a dry grassland in Ontario, buckhorn plantain plants produced an average of 127 seeds in one season, whereas in a fertile agricultural field nearby, plants produced more than 10,000 seeds. In Great Britain, buckhorn produced 4,000 seeds per plant in a year. All three species occasionally produce side shoots that eventually become independent plants, but this is common only in buckhorn plantain. In all three species, the daughter plants usually remain closely clumped with the parent.

Dispersal: All three species spread in forage seed, and buckhorn plantain in particular is a notorious contaminant of red clover seed. Seeds of all three species also spread in soil clinging to tires and machinery. Buckhorn and broadleaf plantain are spread in livestock manure, and blackseed plantain may be as well. They also spread in bird and deer droppings. Seeds of these species contain a mucilage that makes them sticky when wet and may facilitate dispersal on fur and feathers.

Common natural enemies: Many fungi and insects attack plantains, but substantial damage by natural enemies is rare.

Palatability: All three species are palatable to livestock. Domesticated cultivars of buckhorn plantain have been developed. Young leaves of broadleaf plantain have been cooked like spinach and dried leaves brewed as tea.

Note: All three species produce large amounts of windborne pollen that can cause hay fever in sensitive individuals.

FURTHER READING

Cavers, P.B., I.J. Bassett and C.W. Crompton. 1980. The biology of Canadian weeds. 47. *Plantago lanceolata* L. *Canadian Journal of Plant Science* 60: 1269–1282.

Hawthorn, W.R. 1974. The biology of Canadian weeds. 4. *Plantago major* and *P. rugelii. Canadian Journal of Plant Science* 54: 383–396.

Hawthorn, W.R. and P.B. Cavers. 1976. Population dynamics of the perennial herbs *Plantago major* L. and *P. rugelii* Decne. *Journal of Ecology* 64: 511–527.

Sagar, G.R. and J.L. Harper. 1964. Biological flora of the British Isles. *Plantago major* L., *P. media* L. and *P. lanceolata* L. *Journal of Ecology* 52: 189–221.

Prickly lettuce

Lactuca serriola L.

Prickly lettuce seedling
Antonio DiTommaso, Cornell University

Prickly lettuce plant
Antonio DiTommaso, Cornell University

Prickly lettuce spines on midvein of leaf underside
Antonio DiTommaso, Cornell University

IDENTIFICATION

Other common names: wild lettuce, compass plant, milk thistle, horse thistle, wild opium, common wild lettuce, lobed prickly lettuce

Family: Aster family, Asteraceae

Habit: Erect, taprooted, summer or winter annual herb

Description: Seedling cotyledons are round or oval shaped and 0.2–0.4 inch long by 0.1 inch wide, with a tapering base, hairy upper surfaces and a notched tip. Young true leaves form a basal rosette; they are club to oval shaped, widest at the tip, pale green, hairy and toothed or lobed; spines are present on the undersides of the prominent midveins and on leaf edges. **Mature plants** bolt with erect, 1–6 foot-high, hollow, branched, somewhat woody stems. Stems are pale green to white, sometimes with red flecking, and spines are present at the stem base. Leaves are alternate, deeply lobed, blue-green, widest near the pointed leaf tip, 2–12 inches long by 1–4 inches wide, and often oriented vertically. Lobes point backward, form round cavities and clasp the stem at the leaf base. Mature spines on lower leaf surface and leaf edges are larger and more yellow than those found on the seedling. The entire plant, including the large, deep taproot, exudes a milky white sap when wounded. **Flower heads** are yellow but turn blue as they wither. They are 0.1–0.4 inch wide, composed of five to 15 ray flowers, and have three to four rows of hairless, 0.5 inch-long, green bracts covering the base. Flower heads occur in open pyramidal panicles of 13–27 inches at branch ends. One **seed** is produced per ray flower and enclosed in a dry, hard, grayish-yellow, 0.1–0.2 inch-long, ribbed fruit. This dispersal unit is referred to as the seed in the discussion below; it is attached by a 0.13 inch-long stalk to a white, bristly pappus.

Similar species: Sowthistle (*Sonchus*) species lack the characteristic midrib spines of prickly lettuce. Sowthistle leaves also remain parallel to the ground, unlike prickly lettuce.

MANAGEMENT

Prickly lettuce is primarily a problem in winter grains and reduced tillage cropping systems. The species is particularly a problem in cereal grains because it usually flowers at harvest, and the flower buds, which are difficult to separate from the grain, lower quality and raise moisture content. Since established plants can resprout after winter grain harvest operations, post-harvest control is advisable to prevent seed production.

Spring tillage controls plants that established in the fall, and late spring tillage controls most of the spring germinating individuals as well. Tillage that mixes seeds deep into the soil helps eliminate the weed. Since the seeds are small and short lived, few deeply buried seeds will return to emergence depth before they die. Control of populations along field margins may be necessary to prevent re-population by wind dispersed seeds.

The seedlings are weak and suffer high rates of natural mortality. Consequently, they are susceptible to tine weeding or rotary hoeing. They can also be controlled by flaming at the two- to four-leaf stage, but half of plants survive flaming at the six- to eight-leaf stage. Because prickly lettuce often grows taller than the crop at some point in development, it is susceptible to weed pullers, electrocution weeders and raised mowers.

Incorporated white mustard seed meal has the potential for allelopathically controlling prickly lettuce emergence and growth. Soil solarization to 135°F completely eliminates prickly lettuce emergence.

ECOLOGY

Origin and distribution: Prickly lettuce is native to the Mediterranean basin and western Asia. It now occurs in northern Europe, central Asia, North America and Australia. In North America, it occurs throughout southern Canada, the United States and into northern Mexico, except the Florida peninsula.

Seed weight: Mean population seed weight ranges from 0.45–0.62 mg.

Dormancy and germination: Seeds have no primary dormancy and can germinate immediately following dispersal. Optimal temperature for germination in England was 54–75°F, and seeds did not germinate below 46°F. Temperatures from 79–95°F induced dormancy. Seeds from the Czech Republic germinated well at temperatures from 50–86°F. Sunlight promotes germination of fresh seeds, but light filtered through a plant canopy inhibits germination. After exposure to natural alternating temperatures and/or burial for at least eight weeks, however, seeds germinate well in the dark.

Seed longevity: Seeds in the soil normally do not survive longer than three years. In Idaho, prickly lettuce seeds on the soil surface did not survive longer than 12–18 months, whereas those buried 6 inches survived up to 24–33 months. Prickly lettuce seeds' viability was an estimated 15–37% after 12 months burial. In England, the half-life of prickly lettuce seeds in soil was estimated to be 18 months, and mortality after 12 months was estimated to be 40%.

Season of emergence: Most plants emerge in the late fall, but a few emerge in the spring. Natural seedling mortality is much higher for fall germinating plants.

Emergence depth: Seeds emerge best when covered by no more than a thin layer of dust. Emergence at 0.8 inch is less than 25% of that near the soil surface.

Photosynthetic pathway: C_3

Sensitivity to frost: Prickly lettuce is very frost tolerant and commonly overwinters as a rosette, even in the northern parts of its range. During midwinter in Illinois, leaves reached temperatures as much as 18°F above air temperature. This allows continued growth on sunny winter days.

Drought tolerance: Prickly lettuce is highly drought tolerant. Its taproot extends 3 feet or more, and it can therefore tolerate drying of the plow layer. Prickly lettuce has greater taproot length per unit plant weight, number of lateral roots per unit taproot and lateral number near the bottom of the taproot compared to cultivated lettuce. The leaves orient vertically in a north-south plane, which allows maximum photosynthesis early and late in the day while reducing overheating and water stress at mid-day.

Mycorrhiza: Prickly lettuce is mycorrhizal.

Response to fertility: The limited information available seems to indicate that prickly lettuce is favored by additions of P and K but is out-competed by N-responsive species when N or a balanced nutrient source is applied.

Soil physical requirements: The species occurs on all soil textures from gravelly sand to clay. It is most common on dry, well drained soils, but it also occasionally occurs in wet sites.

Response to shade: Prickly lettuce, particularly in the rosette stage, is suppressed by a dense crop leaf canopy and suffers high mortality rates.

Sensitivity to disturbance: If mowed during flowering, the plant will branch from the remaining stem and flower again. Plants cut during winter wheat harvest in Ontario, produced 500–3,000 mature seeds per plant by late October if left unmanaged.

Time from emergence to reproduction: Overwintering plants begin to elongate in May and flower from July to September. Most seeds are shed during August and September. Spring germinating individuals mature only one to two weeks later than plants establishing in the fall. Plants that began flowering in the last week of July produced viable seeds by August 17, indicating approximately three weeks were required for maturation. Low temperature vernalization of rosettes at 40–50°F is necessary to trigger flowering.

Pollination: Prickly lettuce primarily self-pollinates, but insects accomplish some cross pollination.

Reproduction: Seed heads typically contain from 15–22 seeds. Seed production depends on plant height. For plants growing in soybeans and grain stubble in Ontario, 1 foot plants produced about 330 seeds whereas 5 foot plants produced about 55,000 seeds. Mid-sized plants of 40 inches produced 6,700 seeds. In Idaho, plants averaged 4,200–4,900 seeds per plant. In a British study, fall germinating plants averaged 1,550–2,350 seeds each, whereas spring germinating plants averaged 170–1,500 seeds each, with later emerging plants producing fewer seeds.

Dispersal: The seeds have a clump of hairs at the top that provide buoyancy in air and assist in wind dispersal. Although a very few seeds may travel long distances, most fall close to the parent plant. Seeds also disperse by water and can be introduced into fields with surface irrigation water.

Common natural enemies: Downy mildew (*Bremia lactucae*) can infect a high proportion of prickly lettuce populations, but disease severity is usually low relative to that on cultivated lettuce. Powdery mildew (*Erysiphe cichoracearum*) can cause more extensive infection and may reduce seed production under favorable conditions.

Palatability: Although prickly lettuce is the wild ancestor of the domestic lettuce, it is very bitter and unpalatable to people, even when young. Cattle can develop emphysema from feeding on fresh, young plants, but dried or mature plants are apparently not toxic.

FURTHER READING
Marks, M.K. and S.D. Prince. 1982. Seed physiology and seasonal emergence of wild lettuce *Lactuca serriola*. *Oikos* 38: 242–249.

USDA. 1970. *Selected Weeds of the United States*. United States Department of Agriculture, Agricultural Research Service, Agriculture Handbook No. 366. U.S. Government Printing Office: Washington, DC.

Weaver, S.E. and M.P. Downs. 2003. The biology of Canadian weeds. 122. *Lactuca serriola* L. *Canadian Journal of Plant Science* 83: 619–628.

Prickly sida

Sida spinosa L.

Prickly sida seedling
Antonio DiTommaso, Cornell University

Prickly sida plant
Randall Prostak, University of Massachusetts

Prickly sida showing short flower stalks and the presence of leaf stems
Carroll Johnson

Arrowleaf sida showing stemless leaves and long flower stalks
Carroll Johnson

IDENTIFICATION

Other common names: spiny sida, prickly mallow, false mallow, Indian mallow, thistle mallow, teaweed

Family: Mallow family, Malvaceae

Habit: In the temperate zone, it behaves as an upright summer annual; in tropical regions, it is a somewhat shrubby perennial.

Description: Seedlings have notched, round to heart-shaped, 0.3–0.6 inch-long by 0.2–0.3 inch-wide cotyledons. The first pair of true leaves is round, and subsequent leaves are oval to triangular. Long-stalked young leaves are alternate with toothed edges. The seedling stem, cotyledon edges and leaf edges are covered in short hairs. **Mature plant** stems are erect, highly branched, lightly hairy and can reach 0.7–3.3 feet in height. Leaves are alternate, 0.8–2 inches long by 0.4–0.8 inch wide, lanceolate to oval, toothed and have a 0.4–1.2 inch stalk; stipules are linear, and leaves may also have a 0.2–0.3 inch spine at the base. Leaf surfaces and stalks are covered with short hairs. The root system is a long, slender, branching taproot.

Flowers are produced individually or clustered from leaf axils, on 0.08–0.5 inch stalks. Each flower has five pale yellow, 0.16–0.24 inch-long petals. Stamens are fused for most of their length, forming a column. **Seeds** are located in five-chambered capsules that split into five sections when mature; each section contains one seed and has two sharp spines at its apex. Seeds are 0.04–0.1 inch long and reddish brown, with two flat and one rounded side.

Similar species: Arrowleaf sida (*Sida rhombifolia* L.) leaf stalks are nearly absent and are much shorter than the flower stalks, whereas prickly sida leaf stalks are longer than the flower stalks. Virginia fanpettles [*Sida hermaphrodita* (L.) Rusby] can be distinguished from prickly sida by its white to pale pink flowers, lobed leaves and taller height (up to 10 feet) at maturity.

MANAGEMENT

Prickly sida emerges in late spring and summer and grows best in hot conditions. Consequently, it is a poor competitor and is easy to control in winter grains and cool season vegetables. Winter grains and early planted spring grains can be harvested before it goes to seed, thereby interrupting the life cycle. Because the seeds die off relatively quickly in the soil, rotation into hay for a few years can greatly decrease an infestation. Fall seedings of the hay can be timed to prevent reproduction during the establishment year. A spring seeding will require at least two carefully timed mowings: the first to catch the young prickly sida in the two- to four-leaf stage and the second a few days after the weed begins to flower, and an additional mowing may be required to prevent reproduction by late emerging plants.

Prickly sida is most troublesome in summer row crops like cotton, soybeans, corn and sorghum. Since prickly sida seeds die off relatively quickly and seeds must be close to the soil surface to successfully produce seedlings, populations will be easier to manage with annual moldboard plowing than with reduced tillage regimens. This should occur either in the spring or fall but not both, or some seeds will be plowed back up in the spring before they have had a chance to die off. Delaying soybean planting until mid-June will get the crop past the primary peak in prickly sida emergence. Tine weed as long as possible into the June emergence period. Cultivate shallowly to avoid bringing up new

seeds into the near surface emergence zone. Flame weed cotton to destroy late emerging weeds in the crop row. Although close spacing of row crops can reduce light penetration and the growth of prickly sida, the benefits will rarely compensate for the reduced opportunities for cultivation. Prickly sida responds strongly to P fertility, so supply needed N with cover crops or plowed down sod, and minimize manure and compost use.

Prickly sida is among the species most susceptible to solarization. Covering soil with clear plastic for three weeks in mid-summer in Mississippi killed most seeds in the near-surface soil and reduced subsequent emergence by over 90%. The species can also be suppressed with straw mulch.

ECOLOGY

Origin and distribution: Prickly sida is native to the United States. It occurs widely in the Southeast and Midwest, and sporadically in the Southwest and Northeast. It is widely distributed in tropical and subtropical regions.

Seed weight: 2.3 mg.

Dormancy and germination: Most newly matured seeds of prickly sida are dormant due to a water impermeable seed coat. Some newly matured seeds (15–32%), however, will germinate immediately at day/night temperatures of at least 86/59°F, with percentages increasing with temperature up to 104/77°F. Normally, water only enters the seed following weakening of cells in the region where the seed attached to the parent plant. Damage to the seed coat (scarification) also allows the seeds to take up water and results in nearly 100% germination over a wide range of temperatures. Aging intact seeds for 16 weeks in either continuously moist or alternating wet-dry conditions substantially increases germination to as much as 78% when seeds are subsequently germinated in a warm temperature regimen of 104/77°F. Similarly, storing the seeds dry at room temperature for three to nine months results in germination of 42–95%, with percentage germination increasing with the incubation temperature. Studies disagree on whether a few months of cold, moist conditions promote subsequent germination in warm conditions. Light does not promote germination. Seed germination is high at pH 5–8. In short, seeds become permeable and able to germinate after remaining in the soil for several months regardless of moisture conditions, and germination is most rapid and complete at warm temperatures. However, sustained temperatures of over 113°F kill seeds.

Seed longevity: Less than 1% of prickly sida seeds remained viable after 5.5 years of burial. From 22% to 27% of prickly sida seeds remained viable after 1.5 years of burial, giving an estimated annual mortality rate of 60%. When a sowing of prickly sida was tilled annually to various depths, the number of seeds declined by an average of 88% per year over a two-year period. Nevertheless, a few seedlings were still emerging after five years.

Season of emergence: In Kentucky and Mississippi, prickly sida emerges from April through September. Repeated counts through five seasons in Mississippi showed two peak periods of emergence, one in May and another in July–August.

Emergence depth: Emergence is poor at the soil surface, best from 0.2 inch and declines rapidly with greater depths. No seedlings emerge from deeper than 2 inches.

Photosynthetic pathway: C_3

Sensitivity to frost: Prickly sida is sensitive to frost.

Drought tolerance: Prickly sida can maintain leaf function at higher water stress levels than other weeds and had the highest water-use efficiency compared to other C_3 (but not C_4) weeds.

Mycorrhiza: Prickly sida is mycorrhizal.

Response to fertility: Prickly sida is most commonly found on soils with high P, and it grew five-fold larger on soil testing high in P relative to soil testing very low. It grew 40% better on soil testing high in K than on soil testing low or medium. The species occurs most frequently on soils with a pH of 5.5–6.5, and growth is substantially reduced at pH of 5.2 or lower.

Soil physical requirements: The species is typically found on fertile loamy soil. It occurs most frequently on soils with low compaction and good internal drainage, but it tolerates poor drainage.

Response to shade: Prickly sida prefers full or partial sun. High density, narrow row cotton reduced light by 71–81% relative to normal planting but only reduced prickly sida growth by about half. In experiments with shade cloth, prickly sida produced highest growth and reproduction under mild shade (30–50%) and showed suppression only at a high (90%) shade level.

Sensitivity to disturbance: One mowing close to the ground between sorghum rows was sufficient to kill most plants and provided between-row control similar to a broadcast herbicide program. The species tolerates cattle grazing well.

Time from emergence to reproduction: In Illinois, plants flower from mid-summer to early fall. In Mississippi, flowering occurs from June until frost. Seeds become viable 12 days after flowering but continue to dry and develop a hard seed coat for another nine days before becoming ready for dispersal.

Pollination: Prickly sida is primarily self pollinated, and many flowers pollinate without ever opening. The flowers attract bees and butterflies, and it also outcrosses to a small extent.

Reproduction: Prickly sida produced 1,900 seeds per plant in full sunlight but 3,000 seeds per plant under the optimum 30% shade environment. However, it produced 8,100 seeds per plant when exposed to shade early followed by full sun later in the season, suggesting that it is especially adapted to reproduction during crop senescence.

Dispersal: Spines on segments of the seedpods can cling to fur or clothing and thereby distribute the seeds. Capsule segments float, so the species probably also disperses by water.

Common natural enemies: Three-cornered alfalfa hopper (*Spissistilus festinus*) feeds on prickly sida.

Palatability: Prickly sida has better digestibility than common southern forage grasses and has similar levels of crude protein, but it is deficient in P. Inclusion of some prickly sida in hay or silage still provides good animal nutrition. Its palatability is relatively low, however, as indicated by its increasing abundance in response to the intensity of cattle grazing.

FURTHER READING

Baskin, J.M. and C.C. Baskin. 1984. Environmental conditions required for germination of prickly sida (*Sida spinosa*). *Weed Science* 32: 786–791.

Egley, G.H. and R.N. Paul, Jr. 1993. Detecting and overcoming water-impermeable barriers in prickly sida (*Sida spinosa*) seeds. *Seed Science Research* 3: 119–127.

Russian-thistle

Salsola tragus L. = *S. iberica* (Sennen & Pau) Botsch. ex Czerep.

Russian-thistle seedling
Joseph DiTomaso, University of California, Davis

Russian-thistle inflorescence
Joseph DiTomaso, University of California, Davis

IDENTIFICATION

Other common names: Russian tumbleweed, Russian cactus, tumbling Russian-thistle, glasswort, burning bush, saltwort, prickly glasswort, wind witch, tumbleweed

Family: Goosefoot family, Chenopodiaceae

Habit: Erect, branched summer annual herb

Description: The **seedling** stem is striped with reddish-purple streaks. Cotyledons are fleshy, needle-like and 0.8–2 inches long by less than 0.1 inch wide. The first true leaves of the seedling appear opposite and are fleshy, similar in size and shape to the cotyledons, and tipped with spines. The leaves become smaller, flattened and less fleshy as the plant grows. **Mature plants** are 0.5–4 feet tall, often as wide as tall, profusely branched and bush-like in appearance. Stems have short, stiff hairs (or are occasionally smooth) and have reddish-purple streaks. The stems become stiff and dry as the season progresses. Leaves are alternate, hairless or with short hairs, thin and linear or needle-shaped, 0.25–2.5 inches long by 0.04–0.25 inch wide, and have a stiff, prickly spine at the tip. Mature plants may break at the base and become a tumbleweed. The root system is a long taproot. Small, petal-less **flowers** develop in the leaf axils on the upper portions of the stem between a pair of 0.02 inch-long, spine-tipped bracts. As the flower matures, five pale green to red, petal-like, membranous sepals enlarge to 0.15–0.3 inch wide and surround the single developing **seed**. The seed is conical, gray-brown and 0.2–0.3 inch in diameter.

Similar species: Kochia [*Bassia scoparia* (L.) A.J. Scott] has similarly striped stems and a similar growth habit to Russian-thistle. Kochia leaves, however, are broader, lack spines and have dense, greyish hair on all surfaces, while the leaves of Russian-thistle are hairless or have only short hairs and are spine-tipped.

MANAGEMENT

This is the dominant broadleaf weed in dryland grain production areas, and heavy infestations can prevent adoption of less competitive rotational crops such as spring peas or canola. Because Russian-thistle primarily emerges in early spring, rotation with late spring and summer crops allows most of the previous year's seeds to emerge and be killed by tillage before planting. Similarly, due to the very early season emergence of Russian-thistle, a brief fallow (one to two weeks) before planting an early season crop can help control the weed. Because Russian-thistle is intolerant of shade and relatively slow growing, crop competition is an important control mechanism. Thus, choosing a sowing time to maximize early crop growth rate, choosing competitive cultivars, increasing crop density and decreasing the distance between grain rows all help reduce the competitiveness of the weed. Accordingly, Russian-thistle emergence and growth is suppressed more by a crop of established winter wheat than by spring wheat. Also, inclusion of green manure crops such as yellow sweet clover into the rotation can suppress this weed.

Conventional tillage reduces the abundance of this species, whereas it thrives in no-till or reduced tillage fields with surface residue cover.

Russian-thistle can accumulate over 93% of its total dry matter production after grain harvest, so control of this weed after harvest is essential. Shallow undercutting with wide, low-pitch V-blades after grain harvest can completely prevent seed production while retaining 90% of the stubble for prevention of erosion.

Close mowing of newly emerged seedlings will kill most of them. Mowed or rotationally grazed forages compete with the weed, and mowing and treading kills plants and helps prevent seed production by those that remain.

ECOLOGY

Origin and distribution: Russian-thistle is a native of Russia that was introduced into South Dakota in flax seed in the mid 1870s. It presently occurs throughout the United States and southern Canada, except in the Deep South. In Eurasia its range extends from China through most of Europe and into North Africa. It has been introduced into southern Africa, Australia and parts of Central and South America.

Seed weight: Mean population seed weight ranges from 1.1–1.7 mg.

Dormancy and germination: Most seeds are dormant when dispersed from the mother plant and several months are required before they will germinate. Germination of mature seeds collected in fall will only occur under a restricted range of temperatures (68/41°F day/night alternation is optimal), whereas in April, seeds will germinate well under a wide range of fluctuating temperature combinations ranging from 28–86°F. Seeds can germinate, however, at daytime temperatures as low as 36°F with nighttime temperatures below freezing. Winter chilling is not required for after-ripening. In a field study, germination peaked when soil temperatures were 59–77°F during the day and 32–41°F at night. In a lab study, germination was best at day/night temperatures of 86/68°F or 95/77°F. Fluctuating temperatures appear to promote germination, but the species is relatively unaffected by light. Germination of Russian-thistle is not limited by dry soil conditions that would inhibit germination of other species. Germination can occur at high salt concentrations only at higher temperatures (higher than 68°F), but germination of seeds removed from salt conditions is best at lower temperatures (59/41°F). This germination response is adapted to western desert conditions, whereby evaporation and saline conditions increase as seasonal temperatures increase, but intermittent dilution of salt in the soil and cooler temperatures are associated with rainfall. These traits coupled with a capacity to germinate very rapidly (in minutes to hours) allow Russian-thistle to establish in environments where favorable conditions are highly transitory.

Seed longevity: Although a few seeds may remain viable deep in the soil for many years, most seeds remain viable for no more than one to two years. In a Nebraska experiment, seeds did not survive for one year. In a Saskatchewan experiment in which seeds were sown on cultivated soil in the fall, 31% emerged the next spring, less than 0.5% emerged the second year and only two individuals (0.04%) emerged the third year.

Season of emergence: Seedlings mostly emerge in early spring, with emergence continuing through late spring. Russian-thistle also can emerge intermittently following light rainfall.

Emergence depth: Seedlings emerge best from seeds located within the surface 0.4–1 inch of soil. A few seedlings can emerge from 2.4 inches but none from 3.2 inches. Emergence from seeds on the soil surface is low unless crop residue is present and/or the relative humidity of the atmosphere is very high.

Photosynthetic pathway: C_4

Sensitivity to frost: Russian-thistle, both as a seedling and as a mature plant, is killed by hard frost.

Drought tolerance: Established Russian-thistle plants are extremely drought tolerant and highly water efficient. The root system is at least five times as long as shoots and can extend 5 feet laterally and 6 feet vertically, which partially explains the rapid access to and depletion of soil water by this species.

Mycorrhiza: Russian-thistle does not form mycorrhizal associations. Some evidence indicates that growth can be reduced by mycorrhizal fungi.

Response to fertility: Compared with most crops and other weed species, Russian-thistle growth is not responsive to nitrogen. The species is very good at taking up N when it is in short supply, however, and it concentrates high levels of nitrates (more than 5% N) in shoots under high fertility conditions, a capacity that can enhance its competitiveness with other N-requiring plants. Russian-thistle is also remarkably unresponsive to P.

Soil physical requirements: Russian-thistle occurs primarily on dry, sandy soils. It is also common on loam and silty alkaline soils in prairie regions, but it is uncommon on clay soils. It tolerates and thrives on saline soils that inhibit most plant species. It rarely occurs in wet areas. Seedling emergence is poor on compacted soil.

Response to shade: The species does not tolerate shade.

Sensitivity to disturbance: Small plants cut just above the seed leaves do not survive, which indicates that early season mowing may control the species. Older plants recover from mowing by developing prostrate stems below the cutting level, requiring additional mowing or other control measures. After stems are cut during small grain harvest, the root system regrows at a rate faster than the shoots. Overall regrowth, if left unchecked after harvest, can lead to substantial seed production by fall.

Time from emergence to reproduction: Flowering begins in mid-June and can continue until frost. Seed production occurs from August through fall.

Pollination: Russian-thistle is wind pollinated and can self-pollinate.

Reproduction: Estimates of seed production vary greatly from 2,000 to over 100,000 seeds per plant. Very large plants have been reported to produce up to 150,000 seeds, but plants growing in competition with crops typically produce 5,000–17,000 seeds.

Dispersal: Russian-thistle grows in a ball-like shoot structure that frequently breaks off at the base after senescence and rolls in the wind, dropping seeds as it bounces along. Winds with gusts up to 61 mph are required to break off stems of recently senesced plants. Plants tumble an average distance of 2,100 yards and a maximum distance of 2.5 miles until stopped by fence-lines, roadways or other obstacles. Tumbleweeds caught on railroad cars can spread seeds across the landscape for long distances. Plants retain 26–51% of seeds after tumbling approximately one mile, ensuring seed dispersal over large distances. Seeds also can be spread in waterways and contaminated crop seed or straw. Seeds present in the Columbia River and irrigation water laterals had a 56% germination potential.

Common natural enemies: The native caterpillars *Coleophora parthenica* and *C. klimeschiella* are sufficiently destructive to Russian-thistle that they have been released in Canada as biological control agents. *Coleophora parthenica* also was introduced into the Coachella Valley of southern California, but although the larvae infested most Russian-thistle plants, it had little effect on growth or population levels of this species. The rust fungus *Uromyces salsolae*, the anthracnose fungus *Colletotrichum gloeosporioides* and the bare spot fungus *Rhizoctonia solani* have shown potential to suppress growth under controlled conditions.

Palatability: While young and before spines form, Russian-thistle can provide a good source of forage for livestock and native animals. Oxalates and nitrates in the tissue of Russian-thistle can poison sheep. Nitrates become a problem when the plant is growing on well fertilized soils. Salinity enhances forage quality of Russian-thistle, reduces nitrate and oxalate concentrations at full flower, and creates good forage potential on arid lands.

Note: Russian-thistle pollen is an important contributor to summer hay fever problems in regions where it is common.

FURTHER READING

Beckie, H.J. and A. Francis. 2009. The biology of Canadian weeds. 65. *Salsola tragus* L. (updated). *Canadian Journal of Plant Science* 89:7 75–789.

Blackshaw, R.E., J.R. Moyer, R.C. Doram and A.L. Boswell. 2001. Yellow sweetclover, green manure, and its residues effectively suppress weeds during fallow. *Weed Science* 49: 406–413.

Crompton, C.W. and I.J. Bassett. 1985. The biology of Canadian weeds. 65. *Salsola pestifer* A. Nels. *Canadian Journal of Plant Science* 65: 379–388.

Schillinger, W.F. 2007. Ecology and control of Russian-thistle (*Salsola iberica*) after spring wheat harvest. *Weed Science* 55: 381–385.

Young, F.L. 1986. Russian-thistle (*Salsola iberica*) growth and development in wheat (*Triticum aestivum*). *Weed Science* 34: 901–905.

Shepherd's-purse

Capsella bursa-pastoris (L.) Medik.

Shepherd's-purse seedling
Scott Morris, Cornell University

Shepherd's-purse young rosette
Scott Morris, Cornell University

Shepherd's-purse plants in flower
Antonio DiTommaso, Cornell University

Shepherd's-purse flowers and fruit
Scott Morris, Cornell University

IDENTIFICATION

Other common names: pick pocket, pepper plant, case weed, pick purse, shepherd's bag, shepherd's pouch, mother's heart, St. James weed, witches' pouches, toothwort, shovel plant

Family: Mustard family, Brassicaceae

Habit: Erect, winter or summer annual herb arising from a basal rosette

Description: The **seedling** stem is light green to purple. Cotyledons are oval to spatula shaped, 0.1–0.25 inch long by 0.06 inch wide, hairless and long stalked. The first two to four leaves are opposite, round to oval, without teeth or slightly toothed, and densely hairy. Hairs are short and spreading, with branching star-shaped tips. Some hairs may be unbranched. All subsequent leaves are alternate, oval to club-shaped, toothed to deeply lobed (occasionally entire), and form a basal rosette. The rosette leaves have hairs along the underside of the midvein and are hairless to sparsely hairy above. **Mature plants** bolt from the rosette to form a 4–30 inch-tall, occasionally branching flowering stem. The stem is green to purple, hairy towards the base and smooth to sparsely hairy above, and it has leaves only on the lower portions. Rosette leaves are 1.25–6 inches long by up to 1.5 inch wide and tapered towards the base. Leaf shape and hairs are similar to those of the seedling. Stem leaves are lanceolate, 0.4–2.2 inches long by up to 0.6 inch wide, alternate, without lobes, stalkless and may clasp the stem at the base. The taproot is thin and branching, with fibrous secondary roots. White, four-petaled **flowers** continually develop at the tips of the stem and branches. Each flower is up to 0.25 inch wide, with a 0.3–0.6 inch-long, hairless to sparsely hairy stalk. Stalks are initially upright, becoming spreading and nearly perpendicular to the stem as old flowers are replaced by flattened, heart-shaped **seedpods**. The seedpods are notched at the tip, two-segmented and 0.2–0.4 inch long by 0.1–0.3 inch wide. Each seedpod contains 10–20 **seeds**. The seeds are oblong, shiny, pale orange to red-brown, and 0.03–0.04 inch long by 0.01–0.02 inch wide.

Similar species: Spreading, star-shaped hairs and heart-shaped seedpods distinguish shepherd's-purse from similar species in the mustard family. Field pennycress (*Thlaspi arvense* L.), Virginia pepperweed (*Lepidium virginicum* L.) and field pepperweed [*Lepidium campestre* (L.) W.T. Aiton] are all rosette-forming annuals with flattened seedpods. Field pennycress leaves are hairless with wavy margins, while shepherd's-purse leaves are toothed or lobed. The flowering stems of Virginia pepperweed and field pepperweed have leaves throughout, while shepherd's-purse has leaves only on the lower parts of the stem.

MANAGEMENT

Shepherd's-purse germinates primarily in the spring and early fall. A lag between tillage and seedbed preparation at these times of year depletes the seed bank near the soil surface. Since the seedlings are tiny, tine weeding and other in-row and near-row cultivation easily kills them. Since shepherd's-purse is primarily a cool season species, rotation with warm season crops like squash and tomatoes break the life cycle of the weed. Rapid cleanup of the field after short, early season crops like lettuce and radishes will prevent seed set.

Shepherd's-purse is a common weed in winter grain crops. Since the seedlings are small and fragile early in the plant's life cycle and remain very short until spring, they are highly susceptible to fall harrowing. Pre-emergence harrowing will break or dry out white thread stage seedlings, and post-emergence harrowing can bury many of the established seedlings.

Dense planting helps control this weed in crops that will tolerate high density, particularly if the crops are tall, since the weed is always short. A fall forage radish cover crop effectively suppressed establishment of shepherd's-purse in fall and early spring. Emergence and growth may be suppressed by incorporated white mustard or buckwheat cover crops. Organic mulch materials are highly effective for suppressing shepherd's-purse, since the seedlings are tiny and the leaves remain flat on the ground until flowering.

ECOLOGY

Origin and distribution: Shepherd's-purse originated in the eastern Mediterranean but today is a cosmopolitan weed found in most temperate regions of the world and at temperate elevations in the tropics and subtropics. The species occurs throughout the United States and Canada, including the far north. The similarity in genetic variation between three genetic clusters have been identified, the first occurring in the Middle East, a second in Europe and the most recent in eastern Asia. An accession from the United States was aligned with the Middle Eastern cluster.

Seed weight: Mean population seed weights range from 0.09–0.14 mg, with most lying near 0.1 mg. Heavier seeds tended to have higher germination rates.

Dormancy and germination: Seeds of shepherd's-purse are dormant when they ripen and require cold treatment before they are able to germinate. Nitrate can enhance germination, particularly in combination with fluctuating temperatures that include chilling. Once they are non-dormant, a high percentage of seeds will sprout at day/night temperatures ranging from 59/43°F to 86/59°F. Different populations of this species have variable temperature requirements, with some populations having a pronounced temperature optimum and others germinating equally over a wide range of temperatures. Seeds generally require brief exposure to light for germination, except in early spring when some seeds may germinate even in the dark. Warm soil temperatures (e.g., higher than

95/68°F day/night) will induce complete dormancy, but lower summer temperatures will induce dormancy in only a percentage of the seeds. Fall tillage will expose to light seeds that have been chilled the previous winter but were not exposed to dormancy-inducing summer temperatures. This allows prolific germination following tillage in late summer or fall. Stimulation of germination by tillage may also be explained by release of seeds from high carbon dioxide and low oxygen levels, as occurs beneath the soil surface. However, at high latitude locations, chilling fall temperatures maintain dormancy of seeds shed in summer and prevent germination during their first fall.

Seed longevity: Seeds can survive up to 35 years in undisturbed soils. Nearly 100% of seeds survived 30.5 months of burial in Sweden. Seed survival is longer in undisturbed soil. In a five-year study, the number of seeds declined by an average of 43% per year in soil stirred four times each year and 24% per year in uncultivated soil. Likewise, annual mortality was 35–52% in cultivated soil and 11–22% in undisturbed soil over six years in England. Annual mortality rates of 19% in Alaska and 21% in France also have been reported in undisturbed soil.

Season of emergence: Seedlings emerge in the fall and overwinter as rosettes close to the ground, or they emerge in the early spring. When emerging in spring, it is among the earliest emerging species. A few seedlings also come up during the warmer parts of the growing season and are notably more common during relatively cool summers. In Michigan, which has relatively cool summers, shepherd's-purse emerged primarily in late summer and only a small fraction emerged in fall and spring.

Emergence depth: Optimum emergence occurs approximately at the soil surface. Most seedlings emerge from within the top 0.5 inch of soil, and very few emerge from deeper than 1 inch.

Photosynthetic pathway: C_3

Sensitivity to frost: Shepherd's-purse is very frost hardy, with acclimated plants surviving temperatures of 10°F with little damage. Individuals that germinate in the fall commonly persist through the winter.

Drought tolerance: Shepherd's-purse is relatively drought tolerant. For example, it grew better with alternating four days of watering and 10 days of drying than with regimes including more days of watering and fewer days of drying.

Mycorrhiza: Most reports indicate the absence of mycorrhiza on this species in England and Utah, but some samples from England have had mycorrhizal infections.

Response to fertility: Shepherd's-purse is most common on fertile soils. Compared with many other weeds, however, it is only moderately responsive to balanced fertilization with N-P-K, but it does show a growth response to fertility rates beyond typical recommendations. At low P, its growth increases linearly with N fertilization up to, and probably beyond, 133 pounds per acre. At low N, it increased slightly but linearly with increasing P up to 54 pounds per acre of P. The species prefers neutral soils but tolerates soil with a pH of 5.

Soil physical requirements: Shepherd's-purse is common on a wide range of soil textures and drainage classes but is intolerant of flooding. It is highly tolerant of soil compaction.

Response to shade: Shepherd's-purse is relatively shade tolerant and will set seed in 80% shade. It can grow at light levels similar to those at soil level in a grassland with a closed leaf canopy.

Sensitivity to disturbance: Seedlings are tiny, fragile and easily broken. Even small rosettes readily retain soil on the root system, however, and will subsequently reroot after cultivation. Plants survive and regrow well after one cutting at 2 inches but die after three cuts. The species is moderately tolerant of trampling.

Time from emergence to reproduction: Flowering can occur in less than four weeks for spring emerging plants. Flowering time varies greatly both between and within populations. Flowering varied from six to 16 weeks for the majority of European populations. California populations showed a similar range with populations from hot, dry summer climates flowering earlier than those from cool, moist summer climates. Time to first flowering ranged from 23 days for populations from Asia to 44 days for populations from Europe, and early flowering was suggested as a mechanism for avoiding competition. Plants that begin flowering in early May begin producing ripe seeds by late May.

Pollination: Shepherd's-purse is primarily self pollinated but is also insect pollinated.

Reproduction: Shepherd's-purse reproduces only by seed. The flowers open first at the base of the inflorescence and continue to form at the top even as the capsules lower on the plant mature and shed seeds. Consequently, an individual plant produces seeds for many weeks. Seeds are viable when green fruit are full-sized. An average plant produces around 3,000 seeds, but large plants may produce 30,000–60,000 seeds. Immature capsules will continue to ripen and shed seeds even after the plant has been uprooted and killed.

Dispersal: Because the seeds often reach high densities in soil, they are easily spread from one site to another in soil clinging to shoes, tillage machinery and tires. Seed dispersal is facilitated by the mucilaginous seed surface that allows them to cling to animals or machinery. The seeds survive well in the digestive tracts of cows, sheep, horses and deer, and manure is commonly contaminated with shepherd's-purse seeds. Germination is minimally affected by digestion and dispersal by earthworms, which can facilitate deep burial and long-term survival in soils. Shepherd's-purse also can be dispersed in irrigation water.

Common natural enemies: Many species of insects attack shepherd's-purse. The aphids (*Aphis euonymi*) and (*Myzus certus*) cause leaf rolling. Several curculio beetles in the genus *Ceutorhynchus* form galls on the leaves or stem. The fly larvae *Dasineura brassicae* gall the shoots, and *Liriomyza strigara* and *Phytomyza horticola* mine the leaves. Shepherd's-purse is acceptable to all mollusks on which it has been tested, including the slug *Agriolimax caruanae*.

Palatability: Shepherd's-purse has lower protein and digestible matter than alfalfa. The young foliage can be used as a pot herb or to provide a peppery taste in salads. The seeds are edible but are tiny and do not develop synchronously, which makes them hard to collect in quantity. The plant has a wide variety of medicinal uses, and these have contributed to its worldwide spread.

FURTHER READING

Aksoy, A., J.M. Dixon and W.H.G. Hale. 1998. Biological flora of the British Isles No. 199: *Capsella bursa-pastoris* (L.) Medikus (*Thlaspi bursa-pastoris* L., *Bursa bursa-pastoris* (L.) Shull, *Bursa pastoris* (L.) Weber). *Journal of Ecology* 86: 171–186.

Defelice, M.S. 2001. Shepherd's-purse, *Capsella bursa-pastoris* (L.) Medic. *Weed Technology* 15: 892–895.

Didon, U.M.E., A.K. Kolseth, D. Widmark and P. Persson. 2014. Cover crop residues—effects on germination and early growth of annual weeds. *Weed Science* 62: 294–302.

Lawley, Y.E., R.R. Weil and J.R. Teasdale. 2011. Forage radish cover crop suppresses winter annual weeds in fall and before corn planting. *Agronomy Journal* 103: 137–144.

Sicklepod

Senna obtusifolia (L.) Irwin & Barneby = *Cassia obtusifolia* L.

Sicklepod flower
John Gwaltney, SoutheasternFlora.com

Sicklepod plant with curved pods
John Gwaltney, SoutheasternFlora.com

Sicklepod pod
John Gwaltney, SoutheasternFlora.com

IDENTIFICATION

Other common names: coffeebean, javabean, coffeeweed

Family: Legume family, Fabaceae. Some botanists separate this genus and its relatives, most of which are woody, into the Caesalpinia family, Caesalpiniaceae.

Habit: Upright or sprawling summer annual herb or marginally perennial shrub with compound leaves

Description: Seedling cotyledons are large and round, 0.6–0.8 inch wide, with three to five prominent, pale veins radiating from the base in a hand-like pattern. The first true leaves are divided, with two to three pairs of short-stalked, oval to egg-shaped, hairless leaflets. **Mature plants** are 1–6 feet tall and sprawling to upright. Stems are green, angular, hairless and occasionally branching. Older stems may become woody at the base. Leaves similar to those of young plants, with three to four pairs of leaflets and two 0.6 inch-long, leafy bracts at the base. Each leaflet is 1–3.5 inches long by 0.5–1 inch wide; leaflets at the tip of the leaf are generally larger than those at the base. A 0.03–0.1 inch-long, orange/brown gland is present at the base of the lowest pair of leaflets. The branching taproot can reach up to 3.3 feet in length. All parts of the plant give off a distinctive odor when crushed. Yellow, showy **flowers** with 0.3–1 inch-long stalks develop singly or in pairs from the upper leaf axils. Flowers are five-petaled and 0.4–1 inch wide. Petals are 0.3–0.8 inch long, narrowest at the base and have at least three prominent veins. The seedpod is four sided, curved downward, hairless and 3–8 inches long by 0.12–0.25 inch wide. Seedpods split when ripe, releasing 20–40 **seeds**. Seeds are 0.1–0.25 inch long, brown, glossy and rectangular to irregular in shape.

Similar species: Coffee senna [*Senna occidentalis* (L.) Link] and partridge-pea [*Chamaecrista fasciculata* (Michx.) Greene] have similar flowers and foliage to sicklepod. Coffee senna leaflets are larger, up to 3 inches long and 1.5 inch wide, and more numerous (three to seven pairs per leaf) than those of sicklepod, and they have pointed rather than rounded tips. Coffee senna seedpods are cylindrical or flattened rather than angular, and they are curved upward. Partridge-pea cotyledons are spade shaped rather than round and have tapered, bluntly pointed ends. Mature partridge pea leaves have eight to 15 pairs of leaflets that are much smaller, up to 0.7 inch long by 0.3 inch wide, than those of sicklepod. Partridge-pea pods are only slightly curved.

MANAGEMENT

Rotation to winter and spring cereal grains helps suppress sicklepod populations since the grain will be highly competitive by the time the sicklepod germinates and will be harvested before the weed can produce seeds. A tilled fallow after grain harvest can flush seedlings out and kill them.

If sicklepod produces seeds, leave them on or very near the soil surface in the fall to encourage seed predation and weathering of the resistant seed coat. When preparing a seedbed for a spring row crop, till shallowly; deep plowing will bury seeds for the current year, but many will survive and emerge in subsequent years. If deep tillage is necessary, rip with as little inversion as possible. Note also that shallow seedbed preparation will put the sicklepod seeds closer to the soil surface so that they are more vulnerable to in-row weeding after planting.

Plant spring row crops as early as will allow good establishment. Soybeans and corn, in particular, can emerge from substantially cooler soil than sicklepod, and this gives these crops a head start over the weed. High density planting of soybeans, for example 395,000 per acre rather than 264,000 per acre, has been shown to help competitively suppress this weed. Reports vary on the effectiveness of narrow row spacing for suppressing sicklepod, but at least one study has shown substantial reduction in density and dry weight with 7.5-inch row spacing as compared with 38-inch row spacing. Since many sicklepod seedlings emerge from below the normal planting depth of row crops, in-row weeding should focus on breaking or burying young sicklepod, rather than uprooting them. Consequently, a shallow sweep cultivator, rolling cultivator or tine weeder with relatively stiff tines will be more effective than a rotary hoe or tine weeder with highly flexible tines. Cultivate shallow and close to the row, but throw enough soil toward the row to bury late emerging plants if the crop will tolerate this. Cotton rows can be flame weeded to eliminate late emerging sicklepod. Late emerging weeds will not be very competitive, except perhaps in a drought year, but the seeds they produce can pose problems for future crops in the rotation.

The seeds provide sufficient energy to allow seedlings to penetrate even dense layers of straw mulch, but synthetic mulches are effective barriers against sicklepod in vegetable crops. Distillery wastewater applied as a fertility amendment at 1 gallon per square foot suppressed sicklepod emergence by 87% and post-emergence survival by 65%.

Sicklepod can be a problem in poorly managed pastures and a severe problem in high traffic areas such as around hay rings or feed bunks, but a vigorous sward will outcompete the weed. Trampling during intensive rotational grazing will kill sicklepod even though they are not eaten. Mowing the pasture when sicklepod is about to flower will eliminate or greatly retard seed production.

ECOLOGY

Origin and distribution: Sicklepod is native to tropical and subtropical parts of South and North America. It is widely introduced in warmer parts of Africa, Asia and Australia. In the United States it is common throughout the Southeast as far north as Tennessee and Virginia, and it occurs sporadically northward to Wisconsin and New York, westward to Nebraska, and in southern California. It is most problematic in the Southeast.

Seed weight: 23–28 mg.

Dormancy and germination: Sicklepod seeds have a hard, waxy seed coat that prevents absorption of water, thereby preventing germination of most seeds. On average, only about 10% of seeds with intact seed coats will germinate, but germinability varies. Damage to the seed coat promotes germination. In one experiment, incubation of the seeds in wet sand for 12 months increased germination from 5% to 15%, indicating that the seed coat slowly breaks down over time. Fire cracks the seed coat and can prompt high germination following the next rain. Scarified seeds germinate well from 68–97°F but poorly below 59°F or above 104°F. Development of a soybean canopy greatly reduced mid- to late-season emergence of sicklepod seedlings, possibly due to lower soil temperatures or decreased temperature fluctuations under the crop canopy. Rapid seedling emergence occurs at temperatures from 81–97°F. Light does not affect germination if the seed coat is damaged, but light may further inhibit germination of intact seeds. Seeds germinate best at pH 5–8, but a few will germinate at pH 3–9.

Seed longevity: Due to its hard seed, sicklepod forms a

persistent seed bank, and as much as 270 pounds per acre of seeds have been recovered from soil samples. Sicklepod seeds had 46% mortality over a two-year period when soil was tilled annually in late fall and early spring, but only 28% mortality when the soil was undisturbed. Seed mortality was computed to be 40% per year based on seed burial studies for 5.5 years in Mississippi. Annual spring tillage followed by repeated disking through the summer depleted the seed bank more rapidly than annual fall tillage followed by a chemical fallow without soil disturbance during the summer.

Season of emergence: In Arkansas, seedlings emerge from late May to August.

Emergence depth: Sicklepod emerges relatively well from below the planting depth of most crops. In one study, emergence was high at depths to 3 inches, and about 15% of seeds buried at 5 inches produced seedlings. Emergence was faster, however, from shallow depths. In another study seedling emergence was high down to 1.5 inch, and a few seedlings emerged from 4 inches, but the emergence response to burial depth varied with soil texture.

Photosynthetic pathway: C_3

Sensitivity to frost: Day/night temperatures of 84/70°F or lower greatly inhibit growth of sicklepod relative to higher temperatures, and growth ceases at mean daily temperatures below 57°F.

Drought tolerance: Sicklepod is able to germinate at low moisture potential. Although initial root elongation is slowed by low moisture, enough growth occurs even under drought conditions that the roots can grow into deeper soil layers with more moisture. Sicklepod tends to have a higher root-to-shoot ratio than soybeans, which probably gives it a competitive advantage under drought conditions, but several physiological characteristics related to photosynthesis and transpiration are intermediate compared to other C_3 weeds.

Mycorrhiza: Sicklepod is mycorrhizal.

Response to fertility: Although sicklepod is a legume it does not host nitrogen-fixing bacteria. Application of 200 pounds per acre of N to an N-deficient soil increased dry weight of sicklepod by 59% and seed production by 114%. Seeds produced under low N fertility were smaller and grew into less competitive plants than seeds produced under high N. Its response to N-P-K fertility is greater than soybean and about equal to cotton. Although sicklepod growth is somewhat reduced at low and high pH, it grows reasonably well at soil pH as low as 3.2 or as high as 7.9.

Soil physical requirements: The species commonly occurs in soil with textures that range from loams to gravels. It seems particularly associated with coarse soils like sandy agricultural fields and the gravel of railroad ballast. It is highly tolerant of compacted layers in the soil.

Response to shade: Partial shading of 47% or 65% decreased dry weight of sicklepod but induced plant height to increase by 32%, showing that this species can often grow out from under a partial crop canopy. Shade of 80% reduced dry weight more than 75% of plants growing in full sun, and 95% shade suppressed sicklepod almost completely.

Sensitivity to disturbance: If sicklepod plants in a subtropical climate are cut, they may persist as short lived perennials.

Time from emergence to reproduction: Flowering is hastened when days are shorter than 12 hours. Flowering occurs six to 12 weeks after emergence, with northern populations (e.g., Tennessee) flowering more quickly than southern populations (e.g., Florida).

Pollination: Although the flowers are heavily visited by bees, self-pollination usually occurs before the flowers open.

Reproduction: In a comparison of populations from across the South grown with minimal competition, the mean number of seeds per pod was 26 and varied little between populations and years. The number of pods per plant varied from 63–591, with substantial variation between years and with more pods produced by southern (e.g., Louisiana or Florida) than more northern populations (e.g., Tennessee or North Carolina). Consequently, the number of seeds per plant varied from 1,500–16,000. Sicklepod grown without interference produced 11,420 seeds per plant in Arkansas. Plants growing in conventionally tilled soybeans in Alabama averaged 176 pods per plant. A dense stand can produce over 2,700 pounds per acre of seed.

Dispersal: The pods can throw seeds up to 16 feet as they open. Seeds are carried by stream water, overland flow and in mud attached to the feet and fur of animals. They also move in contaminated mulch and in mud on machinery, vehicles and footwear. Although the species is generally unpalatable, livestock nibble on the pods, and seeds will pass through the animals and disperse when the animals are moved.

Common natural enemies: Sicklepod seeds are attacked by the beetle *Sennius fallax*. Caterpillars of several species of sulphur butterflies feed on the foliage, including little sulphur (*Eurema lisa*), sleepy orange (*Eurema nicippe*) and cloudless sulfur (*Phoebis sennae cubule*). *Alternaria cassiae* causes seedling blight in sicklepod and is a potential mycoherbicide. A mixture of spores of the anthracnose fungi *Colletotrichum truncatum* and *C. Gloeosporioides* in corn oil and a surfactant effectively controls sicklepod.

Palatability: Sicklepod is cultivated in Africa for the young shoots, which are eaten as a pot herb. Sicklepod is unpalatable to livestock and can cause poisoning if eaten. The seeds are toxic and pose a serious threat to chickens and pigs when they contaminate feed.

Note: Sicklepod residue inhibits the germination and seedling growth of cotton and other crops.

FURTHER READING

Bararpour, M.T. and L.R. Oliver. 1998. Effect of tillage and interference on common cocklebur (*Xanthium strumarium*) and sicklepod (*Senna obtusifolia*) population, seed production, and seed bank. *Weed Science* 46: 424–431.

Bridges, D.C. and R.H. Walker. 1985. Influence of weed management and cropping systems on sicklepod (*Cassia obtusifolia*) seed in the soil. *Weed Science* 33: 800–804.

Nice, G.R.W., N.W. Buehring and D.R. Shaw. 2001. Sicklepod (*Senna obtusifolia*) response to shading, soybean (*Glycine max*) row spacing, and population in three management systems. *Weed Technology* 15: 155–162.

Smartweeds

Ladysthumb, *Polygonum persicaria* L. = *Persicaria maculosa* Gray
Pennsylvania smartweed, *Polygonum pensylvanicum* L. = *Persicaria pensylvanica* (L.) Small

Ladysthumb seedling
Scott Morris, Cornell University

Ladysthumb flowers
Scott Morris, Cornell University

Pennsylvania smartweed plant
Randall Prostak, University of Massachusetts

Pennsylvania smartweed ocrea
Scott Morris, Cornell University

Pennsylvania smartweed inflorescence
Joseph DiTomaso, University of California, Davis

IDENTIFICATION
Other common names:

Ladysthumb: lady's thumb, smartweed, persicary, spotted smartweed, heartweed, spotted knotweed, red shanks, willow weed, lovers pride, ladythumb, heart's ease

Pennsylvania smartweed: pinkweed, purple head, glandular persicary, heart's ease, swamp persicary, smartweed

Family: Buckwheat family, Polygonaceae

Habit: Upright or sprawling, branched summer annual herbs

Description: Seedlings have hairy edged cotyledons. The first true leaves are lanceolate to ellipse shaped and hairy on the upper surfaces and edges. Seedling stems can be erect or prostrate. Leaf-stem junctions are swollen and covered with transparent, papery membranes (ocreae) that extend up and around the stem.

Ladysthumb: Cotyledons are round tipped, lanceolate and 0.13–0.5 inch long by 0.13–0.3 inch wide. Seedling stems are brown or pink to bright red. Young leaves are lightly hairy on their upper surfaces and pale green, with a black spot sometimes present on the leaf surface. Stems are sometimes lightly hairy, otherwise they are similar to Pennsylvania smartweed stems.

Pennsylvania smartweed: Cotyledons are ellipse to lanceolate and 0.13–1 inch long by 0.13–0.5 inch wide. Seedling stems are pink at the base. Young leaves are hairless, with purple-hued undersides. The hairless, red-purple stems change angle at lower nodes, giving the young plant a zig-zag shape.

Mature plants have green to red, leaning or upright, freely branching stems. Leaf-stem junctions are swollen and covered by an ocrea. Leaves are alternate, smooth to lightly hairy, lanceolate to elliptical and often have a black or purple spot in the center of the blade surface. Roots are fibrous from a small taproot.

Ladysthumb: Stems are 1–3 feet tall. Leaves are 1.25–6 inches long by 0.2–0.7 inch wide and sometimes covered in short, flat hairs; they lack the long stalks of Pennsylvania smartweed. The darkened spot is often present on the blade surface. Ocreae have a row of 0.06 inch-long bristly hairs at the top.

Pennsylvania smartweed: Flat, stiff hairs are present on the 1–4 foot-tall stem. The stalked leaves are 2–6 inches long by 1.25 inch wide and hairless to sparsely hairy. The darkened spot is inconsistently present and is often only a subtly different shade of green. Ocreae are hairless.

Flowers are small, white to red and clustered at stalk ends.

Ladysthumb: Flowers are pink to red-purple and rarely white. Clusters are 1 inch long.

Pennsylvania smartweed: Flower stalks are hairy and occasionally sticky. Flowers are pale to bright pink or white. Clusters are 1.5 inches long.

Fruit and seeds: Seeds are flat, shiny, black, pointy tipped and round to oval shaped.

Ladysthumb: Seeds are 0.1 inch wide and occasionally three sided.

Pennsylvania smartweed: Seeds are 0.13 inch wide.

Similar species: Tufted knotweed [*Polygonum cespitosum* Blume var. *longisetum* (Bruijn) A.N. Steward] ocreae also have bristles, but the bristles are longer, over 0.25 inch, and the plant is shorter, less than 1 foot. Pale smartweed (*Polygonum lapathifolium* L.) has hairless ocreae; spotless, purplish-green, 2–6 inch-long leaves; 3 inch-long nodding flower clusters; and brown seeds.

MANAGEMENT

Crop rotation is an important component in managing Pennsylvania smartweed and ladysthumb. Since nearly all the emergence of these species occurs by late spring, tillage before a summer planted crop like tomatoes or a late planting of soybeans will largely eliminate these species for that year. A winter grain crop will be competitive against these weeds when they emerge in spring. High crop sowing density and close row spacing (4 inches) can suppress ladysthumb in spring wheat, and these tactics are probably helpful in any cereal grain. These species die out of the seed bank at an intermediate rate, and consequently grains following alfalfa were found to have 27–29% as many smartweeds as grains following grains. These species can, however, establish in later years of a forage crop, and the hay must therefore be mowed before seed set.

Tine weed spring grains and row crops pre- and post-emergence to kill seedlings. Since seedlings commonly emerge from up to 2 inches, tine weeding will be more effective than rotary hoeing, though a well-timed rotary hoeing will kill part of the population. If weed density is high, tine weed winter grains in the spring as seedlings emerge. Since the seedlings do not become tall quickly, throwing 2 inches or more of soil into the crop row can effectively bury most seedlings that escape tine weeding in row crops.

Organic and synthetic mulches can effectively suppress these weeds. These species are likely to respond to fertility amendments at least as much as the crop, so avoid over fertilization.

ECOLOGY

Origin and distribution: Ladysthumb is a native of Eurasia, where it occurs from western Europe eastward through central Russia, and the Middle East to northern India and southward across North Africa. It also occurs in Japan and has been introduced into Australia, New Zealand and North and South America. In North America, it occurs throughout the contiguous United States and southern Canada, and northward into the Yukon and Alaska. Pennsylvania smartweed is native to eastern North America. It occurs in most areas from the Atlantic to the prairie states and provinces but is more sporadic in the intermountain states and far west.

Seed weight: ladysthumb, 1.4–4 mg; Pennsylvania smartweed, 3.6–6.8 mg.

Dormancy and germination: The seeds of both species are dormant when shed from the parent plant and require several months of after-ripening before they will germinate. A few weeks of cold, wet conditions break dormancy. Once dormancy has been broken, ladysthumb germinates better at 86°F than at 50°F or 68°F, and temperature fluctuations of at least 14°F greatly increase germination. Similarly, non-dormant seeds of Pennsylvania smartweed germinate best in warm, fluctuating temperatures, for example, 86/59°F or 95/68°F. However, exposure to temperatures over 59°F under conditions unfavorable for germination reinduces dormancy of ladysthumb, and drying also has been reported to induce secondary dormancy. As both species show greatly reduced emergence during summer, induction of secondary dormancy by warm, dry soil conditions probably occurs in Pennsylvania smartweed also. Nitrate stimulates germination of ladysthumb, but the effect is weak and only acts on relatively new seeds. Germination of Pennsylvania smartweed was greater than that of several other common annual weeds under wet conditions but lower under dry conditions.

Seed longevity: In experiments in annually tilled soil, ladysthumb seeds declined at a rate of 20% and 28% per year, but, in a similar experiment, seeds declined at 37% per year. In several five-year experiments in which the top 3 inches of soil was stirred three times per year, seeds of ladysthumb declined by an average of 38–43% per year. In another experiment, the number of seedlings emerging in an annually tilled soil declined 38% per year. However, some seeds of ladysthumb can survive in the soil for 20 years if left undisturbed. Pennsylvania smartweed and ladysthumb seeds can survive up to 30 years of burial. Pennsylvania smartweed seeds buried in late October had 40–63% viability 10 months later. In an Alaskan study, undisturbed seeds declined by an average of 25% per year, and, after burial for 19.7 years, 3.3% of seeds were still viable.

Season of emergence: Both species tend to emerge in early to mid-spring, with some emergence continuing into late spring for Pennsylvania smartweed and early summer for ladysthumb.

Emergence depth: Ladysthumb seedlings emerge best from the top 1.6 inch of soil, but a few seedlings can emerge from as deep as 2.4 inches. Pennsylvania smartweed emerges readily from anywhere in the top 2 inches of soil and a few seedlings can emerge from 4 inches.

Photosynthetic pathway: C_3

Sensitivity to frost: Neither species tolerates temperatures below 32°F. A 27°F frost killed about half the leaves on Pennsylvania smartweed.

Drought tolerance: Ladysthumb tolerates a range of soil moisture, from drought (daily wilting) to continuous flooding. Flexibility in energy allocated to roots and in photosynthesis versus water loss via transpiration explain the adaptive responses of this species to moisture conditions. Pennsylvania smartweed does best in moist but well drained habitats and is more tolerant of overly wet than dry conditions. It maximizes biomass production under saturated soil conditions but maximizes seed production under field capacity conditions. Photosynthesis of Pennsylvania smartweed declines more rapidly than velvetleaf or giant foxtail with increasing drought stress, probably because it closes stomates and maintains a higher plant water potential than the other species.

Mycorrhiza: Mycorrhiza have been reported in some samples of ladysthumb and absent in others.

Response to fertility: Ladysthumb is moderately to highly responsive to fertility, with substantial differences among populations. Plant size increased roughly three- to 10-fold when balanced N-P-K fertility increased from very low to high rates. Excessively high fertility neither helped nor harmed the plants. Moderate fertilization of an agricultural field with 15-30-15 fertilizer increased growth of Pennsylvania smartweed by 43%. The equivalent of 50 pounds per acre each of N, P and K added to a low fertility potting mix increased growth about three-fold, but higher fertility rates did not cause additional growth. Thus, Pennsylvania smartweed appears to be less responsive to fertility than many crops. Pennsylvania smartweed and ladysthumb tolerate a wide range of pH, from 4–8.5.

Soil physical requirements: Pennsylvania smartweed occurs on fine to course textured soils, and ladysthumb is found on a wide variety of soils including muddy riverbanks, silty alluvium, heavy clay, sand, black muck, peat, cinders and manure heaps. Both species are moderately well adapted to anaerobic conditions and commonly occur in wetlands.

Response to shade: Pennsylvania smartweed and ladysthumb are generally considered shade intolerant. Ladysthumb growth and seed production was decreased by shade, however, this species adapts to shade by reducing leaf thickness and by allocating more tissue to leaves. Pennsylvania smartweed growth is significantly reduced by shade, and maximum photosynthesis is only reached in full sunlight.

Sensitivity to disturbance: Ladysthumb tolerates wheel traffic and trampling by livestock. It becomes dwarfed if clipped repeatedly in the vegetative state. If clipped when flowering, it may overwinter in areas with a mild winter. Stem cuttings are capable of regenerating into plants.

Time from emergence to reproduction: Pennsylvania smartweed and ladysthumb begin flowering at six to nine weeks after emergence, and plants can have mature seeds by July. Both species continue producing seeds until fall.

Pollination: Ladysthumb is primarily self pollinated but some cross-pollination by insects probably also occurs. Pennsylvania smartweed is pollinated by insects, including ants, but pollen transfer within a plant can fertilize flowers.

Reproduction: When growing with relatively little competition, ladysthumb typically produces 200–800 seeds per plant, but large plants may produce 1,200–4,500 seeds. In a spring wheat crop, however, plants produced 40–150 seeds per plant, depending on row spacing and seeding rate of the crop. When grown as a monoculture in North Carolina, Pennsylvania smartweed produced 93,000–119,000 seeds per plant, whereas when grown with cotton, it produced 19,000–22,000 seeds per plant. A dense population of Pennsylvania smartweed growing with giant foxtail and common ragweed produced an average of 60 seeds per plant.

Dispersal: Seeds of both species pass through grazing animals intact and thus disperse when animals are moved and when manure is spread. Cottontail rabbits eat both species, but only the smaller seeds of ladysthumb pass through the rabbits unharmed. Seeds of ladysthumb have reportedly been moved in mud on the feet of gulls. In the 1940s, ladysthumb was a common contaminant of forage legume and grass seed, with red clover being particularly prone to contamination. Poorly cleaned forage and cover crop seed probably continues to be a mode of dispersal today. Pennsylvania smartweed is found in irrigation water. Ladysthumb seeds float and are dispersed in irrigation water. As both species are commonly found along the edges of streams and rivers, seeds are likely deposited in lowland fields during flood events.

Common natural enemies: Birds and cottontail rabbits eat seeds of both species. The widespread specialist aphid *Capitophorus hippophaes* infested an experimental population of Pennsylvania smartweed in Illinois but had little influence on growth and reproduction. Pennsylvania smartweed seeds frequently carry pathogenic fungi that damage or kill young plants.

Palatability: Ladysthumb and Pennsylvania smartweed have low forage quality and palatability for grazing animals, although Pennsylvania smartweed was as palatable to sheep as oats. Pennsylvania smartweed is a valued species in wetlands because of its abundant seed production and superior nutritional quality for supporting migratory birds and other wildlife.

FURTHER READING

Lee, H.S., A.R. Zangerl, K. Garbutt and F.A. Bazzaz. 1986. Within and between species variation in response to environmental gradients in *Polygonum pensylvanicum* and *Polygonum virginianum. Oecologia* (Berlin) 68: 606–610.

Mertens, S.K. and J.H. Jansen. 2002. Weed seed production, crop planting pattern, and mechanical weeding in wheat. *Weed Science* 50: 748–756.

Sultan, S.E. and F.A. Bazzaz. 1993a. Phenotypic plasticity in *Polygonum persicaria*. I. Diversity and uniformity in genotypic norms of reaction to light. *Evolution* 47: 1009–1031.

Velvetleaf

Abutilon theophrasti Medik.

Velvetleaf seedling
Antonio DiTommaso, Cornell University

Velvetleaf young plant
Antonio DiTommaso, Cornell University

Velvetleaf flower
Robert Nurse

Velvetleaf mature capsules
Scott Morris, Cornell University

IDENTIFICATION

Other common names: butter print, butter weed, pie marker, Indian mallow, velvet weed, Indian hemp, cotton weed, buttonweed, wild cotton, elephant ear, chingma, velvet leaf butterprint

Family: Mallow family, Malvaceae

Habit: Tall, erect, summer annual herb

Description: The stem of the **seedling** is hairy. **Cotyledons** are heart shaped, 0.13–0.5 inch long and hairy on all surfaces, including the edge. The first true leaves are alternate, heart shaped, covered with soft hairs on both surfaces and have blunt teeth along the edges. **Mature plants** are 2–7 feet tall, with hairy stems that often branch towards the top of the plant. Leaves are alternate, densely covered in soft, fine hairs and are heart shaped with a tapered tip. Each leaf is 4–8 inches long and wide, with a 4–8 inch-long leaf stalk. Leaf edges are bluntly toothed. Prominent veins radiate from the point where the leaf stalk attaches. Leaf and stem tissue give off a distinctive odor when crushed. The root system is a shallow, branching, white taproot. The **flowers** are yellow to yellow-orange and grow from the upper leaf axils on 1–2 inch-long stalks. Each flower is 0.6–1 inch wide and has five green sepals and five shallowly notched petals. Flowers are replaced by circular, cup-shaped capsules composed of a ring of flattened, pointy tipped seedpods. The capsules are light green when young and become brown to black as they mature. Mature capsules may remain on dead stems throughout the winter. Each capsule contains 15–45 grey-brown, kidney to heart-shaped **seeds**. Each seed is 0.04–0.06 inch thick and 0.08–0.11 inch long and wide.

Similar species: Common mallow (*Malva neglecta* Wallr.), prickly sida (*Sida spinosa* L.) and Venice mallow (*Hibiscus trionum* L.) seedlings may sometimes be confused with velvetleaf. The cotyledons of all three species are hairless, however, while those of velvetleaf are covered in fine hairs. The first true leaves of common mallow are rounded rather than heart shaped, like velvetleaf, while prickly sida's true leaves are more oval in shape and have larger teeth than those of velvetleaf. The first true leaves of Venice mallow are irregularly shaped, and all subsequent leaves are divided into three or more distinct lobes.

MANAGEMENT

Velvetleaf is primarily a pest in long season, spring

planted row crops like corn and soybeans, and rotating away from these crops helps manage the weed. Most velvetleaf emerge in the spring, so delaying planting or rotating to a summer planted crop allows destruction of many seedlings by tillage and a reduction in velvetleaf density in the crop. Also, plants are usually not as competitive in summer planted crops due to more rapid initiation of reproduction in response to shortening day length. Velvetleaf generally does not begin setting seeds until late summer. Consequently, early harvested crops like winter or spring grains or peas interrupt this weed's life cycle. In spring/early summer row crops, a good, uniform stand and a vigorous crop are critical, since crop competition can greatly decrease branching and seed production of velvetleaf.

A comparison of several studies on competition between corn and velvetleaf showed that velvetleaf caused little yield loss in years with cool weather (lower than 58°F average temperature) during the first two weeks after planting. In years with warm weather after planting, velvetleaf was only moderately competitive if the weather was wet during the period of rapid corn growth (30–75 days after planting), but it caused large yield losses if the weather was dry during that period. Thus, early season control of velvetleaf is particularly important in years with warm spring weather.

Velvetleaf seedlings can emerge from deep enough in soil that many will survive tine weeding or rotary hoeing. Consequently, you should target these operations against the white thread stage. This may entail more frequent cultivation with these implements than you would use against a more slowly establishing weed like common lambsquarters or redroot pigweed. Very shallow cultivation that cuts the plant just below or even at the soil surface causes high mortality of both seedlings and larger plants.

Both rodents and insects consume large numbers of velvetleaf seeds on the soil surface, and germinating seeds on the soil surface are prone to drying out before they become rooted. Consequently, if you have a year with high seed production of this species, avoid fall tillage to allow these mortality factors time to operate. A large percentage of velvetleaf seeds will germinate immediately following burial, so plowing to place surface seeds too deep for successful emergence will eliminate a substantial portion of the surface seed bank. However, seeds that do not immediately germinate will likely persist for many years and emerge following subsequent tillage events.

Because the seeds are relatively large, velvetleaf can emerge through even thick layers of organic mulch material. However, velvetleaf establishment was consistently suppressed when soybeans were grown in a rye cover crop left on the soil surface without tillage compared to soybeans that were planted after rye was incorporated by tillage. Velvetleaf seeds are exceptionally resistant to heat, and solarization only slightly reduces the survival of velvetleaf seeds.

ECOLOGY

Origin and distribution: Velvetleaf is native to Asia and was introduced to the United States from China as a fiber crop during the colonial era. It subsequently spread throughout the United States and southern Canada. It has also been introduced into Europe.

Seed weight: Seed weights range from 6–12 mg, but most population mean seed weights are near 9 mg.

Dormancy and germination: When shed from the parent plant, 3–62% of velvetleaf seeds are dormant due to a hard seed coat that is impermeable to water. The percentage of hard seeds produced varies greatly between populations. Also, seeds produced in a shady environment have thinner seed coats and a substantially lower percentage dormancy. Although non-dormant seeds could potentially germinate in the fall, most do not due to unfavorable soil conditions, and many of these germinate the following spring. As dormant seeds age, the pore where the seed attaches to the parent plant eventually cracks open and allows water to enter and germination to proceed. This leads to sporadic germination over many years. The seed coat contains germination inhibitors, but these appear to be unimportant in maintaining dormancy in the field. A few seeds will germinate at 46°F, but germination is best at 75–86°F and declines above 95°F. Above 122°F, seed coat permeability increases and so does seed mortality. Temperature fluctuations do not promote germination, but a period of drying at a warm temperature (93°F) after exposure to moisture does. Natural chilling of seeds during the winter has little effect on germination. Light does not affect germination of fresh seeds but promotes germination of seeds that have been buried in the soil. Germination of velvetleaf seeds in the soil is inhibited by volatile organic compounds like ethanol and acetaldehyde that are produced during anaerobic respiration. This may partially explain why tillage, which vents these compounds to the atmosphere, can prompt a flush of emergence. Application of nitrate does not increase seed germination. The large seeds of velvetleaf require good soil-seed contact to germinate, and consequently germination is best in a fine seedbed.

Seed longevity: Velvetleaf seeds can persist in the soil for several decades. Mortality rates for undisturbed seeds range from 11–17% per year over three- to six-year periods. Experiments using locally collected seeds from several Midwestern states, however, found an average seed loss of 41–43% over the first year for shallowly buried seeds. When soil has been tilled annually, seed mortality rates of 32–53% per year have been observed. A demographic study found that 71% of the seeds in the seed bank in the previous fall were lost by the following spring. One year mortality of seeds decreased from an average of 55% at the soil surface to 3% at 6 inches. Mice consume 31–99% of seeds left on the soil surface over the winter, with mortality rate increasing with the amount of cover. Although velvetleaf seeds are highly persistent, a partial draw-down of the seed bank can be achieved over several years if seed production is prevented, and this can result in reduced seedling emergence in the crop.

Season of emergence: Velvetleaf emerges primarily in mid-spring, but a few seedlings emerge sporadically later in the growing season.

Emergence depth: Seedlings emerge best from the top 0.5–1 inch, while emergence is more variable from the top 2–3 inches of soil, and only a few emerge from below 3 inches. Emergence is poor from seeds on the soil surface.

Photosynthetic pathway: C_3

Sensitivity to frost: Velvetleaf is killed by the first hard frost.

Drought tolerance: Velvetleaf is drought tolerant. In drought conditions it loses the lower leaves, which reduces water use and aides survival to reproduction. However, corn uses water more efficiently than velvetleaf and consequently can grow faster when water is in short supply.

Mycorrhiza: Velvetleaf is mycorrhizal, and the importance of mycorrhizae for phosphorus nutrition of this species has been demonstrated.

Response to fertility: Velvetleaf growth is highly responsive to fertilization, especially fertilization with N. Applications of poultry manure compost or blood meal continued to increase productivity up to 320 pounds per acre of N. Adding swine manure at rates of 500–643 pounds per acre of N plus 105 pounds per acre of N chemical fertilizer more than doubled seed production relative to 132 pounds per acre of N of chemical fertilizer alone. In a pot experiment, plants continued to increase in size and seed production up to 440-880-440 pounds per acre of N-P-K. Seedlings from plants grown in highly fertile conditions are larger and more competitive than those from plants grown at lower fertility.

Soil physical requirements: Velvetleaf tolerates poor drainage and a wide range of soil textures.

Response to shade: Velvetleaf is moderately shade tolerant. Shade at 30% only slightly decreased growth, but 76% shade reduced plant weight and seed production by 88% or more. Velvetleaf growth and seed production declined linearly as shading from a corn leaf canopy increased. Because shade and crop competition have relatively small effects on plant height, velvetleaf is often able to grow up into or to overtop crop canopies.

Sensitivity to disturbance: Velvetleaf can tolerate up to 75% defoliation at six weeks after emergence with little effect on plant size or seed production, provided the plants are not shaded. The substantial root system and fibrous stems of large velvetleaf plants make them difficult to uproot with cultivation.

Time from emergence to reproduction: Velvetleaf flowers in response to short days, so spring emerging individuals require more time to flower than summer emerging plants. Spring emerging plants in Wisconsin and Ontario flowered 11–12 weeks after emergence. Flowers pollinate the day they open. A few seeds become viable 12 days after flowering and essentially all are viable within 15 days, but capsules do not open to disperse seeds until 18–23 days after flowering. Flowering and seed production continue until frost.

Pollination: The species is primarily self pollinated, but some cross pollination by insects probably occurs.

Reproduction: The number of seeds produced is proportional to the weight of the plant. Plants typically produce 70–200 capsules, each containing 35–45 seeds. Plants grown without crop competition typically produce 700–17,000 seeds, but much lower seed production has also been observed. Crop competition can substantially reduce seed production, but velvetleaf growing in corn can still produce 1,000–2,000 seeds per plant in favorable circumstances.

Dispersal: Much feed corn is contaminated with velvetleaf seeds. These pass readily through the digestive tracts of livestock and are spread with manure. Seeds also pass intact through the guts of some birds or are regurgitated after being retained for several days, so dispersal by birds seems likely. Dispersal also occurs with soil clinging to tires and tillage implements. Combines probably also spread the weed between fields and farms.

Common natural enemies: Scentless plant bugs (*Niesthrea louisianica*) attack young pods and seeds and can substantially reduce reproduction both by direct damage and by introducing pathogens. Carabid ground beetles, slugs, cutworms and especially mice eat a substantial percentage of the seeds after dispersal,

which reduces seedling density the next year. *Fusarium lateritium* wilt disease reduced velvetleaf growth by as much as 86% and resulted in up to 55% mortality of velvetleaf seedlings. The fungus *Colletotrichum coccodes* reduces plant growth and reproduction and has been tested as a biological control agent. Velvetleaf in organic corn fields in New York are often heavily attacked by a species-specific white fly that greatly damages the leaves, both directly and by introducing a virus. Damage to the velvetleaf may be so extensive that only a few individuals along the field edges reach maturity.

Palatability: The seeds are eaten as food in China and Kashmir. Some sheep find velvetleaf palatable while others reject it.

Note: Velvetleaf has allelopathic effects on some crops.

FURTHER READING

Davis, A.S. and K.A. Renner. 2007. Influence of seed depth and pathogens on fatal germination of velvetleaf (*Abutilon theophrasti*) and giant foxtail (*Setaria faberi*). *Weed Science* 55: 30–35.

Oliver, L.R. 1979. Influence of soybean (*Glycine max*) planting date on velvetleaf (*Abutilon theophrasti*) competition. *Weed Science* 27: 183–188.

Teasdale, J.R. 1998. Influence of corn (*Zea mays*) population and row spacing on corn and velvetleaf (*Abutilon theophrasti*) yield. *Weed Science* 46: 447–453.

Warwick, S.I. and L.D. Black. 1988. The biology of Canadian weeds. 90. *Abutilon theophrasti*. *Canadian Journal of Plant Science* 68: 1069–1085.

Waterhemp

Amaranthus tuberculatus (Moq.) Sauer
Amaranthus rudis Sauer

Waterhemp seedlings
Aaron Hager, University of Illinois

Waterhemp foliage
Lynn Sosnoskie, Cornell University

Waterhemp plant
Antonio DiTommaso, Cornell University

IDENTIFICATION

Other common names: common waterhemp, tall waterhemp

Family: Pigweed family, Amaranthaceae

Habit: Tall, upright, summer annual herb

Taxonomic note: Many sources separate common waterhemp (*Amaranthus rudis*) from tall waterhemp (*Amaranthus tuberculatus*), but recent authorities have considered the two as a single species. Though probably not good species, the two forms may deserve recognition as varieties (*A. tuberculatus* var. *rudis* and *A. tuberculatus* var. *tuberculatus*). Common waterhemp appears to be more common as an agricultural weed, but tall waterhemp also occurs in farm fields. Since the two forms can only be definitively distinguished using minute characteristics of the female flowers, and since these characteristics intergrade extensively in the central Midwest where the species is most problematic, we here treat them as a single species that we refer to simply as waterhemp.

Description: Seedling stems are light green to red-pink and 0.1–0.2 inch tall. Cotyledons are oval to lanceolate, 0.2–0.4 inch long by 0.08–0.16 inch wide, hairless and red-green above and dark red below. The true leaves are oblong, dark green, prominently veined and reddish pink below, shiny above and shallowly notched at the tip. **Mature plants** reach 2.5–8 feet tall on branching, green to reddish, hairless, occasionally ridged stems. The leaves are alternate, shiny, hairless, 0.6–6 inches long by 0.2–1.5 inch wide and oblong to lanceolate with an abruptly tapered, notched tip. Leaf stalks are up to 0.4–2.8 inches long and are usually shorter than the leaf blade. Upper leaves are smaller and more lanceolate than lower leaves. The root system is a taproot with secondary fibrous roots. Male and female **flowers** are produced by separate plants. Flowers of both sexes are green, without petals, 0.07–0.11 inch long and clustered to form upright, 4–8 inch-tall, occasionally branching spikes at branch tips and leaf axils. Male flowers have five green sepals, while female flowers have zero to two green sepals. Female flowers are replaced by 0.06 inch-long oval capsules. Each capsule contains a single shiny, round to oval, 0.03 inch-wide, red-brown to black **seed**.

Similar species: Redroot pigweed (*Amaranthus*

retroflexus L.), smooth pigweed (*A. hybridus* L.) and Powell amaranth (*A. powellii* S. Watson) have hairy stems and leaves, while waterhemp leaves and stems are hairless. These amaranth species also have male and female flowers on a single plant, while waterhemp has separate male and female plants. Like waterhemp, Palmer amaranth (*A. palmeri* S. Watson) has smooth stems and separate male and female plants. The leaf stalks of Palmer amaranth, however, are generally longer than the leaf blade, while waterhemp leaf stalks are usually shorter than the leaf blade. Palmer amaranth leaves may have chevron or V-shaped watermarks, while waterhemp leaves never have such markings.

MANAGEMENT

Waterhemp is a late emerging species, so planting crops as early as possible while still ensuring good establishment will improve the competitiveness of the crop. Plants emerging one month or more after corn or soybeans are substantially suppressed. Tillage greatly reduces waterhemp emergence, and moldboard plowing is more effective than chisel plowing for reducing emergence. Although the seeds are moderately persistent when buried deeply, their small size means that seeds must return close to the soil surface to emerge successfully following any future tillage event, and a large proportion of those buried by tillage will never return. Although inversion tillage can be one component of a waterhemp management strategy, waterhemp seed densities in the soil can be decreased using ridge tillage in row crops if the cultivation program is effective.

Given its late emergence, probably few waterhemp will mature in a winter grain crop, but they may be able to mature in spring oats. If hay is overseeded into the grain, regrowth following grain harvest will create massive seed rain if mowing is delayed long enough to get a hay cutting that fall. An earlier cutting in August at about 4 inches will knock back the waterhemp at a time when the hay species can rebound rapidly from root reserves and shade the waterhemp. Mow with a forage chopper and wagon so that the cut inflorescences are removed; otherwise, they are likely to set seed. The success of this tactic depends on a good stand of hay or cover crop. Since relatively little of the hay leaf area will be removed, a hay cutting in the fall may still be possible. Despite all precautions, anything short of repeated tillage after small grain harvest is likely to result in some waterhemp seed production.

Effective cultivation of corn and soybeans can provide good control of waterhemp. Establishment from near the soil surface and the tiny size of waterhemp seedlings make them highly susceptible to rotary hoeing and tine weeding. Continue in-row weeding as long as possible to get later emerging cohorts out of the row. Begin throwing soil into the row with inter-row cultivation soon after the last in-row operation and mound up around the base of the plants at the final cultivation. Topping waterhemp that emerges above soybeans or other short crops with a mower will not be effective, as this species recovers from clipping rapidly. Pulling plants will prevent competition with the crop, but the rapid flowering habit of the species means that it is unlikely to be effective for preventing seed production unless the plants are removed from the field. The high retention of seeds on plants at harvest provides an opportunity for seed capture and destruction during harvest operations.

The late emergence of waterhemp means that neither a tilled fallow period nor a stale seedbed before planting will be effective for controlling the species, except possibly for a midsummer planted vegetable crop. Straw or other organic mulch materials can completely suppress this species both by preventing germination cues and by blocking emergence of the tiny seedlings.

ECOLOGY

Origin and distribution: Waterhemp is native to the United States. It is most problematic as a weed in the Midwest but occurs through most of the eastern and southern United States, and the southern margin of Canada from Quebec to Saskatchewan. Its scattered occurrence from the Rocky Mountains westward is probably due to introduction. It has also been introduced to Europe.

Seed weight: Population mean seed weights range from 0.19–0.27 mg.

Dormancy and germination: Seeds collected from recently matured plants are dormant. Populations vary in the after-ripening conditions needed to allow germination. Some populations germinate best after 12 weeks of wet, cold (39°F) conditions, whereas others germinate best after a period of warm wet conditions, and a population from a natural habitat germinated well following after-ripening in a wide range of conditions. Seeds begin germination at mean daily temperatures of 41–59°F, depending on the population, but percentage germination is low at cool temperatures. Some germination can occur at day/night temperatures as high as 113/104°F. Peak germination occurs at mean daily temperatures of 68–91°F. Germination at any constant temperature is relatively poor whereas daily temperature fluctuations substantially promote germination, with a daily fluctuation range of 32–43°F maximizing germination. The after-ripening requirements of agricultural populations coupled with peak germination at warm temperatures ensures that most seeds will not germinate until the summer following production. In warm, fluctuating temperature conditions, seeds germinate in as little as one day. Light stimulates germination, especially red light, whereas light that has been depleted in red wavelengths (e.g., by passage through a plant leaf canopy) can inhibit germination. However, high temperatures, for example 97°F, can overcome this latter inhibition and allow germination under plant canopies.

Seed longevity: Waterhemp seeds mixed into the top 2 inches of soil and stirred annually in spring had an average annual mortality of 40%. In Nebraska, the computed annual seed mortality rate over the first three years was 30–44%, but 1–3% of seeds still germinated after 17 years of burial at 8 inches. In several Midwestern states, the average mortality of seed buried 6 inches deep for 12 months was 78%.

Season of emergence: Waterhemp typically emerges from mid-May to late July in Iowa, with peak emergence in mid- to late June. Reduced tillage tends to delay peak emergence. Waterhemp is classified as a late emerging weed with a long duration of emergence in Nebraska. In Ontario, emergence occurs from June through August. Waterhemp required 14–17 days to emerge when planted in May in Missouri and had the slowest emergence of six pigweed species tested. It is generally known for late emergence and a discontinuous germination pattern that extends well into the growing season, thereby allowing waterhemp to emerge and produce seeds after all weed management operations have been completed.

Emergence depth: This has not been studied directly,

but the small seed size and similarity of the seeds to related pigweed species indicate that most seedlings likely emerge from the top 1 inch or less of the soil (see the species chapters "Palmer amaranth" and "Pigweeds: Powell amaranth, redroot and smooth"). The great suppression of emergence by tillage further supports a shallow emergence pattern.

Photosynthetic pathway: C_4

Sensitivity to frost: Flowering and seed set stop at the first frost.

Drought tolerance: Waterhemp is not drought tolerant. Plant growth and seed production declines approximately linearly as soil water content declines and as the number of days without water increases. The pattern of seed production in several short pulses over an extended period appears to be an adaptation to improve reproductive success in the face of unpredictable summer rainfall patterns.

Mycorrhiza: Waterhemp is probably non-mycorrhizal in most circumstances.

Response to fertility: In sand, increased ammonium nitrate raised waterhemp's relative growth rate about five fold, similar to the response of corn. Increasing rates of composted swine manure in a soil mix substantially increased the relative growth rate of waterhemp seedlings while having little effect on the relative growth rates of corn, wheat or soybeans. This effect would make the weed both more competitive against crops and also more difficult to control with cultivation at high compost rates. In soybeans, application of a high rate of composted swine manure increased waterhemp dry weight by 25–50%. In an experiment with corn, application of 24–33 tons per acre of composted swine manure roughly doubled the dry weight of waterhemp. Composted swine manure also inhibited waterhemp emergence but had no effect on emergence of the crops. Waterhemp tolerates a pH from 4.5–8.

Soil physical requirements: The species' natural habitat is the wet soils of marshes and the edges of lakes, ponds and rivers. Consequently, it tolerates wet, anaerobic soils but does best in well drained agricultural soils. It tolerates a wide range of soil textures but grows best in medium to fine textured soils.

Response to shade: Waterhemp is considered shade intolerant. Shade of 68% reduced final plant weight by about 50%, but plants emerging in May still produced 400,000 seeds per plant and June-emerging individuals produced 90,000 seeds per plant. With 99% shade, mortality was substantial, and remaining plants were small and produced few seeds. By comparison, corn in Illinois at the V8 growth stage casts 80–98% shade.

Sensitivity to disturbance: Waterhemp recovers well from clipping. In greenhouse studies, removing half the shoot had no effect on plants 4–8 inches tall and only reduced the weight of 12–16 inch plants by 22%. In the field, removing half the shoot of 6 inch plants reduced final height by 11% but had no effect on seed production. Removing all but the seed leaf node reduced seed production by 78%, but the plants still produced an average of 32,000 seeds. Waterhemp has a taproot, which makes large plants difficult to uproot with a cultivator but relatively susceptible to slicing with shallow cultivating knives.

Time from emergence to reproduction: Waterhemp, as other *Amaranthus* species, begins flowering in response to shortening day length. Populations established in May and June required five to seven weeks after emergence to begin flowering, whereas populations established in July required three to four weeks. Time to initiation of flowering was consistently a few days earlier for male than female plants, ensuring sufficient pollen availability when females began flowering. Some seeds first become viable seven to nine days after pollination, at which time they are brown, but seed weight and percentage viability continues to increase until 12 days after pollination, at which time they are black. Seed maturation occurred similarly in five cohorts emerging from May to July, namely at 20–27 days after flower initiation and six to 13 days after pollination.

Pollination: Since male and female flowers occur on separate plants, the species necessarily outcrosses. It is primarily wind pollinated.

Reproduction: Female waterhemp plants can produce as many as 1 million seeds, but 35,000–200,000

seeds is more typical. Several populations produced 470,000–1.29 million seeds per plant in Indiana when established in May or June but produced 200,000–340,000 seeds when established in July. In soybeans, seed production declined exponentially with an increasing lag between crop planting and waterhemp emergence. Thus, waterhemp establishing simultaneously with soybeans in Iowa produced 300,000 seeds per plant, whereas those emerging 50 days later produced 3,000 seeds per plant. On the other hand, waterhemp seed production per plant also is highly density dependent, so individual plant production will vary among emergence cohorts depending on plant density. Waterhemp produced an average 288,000 seeds per plant in Missouri and had the highest seed production per unit plant weight of six pigweed species tested. Waterhemp plants retained 95–100% of seeds at soybean harvest.

Dispersal: Waterhemp seeds float and probably disperse by overland water flow, in irrigation water and along streams. Although passage through ruminants has not been studied, the seeds are similar to those of other pigweed species that disperse readily in feces and the spreading of manure. Given the persistent seed bank and prolific seed production of the species, dispersal in soil clinging to shoes, tires, animals and machinery seems likely.

Common natural enemies: Seeds are eaten by mourning doves, ducks and songbirds. After dispersal, the seeds are eaten by field crickets, *Gryllus pennsylvanicus*, and several species of carabid ground beetles, including *Amara aeneopolita*, *Anisodactylus rusticus*, *Stenolophus comma* and *Harpalus pennsylvanicus*. These species preferred waterhemp seeds to those of several other prominent weed species. Waterhemp is attacked by several pathogens, including *Albugo bliti* (white rust), *Phymatotrichum omnivorum* (Phymatotrichum root rot), *Cercospora acnidae* (a leaf spot disease) and *Phyllosticta amaranthi* (a leaf spot disease). *Microsphaeropsis amaranthi* has potential as a bioherbicide for control of waterhemp provided formulations can prevent dry leaf surfaces during the first 12 or more hours after application.

Palatability: Palatability of waterhemp for grazing animals is low.

FURTHER READING

Costea, M., S.E. Weaver and F.J. Tardif. 2005. The biology of invasive alien plants in Canada. 3. *Amaranthus tuberculatus* (Moq.) Sauer var. *rudis* (Sauer) Costea & Tardif. *Canadian Journal of Plant Science* 85: 507–522.

Korres, N.E., J.K. Norseworthy, B.J. Young, D.B. Reynolds, W.G. Johnson, S.P. Conley, R.J. Smeda, T.C. Mueller, D.J. Spaunhorst, K.L. Gage, M. Loux, G.R. Kruger and M.V. Bagavathiannan. 2018. Seedbank persistence of Palmer amaranth (*Amaranthus palmeri*) and waterhemp (*Amaranthus tuberculatus*) across diverse geographical regions in the United States. *Weed Science* 66: 446–456.

Leon, R.G. and M.D.K. Owen. 2006. Tillage systems and seed dormancy effects on common waterhemp (*Amaranthus tuberculatus*) seedling emergence. *Weed Science* 54: 1037–1044.

Liebman, M., F.D. Menalled, D.D. Buhler, T.L. Richard, D.N. Sundberg, C.A. Cambardella and K.A. Kohler. 2004. Impacts of composted swine manure on weed and corn nutrient uptake, growth, and seed production. *Weed Science* 52: 365–375.

Wild buckwheat

Polygonum convolvulus L. = *Fallopia convolvulus* (L.) Á. Löve

Wild buckwheat seedlings
Antonio DiTommaso, Cornell University

Wild buckwheat plant
Antonio DiTommaso, Cornell University

Wild buckwheat mature fruits
Antonio DiTommaso, Cornell University

IDENTIFICATION

Other common names: black bindweed, knot bindweed, bear bind, ivy bindweed, climbing bindweed, corn bind, climbing buckwheat, dullseed corn bind, climbing knotweed, devil's bindweed, blackbird bindweed

Family: Buckwheat family, Polygonaceae

Habit: Twining summer annual herb

Description: Seedlings have two elongate, oval cotyledons that are 0.25–1.25 inch long by less than 0.25 inch wide, with round tips and a gritty, waxy surface. Cotyledons are often at a 120 degree angle from one another. The seedling stem is red-purple. Young leaves are arrow shaped with a tapered, pointed tip and basal, backwards pointing lobes. The first true leaves are blue-green on their upper surface and red on their lower surface. All true leaves have a thin, papery covering extending up the stem from the base of the leaf stalk (ocrea). **Mature plants** have smooth, weak, twining stems that branch near the base and reach up to 6.5 feet in length. Leaves are alternate, hairless, 0.75–2.5 inches long, arrow shaped with long pointed tips and backward pointing lobes and attached by

long leaf stalks. Ocreae are 0.13–0.2 inch long sheaths that wrap the stem above leaf junctions. Leaves lower on the stem will be more triangular or heart shaped than younger, narrower leaves near the tip. The root is fibrous. Greenish-pink or greenish-white **flowers** are grouped into clusters of two to six on a long stem arising from axils of upper leaves or on tips of small branches. Individual flowers are 0.2 inch wide. **Seeds** are dull, black, sharply three sided and may be covered by the papery, brown remnants of flower parts.

Similar species: Tartary buckwheat [*Fagopyrum tartaricum* (L.) Gaertn.] and domestic buckwheat (*Fagopyrum esculentum* Moench) do not twine onto other plants and have more compact inflorescences and larger seeds. Hedge bindweed [*Calystegia sepium* (L.) R. Br.] and field bindweed (*Convolvulus arvensis* L.) resemble wild buckwheat, but both are perennials with thick, spreading roots. Leaves of wild buckwheat are more tapered and pointed at the tip than leaves of either of the bindweeds. The lobes of hedge bindweed leaves are squarish in outline when laid flat, whereas wild buckwheat and field bindweed leaf-lobes are pointed. The flowers of both bindweeds are large and either white or pink, and they occur singly. This is unlike wild buckwheat flowers, which are small, green and clustered. Bindweeds lack ocrea.

MANAGEMENT

Wild buckwheat is a severe weed in spring sown cereal grains and a problem in many other field and vegetable crops. In cereal grains, management during preceding crops in the rotation is critical since options are limited in spring grains. A vigorous winter grain crop will tend to suppress the seedlings. Properly timed cultivation can control the weed in row crops like corn and soybeans. If this species is causing severe problems in your spring grains and you rotate with corn, cleaning up plants in the corn rows with a flame weeder may be worthwhile. Including later planted crops like soybeans, dry beans and summer vegetables in the rotation allows a period in the spring for a tilled fallow that will reduce the seed bank going into a subsequent cereal grain the next year. If time permits, work the soil to a depth of about 3 inches at about one-week intervals to ensure that all seeds in the surface soil get exposed to enough warmth to prompt germination.

For this species, effectiveness of tine weeding (harrowing) and rotary hoeing are limited in both grains and row crops because many of the seedlings arise from depths below the seeding depth of the crop. Some seedlings will be buried by a tine weeder, however, so if you can do the operation soon after the first flush of seedlings emerge it can still provide some control. A dense planting of grains helps reduce losses in fields infested with wild buckwheat. Although the species climbs, it does not begin climbing until about a month after emergence. Rapid canopy closure and root proliferation from a dense crop sowing can help suppress this species. Slightly lower than optimal nitrogen fertility suppresses wild buckwheat more than it does grain crops.

Cultivation in row crops should be aimed at uprooting plants while they are small and burying newly emerged seedlings in the row. Rerooting becomes an increasing problem as the wild buckwheat grows. Consequently, tools that cut are more effective against wild buckwheat than are those that dig. If this species is a problem, consider investing in an implement that can work close to the crop when it is still small.

If you do not already have wild buckwheat, inspect any uncertified grain and cover crop seed for the characteristic three-sided seeds of wild buckwheat before you sow it.

ECOLOGY

Origin and distribution: Wild buckwheat originated in Europe but is now found throughout the world's temperate regions and as far north as Greenland and Alaska. The species occurs throughout the United States and Canada.

Seed weight: 4.7–7 mg.

Dormancy and germination: Seeds are dormant when shed from the parent plant, and very few will germinate until subjected to a prolonged period of cold, wet conditions. Dormancy results from a hard seed coat, which, when scarified, allows germination. A cold, moist period of at least two months with daily alternating temperatures between 36–50°F will break dormancy and induce maximum germination similar to mechanical seed coat removal. Non-dormant seeds germinate well at 68–77°F. Most seeds enter secondary dormancy during hot, dry summer weather and then require a second period of cold, wet conditions before germination is again possible. Light and nitrate do not affect germination. In the field, germination and emergence is slow, requiring from nine to 31 days.

Seed longevity: Seed longevity may vary between populations. Most viable seeds germinate the year after production, but some may survive for several decades deep in cool, moist soil. In soil stirred four times per year, the number of wild buckwheat seeds declined by 32–50% per year, whereas in undisturbed soil they declined at 20–25% per year. In Alaska, annual mortality of buried seeds was computed at 52% and at 46% in France. Based on these studies, the seed bank is moderately persistent but susceptible to management.

Season of emergence: Most emergence occurs in early spring with some emergence continuing into summer. Maximum emergence occurs at soil temperatures less than 60°F.

Emergence depth: Most seedlings emerge from anywhere in the top 2 inches of soil, but occasional seedlings emerge from as deep as 5 inches or more. Emergence was often best from 0.4–1.6 inch depths rather than from the soil surface or deeper.

Photosynthetic pathway: C_3

Sensitivity to frost: Wild buckwheat is killed by hard frost.

Drought tolerance: This species absorbs soil moisture efficiently and competes well with crops under dry soil conditions.

Mycorrhiza: Wild buckwheat is not mycorrhizal.

Response to fertility: The growth response of wild buckwheat to N is similar to wheat. Both species continue to increase in size up to application rates beyond 214 pounds per acre of N, but the incremental increase with greater N is moderate. The species is highly responsive to P application rates up to 164 pounds per acre of P_2O_5. Increasing fertility tends to favor wild buckwheat more than field crops like wheat and flax.

Soil physical requirements: Wild buckwheat occurs on a wide range of soil types. But generally, it is best adapted to heavier soil with good moisture holding capacity. Soil compaction does not affect seed germination.

Response to shade: The vining habit of wild buckwheat potentially allows it to avoid shade by climbing up competing crop plants. However, the species does not begin twining onto crops until about a month after emergence, and it can be suppressed by shade if the crop canopy closes quickly.

Sensitivity to disturbance: Since many individuals typically emerge from relatively deep in the soil, shallow disturbance with a tine weeder or rotary hoe leaves many individuals unaffected. In contrast, tools that cut seedlings and young plants at or just below the soil surface kill the weed. The fibrous root system holds soil and helps larger plants reroot after cultivation.

Time from emergence to reproduction: On average, wild buckwheat begins flowering at six to eight weeks after emergence, and the first seeds mature about three weeks later in Saskatchewan. In Wisconsin, 12 weeks were required between emergence and flower initiation. Maturation is faster in warm weather, but the plants are less vigorous. Because of the indeterminant flowering habit of wild buckwheat, flowers, immature seeds and mature seeds may be found on the same plant. Up to half of the seeds of late emerging plants may not reach maturity by the first killing frost.

Pollination: Wild buckwheat primarily self-pollinates, and seed set often occurs even though the flowers remain closed. Even open flowers do not attract pollinating insects.

Reproduction: Wild buckwheat reaches maturity mid-season and continues flowering and releasing seeds for most of the growing season. Large, old plants often lose most of their leaves but still continue maturing seeds. Large plants can produce as many as 12,000–30,000 seeds.

Dispersal: Wild buckwheat spread throughout the world primarily in contaminated seed grain. The seeds float and can spread in surface irrigation water. They also move about on tires, tillage machinery and combines. Wild buckwheat seeds can survive for several months in silage and can survive rumen digestion for 24 hours. Thus, the species probably disperses in manure.

Common natural enemies: The beetle *Gastrophysa polygoni* eats the foliage, and larvae of the fly *Pegomyia setaria* mine the leaves.

Palatability: The seeds are edible but low in oil and protein. The stems and leaves have low forage value for livestock.

FURTHER READING

Forsberg, D.E. and K.F. Best. 1964. The emergence and plant development of wild buckwheat (*Polygonum convolvulus*). *Canadian Journal of Plant Science* 44: 100–103.

Hume, L., J. Martinez and K.F. Best. 1983. The biology of Canadian weeds. 60. *Polygonum convolvulus* L. *Canadian Journal of Plant* Science 63: 959–971.

Mertens, S.K. and J. Jansen. 2002. Weed seed production, crop planting patterns, and mechanical weeding in wheat. *Weed Science* 50: 748–756.

Wild mustard

Sinapis arvensis L. = *Brassica kaber* (DC.) L.C. Wheeler

Wild mustard seedling
Antonio DiTommaso, Cornell University

Wild mustard plant
Scott Morris, Cornell University

Wild mustard seed pods
Scott Morris, Cornell University

IDENTIFICATION

Other common names: charlock, field mustard, field kale, kedlock, common mustard, crunchweed, kraut weed, water cress, yellow flower, herrick, yellow mustard

Family: Mustard family, Brassicaceae

Habit: Highly branched, summer annual herb

Description: Seedlings have cotyledons that are kidney or heart shaped to round, with the indentation at the tip, and are 0.25–0.5 inch long by 0.25–0.75 inch wide. The cotyledons are hairless and have 0.4 inch-long stalks. Young seedling leaves are alternate, egg to club shaped and 0.4–0.8 inch long. Leaf edges are wavy with wide, irregular teeth. The upper surface of the leaves has scattered stiff hairs, a somewhat wrinkled appearance and conspicuous sunken veins. Leaf stalks are 0.1–0.25 inch long and hairy. Early leaves form a basal rosette. **Mature plants** have hairy stems and are 16–40 inches tall. The stem hairs are stiff, sparse, point towards the base and are denser lower on the stem. Leaves are alternate, hairy, 2–10 inches long by 0.5–2.75 inches wide, and larger towards the base of the plant. Lower leaves have long stalks, are oval to egg shaped, deeply lobed and broadest at the tip. Upper leaves are smaller, lanceolate and have short stalks or are stalkless. The root system is a thin and branched taproot with many fibrous secondary roots. Clusters of yellow, four-petaled **flowers** develop at branch tips on 0.1–0.3 inch-long stalks. The flowers are 0.5–1.25 inch wide. Flowers and seedpods are often present simultaneously. Seedpods are upright, cylindrical, 1–1.75 inch long by 0.08–0.12 inch wide, with 0.2–0.3 inch-long stalks and a tapering, two- or four-sided, flat tipped beak. The **seeds** are reddish brown or dark brown to black, round and 0.06 inch wide.

Similar species: White mustard (*Sinapis alba* L.) has stalked upper leaves and hairy seedpods, while the upper leaves of wild mustard are often without stalks and the seedpods are hairless. Wild radish (*Raphanus raphanistrum* L.) has hairier, rougher leaves than those of wild mustard. Wild radish flowers have dark-veined petals, while the flowers of wild mustard lack distinct veins. Wild radish seedpods are larger than those of wild mustard; they appear "beaded" instead of entire or smooth, and they break into segments rather than splitting open when ripe.

MANAGEMENT

This species establishes very quickly, and consequently

it is among the hardest weeds to control with cultivation. If you expect wild mustard to be a problem based on past years, plant large seeded crops a little deeper than usual and blind cultivate aggressively before crop emergence. This practice will be most effective if the crop is planted near its optimum season for establishment. Frequent early cultivations will pay off, especially if your equipment will allow you to get close to the crop row. Throw soil into the crop row as soon as the crop will tolerate it to bury wild mustard seedlings.

Wild mustard reaches its peak emergence following tillage in early to mid-spring, particularly when tillage is preceded by fertilization with inorganic nitrogen. If an area is particularly thick with this species, consider rotating to a late planted crop, and use a spring tilled fallow period to flush out and destroy part of the seed bank.

The growth rate and competitive ability of wild mustard increases greatly with increasing nitrogen fertility, so if this weed is a problem, avoid over fertilizing with nitrogen rich inputs. Wild mustard grows best and is more competitive when nitrogen is side-dressed earlier rather than later, so apply nitrogen as late as possible after the crop is established. This species is sensitive to shading, so plant highly competitive crops in fields where this weed is abundant. High density plantings of cereal grains also are effective for suppressing the weed since it usually is shorter than the grain.

In cereal grains, most pods generally remain closed and seeds are retained on the plant until harvest. Consequently, weed seed collection during combining could substantially reduce population density.

ECOLOGY

Origin and distribution: Wild mustard is native to temperate regions of Europe, Asia and North Africa. It has been introduced into North and South America, Australia and South Africa. It occurs throughout the agricultural regions of North America.

Seed weight: Seed mass ranges from 1–2.3 mg. Light brown colored seeds averaged 1.2 mg, whereas dark brown colored seeds averaged 1.5 mg.

Dormancy and germination: Some seeds can germinate immediately following seed shed (21% in one experiment), but most freshly shed seeds are dormant

with typically about 5% germination. Smaller plants with fewer fruit produced black seeds that had higher dormancy than red seeds produced on larger plants with more fruit. The higher mass and dormancy of dark compared to light colored seeds may be accounted for by the thicker seed coat of dark seeds. Exposure to freezing temperatures over winter can relieve some dormancy, increasing germination from 5% to 28%. When buried deeply in soil, dormancy is induced in seeds, but burial over winter can increase germination and emergence if seeds are brought to the surface the following spring. The degree of light sensitivity apparently varies between populations and environmental conditions, but generally germination is only weakly stimulated by exposure to light alone. Germination is stimulated by a combination of exposure to light and inorganic nitrogen. Under these conditions, optimum temperatures for germination are 50–68°F, with daily alternating temperatures giving maximum response. In favorable conditions, seed germination occurs in two to four days, and emergence in the field occurs five to seven days after initiation of germination.

Seed longevity: When left undisturbed, a few seeds can remain viable in the soil for 60 years, especially if they are deeply buried. However, mortality of seeds buried in packets in the surface 1 inch of soil was 22–45% per year. Similarly, mortality of wild mustard seeds in tilled agricultural soils was 20–52% annually.

Season of emergence: Wild mustard emerges primarily during the spring, beginning when soil temperature exceeds 40°F and recent rainfall has occurred. It continues to emerge sporadically throughout the growing season with a secondary peak of emergence in the fall.

Emergence depth: Emergence is best from at or near the soil surface, is moderate from 2 inches and is negligible from 4 inches or deeper.

Photosynthetic pathway: C_3

Sensitivity to frost: Wild mustard can survive temperatures down to 22°F but suffers some damage. Late germinating plants are commonly winter-killed before they can set seed.

Drought tolerance: Wild mustard has an exceptionally

high density of stomates on the leaves, which would facilitate rapid growth with optimum soil moisture but could make it relatively drought sensitive under low soil moisture conditions. However, wild mustard establishes an extensive root system. For example, a study found that the total length of all roots added together was 3 feet by the fifth day and 394 feet by 21 days after emergence. In one experiment, drought stress reduced plant height by 90% and seed production by 97%, but in another experiment, wild mustard plant height, seed production, seed dormancy and competitive ability were only moderately decreased by drought.

Mycorrhiza: Wild mustard does not form symbiotic relationships with mycorrhizal fungi.

Response to fertility: Wild mustard is highly responsive to N fertility, with plant size increasing in response to application rates of 480 pounds per acre. Increasing nitrogen fertility can enhance the competitiveness of wild mustard relative to selected crops, but barley can outcompete wild mustard even at high N application rates. Delaying N side-dressing from 28 to 56 days after emergence reduced wild mustard biomass and seed production by approximately 50%. The species was highly responsive to P in one experiment but was relatively unresponsive in two other experiments. Plants grew twice as large on soil with a K test of 147 pounds per acre relative to soil with a test of 52 pounds per acre. Wild mustard plants have a high sulfur content and may respond to S. Growth is reduced substantially at a pH of 4.7–4.8 compared to a pH of 5.7–6.5.

Soil physical requirements: Wild mustard grows on a variety of soil types but grows best and is most competitive on clay soils. Emergence is impaired in soils that dry and become crusted.

Response to shade: Shade reduces growth rate and seed production of wild mustard, while time to flowering lengthens.

Sensitivity to disturbance: Because of the rapidly developing root system of wild mustard, plants very quickly become resistant to rotary hoeing or tine weeding, and so these operations should target the weed in the white thread stage. Small plants are susceptible to manual hoeing or cultivation, but plants nearing the flowering stage are prone to rerooting in moist weather.

Time from emergence to reproduction: Wild mustard is a long-day plant that flowers and produces seed when day length exceeds 16 hours. In the northern parts of its range, wild mustard flowers three to six weeks after emergence, with late emerging plants flowering most quickly. Seeds mature five to six weeks later. Time to flowering can be highly variable depending on drought and crop competition, both of which can delay flowering. In warmer parts of the United States fall emerging plants overwinter, flower in the spring and set seeds in early summer.

Pollination: Unlike most annual weeds, wild mustard is self-incompatible. Although it is an introduced species, plants are cross pollinated by a wide variety of native insects. It provides nectar and pollen that attract beneficial insects that prey on crop pests.

Reproduction: Plants in agricultural fields produce from 10–18 seeds per pod and 2,000–3,500 seeds per plant when competing with crops under normal moisture conditions. Plants growing without competition are larger and produce more seeds. Reproduction is substantially reduced when plant density is low and plants are isolated from pollinator habitats. Wild mustard can die as the seeds form or may continue to produce mature pods until frost depending on conditions, but seeds are generally not released until the plant is dead. In Saskatchewan, more than 98% of wild mustard seeds were retained at harvest of field peas and spring wheat, suggesting that this species would be suitable for control by novel harvest practices that collect and destroy weed seeds.

Dispersal: Since the pods are slow to open, wild mustard pods growing in grain fields are often collected and the seeds dispersed by combines and also as contaminants in grain and forage seed. Seeds are also dispersed in manure and in soil clinging to shoes, tires and tillage machinery.

Common natural enemies: Wild mustard frequently shows substantial leaf damage from imported cabbageworm (*Pieris rapae*) and flea beetles (*Phyllotreta* spp.). Despite substantial foliar and root damage from cabbageworms and wireworms, respectively, damaged plants can compensate and still produce similar numbers and mass of seeds as plants that have not been attacked.

Palatability: The young plants are palatable to livestock and are marginally palatable to humans when cooked. The seeds contain toxins, and animals should not be allowed to eat more than small amounts of maturing plants.

FURTHER READING

Edwards, M. 1980. Aspects of the population ecology of charlock. *Journal of Applied Ecology* 17: 151–171.

Mulligan, G.A. and L.G. Bailey. 1975. The biology of Canadian weeds. 8. *Sinapis arvensis* L. *Canadian Journal of Plant Science* 55: 171–183.

Paolini, R., M. Principi, R.J. Froud-Williams, S. Del Puglia and E. Biancardi. 1999. Competition between sugarbeet and *Sinapis arvensis* and *Chenopodium album*, as affected by timing of nitrogen fertilization. *Weed Research* 39: 425–440.

Warwick, S.I., H.J. Beckie, A.G. Thomas and T. McDonald. 2000. The biology of Canadian weeds. 8. *Sinapis arvensis* L. (updated). *Canadian Journal of Plant Science* 80: 939–961.

Wild radish

Raphanus raphanistrum L.

Wild radish seedling
Antonio DiTommaso, Cornell University

Wild radish young plant
Scott Morris, Cornell University

Wild radish flowering plant
Joseph DiTomaso, University of California, Davis

Wild radish flower, yellow phase
Scott Morris, Cornell University

Wild radish mature seed pods
Antonio DiTommaso, Cornell University

IDENTIFICATION

Other common names: charlock, jointed radish, wild turnip, jointed charlock, white charlock, wild kale, cadlock, runch, jointed wild radish

Family: Mustard family, Brassicaceae

Habit: Winter or summer annual herb bolting from a rosette

Description: Seedling stems below cotyledons are purple and have stiff hairs. Cotyledon blades are hairless, kidney shaped to heart shaped with the indentation at the tip, 0.4–0.8 inch long and slightly wider than long, and prominently veined. Cotyledon stalks are 0.4–1 inch long and tapered. First leaves are oval to oblong, alternate, long stalked and strongly veined with wavy, lobed and irregularly toothed edges. The largest lobe occurs at the leaf tip, with two to four smaller lobes near the base. Stiff hairs are scattered on both leaf surfaces and on leaf edges. Early leaves form a basal rosette. **Mature plants** bolt from the rosette, producing a 1–4 foot tall, branched stem. Stems are densely hairy at the base but more sparsely hairy toward the top. Lower leaves are 2–8 inches long by 2 inches wide, oval to oblong and long stalked. Leaf edges are irregularly toothed and lobed like the young leaves. Upper leaves are stalkless or short stalked, lanceolate and smaller (less than 3 inches long). Edges are entire to toothed, with zero to five lobes at base. All leaves have coarse, stiff hairs scattered on edges and on both surfaces. The sturdy taproot has a strong radish scent. **Flowers** occur in branched clusters at the ends of the stem and branches. Individual flowers are 0.4–0.5 inch wide with four petals and are pale yellow to cream colored or white. Petals usually have distinctive purple veins. (Note that purple-pink flowers have been observed and are speculated to be the result of crossing with cultivated radish.) Flower stalks are upright and 0.25–1 inch long. Seedpods form below the flowers and are initially cylindrical, green and fleshy; they are 0.8–3 inches long with a 0.4–2 inch-long beak and contain two to 10 seeds. As seedpods ripen, they become brown, corky and ridged lengthwise, and constricted joints develop between the seeds. Seedpods break into segments at the joints when **seeds** are mature; individual segments do not open. Seeds are grooved, red-brown, 0.1–0.25 inch long by 0.06–0.1 inch wide and kidney shaped.

Similar species: Cultivated radish (*Raphanus sativus* L.) is very similar to wild radish but has pink to purple flowers and unjointed seedpods with only two to three seeds. Wild mustard (*Sinapis arvensis* L.) and other mustard species do not have purple-veined petals and have seedpods that split when mature rather than break into segments. Seeds are almost perfect spheres compared to wild radish seeds. Wild radish leaves are hairier, rougher and more lobed than wild mustard, and the latter does not form a basal rosette. Leaves of yellow rocket (*Barbarea vulgaris* W. T. Aiton) are glossy, dark green and are not hairy like those of wild radish.

MANAGEMENT

If wild radish is a problem, lime to pH 6.8 and avoid applying excess N. Time applications of soluble N sources to correspond to periods of high crop uptake. Because wild radish has a persistent seed bank, the suggestions below for reducing seed density will likely require a few years to have a major impact.

In grain crops, use a tine weeder with stiff tines to break or bury as many wild radish plants as possible since the taproot and deep emergence makes seedlings resistant to uprooting. Since the pods often do not shatter until wheat or canola harvest, consider setting the combine to collect wild radish seeds with the grain, and then clean it out later. This will greatly reduce seed return, and if wild radish is a problem, the grain will likely need secondary cleaning anyway. In warm climates where wild radish acts as a winter annual, till very shallowly (1–2 inches) to stimulate germination and then till again to eliminate seedlings prior to planting grain crops. In corn or soybeans, use a guidance system or belly mounted cultivator to get as close as possible to the row with shallow-pitched sweeps running close to the soil surface. This will cut off the young wild radish with minimum damage to crop roots. Hill up corn and soybeans before wild radish seedlings get too large for complete burial.

In vegetable crops, use side-knives to get as close to the row as possible. In-row weeding machines or hand hoeing will be necessary to obtain good control. Wild radish can pose a severe problem if soil is kept acidic for potato production. Mound about 2 inches of soil over the rows after planting potatoes; then tine weed aggressively when wild radish seedlings appear, and, if necessary, repeat until the potato vines emerge. The aggressive tine weeding will flatten out the field. Hill potatoes in multiple operations to bury successive flushes of wild radish while they are still short. Rather than mowing vines before harvest, use a forage harvester to blow the chopped plants and seedpods into a wagon for disposal.

Where wild radish acts as a summer annual, use a short tilled fallow in spring to reduce seed density before planting a late spring or summer crop. Wild radish may not be very competitive in a summer crop, but this will reduce density in the next spring planted crop. Where wild radish acts as a winter annual, use a late winter to very early spring tilled fallow to flush seeds out of the soil before spring planted crops. This will not help the spring crop because it will have few wild radish anyway, but the fallow will reduce wild radish density in later cool season crops.

If an exceptionally severe infestation of wild radish cannot be prevented from dropping seed, let the seeds weather on the soil surface until the next crop. Then moldboard plow deeply (10 inches). In subsequent years, use direct drilling or shallower tillage (less than 6 inches) to avoid bringing remaining seeds to the surface.

Inspect grain seed for wild radish pod segments before sowing. Avoid uncertified rye seed sold for use as a cover crop or forage.

ECOLOGY

Origin and distribution: Wild radish is native to the Mediterranean regions of Europe, North Africa and the Middle East. It has been introduced widely in Asia, Australia, Latin America and South Africa, and it is a serious weed in grain growing parts of those regions. It is generally rare or absent in the humid tropics. It occurs widely in North America but is rare or absent in much of the center of the continent. As a weed, wild radish causes the greatest problems in the Canadian Maritime provinces, the eastern seaboard and the Pacific coast.

Seed weight: Population mean seed weights for field grown plants range from 5.3–8.6 mg. Seed weight within the same fruit can vary from 1.5–12 mg, and seeds are heaviest in pods with few seeds. Larger seeds had the highest germination, growth and reproductive output.

Dormancy and germination: Freshly produced wild radish seeds are usually dormant, although seeds from plants that emerge in fall are more dormant than those that emerge in spring. Dormancy is caused both by a germination inhibitor in the seed coat and by physical restriction from the woody pod segment that usually remains attached to the seed. After-ripening for six months or burial in the soil over winter generally breaks dormancy. Exposure of seeds to high moisture and fluctuating temperatures appears to facilitate germination by breaking down the pod and seed coat. In some warm climate populations where the species behaves as a winter annual, exposure of seeds on the soil surface during the summer breaks dormancy, and dormancy is lowest in fall but progressively increases from winter to summer months. In this case, cold stratification increases dormancy. Germination in fall is higher from buried seeds than from seeds on the soil surface. Optimum temperature for wild radish germination is 39–68°F and improves with alternating temperatures within this range. Light has minimal effect on germination and can be suppressive at low temperatures.

Seed longevity: Some wild radish seeds remain viable in soil for up to 15–20 years. In a five-year experiment in which the top 3 inches of soil were seeded and then stirred twice each year, the number of seeds declined by 29% per year. This was similar to the 33% per year decline in the seed bank during the first two years under a grass sod. In Australia, the number of viable seeds declined by 32% in one year when buried at 4 inches.

Season of emergence: In northern areas the species emerges primarily in the spring. In the southern parts of the United States, seedlings emerge all year, with peak emergence in fall and winter.

Emergence depth: Most seedlings emerge from the top 1.2 inch of soil, but a few can emerge from as deep as 2.8 inches. Twice as many seedlings emerge from 0.4 inch as from the soil surface. When seeds were sown and tilled into the soil, average depth of seedling emergence was 1.1 inch with a range from 0.3–1.5 inch.

Photosynthetic pathway: C_3

Sensitivity to frost: Seedlings are killed by sub-freezing temperatures, but young rosettes are frost hardy, and the species commonly acts as a winter annual in the southern United States. Frost kills mature plants.

Drought tolerance: The rarity of wild radish in the interior of North America is attributed to a lack of drought tolerance. In California, a moisture deficit in the spring caused more adverse effects than a moisture deficit in the fall.

Mycorrhiza: Wild radish is a poor mycorrhizal host, possibly because of anti-fungal compounds excreted from roots.

Response to fertility: Wild radish is often associated with nitrogen rich soils. In high N soils, wild radish will take up more N than it needs for growth and stores it as nitrate. If N is less available later in the life cycle, plants will use the previously stored nitrate to make proteins. In one chamber experiment, Nitrogen fertilizer increased vegetative growth but not reproductive growth. Liming soil to increase the pH from 6 to 6.8 was a major factor contributing to decreased abundance of wild radish in an 11-year study.

Soil physical requirements: Wild radish can occur on all soil types, including sand, clay, sandy loam and chalky or saline soil, but it appears to do best on acidic sandy soils.

Response to shade: Wild radish grows best in high light. Late emerging plants bypass the rosette stage and bolt early. Wild radish height and the ratio of leaf area to leaf weight are greater under a wheat canopy than in the open, but overall biomass and potential seed production is reduced, especially for later emerging cohorts.

Sensitivity to disturbance: Cutting and grazing have minimal effect on wild radish seed production because of its ability to rapidly produce new flowering stems. Plants have a highly branched stem structure that favors recovery from cutting. Also, multiple stems can form from buds at the base of rosette leaves when resources are adequate.

Time from emergence to reproduction: In Ontario, spring emerging wild radish begins flowering in three to six weeks, but seeds do not mature until August. Plants emerging in Wisconsin took seven weeks to mature. Plants mature most rapidly in warm weather, and plants emerging in June in South Carolina required 44 days to mature whereas plants emerging in November required 231 days. In Western Australia, where the climate is similar to that of California, plants emerging in the fall took 90 days to flower and successively later emerging plants required shorter periods down to 49 days for plants emerging in the spring. Under rapid developmental conditions in Australia, viable seeds can form within pods three weeks after flowering. The flowering period of wild radish in Quebec sufficiently overlaps that of canola for gene flow to potentially occur from herbicide-resistant canola to wild radish.

Pollination: Wild radish cannot self-pollinate. It is cross pollinated primarily by bees, butterflies and syrphid flies.

Reproduction: Plants produce up to 10,000 seeds per plant, depending on region, competition and the time of emergence. Plants in spring wheat in Quebec produced 50–150 seeds per plant, with early emerging individuals producing the most seeds. In northern areas, plants continue to flower and produce seeds until killed by frost. In South Carolina, fertilized, well-watered plants grown with minimal competition produced 8,000–10,000 seeds when emerging in April, October or November, but less than 2,000 when emerging in June or July. Plants grown with wheat in Australia produced up to 1,000 seeds per plant.

Dispersal: Seeds naturally disperse within 1–3 feet of the parent plant. Seeds may move greater distances in soil clinging to tires and machinery or in harvesting equipment. Pod segments, each containing a single seed (sometimes two), are similar in size to cereal grain and are sometimes sown with seed grain. Plants are attractive to grazing mammals, and the seeds pass through to disperse in manure.

Common natural enemies: Wild radish is eaten by grazing mammals and rodents.

Palatability: Some people in the Mediterranean region and Pakistan eat the leaves in salads or cooked. Livestock find young plants palatable, but eating large quantities of wild radish (more than 25% of the diet) can cause sickness. The seeds are especially toxic.

Note: Green manure that included ground-up plants of wild radish suppressed emergence and growth of several crops and weeds in South Carolina.

FURTHER READING

Cheam, A.H. 1986. Seed production and seed dormancy in wild radish (*Raphanus raphanistrum* L.) and some possibilities for improving control. *Weed Research* 26: 405–413.

Cheam, A.H. and G.R. Code. 1995. The biology of Australian weeds. 24. *Raphanus raphanistrum* L. *Plant Protection Quarterly* 10: 2–13.

Del Monte, J.P., J. Dorado and C. Lopez-Fando. 1999. Weed seed bank response to crop rotation and tillage in semiarid agroecosystems. *Weed Science* 47: 67–73.

Warwick, S.I. and A. Francis. 2005. The biology of Canadian weeds. 132. *Raphanus raphanistrum* L. *Canadian Journal of Plant Science* 85: 709–733.

Yellow woodsorrel

Oxalis stricta L.

Yellow woodsorrel seedling
Antonio DiTommaso, Cornell University

Yellow woodsorrel flowering plant
Antonio DiTommaso, Cornell University

Yellow woodsorrel seed pods
Randall Prostak, University of Massachusetts

IDENTIFICATION

Other common names: common yellow woodsorrel, common yellow oxalis, lady's sorrel, lemon clover, sheep's clover, sheep sorrel, sheep sour, sourgrass, tall wood sorrel, toad sorrel, upright wood sorrel

Family: Woodsorrel family, Oxalidaceae

Habit: Short, much branched perennial herb, commonly behaving as an annual in agricultural fields

Description: Cotyledons of the seedling are round to oblong, green to pinkish, hairless and 0.12–0.24 inch long by up to 0.2 inch wide. True leaves are alternate, green to occasionally purplish and are divided into three heart-shaped leaflets. Leaf edges are smooth. The leaflets are smooth on the upper surfaces and have short, scattered hairs on the lower surfaces and a fringe of hairs along the edges. **Mature plants** are 2–15 inches tall and unbranched or branching at the base. Stems are green to purplish and are covered in upward-facing, flattened hairs. Leaf shape and hairiness are similar to the seedling. Mature leaves are 0.5–1.1 inch wide, and individual leaflets are 0.25–0.5 inch wide and long. Leaf stalks are up to 2.5 inches long. The root system is fibrous, but the plant also produces shallow, spreading, white to pinkish rhizomes. **Flowers** grow in clusters of two to six from the leaf axils on stalks, which reach up to 1 inch long. The flowers are yellow, have five notched or rounded petals and five pale green sepals, and are 0.28–0.43 inch wide. Each flower is replaced by an upright, five-sided, pointed, cylindrical **seedpod**. The seedpods are hairy and 0.4–0.6 inch long. When ripe, the seedpods split and eject the **seeds** up to 6.5 feet from the plant. The seeds are flattened, red to brown, ridged transversely and 0.04–0.06 inch long.

Similar species: Creeping woodsorrel (*Oxalis corniculata* L.), a frequent weed in greenhouse and nursery culture, has a more prostrate, spreading habit than yellow woodsorrel and has aboveground runners rather than underground rhizomes. Slender yellow woodsorrel (*Oxalis dillenii* Jacq.) is very similar to yellow woodsorrel but is generally smaller and has a taproot rather than rhizomes. Clovers (*Trifolium* spp.), black medic (*Medicago lupulina* L.) and other trifoliate legumes have similar leaves to yellow woodsorrel. Legume leaflets are not heart shaped, however, and the leaves often have a pair of small bracts at the base of the stalk.

MANAGEMENT

Yellow woodsorrel is relatively non-competitive, but it is sufficiently prolific to make itself a problem in vegetable crops. Even when its density is too low to decrease yield, its rapid, upright growth causes harvest problems in herbs and leafy greens. In addition, it may serve as an alternative host to several diseases (*Puccinia* and *Fusarium*) of field and vegetable crops, including sweet corn and onions. Consequently, rotate these crops to other fields or beds until you have this species under control. This weed is most common in untilled crop fields, so tillage is an effective means of control. For dense infestations, flush the seeds out of the soil with repeated shallow cultivations before planting. Then plant competitive crops like snap beans or short season cabbage that can be repeatedly cultivated shallowly close to the row. Alternatively, grow an early crop like radish or head lettuce and use a tilled fallow during part of the summer. Avoid crops with long, post cultivation periods, as these will allow late emerging yellow woodsorrel to go to seed. Since this weed grows very fast, if you use a summer cover crop, plant it at high density to ensure good suppression. Hay or straw mulch and synthetic barrier mulches effectively suppress this weed. Hand weeding should be done before seed capsules form to prevent dispersal of seeds during the weeding process.

ECOLOGY

Origin and distribution: Yellow woodsorrel is native to eastern North America and probably also eastern Asia. It has been introduced into western North America, Europe, Africa and New Zealand. It now occurs throughout most of the United States and southern Canada, except in the warmer parts of the Pacific coast and Intermountain West, where it is absent or occurs only sporadically.

Seed weight: Population mean seed weight ranges from 0.13–0.15 mg.

Dormancy and germination: Freshly produced seeds of yellow woodsorrel are not dormant and will germinate immediately if sown on warm, moist soil. Seeds germinate at 48–85°F, with optimum temperatures of 60–80°F. Exposure of moist seeds to a high temperature of 97°F will inhibit germination. Seeds require exposure to light after they have taken up water, with only a brief exposure to a low level of light being sufficient. Thus, seeds that get incorporated into the soil will normally wait to germinate until they are exposed to a pulse of light during tillage.

Seed longevity: In undisturbed conditions, seeds persist in the soil for at least five years and probably much longer. Seed viability was 83% after one year. Since the seeds germinate readily in recently disturbed soil, however, they are probably flushed out of the soil relatively quickly in regularly tilled and cultivated fields.

Season of emergence: In temperate climates, yellow woodsorrel emerges from mid-spring through summer. This species can emerge throughout the year in climates with warmer winters such as California.

Emergence depth: This has not been reported, but given the small seed weight, most seedlings probably arise primarily from the top 0.5 inch or less of soil. Sprouts from rhizomes could emerge from deeper in soil, however, unless buried, the rhizomes typically lie just below the soil surface.

Photosynthetic pathway: C_3

Sensitivity to frost: Yellow woodsorrel tolerates light frost but dies back to the ground following hard frost.

Drought tolerance: The species tolerates dry spells of several weeks. Leaflets fold along a center crease in response to stress.

Response to fertility: Yellow woodsorrel tolerates low fertility but is most prolific in highly fertile soils. Plants respond to fertilizer application by producing a flush of new leaves and flowers.

Mycorrhiza: Yellow woodsorrel is mycorrhizal.

Soil physical requirements: Yellow woodsorrel grows on a wide range of soils but thrives in loamy soil. It is an indicator of moist, fertile soils but can tolerate drought-prone sites.

Response to shade: Yellow woodsorrel cannot grow in dense shade, but it tolerates the partial shade cast by many crops.

Sensitivity to disturbance: Plants can be easily uprooted by hand-weeding soon after emergence but can resprout from rhizomes after weeding or cultivation once they are established. The rhizomes lie just below the soil surface and are easily damaged, so resprouting is usually not a major problem. Plants will assume a prostrate growth habit in response to mowing.

Time from emergence to reproduction: Spring emerging plants flower four to six weeks after emergence and set seeds two to four weeks later. Plants emerging in midsummer can set seeds in as few as five weeks.

Pollination: Yellow woodsorrel is often self pollinated but is also cross pollinated by insects.

Reproduction: Yellow woodsorrel reproduces either by seeds and/or by perennating buds on rhizomes. Plants emerging from seeds in the spring and left undisturbed with minimal competition produced an average of 900 capsules each with an average of 23 seeds per capsule, thereby producing approximately 21,000 seeds per plant. Elsewhere, plants have been reported to produce 570–5,000 seeds per plant. Since newly produced seeds lack dormancy, the species can produce two complete generations per year in the northern United States and more in warmer climates. Vegetative reproduction by sprouts from rhizomes is rare in tilled fields but is common in less disturbed habitats.

Dispersal: The mature capsules rupture explosively, scattering seeds up to 13 feet. Seeds pass alive through ruminant digestive tracts and are spread with manure. Seeds may also be transported by rodents. They probably also move with soil on shoes, tires, machinery and by floating in waterways.

Common natural enemies: None of any consequence.

Palatability: Leaves or young plants of yellow woodsorrel are sometimes added to salads or cooked dishes to add a sharp, sour taste. The presence of the toxin oxalic acid, which accumulates in the aerial parts of the plant, gives the shoots their sour taste. Leaves have similar vitamin C content as that found in spinach and oranges.

FURTHER READING

Halverson, W.L. and P. Guertin. 2003. Factsheet for: *Oxalis stricta* L. *USGS Weeds in the West project.* 29 pp.

Lovett Doust, L., A. MacKinnon and J. Lovett Doust. 1985. Biology of Canadian weeds. 71. *Oxalis stricta* L., *O. corniculata* L., *O. dillenii* Jacq. ssp. *dillenii* and *O. dillenii* Jacq. ssp. *filipes* (Small) Eiten. *Canadian Journal of Plant Science* 65: 691–709.

Marshall, G. 1987. A review of the biology and control of selected weed species in the genus *Oxalis*: *O. stricta* L., *O. latifolia* H.B.K. and *O. pes-caprae* L. *Crop Protection* 6:355–364.

Glossary

After-ripening: The transition of a seed from a dormant to a non-dormant state, which takes place after the seed is physiologically independent of the parent plant.

Allelopathy: The poisoning of one plant by another.

Alternate: An arrangement in which leaves, branches or flowers are not paired along the stem.

Annual: A plant that completes its lifespan in less than one year.

Auricle: An extension of a grass leaf on the **collar** that wraps partially around the stem (Figure G.1).

Axil: The upper angle of the point of attachment between a stem and a leaf or branch.

Awn: A thin extension on the chaff of a grass flower that typically sticks out beyond the **spikelet** (Figure G.1).

Biennial: A stationary **perennial** species that usually sets seeds in the second year of life. The term is a misnomer, since most biennials flower when reaching some minimum size, and this may require several years or, in highly favorable circumstances, only a single year.

Biomass: The total weight of tissue of an individual organism, **population** or **community**.

Biotype: A genetically distinct form of a species that persists due to vegetative reproduction or a high level of self-pollination.

Blind cultivation: **Cultivation** of the whole field including the crop row; for example, with a tine weeder or rotary hoe before crop emergence.

Bolt: A change in primary growth habit from a **rosette** of leaves to an upright, reproductive stem that is induced by environmental or developmental cues.

Bunch grass: A grass that grows in discrete clumps or bunches rather than spreading to form a sod.

Bur: A fruit with spines or hooked bristles.

Capsule: A seedpod; capsules often have more than one chamber and usually split open to release the seeds.

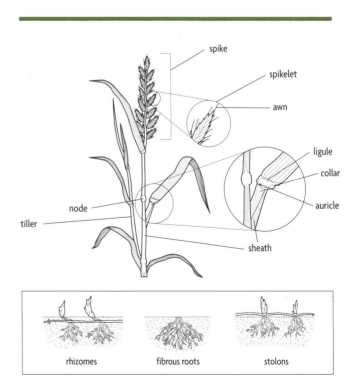

Figure G.1. Anatomy of a grass. Illustration by Vic Kulihin.

Cole crops: Crops in the mustard family (Brassicaceae), such as cabbage and broccoli.

Collar: The point on a grass leaf where the **sheath** transitions to the blade (Figure G.1).

Community: All of the species present in a field. The term is often applied to components of the whole community, for example, the *weed community* or the *insect community*.

Cotyledon: The first leaf-like, aboveground structure present on a newly emerging seedling. Derived from seed tissue, cotyledons usually push up through the soil. They are paired in broadleaf plants and singular in grasses and sedges. They are sometimes called seed leaves.

Creeping perennial: A **perennial** species that spreads by runners, **rhizomes** or horizontal roots that sprout to form new shoots.

Cultivation: Most broadly, any systematic soil disturbance for agricultural purposes, but as used in this book, it refers to mechanized soil disturbance for weed management after crop planting.

Dough stage: The stage in the development of a seed in which the **endosperm** is still soft but is no longer liquid.

Dust mulch: A surface layer of loose soil that dries quickly and thereby prevents emergence of weed seedlings.

Endosperm: The food storage part of a seed.

Entire: Describing a leaf with smooth, uninterrupted, toothless edges.

Fibrous root system: A root system composed of thin, branching roots with no **taproot** (Figure G.1).

Flame weeder: A handheld or tractor mounted burner for killing weeds (Figure 4.16).

Flower head: The "flower" of a plant in the aster family. It is actually made up of many tiny flowers that are tightly clustered together.

Folded in the bud: An arrangement of unopened grass leaves in which the leaves are folded at the **midrib** (Figure G.2).

Growth rate: The rate of addition of tissue to a plant, expressed in units of weight gain per day.

Guess row: The space between two planter passes.

Indeterminate: The growth habit in which new vegetative tissues, flowers and seeds are produced until the plant is killed, usually by frost.

Inflorescence: The flowering part of a plant.

Inoculant for legumes: A mixture of nitrogen-fixing bacteria and a carrier substance (often ground peat) used for coating legume seeds to ensure a high level of **nitrogen fixation**. Various legume species require different types of inoculant.

In the boot: An expression indicating that the **inflorescence** of a grass has formed but has not yet emerged from the surrounding leaf **sheaths**.

Lay-by cultivation: The last **cultivation** before the crop becomes too large for further operations.

Lanceolate: Shaped like the tip of a lance; longer than wide and tapered at each end.

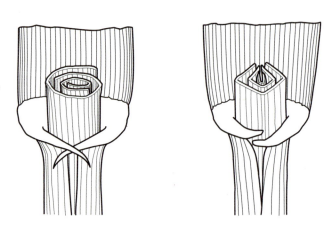

Figure G.2. Left, leaves rolled in the bud; right, leaves folded in the bud. Illustration by Vic Kulihin.

Ligule: An extension of the leaf **sheath** beyond the **collar** of a grass leaf. (Figure G.1)

Mechanical weed management: Tillage and **cultivation** for weed control.

Midrib: The central vein of a leaf.

Milk stage: The stage in the development of a seed in which the **endosperm** is still liquid.

Nitrogen fixation: The conversion of atmospheric nitrogen (N_2 gas) to ammonia by certain species of bacteria. In agricultural systems, most nitrogen fixation occurs in special nodules on the roots of legumes.

Node: A point on a stem at which a leaf or branch arises; or in grasses, a point on a stem at which the leaf blade joins with the **sheath** (Figure G.1).

Ocrea (*pl.* ocreae): A thin membrane, usually papery and translucent, originating at the base of leaves and wrapping around the stem that is present on members of the buckwheat family (Polygonaceae).

Outcrossing: In plants, genetic mixing due to cross-pollination among individuals within a population.

Panicle: An **inflorescence** with multiple, flower-bearing branches that originate from a central stem.

Pappus: A set of hairs, bristles or scales attached to the seeds of many members of the aster family (Asteraceae). These often aid dispersal by catching wind currents or sticking to fur and clothing.

Perennial: A plant that lives for more than one year.

Photosynthetic pathway: The physiological mechanism of carbon dioxide assimilation. C_3 plants fix carbon dioxide into a three-carbon sugar whereas C_4 plants fix carbon dioxide into a four-carbon sugar. The ecology of C_3 plants varies greatly, but they often do well in cool weather and may tolerate shade. C_4 plants usually grow best in hot weather, are intolerant of shade and use water efficiently.

Physical weed management: Weed control methods that attack the body of the weed. These include tillage and **cultivation** but also include methods that attack the weeds without soil disturbance, for example, with flame, electricity or mower blades. Some authors consider mulches, solarization and tarping to be physical weed management, but in this book, we consider them with cultural management methods.

Pinnate (leaf): A compound leaf in which the leaflets are arranged in opposite pairs along the stem.

Pistil: The fruit-producing, female reproductive organ of a flower. Located in the center of the flower.

Population: All of the individuals of a species living in a defined area, such as a field, including the seeds.

Prostrate: Growing along the ground, with shoots spreading horizontally.

PTO: The abbreviation for power take off, a drive shaft on a tractor used to power machinery.

Relative growth rate: The rate of growth relative to the size of the plant, expressed in units of weight gain per plant weight per day. The term is usually applied to seedlings. For example, if a seedling weighed 10 milligrams and gained 5 milligrams per day, then the relative growth rate would be 5/10 = 0.5 milligrams per day per milligram of plant weight.

Rhizome: An underground stem. The term applies to both horizontal, spreading stems of some plants like quackgrass and purple nutsedge and also to the vertical underground stems that sprout up to the soil surface from deep roots, as in Canada thistle (Figure G.1).

Rolled in the bud: An arrangement of unopened grass leaves in which the leaves wrap around each other in arcs (Figure G.2).

Rosette: A circular cluster of leaves that arise from at or near ground level, as in dandelion.

Scarification: Abrasion of the seed coat.

Seed bank: The seeds in the soil but especially those that persist for one or more years.

Seed head: The mature reproductive structure that develops from the composite "flower" of a plant in the aster family (see "flower head").

Sepals: The outermost parts of a flower, typically green, which enclose the petals in the bud. In some flowers, like those of mints, the sepals may be fused into a tubelike structure.

Sheath: The lower part of a grass leaf that encloses the stem (Figure G.1).

Spike: A type of **inflorescence** in which flowers or **spikelets** are attached without stalks to an unbranched central stem (Figure G.1).

Spikelet: A small, tight cluster of grass flowers that share a common attachment to a stem or branch of the **inflorescence** (Figure G.1).

Stale seedbed: A weed management procedure in which a seedbed is prepared, weeds are allowed to emerge and then they are killed without soil disturbance. The crop is then planted and emerges with relatively few weed seedlings.

Stamens: The pollen-producing, male reproductive organs of a flower.

Stationary perennial: A **perennial** species that normally spreads by seeds or bulbs.

Stirrup hoe: A hoe made of a narrow band of steel bent in the shape of a stirrup and attached to a handle or cultivator. It is used for shallow hoeing of small weeds (Figure 4.18).

String trimmer: A motorized device with a whirling string that cuts weeds or grass.

Summer annual: A plant that germinates and completes its entire lifecycle during the warm seasons of the year.

Stratification: A period of wet, cold, but not freezing soil conditions that breaks seed dormancy of some species.

Taproot: A thick, typically straight, vertically growing root from which smaller roots grow laterally.

Tilled fallow: A weed management procedure in which the soil is tilled, weeds are allowed to emerge and then they are killed with subsequent tillage. The weed removal tillage is usually shallow. The sequence may occur only once before planting or be repeated several times.

Tiller: A secondary stem arising from the base of a grass plant (Figure G.1).

Tilth: A desirable quality of the soil in which it is porous and easily broken into small stable crumbs.

Tuber: A distinct, thickened portion of a **rhizome** that stores nutrients and has buds from which new shoots can arise.

Vernalization: The promotion of flowering by exposure to a period of cold temperature.

White thread: The stage in weed growth after the seed germinates and before the seedling emerges.

Whorl: An arrangement of leaves, flowers or other plant parts in which they attach to the stem in a circle or short spiral.

Winter annual: An annual plant that emerges in late summer or fall, lives through the winter and sets seeds the following spring or summer.

Index

organic production, 59–60

corn chamomile, 98, 235–38, 262

corn gluten meal as herbicide, 61–62

costs: in compost application, 29; in disposal of synthetic mulch, 58; of electrical discharge weeders, 99; in energy use, 107; in hand weeding, 102, 124; in high planting density, 45, 46; of larger seeds, 47; of natural product herbicides, 61, 62, 70, 121; in preventive weed management, 63f, 64; in solarization, 61; in stale seedbed, 82; in standard weed management, 63f, 64; in transition to organic production, 58–59, 70; in yield loss from weeds, 63, 63f, 64

cotton, organic, 122–24

cotyledon, 398

cover crops, 49–53, 70; allelopathic compounds in, 31, 51; on Beech Grove Farm, 117, 118, 119; on Martens farm, 111, 112, 113; and mycorrhizal fungi, 29; in onion seedling production, 55; organic matter in, 105–6; on Park Farming Organics, 120; on Pepper farm, 122, 123, 124; planting density, 45–46; in preventive weed management, 64; and row orientation, 47; in tilled fallow, 81–82; in transition to organic production, 58–59; weed seeds in, 36, 66

crabgrass, 21, 38t, 134t–35t, 174–76; large, 21, 32, 38t, 134t–35t, 174–76; in shade, 32, 176; similar species, 160, 163, 174, 208; smooth, 174; southern, 174

creeping bentgrass, 149

creeping perennials, 14, 15t, 127, 398; in field margins, 66; origins, 15, 37; palatability, 37–38; pollination, 34; tillage of, 74, 75t, 76–77; timing of reproduction, 33; vegetative propagation, 17–18, 19f

creeping woodsorrel, 394

crickets, 65

crop density, 45f, 45–46, 70; for hoeing, 103; and plant loss from implements, 83–84

crop diversity, 111, 112–13, 120

crop rotation, 42–44, 70; on Beech Grove Farm, 117f, 117–18, 118f, 119; cover crops in, 50; on Martens farm, 112; on Mugge farm, 115; on Park Farming Organics, 120; in ridge tillage system, 80–81; and seasonal germination, 23–24, 42–43

crops: competitiveness of, 43, 44–49, 70; corn (See corn); field cleanup after harvest, 64; grain (See grain crops); inadvertent loss from cultivation,

83–84, 93, 96, 108, 123; and intercropping, 48–49; leaf canopy (See leaf canopy); length of growing period, 43; organic (See organic production); photosynthesis pathway, 27; planting date, 48, 49; row (See row crops); seed quality for, 44; seed size, 44, 47–48; sequence of mechanical weed management for, 73–74; size difference of weeds and, 40, 47, 83; skip areas in, 44, 48, 49; transplanted (See transplanted crops); uniformity of, 44–45; variety selection, 47; vegetable (See vegetable crops); vigor of, 44–45, 114

crop yield: and compost application, 29; planting density affecting, 45, 45f; in ridge tillage system, 80; seed size affecting, 47; weeds affecting, 63, 63f, 64

cultivation, 42, 73, 82–96, 398; on Beech Grove Farm, 118, 119; blind, 17, 82, 85, 397; depth of, 84–85; dispersal of weed seeds in, 68–69; dust mulch in, 84; energy use in, 106t; inadvertent plant loss in, 83–84, 93, 96, 108, 123; lay-by, 398; on Martens farm, 113; on Mugge farm, 115–16; at night, 104; on Park Farming Organics, 121; on Pepper farm, 122, 123–24; in preventive weed management, 63–64; principles of, 73–74, 82–85, 108; in ridge tillage system, 79–80; of row crops, 17, 43, 46; size of weeds in, 73; of small seeded weeds, 20; and soil tilth, 84, 104–5; timing of, 73, 83, 84, 85, 94, 95, 96; weather affecting, 84, 94; in wet soil, 105

cultivator guidance systems, 85, 99–101, 109; on Martens farm, 113; on Mugge farm, 116; on Park Farming Organics, 121; on Pepper farm, 122, 123–24

cultivators, 85–96; adjustment of, 85, 108; energy use with, 106t; field, 74, 74t; in-row, 82; inter-row, 82, 83; parallel gang, 85–86; rigid, 85; size of, and planter size, 82–83; transport of weed seeds on, 68–69; types of tractor mounting, 85

cultural weed management, 42–72; cover crops in, 49–53; crop competitiveness in, 44–49; crop rotation in, 42–44; livestock in, 62–63; "many little hammers" approach to, 42; mulch in, 53–58; natural product herbicides in, 61–62; preventive, 63–70; solarization in, 60–61; synergistic tactics in, 42; in transition to organic production, 58–60

curly dock, 14, 15t, 22, 66, 280–84

Cynodon dactylon (L.) Pers. (bermudagrass), 27, 38t, 138t–39t, 149–51

Cyperus esculentus L. *See* yellow nutsedge

endosperm, 21, 398, 399

energy use, 106t, 106–7, 107t, 109

entire type of leaf edge, 398

equipment and machinery, 85–104; abrasion weeders, 99; on Beech Grove Farm, 118; cleaning of, 69, 122; in cultivation, 73, 82–96; cultivator guidance systems, 99–101; electrical discharge weeders, 98–99; energy use of, 106–7; in flame weeding (See flame weeding); hoes, 102–4, 109, 400; inadvertent crop loss from, 83–84, 93, 108, 123; on Martens farm, 113–14; matched to task, 73, 101–2; in mechanical weed management, 86t–87t, 86–96; mowers, 96–97; on Mugge farm, 115–16; on Park Farming Organics, 121; on Pepper farm, 122, 123–24; and preventive weed management, 64–65; short-handled tools, 103, 103f; soil compaction from, 75, 105; steam and hot water weeders, 98; in tillage, 74, 74t; weed dispersal from, 68–69, 122; weed pullers, 82, 87t, 99, 99f

Erigeron annuus (L.) Pers. (annual fleabane), 305

Erigeron canadensis L. (horseweed), 132t–33t, 305–8

Erigeron strigosus Muhl. ex Willd. (rough fleabane), 305–6

ethanol in soil, 22

Euphorbia spp. (spurges), 259

Fagopyrum esculentum Moench (domestic buckwheat), 383

Fagopyrum tartaricum (L.) Gaertn. (tartary buckwheat), 383

Fallopia convolvulus (L.) Á. Löve (wild buckwheat), 39t, 134t–35t, 222, 319, 383–85

fallow periods, 117, 118, 119; in tilled fallow, 64, 81–82, 108, 400

fall panicum, 38t, 134t–35t, 146, 155–57, 160, 170, 204, 208

false cleavers, 231–34

fanpettles, Virginia, 354

farm profiles, 111–24; Beech Grove Farm, 116–19; Martens farm, 111–14; Mugge farm, 114–16; Park Farming Organics, 119–22; Pepper farm, 122–24; Walker farm, 55

fertility of soil, 128; and crop tolerance of weed competition, 44–45; and growth rate, 29–30, 30f, 31f, 40; and mineral content of plants, 29, 49; from organic materials, 49; response of crops and weeds to, 29, 40, 49, 70

fescue, tall, 182

fibrous root system, 398

field bindweed, 14, 15t, 38t, 138t–39t, 221–25; depth of storage organs, 74, 75, 224; drought tolerance, 28, 224; at Park Farming Organics, 120; similar species, 222, 319, 383; timing of reproduction, 33, 225; vegetative propagation, 17, 221–22

field brome, 152

field cultivators, 74, 74t

field edges, 66, 69

field pennycress, 27, 38t, 43, 132t–33t, 285–87, 362

field pepperweed, 362

field sandbur, 134t–35t, 186–90

finger millet, 165

fire risk in flame weeding, 98

fixed perennials. *See* stationary perennials

flame weeding, 43, 82, 398; energy use in, 106t, 107; equipment in, 87t, 97f, 97–98; fire risk in, 98; on Mugge farm, 116; on Park Farming Organics, 121; in stale seedbed, 82, 97

fleabane, 305–6

flixweed, 38t, 136t–37t, 288–90

flood water dispersal of weed seeds, 69

flower head, 398

flowering: time from emergence to, 131, 133t–39t; and time to viable seed, 131, 133t–37t

folded in the bud arrangement, 398

foot traffic, soil compaction from, 105

forage, 111, 112; weeds as, 29, 37–39, 62–63; weed seeds in, 66, 67–68; in weed suppression, 43

forage kochia, 313, 316

Four Winds Farm, 55–56

foxtails, 23, 27, 36, 38t, 158–62; giant (See giant foxtail); green, 38t, 134t–35t, 158–62; in ridge till system, 81; seed size, 21, 21f, 160–61; in shade, 33, 160, 161–62; similar species, 146, 155, 160, 174, 188; in tilled fallow, 82; yellow, 38t, 134t–35t, 146, 155, 158–62, 174

frost tolerance, 27–28, 131, 133t–38t

fungi, 36–37, 51; mycorrhizal, 28–29, 131, 133t–39t

Galinsoga ciliata (Raf.) S.F. Blake. *See* hairy galinsoga

Galinsoga parviflora Cav. (smallflower galinsoga), 25, 26, 291–94

Galinsoga quadriradiata Cav. *See* hairy galinsoga

galinsogas, 25, 26, 38t, 132t–33t, 291–94; hairy (See hairy galinsoga); smallflower, 25, 26, 291–94

Galium aparine L. (catchweed bedstraw), 35, 96, 136t–37t, 231–34

Galium mollugo L. (smooth bedstraw), 232

Galium spurium L. (false cleavers), 231–34

Gallandt, Eric, 42

garden hoes, 102f

garden huckleberry, 39t, 328

garden rakes, 102f, 103

garlic: domestic, 196; wild, 74, 138t–39t, 195–99

geese, 62–63, 234

germination, 21–25; allelopathic compounds affecting, 31, 32f; of crops, 44; cues for, 20, 21–22, 23, 40, 56, 77; cultivation affecting, 84–85; and emergence season, 23–24, 42; light affecting, 22, 104; mulch affecting, 56; and seed depth in soil, 24, 25–26, 44; seed weight affecting, 20; in stale seedbed, 82; summary tables on, 130, 132t–39t; in tarping, 58; temperature affecting, 22, 23, 48; tillage affecting, 20, 21–22, 24, 73, 77, 78; tillage after, 77; in tilled fallow, 81, 82

giant foxtail, 14, 32, 33, 38t, 134t–35t, 158–62; on Marten farm, 112; nutrient use and growth rate, 30f, 31f; seed size, 21, 21f, 160

giant hogweed, 69

giant ragweed. *See* ragweed, giant

Glean, 107, 107t

Glechoma hederacea L. (ground ivy), 276

global positioning system guidance. *See* GPS guidance

glyphosate, 106, 107, 107t

goats in weed management, 62

goosefoot, 28

goosefoot cultivator shovels, 87, 88f

goosegrass, 38t, 134t–35t, 163–65, 174

GPS guidance: for cultivators, 100, 113, 116, 121, 122, 123–24; on Martens farm, 113–14; on Mugge farm, 116; on Park Farming Organics, 121; on Pepper farm, 122, 123–24; for planting, 113–14; for wheel traffic, 105

grain crops: contamination with weed seeds, 66, 67–68; of Martens farm, 111, 112, 114; of Mugge farm, 115, 116; of Park Farming Organics, 119–22; planting date and tillage timing for, 77; transition to organic production, 59–60

grasses. *See also specific species*.: in cover crop mixtures, 51; electrical discharge weeding of, 98; emergence depth, 26; flame weeding of, 98; geese in management of, 62–63; mycorrhizal associations, 28; palatability, 38t; perennial, 138t–39t; photosynthesis pathway, 27; seed longevity, 25; solarization affecting, 60, 60f, 61;

summer annual, 134t–35t; tillage and drying of, 76–77; winter annual, 136t–37t

grazing, 37–38, 62–63

great burdock, 243–44, 281

green foxtail, 38t, 134t–35t, 158–62

green manure, 49, 51

ground beetles, 25, 42

groundcherry, clammy, 302

ground ivy, 276

groundsel, common, 37, 98, 132t–33t, 247–50

growth habits, 130, 132t–38t

growth rate, 14, 398; crop variety selection for, 47; crop vigor and uniformity in, 44–45; difference between crops and weeds in, 26–27; and nutrient use, 29–30, 30f, 31f, 40; and photosynthesis pathway, 27; relative, 26, 26t, 399; and root size, 29; and seed size, 20, 26, 29, 47; and soil physical conditions, 31–32

guess row, 398

guidance systems. *See* GPS guidance

hairy fleabane, 305

hairy galinsoga, 26, 27, 119, 291–94; seed survival, 25, 79, 292; tillage of, 77, 79, 292–93; timing of reproduction, 33, 294

hairy nightshade, 47, 323–28

Halberdleaf orach, 251

hand weeding, 56, 78, 102–4, 109; on Beech Grove Farm, 119; on Mugge farm, 116; on Park Farming Organics, 121; on Pepper farm, 124

hard seed coat dormancy, 23

Harrington Seed Destructor, 64

harrows, 87t, 95–96

harvest: field cleanup after, 64; preventive weed management in, 64–65; and seed predation, 65; soil compaction in, 105; transport of weed seeds in, 69

hay, 53–54, 68, 80, 81

healall, 276

hedge bindweed, 23, 53, 138t–39t, 221–25, 383

Helianthus spp. *See* sunflowers

Helianthus annuus L., 266–69. *See also* sunflowers

Helianthus ciliaris DC. (Texas blueweed), 267

Helianthus petiolaris Nutt. (prairie sunflower), 266

Helianthus tuberosus L. (Jerusalem artichoke), 267

hemp dogbane, 255

hemp sesbania, 39t, 132t–33t, 299–301

henbit, 28, 37, 39t, 136t–37t, 275–79

herbicides: and bioherbicides, 36–37; in cover crop termination, 51; and energy use, 106–7, 107t; for mulch edges, 57; natural product, 57, 61–62, 70, 121; in no-till system, 78, 106; resistance to, 34, 107; in stale seedbed, 82

Hibiscus trionum L. (Venice mallow), 374

hoary plantain, 346

hoes, 102–4, 109, 400

hogweed, giant, 69

Holcus lanatus L. (velvetgrass), 152

horizontal disk cultivators, 86t, 88, 101, 108

horsenettle, 138t–39t, 302–4, 325

horse purslane, 60f, 61, 259

horsetail, 62

horseweed, 132t–33t, 305–8

hot water weeders, 87t, 98

huckleberry, garden, 39t, 328

Hypochaeris radicata L. (common catsear), 271

implements. *See* equipment and machinery

indeterminate growth, 398

inflorescence, 398, 399, 400

inoculant, 398

in-row cultivators, 82

insects as seed predators, 37

intercropping, 48–49

inter-row cultivators, 82, 83

inter-row mowers, 86t

in-the-boot, 398

Ipomoea hederacea Jacq. (ivyleaf morningglory), 39t, 79, 317–22

Ipomoea lacunosa L. (pitted morningglory), 317–22

Ipomoea purpurea (L.) Roth (tall morningglory), 22, 39t, 317–22

irrigation, 28, 32, 49, 69, 120

Italian ryegrass, 38t, 136t–37t, 142, 166–69

ivy, ground, 276

ivyleaf morningglory, 39t, 79, 317–22

Jacquemontia tamnifolia (L.) Griseb. (smallflower morningglory), 319

Japanese brome, 152

Japanese millet, 148

Jerusalem artichoke, 267

jimsonweed, 33, 37, 132t–33t, 243–44, 309–12

johnsongrass, 38t, 76, 120, 138t–39t, 170–73, 192

junglerice, 145

Kentucky bluegrass, 141–42

knives, vegetable, 88

knotweed: prostrate, 259; tufted, 370

kochia: annual, 32, 37, 39t, 132t–33t, 313–16, 357; perennial forage, 313, 316

Kochia scoparia (L.) Schrad. (annual kochia), 32, 37, 39t, 132t–33t, 313–16, 357

Kovar weeder, 113

kraft paper mulch, 57

Lactuca serriola L. (prickly lettuce), 39t, 132t–33t, 217, 334–35, 350–52

ladysthumb smartweed, 134t–35t, 369–73

Lakeview Organic Grain, 111, 114

lambsquarters, common, 14, 15t, 38, 38t, 132t–33t, 251–54; crop rotation affecting, 43, 252; dispersal, 36, 254; dormancy and germination, 22, 23, 156, 252–53; flame weeding of, 98; growth rate, 26t, 253; on Marten farm, 112; mortality rate, 37, 253; nutrient use, 30f, 31f, 252, 253; at Park Farming Organics, 120; photosynthesis pathway, 27, 253; seed collection at harvest time, 65, 252; seed size, 20, 21, 21f, 26t, 252; seed survival, 25, 253; in shade, 33, 253; similar species, 251–52, 313, 340; tilled fallow reducing, 82; timing of reproduction, 33, 253–54

Lamium amplexicaule L. (henbit), 28, 37, 39t, 136t–37t, 275–79

Lamium maculatum L. (spotted deadnettle), 276

Lamium purpureum L. (purple deadnettle), 23, 39t, 136t–37t, 275–79

lanceolate shape, 398

landscape fabrics as mulch, 56–57

large crabgrass, 21, 32, 38t, 134t–35t, 174–76

late watergrass, 145

lay-by cultivation, 398

leaf canopy, 45, 46; and seedling growth, 20; shade from, 32; soil covered by, 106

leaf sheath, 398, 399, 400

legumes, 28, 51, 398

Lepidium campestre (L.) W.T. Aiton (field pepperweed), 362

Lepidium virginicum L. (Virginia pepperweed), 362

lettuce, domestic, 30f, 352

lettuce, prickly, 39t, 132t–33t, 217, 334–35, 350–52

Liebman, M., 13, 42

life cycle stages, 16, 16f, 42

light: competition for, 26–27, 32; and day length

sensitivity, 33; as germination cue, 22, 104; and photosynthesis pathways, 27; row orientation affecting crop capture of, 46–47

ligule, 399

little starwort, 239

livestock, 62–63, 70, 112, 178; new, as source of weed seeds, 68; transition to organic production, 59; weeds as food for, 29, 37–39, 62–63; and weed seeds in feed and forages, 67–68; and weed seeds in manure, 36, 67, 68

livid amaranth, 340

Lolium spp., 182

Lolium multiflorum Lam. (Italian ryegrass), 38t, 136t–37t, 142, 166–69

Lolium perenne L. (perennial ryegrass), 166

Lolium perenne L. ssp. *multiflorum* (Lam.) Husnot (Italian ryegrass), 38t, 136t–37t, 142, 166–69

longevity of seeds, 24t, 24–25, 79

longspine sandbur, 134t–35t, 186–90

machinery. *See* equipment and machinery

machine vision systems in cultivator guidance, 100, 101, 113, 114

mallow: common, 374; Venice, 374

Malva neglecta Wallr. (common mallow), 374

manure, 29–30, 30f; composted, 29, 30f, 36, 49, 55; on Martens farm, 112; organic regulations on, 62; on Pepper farm, 123; rapid release of nutrients from, 70; in ridge tillage, 80; weed seeds in, 36, 67, 68

Martens farm, 111–14

Matricaria discoidea DC. (pineapple-weed), 236, 248

Matricaria matricarioides (Less.) Porter (pineapple weed), 236, 248

Matricaria perforata Mérat (scentless chamomile), 236

mayweed chamomile, 23, 68, 235–38

mechanical weed management, 73–110; definition of, 399; energy use in, 106–7; equipment in, 86–96; essential concepts of, 73–74; matching implement to tasks in, 73, 101–2; principles of, 82–85; tillage in, 73, 74–82

medic, black, 394

Medicago lupulina L. (black medic), 394

mice, 25

Microthlaspi perfoliatum (L.) F.K. Mey. (thoroughwort pennycress), 285

midrib, 398, 399

milk stage, 399

milkweed, common, 38t, 125, 138t–39t, 255–58; allelopathic compounds in, 31, 258; depth of storage organs, 74, 75; dispersal, 35, 35f, 257

millet: domesticated proso, 15, 205, 206, 207; finger, 165; foxtail, 162; Japanese, 148; wild-proso, 15, 34, 38t, 134t–35t, 155, 204–7, 208

minerals, 29

Mohler, C. L., 13

moldboard plows, 18, 74, 74t, 108; energy use with, 106, 106t; seed redistribution from, 78, 78f, 108; in severe weed problems, 76

Mollugo verticillata L. (carpetweed), 232

morningglories, 36, 39t, 132t–33t, 317–22; dormancy and germination, 22, 79, 126, 320; ivyleaf, 39t, 79, 317–22; pitted, 317–22; seed size, 20, 320; smallflower, 319; solarization affecting, 60f, 61, 319; tall, 22, 39t, 317–22; tillage of, 79, 319

mortality rate of weeds, 15–17, 37; as seeds (*See* seed mortality)

mouseear chickweed, 239

mowing, 82, 96–97; of cover crops, 51; of field margins, 66; inter-row, 86t; weed topping, 87t, 96–97, 114, 121

Mugge, Paul, 114–16

mugwort, 248, 262–63

mulch, 53–58, 70; allelopathic effect of, 31, 32f, 55; in continous no-till vegetable production, 55–56, 78; cover crop residue as, 51, 52, 53; dust, 81, 84, 108, 119, 398; organic, 53–56, 70, 106; for small seeded weeds, 20, 53; synthetic materials in, 56–58, 59, 61, 70; in transition to organic production, 59

mullein, common, 33

multivators, 89

mustard, white, 98, 386

mustard, wild, 39t, 134t–35t, 386–89, 391; frost sensitivity, 28, 387; mycorrhizal associations, 28, 388; seed collection in combine harvesting, 65, 387

mustard cover crop, 50, 82

mustard seed meal, 62, 351

mycorrhizal fungi, 28–29, 131, 133t–39t

natural enemies of weeds, 36–37; seed predators in, 25, 37, 65–66

natural product herbicides, 57, 61–62, 70, 121

New York Certified Organic, 114

nightshades, 29, 36, 39t, 132t–33t, 323–28;

bittersweet, 325; black, 29, 36, 39t, 101, 323–28; and crop competition, 47, 325, 327; Eastern black, 323–28; hairy, 47, 323–28; horsenettle, 302–4; jimsonweed, 33, 37, 132t–33t, 243–44, 309–12; and machine vision systems, 101; mycorrhizal associations, 28, 326; silverleaf, 124, 325

night-time cultivation and tillage, 104

nitrate, 22, 29, 37, 58

nitrogen fixation, 28, 398, 399

node, 399

Nordell, Eric and Anne, 116–19

no-till production: continuous, with organic mulch, 55–56, 78; cover crops in, 51–53; herbicides in, 78, 106; seed bank in, 108

nutrient use, 29–30, 128, 131, 133t–39t; of crops and weeds, 29, 40, 49, 70; and growth rate, 29–30, 30f, 31f, 40; and mineral content of plants, 29, 49; from organic materials, 49; in slow and rapid release, 49, 70

nutsedge: pigs in management of, 62, 178, 181, 212, 214; purple (*See* purple nutsedge); yellow (*See* yellow nutsedge)

oats: cultivated, 201; wild, 38t, 43, 48, 64, 134t–35t, 200–203

ocrea (ocreae), 399

onion: as crop, 55, 117, 118; wild, 196

onion hoes, 102, 102f

orach, Halberdleaf, 251

orchardgrass, 163

organic matter, 49, 105–6, 123; as mulch, 53–56, 70, 106; slow and rapid release of nutrients from, 49, 70

organic production: on Beech Grove Farm, 116–19; and chickens in weed management, 62; manure applications in, 62; on Martens farm, 111–14; mechanical weed management in, 101; on Mugge farm, 114–16; natural product herbicides in, 61; of onion seedlings, 55; on Park Farming Organics, 119–22; on Pepper farm, 122–24; planting date and tillage timing in, 77; synthetic mulch materials in, 57, 58, 59; transition to, 58–60, 70

Ornithogalum umbellatum L. (star of Bethlehem), 196

outcrossing, 399

Oxalis corniculata L. (creeping woodsorrel), 394

Oxalis dillenii Jacq. (slender yellow woodsorrel), 394

Oxalis stricta L. (yellow woodsorrel), 35, 138t–39t, 394–96

oxtongue, 217

oxygen levels in soil, 22

palatability of weeds, 37–39, 38t–39t, 62

paleseed plantain, 346

pale smartweed, 370

Palmer amaranth, 28, 32, 33, 39t, 132t–33t, 329–33, 340, 379

panicle, 399

panicum, fall, 38t, 134t–35t, 146, 155–57, 160, 170, 204, 208

Panicum capillare L. (witchgrass), 35, 134t–35t, 155, 174, 204, 208–10

Panicum dichotomiflorum Michx. (fall panicum), 38t, 134t–35t, 146, 155–57, 160, 170, 204, 208

Panicum miliaceum L. (wild-proso millet), 15, 34, 38t, 134t–35t, 155, 204–7, 208

paper mulch, 57, 58

pappus, 399

parallel gang cultivators, 85–86, 88f

Park Farming Organics, 119–22

partridge-pea, 299, 365

Pennsylvania smartweed, 134t–35t, 369–73

pennycress: field, 27, 38t, 43, 132t–33t, 285–87, 362; thoroughwort, 285

Pepper, Carl, 122–24

pepperweed: field, 362; Virginia, 362

perennial weeds, 399; creeping (*See* creeping perennials); mulch limitations in control of, 53, 56; origins of, 37; palabability, 37–38; stationary (*See* stationary perennials); storage organs of (*See* storage organs of perennials); summary tables on, 131, 138t–39t; tarping resistance, 58; tillage of, 74–77; tilled fallow in management of, 81; timing of reproduction, 33; vegetative propagation, 17–19

permanent beds, 105, 120

Persian speedwell, 276

Persicaria maculosa Gray (ladysthumb smartweed), 134t–35t, 369–73

Persicaria pensylvanica (L.) Small (Pennsylvania smartweed), 134t–35t, 369–73

phosphorus, 29, 30, 30f

photosynthesis, 26; pathways in, 27, 131, 133t–38t, 399

Physalis heterophylla Nees (clammy groundcherry), 302

physical weed management, 73–110, 399

Picris (oxtongue), 217

pigs, 62, 178, 181, 212, 214

pigweeds, 27, 33, 36, 132t–33t, 338–43; crop rotation affecting, 43, 340; dormancy and germination, 23, 104, 341; mycorrhizal associations, 28, 341; nutrient use, 29, 340, 341–42; palatability, 37, 38, 39t, 342–43; Palmer amaranth, 28, 32, 33, 39t, 132t–33t, 329–33, 340, 379; at Park Farming Organics, 120; Powell amaranth, 25f, 30f, 31f, 39t, 330, 338–43, 379; prostrate, 259, 340; redroot (*See* redroot pigweed); seed size, 20, 21, 21f, 341; similar species, 251, 259, 340; smooth, 330, 338–43, 379; solarization affecting, 60f, 61; tumble, 340; waterhemp (*See* waterhemp)

pineapple-weed, 236, 248

pinnate leaves, 399

pinnate tansymustard, 288

pistil, 399

pitted morningglory, 317–22

Plantago aristata Michx. (bracted plantain), 346

Plantago media L. (hoary plantain), 346

Plantago virginica L. (paleseed plantain), 346

plantain, 39t, 138t–39t, 344–49; blackseed, 138t–39t, 344–49; bracted, 346; broadleaf, 39t, 75, 138t–39t, 344–49; buckhorn, 39t, 138t–39t, 344–49; hoary, 346; paleseed, 346; as rosette, 19; in shade, 32, 348; as stationary perennial, 19, 75; tillage of, 75

plant density, 47–48; of crops, 45–46, 70, 83–84, 103; of weeds, 16–17, 44, 63, 77–79, 83

planters, 79–80, 82–83, 113, 115

plant identification, 67, 126, 127–28

planting date, 48, 77; and cover crop management, 51, 52–53; in intercropping, 49; in ridge tillage system, 80

plant size: of crops and weeds, 40, 47, 83; and removal of largest weeds, 35; and seed production, 35, 40

plastic mulch, 56, 56f, 57

plastic tarp for solarization, 60

plows, 74, 74t; energy use with, 106t; seed redistribution from, 78, 108; seed transport on, 69

pneumatic weeders, 94

Poa annua L. (annual bluegrass), 38t, 60, 75, 136t–37t, 141–44

Poa compressa L. (Canada bluegrass), 141

Poa pratensis L. (Kentucky bluegrass), 141–42

Poa trivialis L. (roughstalk bluegrass), 142

poisoning. *See* toxicity

pollination, 34, 131, 133t–37t

Polygonum aviculare L. (prostrate knotweed), 259

Polygonum cespitosum Blume var. *longisetum* (Bruijn) A.N. Steward (tufted knotweed), 370

Polygonum convolvulus L. (wild buckwheat), 39t, 222, 319, 383–85

Polygonum lapathifolium L. (pale smartweed), 370

Polygonum pensylvanicum L. (Pennsylvania smartweed), 134t–135, 369–73

Polygonum persicaria L. (ladysthumb smartweed), 134t–135, 369–73

population of species, 15–17, 399

Portulaca oleracea L. *See* purslane, common

Powell amaranth, 25f, 30f, 31f, 39t, 330, 338–43, 379

Practical Farmers of Iowa, 116

prairie sunflower, 266

predators of seeds, 25, 37, 65–66

preventive weed management, 63–70, 71

prickly lettuce, 39t, 132t–33t, 217, 334–35, 350–52

prickly sida, 39t, 132t–33t, 353–56, 374

proso millet: domesticated, 15, 205, 206, 207; wild, 15, 34, 38t, 134t–35t, 155, 204–7, 208

prostrate growth, 399

prostrate knotweed, 259

prostrate pigweed, 259, 340

Prunella vulgaris L. (healall), 276

Pseudoroegneria spicata (Pursh) Á. Löve ssp. *spicata* (bluebunch wheatgrass), 153

PTO (power take off), 399

purple deadnettle, 23, 39t, 136t–37t, 275–79

purple nutsedge, 17, 38t, 138t–39t, 177–81, 211; in solarization, 60f, 61, 178; tillage of, 74, 178; timing of reproduction, 33, 180

purslane, common, 27, 36, 38t, 132t–33t, 259–61; dormancy and germination, 22, 23, 260; drought tolerance, 28, 32, 260; flame weeding, 98, 260; in shade, 32, 260, 261; in summer planted crops, 23, 77; tilled fallow in management of, 81; timing of reproduction, 33, 261

purslane, horse, 60f, 61, 259

quackgrass, 14, 15t, 38t, 138t–39t, 182–85; cover crops affecting, 50, 183; intercropping affecting, 49; livestock in management of, 62; on Marten farm, 112; photosynthesis pathway, 27, 184; similar species, 166, 182; tillage of, 74, 76, 183; timing of reproduction, 33, 185; vegetative propagation, 17, 18, 185

fertility of soil); nitrate in, 22, 58; organic matter in, 49, 105–6; oxygen levels in, 22; seed depth in, 24, 25–26, 78–79; seed distribution in, 77–78, 78f, 108; solarization, 60–61, 70; temperature, 22, 23, 48, 77; texture, 31–32; tilth (*See* tilth)

soil preparation: corn gluten as herbicide in, 61–62; for cover crop, 52; and effective cultivation, 84; for onion seedling production, 55; solarization in, 60–61, 70; tarping in, 57–58; timing of, 77; in transition to organic production, 58–60; for transplants, 48

Solanum americanum Mill. (American black nightshade), 325

Solanum carolinense L. (horsenettle), 138t–39t, 302–4, 325

Solanum dulcamara L. (bittersweet nightshade), 325

Solanum elaeagnifolium Cav. (silverleaf nightshade), 325

Solanum nigrum L. (black nightshade), 29, 36, 39t, 101, 323–28

Solanum physalifolium Rusby (hairy nightshade), 47, 323–28

Solanum ptycanthum Dunal (Eastern black nightshade), 323–28

Solanum rostratum Dunal (buffalobur), 302

Solanum sarachoides auct. non Sendtn. (hairy nightshade), 47, 323–28

solarization, 60–61, 70

Sonchus arvensis L. *See* sowthistle, perennial

Sonchus asper (L.) Hill (spiny sowthistle), 216–20, 334–35

Sonchus oleraceus L. (annual sowthistle), 134t–35t, 216–20, 271, 334–35, 350

sorghum, 15, 191–94

Sorghum bicolor (L.) Moench nothosubsp. *drummondii* (Steud.) de Wet ex. Davidse, 191

Sorghum bicolor (L.) Moench ssp. *bicolor,* 191

Sorghum bicolor (L.) Moench ssp. *drummondii* (Nees ex Steud.) de Wet & Harlan, 191

Sorghum bicolor (L.) Moench ssp. *verticilliflorum* (Steud.) de Wet ex Wiersema & J. Dahlb. *See* shattercane

Sorghum halepense (L.) Pers. (johnsongrass), 38t, 76, 120, 138t–39t, 170–73, 192

sorghum-sudangrass cover crops, 50–51

southern crabgrass, 174

southern sandbur, 134t–35t, 186–90

sowthistle, annual, 134t–35t, 216–20, 271, 334–35,

350

sowthistle, perennial, 39t, 62, 138t–39t, 334–37; recovery of root fragments, 19f, 336; similar species, 217, 271, 334–35, 350; tillage of, 74, 76, 335

sowthistle, spiny, 216–20, 334–35

soybeans: growth rate, 26t; on Martens farm, 111, 112, 113; mechanical weed management for, 101; on Mugge farm, 115, 116; planting density, 45; in ridge tillage system, 81, 115; row spacing, 46; seed contamination with weed seeds, 66; seed size, 21, 21f, 26t; tine weeding of, 95; transition to organic production, 59

spading machines, 18

speedwell: Persian, 276; thymeleaf, 239

spider gangs, rolling cultivators with, 86t, 88

spike harrows, 87t, 95

spike inflorescence, 400

spikelet, 397, 400

spinners, 86t, 92–93, 93f

spiny amaranth, 340

spiny cocklebur, 243–44

spiny sowthistle, 216–20, 334–35

spotted deadnettle, 276

spring hoes, 86t, 91–92, 92f

spurges, 259

spyders, 86t, 89–90, 90f

stale seedbed, 81, 82, 108, 400; flame weeding of, 82, 97; natural product herbicides in, 61; tarping of, 57

stamens, 400

star of Bethlehem, 196

starwort, little, 239

stationary perennials, 14, 15t, 400; and biennials, 397; pollination of, 34; taproots of, 18–19; tillage of, 75

Staver, C. P., 13

steam weeders, 98

Stellaria graminea L. (little starwort), 239

Stellaria media (L.) Vill. *See* chickweed, common

stirrup hoes, 102f, 102–3, 400

storage organs of perennials, 17–18; drying as control method, 18, 76–77; exhausting food reserves of, 75–76; freeezing temperatures affecting, 28; removal of, 18, 76; summary tables on, 131, 138t; tillage affecting, 74, 75t, 75–77

stratification, 400

straw mulch, 31, 32f, 53f, 53–54, 55, 68

string trimmers, 400

sulfonyl urea herbicides, 107, 107t

summer annuals, 130, 132t–35t, 400

summer cover crops, 50–51

sunflowers, 26t, 38t, 132t–33t, 266–69, 295; nutrient use, 29, 268; origin, 15, 267; prairie, 266

swan neck weeders, 103, 103f

sweep hoes, 102, 102f

sweep plows, 74, 74t

sweeps, 86t, 86–88, 88f; in ridge tillage, 80; rolling cultivators with, 88–89

synthetic mulch, 56–58, 59, 61, 70

tall fescue, 182

tall morningglory, 22, 39t, 317–22

tall waterhemp, 34, 378

tansymustard, pinnate, 288

taproots, 18–19, 75, 398, 400

Taraxacum laevigatum (Willd.) DC. (redseed dandelion), 271

Taraxacum officinale F.H. Wigg. *See* dandelion

tarping, 57–58

teasle, common, 33

temperature: in composting, 68; and crop planting date, 48; in flame weeding, 97; and frost sensitivity, 27–28; and germination, 22, 23, 48, 77; in soil solarization, 60

Texas blueweed, 267

thistle: bull, 75, 226; Canada, 18, 27, 68, 138t–39t, 226–30; Russian, 32, 35, 39t, 132t–33t, 313, 357–60

Thlaspi arvense L. (field pennycress), 27, 38t, 43, 132t–33t, 285–87, 362

Thompson, Dick and Sharon, 81

thoroughwort pennycress, 285

thymeleaf sandwort, 239

thymeleaf speedwell, 239

tillage, 73, 74–82; on Beech Grove Farm, 118, 118f; of cover crops, 51–52; depth of, 26, 48, 74, 77, 79; drying of storage organs in, 76–77; and emergence depth, 26, 48, 79; energy use in, 106, 106t; exhausting storage organs in, 75–76; and fallow periods, 81–82; and germination, 20, 21–22, 24, 73, 77, 78; implements in, 74, 74t; matched to weed population, 79; and mycorrhizal fungi, 29; at night, 104; on Park Farming Organics, 121; on Pepper farm, 122, 123, 124; of perennial weeds, 19, 74–77; principles of, 73; prior to cover crop planting, 52; prior to transplanting, 48; removal of storage organs in, 76; ridge (*See* ridge tillage);

seed redistribution in, 77–78, 78f, 108; and seed survival, 25, 79; in stale seedbed, 82; timing of, 73, 77, 79, 107, 108; and vegetative propagation, 18, 18f, 19; and weed density, 77–79; weed seeds on machinery for, 68–69; weed size in, 73; of wet soil, 105

tilled fallow, 64, 81–82, 108, 400

tiller of grass plant, 400

tilth, 59, 104–6, 109, 400; and effective cultivation, 84, 104–5, 108; in no-till production and mulch, 56; tillage affecting, 77, 81; in transition to organic production, 58, 59; in weed management, 32

tine weeders, 87t, 95f, 95–96; in blind cultivation, 17, 85; planting density increased for, 83; in sequence of mechanical weed management, 73, 74; for small seeded weeds, 20; vertical axis, 86t, 92

tires: soil compaction from, 105; weed transport on, 69

torsion weeders, 86t, 91–92, 92f

toxicity: of allelopathic compounds, 31; of black nightshade, 29; of catchweed bedstraw, 234; of cocklebur, 246; of field pennycress, 287; of flixweed, 290; of garlic, wild, 199; of groundsel, 249–50; of hemp sesbania, 301; of horsenettle, 304; of Italian ryegrass, 169; of jimsonweed, 37, 311; of johnsongrass, 173; of kochia, 316; of lambsquarters, 254; of milkweed, 257; of nightshades, 29, 327; nitrate levels in, 29, 37; of pigweeds, 29, 37; of proso millet, wild, 207; of purslane, 261; of radish, wild, 393; of Russian-thistle, 359; of shattercane, 194; of witchgrass, 210

tractors: cultivator mounted to, 85; weed transport on, 69

transplanted crops, 44; competitiveness of, 70; corn gluten as herbicide for, 61–62; of early crops, 48; mulching of, 31, 54; production of onion seedlings for, 55; of small seeded crops, 48

tree leaves, 54–55, 68

Trianthema portulacastrum L. (horse purslane), 60f, 61, 259

Trifolium spp. (clovers), 394

Tripleurospermum perforatum (Mérat) M. Lainz (scentless chamomile), 236

tubers, 74–75, 76, 400

tufted knotweed, 370

tumble pigweed, 340

tumbleweeds, 35

vegetable crops: at Beech Grove Farm, 116–19; at